Mathematical Methods, Modelling and Applications

Mathematical Methods, Modelling and Applications

Editors

Lucas Jódar
Rafael Company

MDPI • Basel • Beijing • Wuhan • Barcelona • Belgrade • Manchester • Tokyo • Cluj • Tianjin

Editors
Lucas Jódar
Universitat Politècnica de
València
Spain

Rafael Company
Universitat Politècnica de
València
Spain

Editorial Office
MDPI
St. Alban-Anlage 66
4052 Basel, Switzerland

This is a reprint of articles from the Special Issue published online in the open access journal *Mathematics* (ISSN 2227-7390) (available at: https://www.mdpi.com/journal/mathematics/special_issues/Mathematical_Methods_Modelling_Applications).

For citation purposes, cite each article independently as indicated on the article page online and as indicated below:

LastName, A.A.; LastName, B.B.; LastName, C.C. Article Title. *Journal Name* **Year**, *Volume Number*, Page Range.

ISBN 978-3-0365-4357-4 (Hbk)
ISBN 978-3-0365-4358-1 (PDF)

© 2022 by the authors. Articles in this book are Open Access and distributed under the Creative Commons Attribution (CC BY) license, which allows users to download, copy and build upon published articles, as long as the author and publisher are properly credited, which ensures maximum dissemination and a wider impact of our publications.

The book as a whole is distributed by MDPI under the terms and conditions of the Creative Commons license CC BY-NC-ND.

Contents

About the Editors vii

Preface to "Mathematical Methods, Modelling and Applications" ix

Lucas Jódar and Rafael Company
Preface to "Mathematical Methods, Modelling and Applications"
Reprinted from: *Mathematics* **2022**, *10*, 1607, doi:10.3390/math10091607 1

Carlos Fuentes, Carlos Chávez, Antonio Quevedo, Josué Trejo-Alonso and Sebastián Fuentes
Modeling of Artificial Groundwater Recharge by Wells: A Model Stratified Porous Medium
Reprinted from: *Mathematics* **2020**, *8*, 1764, doi:10.3390/math8101764 3

Francisco Pedroche and J. Alberto Conejero
Corrected Evolutive Kendall's τ Coefficients for Incomplete Rankings with Ties: Application to Case of Spotify Lists
Reprinted from: *Mathematics* **2020**, *8*, 1828, doi:10.3390/math8101828 15

Hao Chen, Ling Liu and Junjie Ma
Analysis of Generalized Multistep Collocation Solutions for Oscillatory Volterra Integral Equations
Reprinted from: *Mathematics* **2020**, *8*, 2004, doi:10.3390/math8112004 45

Carlos Fuentes, Carlos Chávez and Fernando Brambila
Relating Hydraulic Conductivity Curve to Soil-Water Retention Curve Using a Fractal Model
Reprinted from: *Mathematics* **2020**, *8*, 2201, doi:10.3390/math8122201 61

Chin-Tsai Lin and Cheng-Yu Chiang
Developing and Applying a Selection Model for Corrugated Box Precision Printing Machine Suppliers
Reprinted from: *Mathematics* **2021**, *9*, 68, doi:10.3390/math9010068 75

Alicia Cordero, Eva G. Villalba, Juan R. Torregrosa and Paula Triguero-Navarro
Convergence and Stability of a Parametric Class of Iterative Schemes for Solving Nonlinear Systems
Reprinted from: *Mathematics* **2021**, *9*, 86, doi:10.3390/math9010086 95

Rafael Company, Vera N. Egorova and Lucas Jódar
Quadrature Integration Techniques for Random Hyperbolic PDE Problems
Reprinted from: *Mathematics* **2021**, *9*, 160, doi:10.3390/math9020160 113

Juan-Carlos Cortés, Elena López-Navarro, José-Vicente Romero and María-Dolores Roselló
Approximating the Density of Random Differential Equations with Weak Nonlinearities via Perturbation Techniques
Reprinted from: *Mathematics* **2021**, *9*, 204, doi:10.3390/math9030204 129

María Consuelo Casabán, Rafael Company and Lucas Jódar
Reliable Efficient Difference Methods for Random Heterogeneous Diffusion Reaction Models with a Finite Degree of Randomness
Reprinted from: *Mathematics* **2021**, *9*, 206, doi:10.3390/math9030206 147

Michael M. Tung
The Relativistic Harmonic Oscillator in a Uniform Gravitational Field
Reprinted from: *Mathematics* **2021**, *9*, 294, doi:10.3390/math9040294 163

Benito Chen-Charpentier
Stochastic Modeling of Plant Virus Propagation with Biological Control
Reprinted from: *Mathematics* 2021, 9, 456, doi:10.3390/math9050456 175

Jitraj Saha and Andreas Bück
Conservative Finite Volume Schemes for MultidimensionalFragmentation Problems
Reprinted from: *Mathematics* 2021, 9, 635, doi:10.3390/math9060635 191

Iván Alhama, Gonzalo García-Ros and Matteo Icardi
Non-Stationary Contaminant Plumes in the Advective-Diffusive Regime
Reprinted from: *Mathematics* 2021, 9, 725, doi:10.3390/math9070725 219

Sagrario Lantarón and Susana Merchán
The Dirichlet-to-Neumann Map in a Disk with a One-Step Radial Potential: An Analytical and Numerical Study
Reprinted from: *Mathematics* 2021, 9, 794, doi:10.3390/math9080794 245

Elena de la Poza, Lucas Jódar and Paloma Merello
Modeling Political Corruption in Spain
Reprinted from: *Mathematics* 2021, 9, 952, doi:10.3390/math9090952 263

Aleksandr Shirokanev, Nataly Ilyasova, Nikita Andriyanov, Evgeniy Zamytskiy, Andrey Zolotarev and Dmitriy Kirsh
Modeling of Fundus Laser Exposure for Estimating Safe Laser Coagulation Parameters in the Treatment of Diabetic Retinopathy
Reprinted from: *Mathematics* 2021, 9, 967, doi:10.3390/math9090967 281

Elías Berriochoa, Alicia Cachafeiro, Alberto Castejón and José Manuel García-Amor
Mechanical Models for Hermite Interpolation on the Unit Circle
Reprinted from: *Mathematics* 2021, 9, 1043, doi:10.3390/math9091043 297

Xue Li, Jun-Yi Sun, Xiao-Chen Lu, Zhi-Xin Yang and Xiao-Ting He
Steady Fluid–Structure Coupling Interface of Circular Membrane under Liquid Weight Loading: Closed-Form Solution for Differential-Integral Equations
Reprinted from: *Mathematics* 2021, 9, 1105, doi:10.3390/math9101105 317

Oksana Mandrikova, Nadezhda Fetisova and Yuriy Polozov
Hybrid Model for Time Series of Complex Structure with ARIMA Components
Reprinted from: *Mathematics* 2021, 9, 1122, doi:10.3390/math9101122 341

Eduardo Diz-Mellado, Samuele Rubino, Soledad Fernández-García, Macarena Gómez-Mármol, Carlos Rivera-Gómez and Carmen Galán-Marín
Applied Machine Learning Algorithms for Courtyards Thermal Patterns Accurate Prediction
Reprinted from: *Mathematics* 2021, 9, 1142, doi:10.3390/math9101142 359

Javier Ibáñez, José M. Alonso, Jorge Sastre, Emilio Defez and Pedro Alonso Jordá
Advances in the Approximation of the Matrix Hyperbolic Tangent
Reprinted from: *Mathematics* 2021, 9, 1219, doi:10.3390/math9111219 379

About the Editors

Lucas Jódar

Lucas Jódar is a Full Professor in Applied Mathematics of the Universitat Politècnica de València (Spain). His research interests include mathematical modelling in sciences, engineering, and a wide range of fields in social sciences. Apart from his activity as Editor of several reputed journals, his works include areas such as numerical methods for ordinary and partial differential equations, computing and numerical analysis, and random differential equations.

Rafael Company

Rafael Company is a Full Professor in Applied Mathematics of the Universitat Politècnica de València (Spain). His research interest is focused on the numerical analysis and computing of numerical methods for solving partial differential equations modelling a wide range of scientific fields, especially mathematical finance and physics.

Preface to "Mathematical Methods, Modelling and Applications"

Reality is more complex than it seems. The segmentation of science does not help capture reality; each scientific point of view seems to be a partial mirror of the problem under consideration. A model is an approximation to represent an actual phenomenon in a simplified way, disregarding some factors but considering enough of them to achieve an acceptable answer. A mathematical model is an idealization of the phenomenon one wishes to represent in mathematical terms, typically an equation. The modelling process is divided in several parts:

i. Observations obtaining data and pattern recognition.

ii. Hypothesis, identification of variables. Building a mathematical model.

iii. Resolution of the model and applications.

The present book contains 21 articles accepted for publication in the Special Issue "Mathematical Methods, Modelling and Applications" of the MDPI Mathematics journal. The contents of the book are organized in the following way. Some papers are concerned with step (i) of the modelling process. Other papers are linked to step (ii). All the remaining papers are related to step (iii), covering a wide spectrum of methods, deterministic and random, algebraic and differential, in different fields of hydrodynamics, physics, and health sciences.

We would like to thank the MDPI publishing editorial team, the scientific peer reviewers and all the authors who contributed to this book. We are confident that the contents will be of value to researchers, academics and professionals involved in the resolution of real-world nature and social problems

Lucas Jódar and Rafael Company
Editors

Editorial

Preface to "Mathematical Methods, Modelling and Applications"

Lucas Jódar and Rafael Company *

Instituto de Matemàtica Multidisciplinar, Universitat Politècnica de València, 46022 Valencia, Spain; ljodar@imm.upv.es
* Correspondence: rcompany@imm.upv.es

The reality is more complex than it seems. The segmentation of science does not help capture the reality; each scientific point of view seems to be a partial mirror of the problem under consideration. A model is an approximation to represent an actual phenomenon in a simplified way, disregarding some factors but considering enough of them to achieve an acceptable answer. A mathematical model is an idealization of the phenomenon one wishes to represent in mathematical terms, typically an equation. The modelling process is divided in several parts:

i. Observations obtaining data and Pattern Recognition.
ii. Hypothesis, identification of variables. Building the Mathematical Model.
iii. Resolution of the Model and applications.

The present book contains the 21 articles accepted for publication in the Special Issue "Mathematical Methods, Modelling and Applications" of the MDPI "Mathematics" journal. The contents of the book are organized in the following way. Papers [1–3] are concerned with step (i) of the modelling process. Papers [4–6] are linked to step (ii). All the remaining papers [7–21] are related to step (iii) covering a wide spectrum of methods, deterministic and random, algebraic and differentials, in different fields Hydrodynamic, Physics, Health Sciences.

We would like to thank the MDPI publishing editorial team, the scientific peer reviewers and all the authors who contributed to this book. We are confident that the contents will be of value to researchers, academics and professionals involved in the resolution of real-world nature and social problems.

Funding: This research received no external funding.

Conflicts of Interest: The authors declare no conflict of interest.

Citation: Jódar, L.; Company, R. Preface to "Mathematical Methods, Modelling and Applications". *Mathematics* **2022**, *10*, 1607. https://doi.org/10.3390/math10091607

Received: 3 May 2022
Accepted: 6 May 2022
Published: 9 May 2022

Publisher's Note: MDPI stays neutral with regard to jurisdictional claims in published maps and institutional affiliations.

Copyright: © 2022 by the authors. Licensee MDPI, Basel, Switzerland. This article is an open access article distributed under the terms and conditions of the Creative Commons Attribution (CC BY) license (https://creativecommons.org/licenses/by/4.0/).

References

1. Mandrikova, O.; Fetisova, N.; Polozov, Y. Hybrid Model for Time Series of Complex Structure with ARIMA Components. *Mathematics* **2021**, *9*, 1122. [CrossRef]
2. Lin, C.-T.; Chiang, C.-Y. Developing and Applying a Selection Model for Corrugated Box Precision Printing Machine Suppliers. *Mathematics* **2021**, *9*, 68. [CrossRef]
3. Pedroche, F.; Conejero, J.A. Corrected Evolutive Kendall's τ Coefficients for Incomplete Rankings with Ties: Application to Case of Spotify Lists. *Mathematics* **2020**, *8*, 1828. [CrossRef]
4. Shirokanev, A.; Ilyasova, N.; Andriyanov, N.; Zamytskiy, E.; Zolotarev, A.; Kirsh, D. Modeling of Fundus Laser Exposure for Estimating Safe Laser Coagulation Parameters in the Treatment of Diabetic Retinopathy. *Mathematics* **2021**, *9*, 967. [CrossRef]
5. de la Poza, E.; Jódar, L.; Merello, P. Modeling Political Corruption in Spain. *Mathematics* **2021**, *9*, 952. [CrossRef]
6. Chen-Charpentier, B. Stochastic Modeling of Plant Virus Propagation with Biological Control. *Mathematics* **2021**, *9*, 456. [CrossRef]
7. Ibáñez, J.; Alonso, J.; Sastre, J.; Defez, E.; Alonso-Jordá, P. Advances in the Approximation of the Matrix Hyperbolic Tangent. *Mathematics* **2021**, *9*, 1219. [CrossRef]
8. Diz-Mellado, E.; Rubino, S.; Fernández-García, S.; Gómez-Mármol, M.; Rivera-Gómez, C.; Galán-Marín, C. Applied Machine Learning Algorithms for Courtyards Thermal Patterns Accurate Prediction. *Mathematics* **2021**, *9*, 1142. [CrossRef]
9. Li, X.; Sun, J.-Y.; Lu, X.-C.; Yang, Z.-X.; He, X.-T. Steady Fluid–Structure Coupling Interface of Circular Membrane under Liquid Weight Loading: Closed-Form Solution for Differential-Integral Equations. *Mathematics* **2021**, *9*, 1105. [CrossRef]
10. Berriochoa, E.; Cachafeiro, A.; Rábade, H.G.; García-Amor, J. Mechanical Models for Hermite Interpolation on the Unit Circle. *Mathematics* **2021**, *9*, 1043. [CrossRef]
11. Lantarón, S.; Merchán, S. The Dirichlet-to-Neumann Map in a Disk with a One-Step Radial Potential: An Analytical and Numerical Study. *Mathematics* **2021**, *9*, 794. [CrossRef]
12. Alhama, I.; García-Ros, G.; Icardi, M. Non-Stationary Contaminant Plumes in the Advective-Diffusive Regime. *Mathematics* **2021**, *9*, 725. [CrossRef]
13. Saha, J.; Bück, A. Conservative Finite Volume Schemes for Multidimensional Fragmentation Problems. *Mathematics* **2021**, *9*, 635. [CrossRef]
14. Tung, M. The Relativistic Harmonic Oscillator in a Uniform Gravitational Field. *Mathematics* **2021**, *9*, 294. [CrossRef]
15. Casabán, M.; Company, R.; Jódar, L. Reliable Efficient Difference Methods for Random Heterogeneous Diffusion Reaction Models with a Finite Degree of Randomness. *Mathematics* **2021**, *9*, 206. [CrossRef]
16. Cortés, J.-C.; López-Navarro, E.; Romero, J.-V.; Roselló, M.-D. Approximating the Density of Random Differential Equations with Weak Nonlinearities via Perturbation Techniques. *Mathematics* **2021**, *9*, 204. [CrossRef]
17. Company, R.; Egorova, V.N.; Jódar, L. Quadrature Integration Techniques for Random Hyperbolic PDE Problems. *Mathematics* **2021**, *9*, 160. [CrossRef]
18. Cordero, A.; Villalba, E.G.; Torregrosa, J.R.; Triguero-Navarro, P. Convergence and Stability of a Parametric Class of Iterative Schemes for Solving Nonlinear Systems. *Mathematics* **2021**, *9*, 86. [CrossRef]
19. Fuentes, C.; Chávez, C.; Brambila, F. Relating Hydraulic Conductivity Curve to Soil-Water Retention Curve Using a Fractal Model. *Mathematics* **2020**, *8*, 2201. [CrossRef]
20. Chen, H.; Liu, L.; Ma, J. Analysis of Generalized Multistep Collocation Solutions for Oscillatory Volterra Integral Equations. *Mathematics* **2020**, *8*, 2004. [CrossRef]
21. Fuentes, C.; Chávez, C.; Quevedo, A.; Trejo-Alonso, J.; Fuentes, S. Modeling of Artificial Groundwater Recharge by Wells: A Model Stratified Porous Medium. *Mathematics* **2020**, *8*, 1764. [CrossRef]

Article

Modeling of Artificial Groundwater Recharge by Wells: A Model Stratified Porous Medium

Carlos Fuentes [1], Carlos Chávez [2,*], Antonio Quevedo [1], Josué Trejo-Alonso [2] and Sebastián Fuentes [2]

[1] Mexican Institute of Water Technology, Paseo Cuauhnáhuac Núm. 8532, Jiutepec, Morelos 62550, Mexico; cbfuentesr@gmail.com (C.F.); jose_quevedo@tlaloc.imta.mx (A.Q.)
[2] Water Research Center, Department of Irrigation and Drainage Engineering, Autonomous University of Queretaro, Cerro de las Campanas SN, Col. Las Campanas, Queretaro 76010, Mexico; josue.trejo@uaq.mx (J.T.-A.); sefuca.1196@gmail.com (S.F.)
* Correspondence: chagcarlos@uaq.mx; Tel.: +52-442-192-1200 (ext. 6036)

Received: 21 September 2020; Accepted: 8 October 2020; Published: 13 October 2020

Abstract: In recent years, groundwater levels have been decreasing due to the demand in agricultural and industrial activities, as well as the population that has grown exponentially in cities. One method of controlling the progressive lowering of the water table is the artificial recharge of water through wells. With this practice, it is possible to control the amount of water that enters the aquifer through field measurements. However, the construction of these wells is costly in some areas, in addition to the fact that most models only simulate the well as if it were a homogeneous profile and the base equations are restricted. In this work, the amount of infiltrated water by a well is modeled using a stratified media of the porous media methodology. The results obtained can help decision-making by evaluating the cost benefit of the construction of wells to a certain location for the recharge of aquifers.

Keywords: mathematical modeling; infiltration well; differential equations; porous medium; fractal conductivity model

---

1. Introduction

Infiltration wells are used to contribute to the evacuation of rains in urban areas and also as a mechanism to recharge aquifers in regions where they present an unsustainable abatement [1–4]. Their construction must be analyzed from several angles: objectives of artificial recharge, available technological options, chemical quality of the water, social factors, place, quantity of water to contribute, among others [5–9].

In the literature, several numerical and analytical solutions can be found to model the flow of water in the porous medium, however, the models present restrictions to estimating the properties of soils, in addition to considering the stratum of the soil well profile as a homogeneous medium [1,10–14].

The artificial recharge capacity in a well is measured as the amount of water that infiltrates the soil during a specific period of time, and varies depending on the number of strata in the soil in which it was built. In this way, if you want to know the amount of water that the entire well contributes, you must evaluate the infiltration rate in all the strata to have a better knowledge about the contributions to the aquifer and the behavior of the system as a whole.

The phenomenon of infiltration in porous media can be studied from the general principles of the conservation of mass and momentum. The equation that results from the application of the first principle is:

$$\frac{\partial \theta}{\partial t} = -\nabla \cdot \vec{q}. \tag{1}$$

Darcy's law generalized to partially saturated porous media is used as a dynamic equation [15]:

$$\vec{q} = -K\nabla H, \quad (2)$$

where H is the hydraulic potential and is the sum of the pressure potential (ψ) and the gravitational potential assimilated to the vertical coordinate (z) oriented, in this case, as positive upwards. The pressure potential is positive in the saturated zone and negative in the unsaturated zone, since it is agreed that the zero pressure corresponds to the atmospheric pressure; $\theta = \theta(\psi)$ is the volumetric water content, also called moisture content, and is a function of the water pressure, $\theta(\psi)$ is known as the retention curve or soil moisture characteristic; $\vec{q} = (q_x, q_y, q_z)$ is the flow of water per unit of soil surface or Darcy flow, with its components in a rectangular system; (x, y, z) are the spatial coordinates in a rectangular or Cartesian system, t is time; ∇ is the gradient operator; $K = K(\psi)$ is the hydraulic conductivity as a function of the water pressure.

Thus, the general equation of flow in a porous medium results from the combination of Equations (1) and (2):

$$\frac{\partial \theta}{\partial t} = \nabla \cdot [K(\psi)\nabla(\psi + z)]. \quad (3)$$

This equation presents two independent variables, θ and ψ, but since there is a relationship between them, the specific capacity defined as the slope of the retention curve is introduced. The chain rule is applied and the equation with the dependent variable pressure is established, known as the Richards equation [16]:

$$C(\psi)\frac{\partial \psi}{\partial t} = \nabla \cdot [K(\psi)\nabla \psi] + \frac{\partial K}{\partial \psi}\frac{\partial \psi}{\partial z}; \quad C(\psi) = \frac{\partial \theta}{\partial \psi}. \quad (4)$$

In this work a methodology is presented to obtain the water infiltration rate by partially or totally filled artificial recharge wells. The equations have been adapted to be used in a homogeneous stratified medium, taking into account the soil characteristics of each strata in the profile of the well.

2. Materials and Methods

2.1. The Richards and Kirchhoff Equations in Spherical and Cylindrical Coordinates

In some problems the analysis is simplified if Equation (4) is written in cylindrical or spherical coordinates. The Richards equation [16] in cylindrical coordinates (r, φ, z) is as follows:

$$C(\psi)\frac{\partial \psi}{\partial t} = \frac{1}{r}\frac{\partial}{\partial r}\left[rK(\psi)\frac{\partial \psi}{\partial r}\right] + \frac{1}{r^2}\frac{\partial}{\partial \varphi}\left[K(\psi)\frac{\partial \psi}{\partial \varphi}\right] + \frac{\partial}{\partial z}\left[K(\psi)\frac{\partial \psi}{\partial z}\right] + \frac{\partial K}{\partial \psi}\frac{\partial \psi}{\partial z}, \quad (5)$$

where r is the radius and φ is the azimuth: $r^2 = x^2 + y^2$, $x = r\cos\varphi$, $y = r\sin\varphi$.

In spherical coordinates (ϱ, ϑ, φ) the Richards equation is written as:

$$C(\psi)\frac{\partial \psi}{\partial t} = \frac{1}{\varrho^2}\frac{\partial}{\partial \varrho}\left[\varrho^2 K(\psi)\frac{\partial \psi}{\partial \varrho}\right] + \frac{1}{\varrho^2 \sin\vartheta}\frac{\partial}{\partial \vartheta}\left[\sin\vartheta K(\psi)\frac{\partial \psi}{\partial \vartheta}\right] \\ + \frac{1}{\varrho^2 \sin^2\vartheta}\frac{\partial}{\partial \varphi}\left[K(\psi)\frac{\partial \psi}{\partial \varphi}\right] + \frac{\partial K}{\partial \psi}\frac{\partial \psi}{\partial z} \quad (6)$$

where ϱ is the radio, ϑ is the polar angle and φ is the azimuth: $\varrho^2 = x^2 + y^2 + z^2$, $x = \varrho \sin\vartheta\cos\varphi$, $y = \varrho\sin\vartheta\sin\varphi$, $z = \varrho\cos\vartheta$.

In a symmetric well with respect to the z axis, the radius r takes this axis as its origin, and Equation (5) is very useful for the infiltration analysis when it is assumed that the pressure does not depend on the azimuth, that is, when the heterogeneity is presented by layers. In this case the equation simplifies to the following:

$$C(\psi)\frac{\partial \psi}{\partial t} = \frac{1}{r}\frac{\partial}{\partial r}\left[rK(\psi)\frac{\partial \psi}{\partial r}\right] + \frac{\partial}{\partial z}\left[K(\psi)\frac{\partial \psi}{\partial z}\right] + \frac{\partial K}{\partial \psi}\frac{\partial \psi}{\partial z}, \qquad (7)$$

which has only two spatial coordinates (r, z).

The equation in spherical coordinates presents a very particular importance when, in the analysis of a problem, it is considered that the medium is homogeneous and isotropic, that is, when the phenomenon does not depend on either the colatitude or the azimuth:

$$C(\psi)\frac{\partial \psi}{\partial t} = \frac{1}{\varrho^2}\frac{\partial}{\partial \varrho}\left[\varrho^2 K(\psi)\frac{\partial \psi}{\partial \varrho}\right] + \frac{\partial K}{\partial \psi}\frac{\partial \psi}{\partial z}, \qquad (8)$$

in which only two spatial coordinates are presented (ϱ, z).

Furthermore, in some particular problems involving homogeneous porous media, the analysis is simplified if the potential Kirchhoff flow is defined by:

$$\Phi = \int_{-\infty}^{\psi} K(\overline{\psi})d\overline{\psi} = \int_{\theta_r}^{\theta} D(\overline{\theta})d\overline{\theta}, \qquad (9)$$

from which it follows that:

$$\frac{d\Phi}{d\psi} = K(\psi); \quad \frac{d\Phi}{d\theta} = D(\theta), \qquad (10)$$

where D(θ) is the hydraulic diffusivity, in analogy with the diffusion of gases, which is expressed as D(θ) = K(θ)/C(θ), considering, now, that both the hydraulic conductivity and the specific capacity are functions of the volumetric content moisture.

The water transfer equation in porous media as a dependent variable for moisture content, Equation (3), is as follows:

$$\frac{\partial \theta}{\partial t} = \nabla \cdot [D(\theta)\nabla \theta] + \frac{dK}{d\theta}\frac{\partial \theta}{\partial z}, \qquad (11)$$

which presents the structure of a nonlinear Fokker–Planck equation [17], the linear version of which is widely known in diffusion problems.

In terms of the potential Kirchhoff flow, Equation (11) becomes:

$$\frac{1}{D(\Phi)}\frac{\partial \Phi}{\partial t} = \nabla^2 \Phi + \frac{dK}{d\Phi}\frac{\partial \Phi}{\partial z}. \qquad (12)$$

Kirchhoff's equation in cylindrical coordinates (r, φ, z), where r is the radius and φ is the azimuth, is:

$$\frac{1}{D(\Phi)}\frac{\partial \Phi}{\partial t} = \frac{1}{r}\frac{\partial}{\partial r}\left(r\frac{\partial \Phi}{\partial r}\right) + \frac{1}{r^2}\frac{\partial^2 \Phi}{\partial \varphi^2} + \frac{\partial^2 \Phi}{\partial z^2} + \frac{dK}{d\Phi}\frac{\partial \Phi}{\partial z}. \qquad (13)$$

In spherical coordinates (ϱ, ϑ, φ) the Kirchhoff equation is written as follows:

$$\frac{1}{D(\Phi)}\frac{\partial \Phi}{\partial t} = \frac{1}{\varrho^2}\frac{\partial}{\partial \varrho}\left(\varrho^2 \frac{\partial \Phi}{\partial \varrho}\right) + \frac{1}{\varrho^2 \sin \vartheta}\frac{\partial}{\partial \vartheta}\left(\sin \vartheta \frac{\partial \Phi}{\partial \vartheta}\right) + \frac{1}{\varrho^2 \sin^2 \vartheta}\frac{\partial^2 \Phi}{\partial \varphi^2} + \frac{dK}{d\Phi}\frac{\partial \Phi}{\partial z}. \qquad (14)$$

These formulations are only applicable in homogeneous media, and, if these are isotropic, they are written respectively in cylindrical coordinates as follows:

$$\frac{1}{D(\Phi)}\frac{\partial \Phi}{\partial t} = \frac{1}{r}\frac{\partial}{\partial r}\left(r\frac{\partial \Phi}{\partial r}\right) + \frac{\partial^2 \Phi}{\partial z^2} + \frac{dK}{d\Phi}\frac{\partial \Phi}{\partial z}, \qquad (15)$$

and in spherical coordinates:

$$\frac{1}{D(\Phi)}\frac{\partial \Phi}{\partial t} = \frac{1}{\varrho^2}\frac{\partial}{\partial \varrho}\left(\varrho^2 \frac{\partial \Phi}{\partial \varrho}\right) + \frac{dK}{d\Phi}\frac{\partial \Phi}{\partial z}. \tag{16}$$

2.2. The Hydrodynamic Characteristics of Porous Media

In order to solve the mass or energy transfer equations of water in porous media, aside from specifying the limit conditions, it is necessary to know the hydrodynamic characteristics formed by the water retention curve θ (ψ) and the hydraulic conductivity curve either as a function of the water pressure, K(ψ), or as a function of the moisture content K(θ). The analysis is greatly simplified if these curves are represented with analytical functions.

The retention curve can be represented with the equation of van Genuchten [18]:

$$\Theta(\psi) = \left[1 + \left(\frac{\psi}{\psi_d}\right)^n\right]^{-m}, \tag{17}$$

where m > 0 and n > 0 are two shape parameters (dimensionless), ψ_d is a characteristic value of the water pressure and Θ is the effective degree of saturation defined by:

$$\Theta = \frac{\theta - \theta_r}{\theta_s - \theta_r}, \tag{18}$$

in which θ_r is the residual moisture content defined such that $K(\theta_r) = 0$ and $\theta(\psi \to -\infty) = \theta_r$ [19]; θ_s is the moisture content at saturation, assimilated to the total porosity of the soil (ϕ), when under saturation conditions no air is trapped in the interstices of the porous medium: $\theta_s = \phi$. In general, $\theta_r = 0$ can be assumed [20].

A closed way to represent the conductivity curve can be obtained using prediction models of the same from the retention curve. In the literature we can find various works, but given the condition of the phenomenon we are studying, this work uses one of the fractal models proposed, calibrated and validated by Fuentes et al. [20]. The results found by [20] shows a better adjustment between observed and estimated data by the function given by:

$$K(\Theta) = K_s \left[\int_0^\Theta \frac{\vartheta^{s-1} d\vartheta}{|\psi(\vartheta)|^{2s}} \bigg/ \int_0^1 \frac{\vartheta^{s-1} d\vartheta}{|\psi(\vartheta)|^{2s}}\right]^2, \tag{19}$$

where K_s is the hydraulic saturation conductivity and s = D/E, with D the fractal dimension of the porous medium and E = 3 the Euclid dimension of the physical space where the medium is embedded, related to the total porosity through the relation:

$$(1 - \phi)^s + \phi^{2s} = 1. \tag{20}$$

The introduction of Equation (17) in Equation (19) leads to the following equation to represent the hydraulic conductivity curve, accepting the relationship between the parameters as indicated:

$$K(\Theta) = K_s\left[1 - \left(1 - \Theta^{1/m}\right)^{sm}\right]^2; \ 0 < sm = 1 - 2s/n < 1. \tag{21}$$

The solution of the transfer equation in its different forms is generally numerical [21,22]. However, in some simplified cases, characteristics of the solution can be obtained analytically [23–25].

3. Results and Discussion

3.1. Conceptual Model

To analyze the infiltration in steady-state wells, it is necessary to write the Darcy flows in the radial and vertical directions:

$$\vec{q}_r = -K_s \frac{\partial \psi}{\partial r} \hat{r}, \tag{22}$$

$$\vec{q}_z = -\left(K_s \frac{\partial \psi}{\partial z} + K_s\right) \hat{k}, \tag{23}$$

where $\hat{r}$ and $\hat{k}$ are unitary vectors in the r and z directions respectively.

Flow through the wall and bottom of the well is defined by:

$$Q_s = \int_{A_p} \vec{q}_r \cdot dA_p + \int_{A_b} \vec{q}_z \cdot dA_b, \tag{24}$$

where dA_p y dA_b are, respectively, the differential areas in the wall and at the bottom of the well defined by:

$$dA_p = (2\pi R dz)\hat{r}, \tag{25}$$

$$dA_b = (2\pi r Dr)(-\hat{k}). \tag{26}$$

Equation (24), considering Equations (22), (23), (25) and (26), is written as follows:

$$Q_s = -2\pi R K_s \int_0^H \left.\frac{\partial \psi}{\partial r}\right|_{r=r} dz + 2\pi K_s \int_0^R \left.\frac{\partial \psi}{\partial z}\right|_{z=0} r dr + \pi K_s R^2. \tag{27}$$

Introducing the dimensionless variables:

$$z^* = \frac{z}{H}; \ r^* = \frac{r}{R}; \ \psi^* = \frac{\psi}{H}; \tag{28}$$

Equation (27) is written as follows:

$$Q_s = Q_o + \pi K_s R^2; \ Q_o = \frac{2\pi K_s H^2}{C}; \tag{29}$$

where C is a coefficient defined as:

$$\frac{1}{C} = -\int_0^1 \left.\frac{\partial \psi^*}{\partial r^*}\right|_{r^*=1} dz^* + \left(\frac{R}{H}\right)^2 \int_0^1 \left.\frac{\partial \psi^*}{\partial z^*}\right|_{z^*=0} r^* dr^*. \tag{30}$$

To find this coefficient it is necessary to know ψ (r, z).

3.2. The Glover Model

According to Glover [26], in a first approximation, the steady-state pressure flow through an infiltration well in a homogeneous and isotropic porous medium can be described with Laplace's equation in spherical coordinates that describes the pressure in the absence of gravitational gradients. From Equation (8) we have:

$$\nabla^2 \psi = \frac{1}{\varrho^2} \frac{\partial}{\partial \varrho}\left(\varrho^2 \frac{\partial \psi}{\partial \varrho}\right) = 0, \tag{31}$$

which must be subject to border conditions:

$$\psi = \psi_R; \varrho = R \tag{32}$$

$$\psi = 0; \varrho \to \infty. \tag{33}$$

Integration of Equation (31) leads to $\psi = -c_1 \varrho^{-1} + c_2$, where c_1 and c_2 are integration constants; the Equation (33) implies $c_2 = 0$ and the Equation (32) $c_1 = -\psi_R R$, ergo $\psi = \psi_R (R/\varrho)$. The Darcy flux is $q_\varrho = -K_s \partial\psi/\partial\varrho = K_s \psi_R (R/\varrho^2)$, when $\varrho = R$, $q_R = K_s \psi_R/R$; the flow through the surface of the sphere of radius R is $q_o = 4\pi R^2 q_R = 4\pi K_s R \psi_R$; this flow from the point source in the center of the sphere is the variable of interest. Since $\psi_R R = q_o/4\pi K_s$, it is better to set the pressure variation around the source flow to continue with the Glover approach:

$$\psi = \frac{q_o}{4\pi K_s \varrho}. \tag{34}$$

If h represents the position of the center of the sphere from the base, then the spherical coordinate (ϱ) and the cylindrical coordinate (r) are related by:

$$\varrho = \sqrt{r^2 + (z-h)^2}. \tag{35}$$

The pressure in terms of the cylindrical coordinates is obtained by introducing Equation (35) into Equation (34):

$$\psi = \frac{q_o}{4\pi K_s \sqrt{r^2 + (z-h)^2}}. \tag{36}$$

To provide a series of point sources whose magnitude increases with depth, an expression similar to that originally proposed by Glover, we have:

$$dq_o = B(h_c - h)dh, \tag{37}$$

where B is a parameter to be determined and h_c defines the range of the sources $h_o \le h \le h_c$ and sinks $h_c < h \le h_s$.

The total flow is found by integrating Equation (37):

$$Q_o = B \int_{H_o}^{h_s} (h_c - h)dh = \frac{1}{2} B H^2 \left[(h_c^* - h_o^*)^2 - (h_c^* - h_s^*)^2 \right] \tag{38}$$

hence parameter B is deduced:

$$B = \frac{2Q_o}{H^2 \left[(h_c^* - h_o^*)^2 - (h_c^* - h_s^*)^2 \right]} \tag{39}$$

where $h^* = h/H$ for all subscripts.

From Equations (36), (37) and (39) we have:

$$d\psi = \frac{Q_o (h_c - h)}{2\pi K_s H^2 \left[(h_c^* - h_o^*)^2 - (h_c^* - h_s^*)^2 \right] \sqrt{r^2 + (z-h)^2}} dh \tag{40}$$

the integration of which leads to:

$$\psi = \frac{Q_o}{2\pi K_s H^2 \left[(h_c^* - h_o^*)^2 - (h_c^* - h_s^*)^2\right]} \left[(h_c - z)\mathrm{asinh}\left(\frac{z-h}{r}\right) + \sqrt{r^2 + (z-H)^2}\right]_{h=h_s}^{h=h_o}, \quad (41)$$

ergo:

$$\psi = \frac{Q_o}{2\pi K_s H^2} \frac{\left[\begin{array}{c}(h_c - z)\mathrm{asinh}\left(\frac{z-h_o}{r}\right) - (h_c - z)\mathrm{asinh}\left(\frac{z-h_s}{r}\right) \\ + \sqrt{r^2 + (z-h_o)^2} - \sqrt{r^2 + (z-h_s)^2}\end{array}\right]}{\left[(h_c^* - h_o^*)^2 - (h_c^* - h_s^*)^2\right]}. \quad (42)$$

At the point on the boundary $(r, z) = (R, 0)$ we have $\psi = H$, which allows obtaining the expression of the flow, Equation (29):

$$Q_o = \frac{2\pi K_s H^2}{C}, \quad (43)$$

where the form coefficient is defined by:

$$C = \frac{h_c^*\left[\mathrm{asinh}\left(\frac{H}{R}h_s^*\right) - \mathrm{asinh}\left(\frac{H}{R}h_o^*\right)\right] + \sqrt{\left(\frac{R}{H}\right)^2 + h_o^{*2}} - \sqrt{\left(\frac{R}{H}\right)^2 + h_s^{*2}}}{(h_c^* - h_o^*)^2 - (h_c^* - h_s^*)^2}. \quad (44)$$

Glover formula is derived from Equation (44) by making $h_c^* = 1$, $h_o^* = 0$ y $h_s^* = 1$:

$$C = \mathrm{asinh}\left(\frac{H}{R}\right) + \frac{R}{H} - \sqrt{\left(\frac{R}{H}\right)^2 + 1}. \quad (45)$$

3.3. The Reynolds and Elrick Model

This model proposed by Reynolds and Elrick [27] assumes $h_c^* = 1/2$, $h_i^* = 0$ y $h_s^* = 1/2$:

$$C = 4\left[\frac{1}{2}\mathrm{asinh}\left(\frac{H}{2R}\right) + \frac{R}{H} - \sqrt{\left(\frac{R}{H}\right)^2 + \frac{1}{4}}\right]. \quad (46)$$

3.4. A Model for Stratified Porous Media

Glover's model can be adapted for the case of stratified porous media. The well is considered to be in a medium composed of N layers of thickness P_j, $j = 1, 2, \ldots, N$; the total hydraulic head, denoted as H_T, is the height of the water column counted from the base of the well to the upper border of the N-th stratum.

The flow infiltrated by the walls of the j-th stratum is provided by Equation (43) modified as:

$$Q_{oj} = \frac{2\pi K_{sj} P_j^2}{C_j}, \quad (47)$$

where K_{sj} and C_j are the saturated hydraulic conductivity and the shape coefficient of the j-th stratum, respectively.

The shape coefficient is derived from Equation (44) denoting by H_j the hydraulic head at the base of the j-th stratum:

$$C_j = h_{pj}^* \frac{h_{Cj}^*\left[\mathrm{asinh}\left(\frac{P_j}{R}h_{sj}^*\right) - \mathrm{asinh}\left(\frac{P_j}{R}h_{oj}^*\right)\right] + \sqrt{\left(\frac{R}{P_j}\right)^2 + h_{oj}^{*2}} - \sqrt{\left(\frac{R}{P_j}\right)^2 + h_{sj}^{*2}}}{\left(h_{cj}^* - h_{oj}^*\right)^2 - \left(h_{cj}^* - h_{sj}^*\right)^2}, \quad (48)$$

where $h^*_{Pj} = P_j/H_j$, $h^*_{cj} = h_{cj}/P_j$, $h^*_{oj} = h_{oj}/P_j$, $h^*_{sj} = h_{sj}/P_j$. It is noted that h_{cj}, h_{oj} y h_{sj} are calculated from the base of the j-th stratum.

The Reynolds and Elrick model assumes $h^*_c = 1/2$, $h^*_o = 0$ y $h^*_s = 1/2$ and therefore:

$$C_j = 4\left[\frac{1}{2}\mathrm{asinh}\left(\frac{P_j}{2R}\right) + \frac{R}{P_j} - \sqrt{\left(\frac{R}{P_j}\right)^2 + \frac{1}{4}}\right]h_{Pj}. \tag{49}$$

The total flow is obtained as:

$$Q = \sum_{j=1}^{N} Q_{oj} + \pi R^2 K_{s1} \tag{50}$$

where the flow at the bottom of the well has been added.

3.5. Aplications

To show the versatility of the solution, data obtained from an infiltration well built on the Queretaro Valley aquifer of radio are used: R = 0.3937 m (15.5″) and depth P_T = 36 m; Five strata were located in the profile (Figure 1). As drilling was carried out, the infiltration tests per stratum were carried out until the permanent regime was reached. The measured data are concentrated in Table 1, and the saturated hydraulic conductivity calculated from Equation (43) is also shown.

Table 1. Calculation of hydraulic conductivity per stratum, Equation (43).

Stratum	H (m/s)	Q (L/s)	C	K_s (m/d)
1	12	0.02841	4.9632	0.0134
2	4	0.33681	3.0113	0.8592
3	10	0.81158	4.6239	0.5142
4	6	1.01448	3.7017	1.4231
5	4	1.62317	3.0113	4.1405

Table 2 shows the flow rates calculated for each stratum when the well is full. In the last row is the total flow, Equation (50), and the saturated hydraulic conductivity corresponding to an equivalent homogeneous stratum.

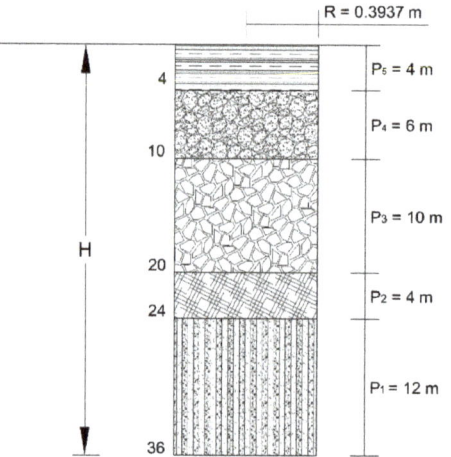

Figure 1. Simplified well scheme.

Table 2. Calculation of the flow per stratum corresponding to a full well, Equations (47) and (49).

Stratum	P (m)	K_s (m/d)	H (m)	C	Q (L/s)
1	12	0.0134	36	1.6544	0.085
2	4	0.8592	24	0.5019	1.992
3	10	0.5142	20	2.3119	1.617
4	6	1.4231	10	2.2210	1.677
5	4	4.1405	4	3.0113	1.600
Equivalent	36	0.5231	36	7.0749	6.972

The artificial volume that can be recharged to the aquifer is 6972 L/s (602.3808 m^3/day), however, stratum 1 only contributes 1.22% of the entire volume and is the deepest layer of the entire well (12 m). With these data, the cost benefit is analyzed and the decision is made to drill wells to a depth of 24 m with the understanding that we would contribute only 595.0318 m^3/day to the aquifer but we reduce time and money.

The time to drill the well up to 36 m is 27 days at a cost of 25,457 USD. Therefore, if you choose to build two, the time taken would be 60 days with a total of 50,915 USD. Conversely, if you only drill to a depth of 24 m, the time taken is 7 days and a cost of 13,376 USD, which gives us a total cost of 53,504 USD for four wells drilled in 30 days. This is due to the fact that the material in the last 12 m is basalt and drilling progress is slower.

Regarding the volumes of recharge to the aquifer, with the four wells in the same area it would be 2380.1272 m^3/day compared to 1204.7616 m^3/day that we would obtain with only two at a depth of 36 m. Finally, the cost-benefit of annual recharge in the aquifer would be 16.24 m^3/USD invested in four wells, compared to the 8.64 m^3/USD that you have if you choose two wells.

4. Conclusions

In recent years the construction of artificial wells to recharge aquifers has been very popular in Mexico, however, as has been demonstrated in this work, the construction of a well at a greater depth does not necessarily give us a greater volume of recharge. The lack of information to calculate the total volumes has led to decisions being made with unscientific bases and, on several occasions, it has resulted in not achieving the expectations for which they were built.

This work provides a tool for knowing the behavior of the water infiltration rate in the porous medium in stratified media to artificially recharge an aquifer through wells. The analysis takes into account all the characteristics of the soil profiles that construct the well, resulting in the cost-benefit analysis of the complete operation to make a better decision.

It is widely demonstrated in the literature that several experimental tests are needed to know the behavior of this phenomenon, and that in order to know the process in detail, other factors that are not analyzed here must be taken into account: preferential flow in heterogeneous soils, trapped air, sediment deposit, among others.

When exploration drilling is done to propose a series of wells to recharge the aquifer, the data from the first layer is usually measured to simulate the behavior of the entire profile as a homogeneous stratum. However, as verified in this work, it is necessary to know the behavior of the entire well by stratum so that pertinent decisions are made, since a deeper well does not necessarily imply a greater volume of recharge to the aquifer.

Author Contributions: Conceptualization, C.F.; methodology, C.F.; software, C.F., A.Q., J.T.-A. and S.F.; validation, C.F., C.C. and A.Q.; formal analysis, C.F., C.C. and A.Q.; investigation, C.F.; data curation, J.T.-A. and S.F.; writing—original draft preparation, C.F.; writing—review and editing, C.C. and J.T.-A. All authors have read and agreed to the published version of the manuscript

Funding: This research received no external funding.

Conflicts of Interest: The authors declare no conflict of interest.

References

1. Chitsazan, M.; Movahedian, A. Evaluation of Artificial Recharge on Groundwater Using MODFLOW Model (Case Study: Gotvand Plain-Iran). *J. Geos. Environ. Prot.* **2015**, *3*, 1221–1232. [CrossRef]
2. Karim, I. Artificial Recharge of Groundwater by Injection Wells (Case Study). *Int. J. Sci. Eng. Technol. Res.* **2018**, *6*, 6193–6196.
3. Salem, S.B.; Chkir, N.; Zouari, K.; Cognard-Plancq, A.L.; Valles, V.; Marc, V. Natural and artificial recharge investigation in the Zeroud Basin, Central Tunisia: Impact of Sidi Saad Dam storage. *Environ. Earth Sci.* **2012**, *66*, 1099–1110. [CrossRef]
4. Bouwer, H. Artificial recharge of groundwater: Hydrogeology and engineering. *Hydrogeol. J.* **2002**, *10*, 121–142. [CrossRef]
5. Sun, Y.; Xu, S.; Wang, Q.; Hu, S.; Qin, G.; Yu, H. Response of a Coastal Groundwater System to Natural and Anthropogenic Factors: Case Study on East Coast of Laizhou Bay, China. *Int. J. Environ. Res. Public Health* **2020**, *17*, 5204.
6. Loizeau, S.; Rossier, Y.; Gaudet, J.P.; Réfloch, A.; Besnard, K.; Angulo-Jaramillo, R.; Lassabatere, L. Water infiltration in an aquifer recharge basin affected by temperature and air entrapment. *J. Hydrol. Hydromech.* **2017**, *65*, 222–233. [CrossRef]
7. Zhang, G.; Feng, G.; Li, X.; Xie, C.; Pi, X. Flood effect on groundwater recharge on a typical silt loam soil. *Water* **2017**, *9*, 523. [CrossRef]
8. Edwards, E.C.; Harter, T.; Fogg, G.E.; Washburn, B.; Hamad, H. Assessing the effectiveness of drywells as tools for stormwater management and aquifer recharge and their groundwater contamination potential. *J. Hydrol.* **2016**, *539*, 539–553. [CrossRef]
9. Jarraya Horriche, F.; Benabdallah, S. Assessing Aquifer Water Level and Salinity for a Managed Artificial Recharge Site Using Reclaimed Water. *Water* **2020**, *12*, 341. [CrossRef]
10. Ward, J.D.; Simmons, C.T.; Dillon, P.J. A theoretical analysis of mixed convection in aquifer storage and recovery: How important are density effects? *J. Hydrol.* **2007**, *343*, 169–186. [CrossRef]
11. Maples, S.R.; Fogg, G.E.; Maxwell, R.M. Modeling managed aquifer recharge processes in a highly heterogeneous, semi-confined aquifer system. *Hydrogeol. J.* **2019**, *27*, 2869–2888.
12. Xu, Y.; Shu, L.; Zhang, Y.; Wu, P.; Atlabachew Eshete, A.; Mabedi, E.C. Physical Experiment and Numerical Simulation of the Artificial Recharge Effect on Groundwater Reservoir. *Water* **2017**, *9*, 908.
13. Ringleb, J.; Sallwey, J.; Stefan, C. Assessment of Managed Aquifer Recharge through Modeling—A Review. *Water* **2016**, *8*, 579.
14. Händel, F.; Liu, G.; Dietrich, P.; Liedl, R.; Butler, J.J. Numerical assessment of ASR recharge using small-diameter wells and surface basins. *J. Hydrol.* **2014**, *517*, 54–63. [CrossRef]
15. Darcy, H. Dètermination des lois d'ècoulement de l'eau à travers le sable. In *Les Fontaines Publiques de la Ville de Dijon*; Dalmont, V., Ed.; Victor Dalmont: Paris, France, 1856; pp. 590–594.
16. Richards, L.A. Capillary conduction of liquids through porous mediums. *Physics* **1931**, *1*, 318–333. [CrossRef]
17. Fuentes, C.; Chávez, C.; Saucedo, H.; Zavala, M. On an exact solution of the non-linear Fokker-Planck equation with sink term. *Water Technol. Sci.* **2011**, *2*, 117–132.
18. Van Genuchten, M.T. A closed-form equation for predicting the hydraulic conductivity of unsaturated soils. *Soil Sci. Soc. Am. J.* **1980**, *44*, 892–898. [CrossRef]
19. Brooks, R.H.; Corey, A.T. Hydraulic properties of porous media. In *Hydrology Papers*; Colorado State University: Colorado, CO, USA, 1964; Volume 3.
20. Fuentes, C.; Antonino, A.C.D.; Sepúlveda, J.; Zataráin, F.; De León, B. Prediction of the relative soil hydraulic conductivity with fractal models. *Hydra. Engine. Mex.* **2003**, *18*, 31–40.
21. Fuentes, S.; Trejo-Alonso, J.; Quevedo, A.; Fuentes, C.; Chávez, C. Modeling Soil Water Redistribution under Gravity Irrigation with the Richards Equation. *Mathematics* **2020**, *8*, 1581. [CrossRef]
22. Saucedo, H.; Fuentes, C.; Zavala, M. The Saint-Venant and Richards equation system in surface irrigation: 2) Numerical coupling for the advance phase in border irrigation. *Ing. Hidraul. Mex.* **2005**, *20*, 109–119.
23. Chen, X.; Dai, Y. An approximate analytical solution of Richards equation with finite boundary. *Bound. Value Probl.* **2017**, *2017*, 167. [CrossRef]
24. Tracy, F.T. Clean two and three-dimensional analytical solution of Richards' equation for testing numerical solvers. *Water Resour. Res.* **2006**, *42*, 85038–85513. [CrossRef]

25. Baiamonte, G. Analytical solution of the Richards equation under gravity-driven infiltration and constant rainfall intensity. *J. Hydrol. Eng.* **2020**, *25*, 04020031. [CrossRef]
26. Glover, R.E. Flow from a test-hole located above groundwater level, in Theory and problems of water percolation. *U.S. Bur. Rec. Eng. Monogr.* **1953**, *8*, 69–71.
27. Reynolds, W.D.; Elrick, D.E. In situ measurement of field—Saturated hydraulic conductivity, sorptivity and the-parameter using the Guelph permeameter. *Soil Sci.* **1985**, *140*, 292–302. [CrossRef]

© 2020 by the authors. Licensee MDPI, Basel, Switzerland. This article is an open access article distributed under the terms and conditions of the Creative Commons Attribution (CC BY) license (http://creativecommons.org/licenses/by/4.0/).

Article

Corrected Evolutive Kendall's τ Coefficients for Incomplete Rankings with Ties: Application to Case of Spotify Lists

Francisco Pedroche [1,*] and J. Alberto Conejero [2]

1. Institut de Matemàtica Multidisciplinària, Universitat Politècnica de València, Camí de Vera s/n, 46022 València, Spain
2. Instituto Universitario de Matemática Pura y Aplicada, Universitat Politècnica de València, Camí de Vera s/n, 46022 València, Spain; aconejero@upv.es
* Correspondence: pedroche@mat.upv.es

Received: 23 September 2020; Accepted: 11 October 2020; Published: 18 October 2020

Abstract: Mathematical analysis of rankings is essential for a wide range of scientific, public, and industrial applications (e.g., group decision-making, organizational methods, R&D sponsorship, recommender systems, voter systems, sports competitions, grant proposals rankings, web searchers, Internet streaming-on-demand media providers, etc.). Recently, some methods for incomplete aggregate rankings (rankings in which not all the elements are ranked) with ties, based on the classic Kendall's tau coefficient, have been presented. We are interested in ordinal rankings (that is, we can order the elements to be the first, the second, etc.) allowing ties between the elements (e.g., two elements may be in the first position). We extend a previous coefficient for comparing a series of complete rankings with ties to two new coefficients for comparing a series of incomplete rankings with ties. We make use of the newest definitions of Kendall's tau extensions. We also offer a theoretical result to interpret these coefficients in terms of the type of interactions that the elements of two consecutive rankings may show (e.g., they preserve their positions, cross their positions, and they are tied in one ranking but untied in the other ranking, etc.). We give some small examples to illustrate all the newly presented parameters and coefficients. We also apply our coefficients to compare some series of Spotify charts, both Top 200 and Viral 50, showing the applicability and utility of the proposed measures.

Keywords: incomplete rankings; Kendall's tau; permutation graph; competitive balance; Spotify

1. Introduction

The analysis of rankings of scores (*cardinal rankings*) or, particularly, rankings composed of natural numbers (*ordinal rankings*), have been studied from different perspectives attending to the ultimate goal of the researchers or practitioners (see [1]). When the interest is on obtaining a *consensus* score that summarizes the opinion of various judges, the used mathematical tools are usually aimed to find a ranking that minimizes a given *distance metric* (see the seminal paper [2,3] for some properties of different metrics). In such a case, we say that a *distance metric* minimizes disagreement. We can place in this area the methods called *voter systems*, *ranking aggregation*, and others (see the detailed review in [4]).

When the interest is focused on comparing two series of rankings, one of the key points is to obtain a measure that describes the *evolution* of the series. In this case, we have a series of rankings such that each one of them prioritizes the elements based on the scores obtained at a particular time (see [5]). For example, *sports rankings* belong to this category. Obviously, at the end of a season, there is no need to find a consensus ranking since, by the nature of sports leagues, it is the last ranking that serves to summarize the result of the overall season. The same happens with the Stock Market, the richest people rankings made by the Fortune magazine [6], university rankings (e.g., [7,8]), songs rankings based on the number of downloads, streaming, or sales (see [9]), etc. Our work is focused on a series of rankings behavior.

The terminology applied to rankings is not unique. For example, in [10] the term *partial* is used to indicate rankings in which ties are presented, while in [11] the term *partial* indicates that not all the objects are compared. In this paper, we use the terminology coined in [4,12]. We talk of *complete* rankings when all the objects are compared (as in a football league) and *incomplete* when there are absent objects (as in a Top k ranking). We explicitly use the terms *with ties* or *without ties* to indicate whether we consider the presence of tied objects in the rankings. We recall that in [11] the term *linear order* is used when all objects are compared and no ties are allowed (that is, for us, *complete rankings with no ties*) and the term *weak ordering* when all objects are compared, but ties are allowed (that is, for us, *complete rankings with ties*).

Incomplete rankings appear in multiple areas. For example, in national or European grant calls, judges evaluate only a subset of the applications, and therefore each judge handles an incomplete ranking. The same happens in literary contests, where each judge only reads a small number of manuscripts. In the case of the results shown by search engines, it is clear that only the first Top k web pages are displayed, being, as a consequence, an incomplete ranking.

We use, and extend, the results of some previous papers. Some concepts are taken from [5], where a method to compare series of complete rankings with no ties was presented, and from [13], where a method to compare series of complete rankings with ties was analyzed. We also make reference to [14], where some theoretical aspects where studied. In all these works, there are two main ingredients:

1. The use of generalizations of the classical concept of Kendall's τ coefficient of disagreement [15–17];
2. The use of graphs associated to the series of rankings as a tool to visualize and also to help in the definition the coefficients that summarize the "behaviour" of the series of rankings.

Regarding to extensions of Kendall's τ coefficient, the first attempt to incorporate an axiomatic distance metric was in [2], followed by the works [11,18,19].

More recently, in [4] these previous works were revised and a new axiomatic framework for incomplete rankings was introduced. To the best of our knowledge, the last paper devoted to an axiomatic study for incomplete rankings is [12], where it is shown as an extension of Kendall's τ coefficient to the case of incomplete rankings with ties.

Kendall's τ has been extensively used, and some extensions can be found in the literature up to the present day on [10,12,20]. In particular, Kendall's τ has been recently reviewed for ophthalmic research in [21] and it is a tool used in neuroscience studies—e.g., [22]—and in bioinformatics [23].

Regarding the use of graphs to represent a series of rankings, we recall, in particular, that a graph can be used to describe the crossings between two rankings. This graph is called a *permutation graph* (see [24,25]). When a graph is defined to show the *consecutive crossings* between a series of m rankings, it is called a *Competitivity graph* [5]. This concept corresponds to that of *intersection graph of a concatenation of permutation diagrams* in graph theory (see [26]). For more relations on graphs associated with rankings, see [14].

In this paper, we take some results of [4,12] as our starting point to develop two coefficients to describe the evolution of a series of $m \geq 2$ incomplete rankings with ties. When applied to the case of only two rankings, our measures reduce to the measures given in [4,12].

We also extend the study of a series of complete rankings with ties developed in [13] to the case of incomplete rankings with ties. We make use of the standard modern notation in the field of rankings mainly based on [10,12,27], among others.

We take as our starting point the definition of τ_x of [12] that is based on the computation of a certain sum of the form $\sum_{i=1}^{n} \sum_{j=1}^{n} A_{ij} B_{ij}$ that involves the terms of some matrix A and B that indicate the relative positions of the elements of two rankings. In Theorem 1, we give an expression of this sum as a function of the type of interactions between a pair of elements $\{i,j\}$ from one ranking to the next one (e.g., interchanges from tie to untie, absence of one of the elements in one ranking, crossings, etc.). This result allows for writing τ_x (and $\hat{\tau}_x$) in terms of the interactions of the elements of the rankings.

On the one hand, this theoretical result also allows a computation of the sum $\sum_{i=1}^{n}\sum_{j=1}^{n}A_{ij}B_{ij}$ without computing explicitly the involved matrices. On the other hand, it allows for interpreting the interactions of a series of rankings by using a permutation graph or, more generally speaking, a competitivity graph. The edges are weighted to represent the weight of the corresponding interactions and the whole series of rankings.

We define two coefficients $\tau_{ev}^{\bullet}$ and $\hat{\tau}_{ev}^{\bullet}$ for series of incomplete rankings with ties by using an analogy based on previous well-established definitions. We recall that, in the field of incomplete rankings, "intuition" is usually used for some measures over others since when you handle an incomplete ranking, there is no unique form to interpret the results (see this kind of reasoning in [4,12]). In our case, our measures' behaviour is checked by ensuring that they are well normalized and that they reduce to well-known cases in limit situations.

Finally, other contributions of the paper are placed on a practical field. We give a methodology to study the movements of rankings (of songs) in Spotify by using two different approaches: the cases of series of incomplete rankings without ties and series of incomplete rankings with ties.

The structure of the paper is as follows. In Section 2, we recall Kendall's τ and give the fundamental relations that will be useful throughout the paper. In Section 3, we recall the notation and basic results for the case of two incomplete rankings with ties allowed.

In Section 4, we give the fundamental theoretical result of the paper and some remarks that give insight both into the validity and application of this result. In Section 5, we recall some definitions from [13] to measure the evolution of m complete rankings with ties. In Section 6, we present two coefficients, denoted as $\tau_{ev}^{\bullet}$ and $\hat{\tau}_{ev}^{\bullet}$ to characterize the evolution of m incomplete rankings with ties and some examples are given. In Section 7, we illustrate the applicability of the new coefficients by using some real data obtained from Spotify charts. Finally, in Section 8, we outline the main conclusions of the paper.

2. Preliminaries

In [16] it is shown that Kendall's τ coefficient (also called *measure of disarray*) associated with two rankings with the same number of elements n, can be written in the form

$$\tau = 1 - \frac{2s}{\frac{1}{2}n(n-1)} \qquad (1)$$

where s is the minimum number of interchanges required to transform one ranking into the other. This coefficient is a measure of the intensity of rank correlation. The coefficient can also be written as

$$\tau = \frac{P-Q}{\frac{1}{2}n(n-1)} \qquad (2)$$

where P is the number of pair of elements that maintain its relative order when passing from the first ranking to the second one (that is, the first element is above or below the second in both rankings) and Q is the number of pairs of elements that interchange its order (that is, in one ranking, the first element is above the second and, in the other ranking, the first element is below the second, or vice-versa).

Note that Q and s are equal. Furthermore, this quantity can be identified with the number of *crossings* or *inversions* when passing from the first ranking to the second. For this reason, throughout the paper, we will keep in mind that Equation (1) gives the equivalence between the number of crossings and the associated τ. This will be important in what follows since we will deal with different extensions of Kendall's τ coefficient and since one of our preferred tools will be counting the number of crossings, as in [5].

We recall from [27] that a *distance metric* $d(\mathbf{a}, \mathbf{b})$ can be transformed into a correlation coefficient $\tau(\mathbf{a}, \mathbf{b})$ by the formula

$$\tau(\mathbf{a}, \mathbf{b}) = 1 - \frac{2d(\mathbf{a}, \mathbf{b})}{d_{max}(\mathbf{a}, \mathbf{b})} \tag{3}$$

where $d_{max}(\mathbf{a}, \mathbf{b})$ is the maximum possible distance between two rankings. We recall that a distance metric between two rankings $\mathbf{a}$ and $\mathbf{b}$ is a non-negative real function f, such that it is *symmetric* ($f(\mathbf{a}, \mathbf{b}) = f(\mathbf{b}, \mathbf{a})$, for any pair of rankings), *regular* ($f(\mathbf{a}, \mathbf{b}) = 0 \leftrightarrow \mathbf{a} = \mathbf{b}$) and satisfying the *triangle inequality* ($f(\mathbf{a}, \mathbf{c}) \leq f(\mathbf{a}, \mathbf{b}) + f(\mathbf{b}, \mathbf{c})$, for any rankings $\mathbf{a}$, $\mathbf{b}$, and $\mathbf{c}$). Note that Equation (1) is of this form, since $n(n-1)/2$ is the maximum number of crossings between two given rankings. The same happens with the Spearman's ρ coefficient. In [16] the Spearman's ρ for two ordinal complete rankings $\mathbf{x} = (x_1, x_2, \ldots, x_n)$ and $\mathbf{y} = (y_1, y_2, \ldots, y_n)$ with $x_i, y_i \in \mathbb{N}$ is defined by

$$\rho = 1 - \frac{6\sum_{i=1}^{n}(x_i - y_i)^2}{n^3 - n}$$

and this is of the form (3) since it is easy to show that the maximum value of $\sum_{i=1}^{n}(x_i - y_i)^2$ occurs when one ranking is the reverse of the other and, as a consequence, the maximum value of the *distance metric* $d(\mathbf{x}, \mathbf{y}) = \sum_{i=1}^{n}(x_i - y_i)^2$ is $\frac{1}{3}(n^3 - n)$ (see [3] for this and other properties of distance metrics).

We also recall that a permutation graph (called *competitivity graph* in [5]) is associated with two rankings over the same elements in such a way that the nodes represent the elements and two nodes are connected with an edge if they cross their positions when passing from one ranking to the other.

In this way, it is clear that the number of edges of this graph is, precisely, s. Furthermore, another quantity (borrowed from graph theory) is also introduced in [5]: the *Normalized Mean Strength NS*; that is, the normalized sum of the weights of the edges of a weighted graph. When considering only two rankings and its corresponding *competitivity graph*, we have the following relation

$$NS = \frac{1 - \tau}{2} \tag{4}$$

that gives the equivalence between the *Normalized Mean Strength* and Kendall's τ for two rankings. Note that $\tau \in [-1, 1]$ and $NS \in [0, 1]$. We consider that the measure NS is more intuitive than τ since it allows us to interpret the movements or *activity* of a series of rankings as a percentage.

3. Coefficients for Two Incomplete Rankings with Ties

In this section, we recall some definitions used in [4,12]. We will use the next three ingredients in order to define a coefficient to compare two rankings:

1. A vector to define the ordinal ranking (including the description of absent elements and tied elements);
2. A matrix to indicate the relative positions of the elements of the ranking (including absent and tied elements);
3. A formula to define the coefficients for a pair of rankings by using the entries of their associate matrices defined in the previous step.

Let $V = \{v_1, v_2, \cdots, v_n\}$ be the objects to be ranked, with $n > 1$. The ranking is given by

$$\mathbf{a} = [a_1, a_2, \cdots, a_n] \tag{5}$$

where a_i is the position of v_i in the ranking. Note that if $a_i = a_j$, then v_i and v_j are tied. If v_i is not ranked, then it is denoted as $a_i = \bullet$. We also define the set

$$V_{\mathbf{a}} = \{v_i \in V \mid a_i \neq \bullet\}.$$

We define an $n \times n$ matrix $A = (A_{ij})$, with entries A_{ij} associated to $\mathbf{a}$ as follows:

$$A_{ij} = \begin{cases} 1 & \text{if } a_i \leq a_j \\ -1 & \text{if } a_i > a_j \\ 0 & \text{if } i = j, a_i = \bullet, \text{ or } a_j = \bullet \end{cases} \tag{6}$$

According to [12], we define the coefficients

$$\tau_x(\mathbf{a}, \mathbf{b}) = \frac{\sum_{i=1}^n \sum_{j=1}^n A_{ij} B_{ij}}{n(n-1)} \tag{7}$$

and, when $\bar{n} > 1$

$$\hat{\tau}_x(\mathbf{a}, \mathbf{b}) = \frac{n(n-1)}{\bar{n}(\bar{n}-1)} \tau_x(\mathbf{a}, \mathbf{b}) \tag{8}$$

where $\bar{n}$ is the number of common ranked elements v_i to $\mathbf{a}$ and $\mathbf{b}$. That is:

$$\bar{n} = |V_a \cap V_b| \tag{9}$$

Example 1. *Let $V = \{1, 2, 3, 4, 5, 6, 7, 8\}$, and let us consider two rankings $\mathbf{a}$ and $\mathbf{b}$. Then, $\mathbf{a} = [6, 4, 5, 5, \bullet, 2, 1, 3]$ represents the incomplete ranking with ties $(7, 6, 8, 2, 3-4, 1)$, where $3-4$ indicate tied elements. Analogously, $\mathbf{b} = [3, 3, 2, 2, \bullet, 1, \bullet, 4]$ represents the ranking $(6, 3-4, 1-2, 8)$. Note that $n = 8$ and $\bar{n} = 6$.*

Note that τ_x with complete rankings and no ties reduces to the classic Kendall's τ given by (1), while $\hat{\tau}_x$ is a renormalization of τ_x, verifying $|\hat{\tau}_x| \geq |\tau_x|$.

As we will see, Definition 6 in Section 6, is based on an analogy with Equation (1). To that end, it will be necessary to count all the possible cases when passing from $\mathbf{a}$ to $\mathbf{b}$ (interactions between the relative positions of pair of elements such as crossings, pass from tie to untie, from being in the ranking to quitting it, etc.). We do this in the next section.

4. Main Result

The following result is the fundamental theoretical result of this paper. This result will allow us to write τ_x and $\hat{\tau}_x$ in terms of the interactions of the rankings' elements. It opens the possibility of giving weights to the interactions, as is a common practice in modern definitions of Kendall's tau [10]. This result also constitutes our starting point to define a coefficient for a series of more than two incomplete rankings. This theorem also allows giving insight into the differences between τ_x and $\hat{\tau}_x$. Some other consequences are detailed in the remarks below and in Corollary 1.

Theorem 1. *Given two vectors $\mathbf{a}, \mathbf{b}$ representing incomplete rankings of n elements with ties, represented as in (5), and their corresponding matrices $A = (A_{ij})$ and $B = (B_{ij})$ defined by (6), it holds that*

$$\sum_{i=1}^n \sum_{j=1}^n A_{ij} B_{ij} = n(n-1) - 4s - 2n_{tu} - 2N_{inc} \tag{10}$$

where

$$N_{inc} = \binom{n_{\bullet\bullet}}{2} + \binom{n_{*\bullet}}{2} + \binom{n_{\bullet*}}{2} + n_{\bullet\bullet}(n_{*\bullet} + n_{\bullet*} + n_{**}) + n_{**}(n_{*\bullet} + n_{\bullet*}) + n_{*\bullet}n_{\bullet*} \quad (11)$$

s is the number of crossings—that is, the number of pairs $\{i, j\}$—such that $a_i < a_j$ and $b_i > b_j$, or $a_i > a_j$ and $b_i < b_j$.

n_{tu} is the number of pairs that are tied in only one ranking (from tie to untie or viceversa), that is, such that $a_i = a_j$ and $b_i \neq b_j$, or $a_i \neq a_j$ and $b_i = b_j$.

In the definitions of s, and n_{tu}, it is assumed that a_i and b_i are different from $\bullet$. For the cases when one or more $\bullet$ may appear, the following notation holds:

$n_{\bullet\bullet}$ is the number of entries such that $a_i = b_i = \bullet$;
$n_{\bullet*}$ is the number of entries, such that $a_i = \bullet$ and $b_i \neq \bullet$;
$n_{*\bullet}$ is the number of entries, such that $a_i \neq \bullet$ and $b_i = \bullet$.

Finally, it is also needed to define n_{**} as the number of entries, such that $a_i \neq \bullet$ and $b_i \neq \bullet$.

Proof of Theorem 1. For each pair $\{i, j\}$ we will evaluate each term $A_{ij}B_{ij} + A_{ji}B_{ji}$ in the expression $\sum_{i=1}^{n}\sum_{j=1}^{n} A_{ij}B_{ij}$. The case $i = j$ gives $A_{ii}B_{ii} + A_{ii}B_{ii} = 0$.

Thus, we focus on pairs $\{i, j\}$ with $i \neq j$. There is a total number of $n(n-1)/2$ of these pairs. It is useful to consider the basic cell of the pair $\{i, j\}$ with $i < j$.

$$\begin{pmatrix} a_i & b_i \\ a_j & b_j \end{pmatrix}$$

where a_k and b_k can be natural numbers or a $\bullet$ if the element k is not ranked in **a** or **b**.

Let us study first the cases that can appear when no $\bullet$ is present in the basic cell.

The Complete Case (C):

That is $a_k \neq \bullet, b_k \neq \bullet$, for all $k \in \{1, 2, \ldots n\}$. We distinguish four types of basic cells.

*Type C.1: Not crossing, and no ties in **a** nor in **b**.*

For example:

$$\begin{pmatrix} 1 & 3 \\ 2 & 4 \end{pmatrix} \quad \text{or} \quad \begin{pmatrix} 2 & 4 \\ 1 & 3 \end{pmatrix}.$$

So that, we have $a_i \neq a_j$ and $b_i \neq b_j$ and two cases can appear:

C.1.1. If $a_i < a_j$ and $b_i < b_j$, then $A_{ij}B_{ij} + A_{ji}B_{ji} = 1 \cdot 1 + (-1) \cdot (-1) = 2$.
C.1.2. If $a_i > a_j$ and $b_i > b_j$, then $A_{ij}B_{ij} + A_{ji}B_{ji} = (-1) \cdot (-1) + 1 \cdot 1 = 2$.

Type C.2: Crossing.

For example:

$$\begin{pmatrix} 1 & 4 \\ 2 & 3 \end{pmatrix} \quad \text{or} \quad \begin{pmatrix} 2 & 3 \\ 1 & 4 \end{pmatrix}.$$

Again, we have $a_i \neq a_j$ and $b_i \neq b_j$ and two more cases can appear:

C.2.1. If $a_i < a_j$ and $b_i > b_j$, then $A_{ij}B_{ij} + A_{ji}B_{ji} = 1 \cdot (-1) + (-1) \cdot 1 = -2$.
C.2.2. If $a_i > a_j$ and $b_i < b_j$, then $A_{ij}B_{ij} + A_{ji}B_{ji} = (-1) \cdot (1) + 1 \cdot (-1) = -2$.

Type C.3: From tie to untie or viceversa.

For example:

$$\begin{pmatrix} 1 & 3 \\ 1 & 4 \end{pmatrix}, \quad \begin{pmatrix} 1 & 4 \\ 1 & 3 \end{pmatrix}, \quad \begin{pmatrix} 3 & 1 \\ 4 & 1 \end{pmatrix}, \quad \text{or} \quad \begin{pmatrix} 4 & 1 \\ 3 & 1 \end{pmatrix}$$

We have $a_i = a_j$ and $b_i \neq b_j$ or $a_i \neq a_j$ and $b_i = b_j$. Therefore, four cases can appear:

C.3.1. If $a_i = a_j$ and $b_i < b_j$ then $A_{ij}B_{ij} + A_{ji}B_{ji} = 1 \cdot 1 + 1 \cdot (-1) = 0$.
C.3.2. If $a_i = a_j$ and $b_i > b_j$ then $A_{ij}B_{ij} + A_{ji}B_{ji} = 1 \cdot (-1) + 1 \cdot 1 = 0$.
C.3.3. If $a_i < a_j$ and $b_i = b_j$ then $A_{ij}B_{ij} + A_{ji}B_{ji} = 1 \cdot 1 + (-1) \cdot 1 = 0$.
C.3.4. If $a_i > a_j$ and $b_i = b_j$ then $A_{ij}B_{ij} + A_{ji}B_{ji} = (-1) \cdot 1 + 1 \cdot 1 = 0$.

Type C.4: From tie to tie.

For example:

$$\begin{pmatrix} 1 & 2 \\ 1 & 2 \end{pmatrix}$$

That is, we have: $a_i = a_j$ and $b_i = b_j$, and then $A_{ij}B_{ij} + A_{ji}B_{ji} = 1 \cdot 1 + 1 \cdot 1 = 2$.

We denote the number of pairs of each case using the terminology of Table 1. Note that n_{tt} is *the number of pairs that are tied in both rankings*, that is, such that $a_i = a_j$ and $b_i = b_j$. Note also that n_{tu} is the number of pairs that go from tie to untie or viceversa.

Table 1. Number of pairs $\{i, j\}$ corresponding to each type for the complete cases.

Type	Number of Pairs
C.1	n_{nc}
C.2	s
C.3	n_{tu}
C.4	n_{tt}

The Incomplete Case (I):

There is at least one • in the basic cell. In other words, there is some k such that $a_k = \bullet$, or $b_k = \bullet$, or both. We distinguish seven cases:

Type I.1: Four •. That is $a_i = a_j = b_i = b_j = \bullet$, or graphically

$$\begin{pmatrix} \bullet & \bullet \\ \bullet & \bullet \end{pmatrix}$$

Then $A_{ij}B_{ij} + A_{ji}B_{ji} = 0 \cdot 0 + 0 \cdot 0 = 0$. Let us denote by $n_{\bullet\bullet}$ the number of null rows that appear in the matrix with columns **a** and **b**. Therefore, we have $\binom{n_{\bullet\bullet}}{2}$ pairs $\{i, j\}$ of this type.

Type I.2: Three •. That is, a cell of one of these forms

$$\begin{pmatrix} \bullet & \bullet \\ * & \bullet \end{pmatrix}, \quad \begin{pmatrix} * & \bullet \\ \bullet & \bullet \end{pmatrix}, \quad \begin{pmatrix} \bullet & \bullet \\ \bullet & * \end{pmatrix}, \quad \text{or} \quad \begin{pmatrix} \bullet & * \\ \bullet & \bullet \end{pmatrix}$$

where $*$ is a number (not a •). Therefore, we have four cases, but all are similar to this one: $a_i \neq \bullet$ and $a_j = b_i = b_j = 0$. Then, $A_{ij}B_{ij} + A_{ji}B_{ji} = 0 \cdot 0 + 0 \cdot 0 = 0$.

Denoting $n_{*\bullet}$ the number of rows of the form $(* \ \bullet)$ in the $n \times 2$ matrix $(\mathbf{a} \ \mathbf{b})$, and $n_{\bullet*}$ the number of rows of the form $(\bullet \ *)$ in the same matrix, it is clear that the number of pairs $\{i, j\}$ of this type is: $n_{\bullet\bullet}(n_{*\bullet} + n_{\bullet*})$.

Type I.3: Two •, one on each ranking. That is, any cell of one of these forms

$$\begin{pmatrix} \bullet & \bullet \\ * & * \end{pmatrix}, \begin{pmatrix} * & * \\ \bullet & \bullet \end{pmatrix}, \begin{pmatrix} \bullet & * \\ * & \bullet \end{pmatrix}, \text{ or } \begin{pmatrix} * & \bullet \\ \bullet & * \end{pmatrix}$$

These four cases can be reduced to two:

I.3.1. If $a_i = b_i = \bullet, a_j \neq \bullet$ and $b_j \neq \bullet$, then $A_{ij}B_{ij} + A_{ji}B_{ji} = 0 \cdot 0 + 0 \cdot 0 = 0$.
I.3.2. If $a_i = \bullet, a_j \neq \bullet, b_i \neq \bullet$ and $b_j = \bullet$, then $A_{ij}B_{ij} + A_{ji}B_{ji} = 0 \cdot 0 + 0 \cdot 0 = 0$.

Denoting by n_{**} the number of rows of the form $(* \ *)$ in the $n \times 2$ matrix $(\mathbf{a} \ \mathbf{b})$, it is clear that the number of pairs $\{i,j\}$ of this type is $n_{\bullet\bullet} n_{**} + n_{*\bullet} n_{\bullet*}$.

Type I.4: Tied in one ranking and two • in the other. For example,

$$\begin{pmatrix} 1 & \bullet \\ 1 & \bullet \end{pmatrix}, \begin{pmatrix} \bullet & 1 \\ \bullet & 1 \end{pmatrix}$$

That is, we have two cases, which are similar to this $a_i = a_j$ and $b_i = b_j = \bullet$, and then $A_{ij}B_{ij} + A_{ji}B_{ji} = 0 \cdot 0 + 0 \cdot 0 = 0$.

Let us denote by n_a the number of different natural numbers in $\mathbf{a}$ and by n_b be the number of different natural numbers in $\mathbf{b}$. Let $n_{i\bullet}$ be the number of rows of the form $(i, \bullet)$ in that matrix, for $i = 1, \ldots, n_a$ and, analogoulsly, let $n_{\bullet i}$ be the number of rows of the form $(\bullet, i)$ in the matrix $(\mathbf{ab})$ for $i = 1, \ldots, n_b$. Then, it is straightforward to see that the number of cases of this type is given by

$$\sum_{i=1}^{n_a} \binom{n_{i\bullet}}{2} + \sum_{i=1}^{n_b} \binom{n_{\bullet i}}{2}.$$

Type I.5: Tied in one ranking, one • in the other. For example

$$\begin{pmatrix} 1 & \bullet \\ 1 & 2 \end{pmatrix}, \begin{pmatrix} 1 & 2 \\ 1 & \bullet \end{pmatrix}, \begin{pmatrix} \bullet & 1 \\ 2 & 1 \end{pmatrix}, \begin{pmatrix} 2 & 1 \\ \bullet & 1 \end{pmatrix}.$$

We have the following 4 cases:

I.5.1. If $a_i = a_j$ and $b_i = \bullet$ and $b_j \neq \bullet$, then $A_{ij}B_{ij} + A_{ji}B_{ji} = 0 \cdot 0 + 0 \cdot 0 = 0$.
I.5.2. If $a_i = a_j$ and $b_i \neq \bullet$ and $b_j = \bullet$, then $A_{ij}B_{ij} + A_{ji}B_{ji} = 0 \cdot 0 + 0 \cdot 0 = 0$.
I.5.3. If $a_i = \bullet$ and $a_j \neq \bullet$ and $b_i = b_j$, then $A_{ij}B_{ij} + A_{ji}B_{ji} = 0 \cdot 0 + 0 \cdot 0 = 0$.
I.5.4. If $a_i \neq \bullet$ and $a_j = \bullet$ and $b_i = b_j$, then $A_{ij}B_{ij} + A_{ji}B_{ji} = 0 \cdot 0 + 0 \cdot 0 = 0$.

Let n_{i*} be the number of rows of the form $(i, *)$ (where $*$ can be i) in the same matrix, with $i \in \{1, 2, \ldots n_a\}$.

Analogously, let n_{*i} be the number of rows of the form $(*, i)$ (where $*$ can be i) in the matrix $(\mathbf{a} \ \mathbf{b})$. Then, it is straightforward to see that the number of cases of this type is given by

$$\sum_{i=1}^{n_a} n_{i*} n_{i\bullet} + \sum_{i=1}^{n_b} n_{*i} n_{\bullet i}.$$

Type I.6: Two • in one ranking and different numbers in the other.

For example

$$\begin{pmatrix} 1 & \bullet \\ 2 & \bullet \end{pmatrix}, \begin{pmatrix} \bullet & 2 \\ \bullet & 1 \end{pmatrix}$$

We have here only two cases:

I.6.1. If $a_i \neq a_j$ and $b_i = b_j = \bullet$ then $A_{ij}B_{ij} + A_{ji}B_{ji} = (\pm 1) \cdot 0 + (\pm 1) \cdot 0 = 0$.
I.6.2. If $a_i = a_j = \bullet$ and $b_i \neq b_j$ then $A_{ij}B_{ij} + A_{ji}B_{ji} = 0 \cdot (\pm 1) + 0 \cdot (\pm 1) = 0$.

Then, it is easy to see that the number of pairs $\{i,j\}$ of this type is

$$\binom{n_{*\bullet}}{2} + \binom{n_{\bullet*}}{2} - \sum_{i=1}^{n_a} n_{i*}n_{i\bullet} - \sum_{i=1}^{n_b} n_{*i}n_{\bullet i}$$

where we have subtracted the number of cases of the type I.4.

Type I.7: Only one $\bullet$ and no ties.

For example, they are cases of the form

$$\begin{pmatrix} 1 & 1 \\ 2 & \bullet \end{pmatrix}, \begin{pmatrix} 1 & \bullet \\ 2 & 1 \end{pmatrix}, \begin{pmatrix} 1 & 1 \\ \bullet & 2 \end{pmatrix}, \begin{pmatrix} \bullet & 2 \\ 1 & 1 \end{pmatrix}$$

We can have four cases that are similar to these

If $a_i < a_j$ and $b_i \neq \bullet, b_j = \bullet$ then $A_{ij}B_{ij} + A_{ji}B_{ji} = 1 \cdot 0 + (-1) \cdot 0 = 0$.
If $a_i > a_j$ and $b_i \neq \bullet, b_j = \bullet$ then $A_{ij}B_{ij} + A_{ji}B_{ji} = (-1) \cdot 0 + 1 \cdot 0 = 0$.

Let n_{i*} be number of rows of the form $(i,*)$ (where $*$ can be i) in the same matrix, with $i \in \{1,2,\ldots,n_a\}$ and, analogously, let n_{*i} be the number of rows of the form $(*,i)$ (where $*$ can be i) in the matrix $(\mathbf{a}\,\mathbf{b})$, with $i \in \{1,2,\ldots,n_a\}$. Then, the number of pairs $\{i,j\}$ of this type is given by

$$n_{**}(n_{*\bullet} + n_{\bullet*}) - \sum_{i=1}^{n_a} n_{i*}n_{i\bullet} - \sum_{i=1}^{n_b} n_{*i}n_{\bullet i}$$

where we have subtracted the number of cases of the type I.5.

In Table 2 we overview the number of cases for each type of the incomplete case.

Table 2. Number of pairs $\{i,j\}$ corresponding to each type for the incomplete cases.

Type	Number of Pairs $\{i,j\}$
I.1	$\binom{n_{\bullet\bullet}}{2}$
I.2	$n_{\bullet\bullet}(n_{*\bullet} + n_{\bullet*})$
I.3	$n_{\bullet\bullet}n_{**} + n_{*\bullet}n_{\bullet*}$
I.4	$\sum_{i=1}^{n_a} \binom{n_{i\bullet}}{2} + \sum_{i=1}^{n_b} \binom{n_{\bullet i}}{2}$
I.5	$\sum_{i=1}^{n_a} n_{i*}n_{i\bullet} + \sum_{i=1}^{n_b} n_{*i}n_{\bullet i}$
I.6	$\binom{n_{*\bullet}}{2} + \binom{n_{\bullet*}}{2} - \sum_{i=1}^{n_a} \binom{n_{i\bullet}}{2} - \sum_{i=1}^{n_b} \binom{n_{\bullet i}}{2}$
I.7	$n_{**}(n_{*\bullet} + n_{\bullet*}) - \sum_{i=1}^{n_a} n_{i*}n_{i\bullet} - \sum_{i=1}^{n_b} n_{*i}n_{\bullet i}$

To end the proof, we add the contributions for all the cases, complete (C) and incomplete (I), to the sum $\sum_{i=1}^{n}\sum_{j=1}^{n} A_{ij}B_{ij}$ and we obtain

$$\sum_{i=1}^{n}\sum_{j=1}^{n} A_{ij}B_{ij} = 2n_{nc} - 2s + 2n_{tt} \tag{12}$$

Now, taking into account that all the cases must amount up to the total number of pairs we have

$$\frac{n(n-1)}{2} = n_{nc} + s + n_{tt} + n_{tu} + N_{inc} \tag{13}$$

where N_{inc} is the sum of all the cases in Table 2. By plugging $n_{nc} = \frac{n(n-1)}{2} - s - n_{tt} - n_{tu} - N_{inc}$ into (12), we finally get

$$\sum_{i=1}^{n}\sum_{j=1}^{n} A_{ij}B_{ij} = n(n-1) - 4s - 2n_{tu} - 2N_{inc}$$

where

$$N_{inc} = \binom{n_{\bullet\bullet}}{2} + \binom{n_{*\bullet}}{2} + \binom{n_{\bullet*}}{2} + n_{\bullet\bullet}(n_{*\bullet} + n_{\bullet*} + n_{**}) + n_{**}(n_{*\bullet} + n_{\bullet*}) + n_{*\bullet}n_{\bullet*}$$

□

In the next example, we illustrate the previous result.

Example 2. *Given the rankings* $\mathbf{a} = [1, \bullet, 2, \bullet, 3, 2, \bullet, \bullet, \bullet, 1]$ *and* $\mathbf{b} = [2, \bullet, 4, 2, \bullet, 1, 3, 3, \bullet, 2]$, *then* $n = 10$, $n_{\bullet\bullet} = 2$, $n_{\bullet*} = 3$, $n_{*\bullet} = 1$, $n_{**} = 4$, $s = 2$ *(corresponding to the pairs* $\{1,6\}$ *and* $\{6,10\}$*),* $n_{tu} = 1$ *(corresponding to the pair* $\{3,6\}$*)*, $n_{tt} = 1$ *(corresponding to the pair* $\{1,10\}$*),* $n_a = 3$, $n_b = 4$, $n_{1\bullet} = n_{2\bullet} = 0$, $n_{3\bullet} = 1$, $n_{\bullet 1} = 0$, $n_{\bullet 2} = 1$, $n_{\bullet 3} = 2$, $n_{\bullet 4} = 0$, $n_{1*} = 2$, $n_{2*} = 2$, $n_{3*} = 0$, $n_{*1} = 1$, $n_{*2} = 2$, $n_{*3} = 0$, *and,* $n_{*4} = 1$.

From the parameters of Table 3, we obtain $N_{inc} = 39$. *Thus, it is easy to check that* $\sum_{i=1}^{n}\sum_{j=1}^{n} A_{ij}B_{ij} = n(n-1) - 4s - 2n_{tu} - 2N_{inc} = 2$ *as stated in Theorem 1.*

The number of pairs $\{i,j\}$ *is 45, corresponding to the following cells*

$$\begin{pmatrix} 1 & 2 \\ \bullet & \bullet \end{pmatrix}, \begin{pmatrix} 1 & 2 \\ 2 & 4 \end{pmatrix}, \begin{pmatrix} 1 & 2 \\ \bullet & 2 \end{pmatrix}, \begin{pmatrix} 1 & 2 \\ 3 & \bullet \end{pmatrix}, \begin{pmatrix} 1 & 2 \\ 2 & 1 \end{pmatrix}, \begin{pmatrix} 1 & 2 \\ \bullet & 3 \end{pmatrix}, \begin{pmatrix} 1 & 2 \\ \bullet & 3 \end{pmatrix}, \begin{pmatrix} 1 & 2 \\ \bullet & \bullet \end{pmatrix}$$

$$\begin{pmatrix} 1 & 2 \\ 1 & 2 \end{pmatrix}, \begin{pmatrix} \bullet & \bullet \\ 2 & 4 \end{pmatrix}, \begin{pmatrix} \bullet & \bullet \\ \bullet & 2 \end{pmatrix}, \begin{pmatrix} \bullet & \bullet \\ 3 & \bullet \end{pmatrix}, \begin{pmatrix} \bullet & \bullet \\ 2 & 1 \end{pmatrix}, \begin{pmatrix} \bullet & \bullet \\ \bullet & 3 \end{pmatrix}, \begin{pmatrix} \bullet & \bullet \\ \bullet & 3 \end{pmatrix}, \begin{pmatrix} \bullet & \bullet \\ \bullet & \bullet \end{pmatrix}$$

$$\begin{pmatrix} \bullet & \bullet \\ 1 & 2 \end{pmatrix}, \begin{pmatrix} 2 & 4 \\ \bullet & 2 \end{pmatrix}, \begin{pmatrix} 2 & 4 \\ 3 & \bullet \end{pmatrix}, \begin{pmatrix} 2 & 4 \\ 2 & 1 \end{pmatrix}, \begin{pmatrix} 2 & 4 \\ \bullet & 3 \end{pmatrix}, \begin{pmatrix} 2 & 4 \\ \bullet & 3 \end{pmatrix}, \begin{pmatrix} 2 & 4 \\ \bullet & \bullet \end{pmatrix}, \begin{pmatrix} 2 & 4 \\ 1 & 2 \end{pmatrix}$$

$$\begin{pmatrix} \bullet & 2 \\ 3 & \bullet \end{pmatrix}, \begin{pmatrix} \bullet & 2 \\ 2 & 1 \end{pmatrix}, \begin{pmatrix} \bullet & 2 \\ \bullet & 3 \end{pmatrix}, \begin{pmatrix} \bullet & 2 \\ \bullet & 3 \end{pmatrix}, \begin{pmatrix} \bullet & 2 \\ \bullet & \bullet \end{pmatrix}, \begin{pmatrix} \bullet & 2 \\ 1 & 2 \end{pmatrix}, \begin{pmatrix} 3 & \bullet \\ 2 & 1 \end{pmatrix}, \begin{pmatrix} 3 & \bullet \\ \bullet & 3 \end{pmatrix}$$

$$\begin{pmatrix} 3 & \bullet \\ \bullet & 3 \end{pmatrix}, \begin{pmatrix} 3 & \bullet \\ \bullet & \bullet \end{pmatrix}, \begin{pmatrix} 3 & \bullet \\ 1 & 2 \end{pmatrix}, \begin{pmatrix} 2 & 1 \\ \bullet & 3 \end{pmatrix}, \begin{pmatrix} 2 & 1 \\ \bullet & 3 \end{pmatrix}, \begin{pmatrix} 2 & 1 \\ \bullet & \bullet \end{pmatrix}, \begin{pmatrix} 2 & 1 \\ 1 & 2 \end{pmatrix}, \begin{pmatrix} \bullet & 3 \\ \bullet & 3 \end{pmatrix}$$

$$\begin{pmatrix} \bullet & 3 \\ \bullet & \bullet \end{pmatrix}, \begin{pmatrix} \bullet & 3 \\ 1 & 2 \end{pmatrix}, \begin{pmatrix} \bullet & 3 \\ \bullet & \bullet \end{pmatrix}, \begin{pmatrix} \bullet & 3 \\ 1 & 2 \end{pmatrix}, \begin{pmatrix} \bullet & \bullet \\ 1 & 2 \end{pmatrix}$$

and the number of cases of each type for the incomplete case appearing on Theorem 1 are shown in Table 3.

Table 3. Number of pairs $\{i,j\}$ that have some •, corresponding to Example 2. Note that the sum of all the types is, by definition in (11), N_{inc}.

Type	Number of Pairs $\{i,j\}$
I.1	$\binom{n_{\bullet\bullet}}{2} = 1$
I.2	$n_{\bullet\bullet}(n_{*\bullet} + n_{\bullet *}) = 8$
I.3	$n_{\bullet\bullet}n_{**} + n_{*\bullet}n_{\bullet *} = 11$
I.4	$\sum_{i=1}^{n_a} \binom{n_{i\bullet}}{2} + \sum_{i=1}^{n_b} \binom{n_{\bullet i}}{2} = 1$
I.5	$\sum_{i=1}^{n_a} n_{i*}n_{i\bullet} + \sum_{i=1}^{n_b} n_{*i}n_{\bullet i} = 2$
I.6	$\binom{n_{*\bullet}}{2} + \binom{n_{\bullet *}}{2} - \sum_{i=1}^{n_a} \binom{n_{i\bullet}}{2} - \sum_{i=1}^{n_b} \binom{n_{\bullet i}}{2} = 2$
I.7	$n_{**}(n_{*\bullet} + n_{\bullet *}) - \sum_{i=1}^{n_a} n_{i*}n_{i\bullet} - \sum_{i=1}^{n_b} n_{*i}n_{\bullet i} = 14$

Remark 1. *By using (10) and (7) we obtain*

$$\tau_x = 1 - \frac{4(s + \frac{1}{2}n_{tu}) + 2N_{inc}}{n(n-1)} \tag{14}$$

that can be thought of an extension of (1) to the case of two incomplete rankings with ties. This formula is one of the original contributions of this paper. Note that the term N_{inc} is known since it is given by (11). This formula will be useful in Section 6 to define our measure of correlation for a series of incomplete rankings with ties.

Remark 2. *For two complete rankings with ties allowed, Equation (10) simplifies to*

$$\sum_{i=1}^{n}\sum_{j=1}^{n} A_{ij}B_{ij} = n(n-1) - 4s - 2n_{tu} \tag{15}$$

If we recall the definition of the distance of Kemeny and Snell [2] depending on a matrix $C(\mathbf{a}) = C_{ij}(\mathbf{a})$ such that

$$C_{ij}(\mathbf{a}) = \begin{cases} 1 & \text{if element } i \text{ is preferred to element } j \\ -1 & \text{if element } j \text{ is preferred to element } i \\ 0 & \text{if } i = j, \text{ or if both elements } i \text{ and } j \text{ are tied} \end{cases} \tag{16}$$

by following a similar procedure as in the proof of Theorem 1 it is easy to show that

$$\sum_{ij} |C_{ij}(\mathbf{a}) - C_{ji}(\mathbf{b})| = 4s + 2n_{tu} \tag{17}$$

and by using (15) we get

$$\sum_{i=1}^{n}\sum_{j=1}^{n} A_{ij}B_{ij} = n(n-1) - \sum_{ij} |C_{ij}(\mathbf{a}) - C_{ji}(\mathbf{b})| \tag{18}$$

that it is in agreement with the results shown in [27], but we obtain it as a particular case of Theorem 1.

Remark 3. *The common number of ranked elements in $\mathbf{a}$ and $\mathbf{b}$ that we denote as $\bar{n}$ in (9) is precisely n_{**}. Moreover, by using that*

$$n_{\bullet *} + n_{*\bullet} + n_{\bullet\bullet} = n - \bar{n}$$

Let us check that N_{inc} given by (11) can be rewritten as

$$N_{inc} = \binom{n}{2} - \binom{\overline{n}}{2} \tag{19}$$

To that end, it is needed to use that $n_{**} = \overline{n}$ and

$$n_{\bullet *} + n_{*\bullet} + n_{\bullet\bullet} = n - \overline{n} \tag{20}$$

To see how it is, we first note that

$$\begin{aligned}\binom{n_{\bullet\bullet}}{2} + \binom{n_{*\bullet}}{2} + \binom{n_{\bullet *}}{2} &= \frac{1}{2}\left[n_{\bullet\bullet}^2 + n_{*\bullet}^2 + n_{\bullet *}^2 - (n_{\bullet\bullet} + n_{*\bullet} + n_{\bullet *})\right] \\ &= \frac{1}{2}\left[n_{\bullet\bullet}^2 + n_{*\bullet}^2 + n_{\bullet *}^2 - n + \overline{n}\right]\end{aligned} \tag{21}$$

Second, we can simplify, by using (20)

$$n_{\bullet\bullet}(n_{*\bullet} + n_{\bullet *} + n_{**}) = n_{\bullet\bullet}(n - n_{\bullet\bullet}) \tag{22}$$

Third, note that, by using (20),

$$n_{**}(n_{*\bullet} + n_{\bullet *}) = \overline{n}n - \overline{n}^2 - \overline{n}n_{\bullet\bullet} \tag{23}$$

Now, by using (21)–(23) we have that N_{inc} given by (11) becomes

$$N_{inc} = \frac{1}{2}n_{\bullet\bullet}^2 + \frac{1}{2}(n_{*\bullet} + n_{\bullet *})^2 + \frac{1}{2}(\overline{n} - n) + n_{\bullet\bullet}(n - n_{\bullet\bullet} - \overline{n}) + \overline{n}(n - \overline{n})$$

and since

$$\frac{1}{2}(n_{*\bullet} + n_{\bullet *})^2 = \frac{1}{2}\left(n^2 - 2n\overline{n} + \overline{n}^2 + 2\overline{n}n_{\bullet\bullet} - 2nn_{\bullet\bullet} + n_{\bullet\bullet}^2\right)$$

we get

$$N_{inc} = \frac{1}{2}(\overline{n} - n) + \frac{1}{2}\left(n^2 - 2n\overline{n} + \overline{n}^2\right) + \overline{n}n - \overline{n}^2 = \frac{n(n-1)}{2} + \frac{\overline{n} - \overline{n}^2}{2}$$

that is to say

$$N_{inc} = \binom{n}{2} - \binom{\overline{n}}{2}$$

and the proof is done. Note also that, by using (13), we have: $\binom{\overline{n}}{2} = n_{nc} + s + n_{tt} + n_{tu}$.

This last remark motivates the next result.

Corollary 1. *Given two vectors* **a**, **b** *representing incomplete rankings of n elements with ties and their corresponding matrices $A = (A_{ij})$ and $B = (B_{ij})$, it holds that*

$$\sum_{i=1}^{n}\sum_{j=1}^{n} A_{ij}B_{ij} = \overline{n}(\overline{n} - 1) - 4s - 2n_{tu} \tag{24}$$

where $\overline{n}$ is the number of common ranked elements in both rankings—see (9)—s is the number of crossings, that is, the number of pairs $\{i,j\}$, such that $a_i < a_j$ and $b_i > b_j$ or $a_i > a_j$ and $b_i < b_j$, and n_{tu} is the number of pairs that are tied in only one ranking (from tie to untie or viceversa), that is, such that $a_i = a_j$ and $b_i \neq b_j$, or $a_i \neq a_j$ and $b_i = b_j$.

With (24), it is easy to obtain the maximum and minimum of the expression $\sum_{i=1}^{n}\sum_{j=1}^{n} A_{ij}B_{ij}$. When $s = 0$ and $n_{tu} = 0$ we have

$$\sum_{i=1}^{n}\sum_{j=1}^{n} A_{ij}B_{ij} = \bar{n}(\bar{n}-1)$$

that is the maximum value of $\sum_{i=1}^{n}\sum_{j=1}^{n} A_{ij}B_{ij}$. Analogously, by taking $s = \binom{\bar{n}}{2}$, that is the maximum number of crossings and consequently $n_{tu} = 0$, we obtain from (24)

$$\sum_{i=1}^{n}\sum_{j=1}^{n} A_{ij}B_{ij} = \bar{n}(\bar{n}-1) - 4\binom{\bar{n}}{2} = -\bar{n}(\bar{n}-1)$$

that is the minimum value of $\sum_{i=1}^{n}\sum_{j=1}^{n} A_{ij}B_{ij}$. These facts, that are in agreement with the results shown in [12], explain why $\hat{\tau}_x$ defined by (8) takes values in $[-1, 1]$.

Remark 4. *By using (7) and (24) we obtain*

$$\tau_x = \frac{\bar{n}(\bar{n}-1)}{n(n-1)} - \frac{4s + 2n_{tu}}{n(n-1)} \qquad (25)$$

and from (8) and (25) we get

$$\hat{\tau}_x = 1 - \frac{4s + 2n_{tu}}{\bar{n}(\bar{n}-1)} \qquad (26)$$

Remark 5. *As we have pointed out in (3), a distance metric $d(\mathbf{a}, \mathbf{b})$ can be transformed into a correlation coefficient $\tau(\mathbf{a}, \mathbf{b})$ by the formula*

$$\tau(\mathbf{a}, \mathbf{b}) = 1 - \frac{2d(\mathbf{a}, \mathbf{b})}{d_{max}(\mathbf{a}, \mathbf{b})} \qquad (27)$$

Note that in expression (14), when $N_{inc} \neq 0$, the quantity $n(n-1)$ is not the maximum value of the distance metric $d(\mathbf{a}, \mathbf{b}) = 2s + n_{tu} + N_{inc}$ (see Example 6). This problem does not appear with the use of $\hat{\tau}_x$ since, by using (26) we can identify a "distance metric" given by $\hat{d}(\mathbf{a}, \mathbf{b}) = 2s + n_{tu}$ and its maximum value is achieved when $s = \bar{n}(\bar{n}-1)/2$ (and consequently $n_{tu} = 0$) and has the value of

$$\hat{d}_{max} = \bar{n}(\bar{n}-1)/2$$

Therefore, $\hat{\tau}_x$ should be preferred over τ_x in terms of normalization (see [12] for other considerations). This fact will be useful for the definition that we will introduce in Section 6.

In the next examples, we illustrate the two previous remarks. Note that when $s = 0$ and $n_{tu} = 0$ then, by (26), $\hat{\tau}_x = 1$ and it is not affected by the presence of • in the rankings. By analogy with (4), we denote the *Normalized Mean Strength* of **a** and **b** as

$$NS(\mathbf{a}_1, \mathbf{a}_2) = \frac{(1 - \tau_x)}{2}, \quad \text{and} \quad \widehat{NS}(\mathbf{a}_1, \mathbf{a}_2) = \frac{(1 - \hat{\tau}_x)}{2}.$$

Example 3. *Let $\mathbf{a}_1 = [1,2,3,\bullet,\bullet,\bullet]$ and $\mathbf{a}_2 = [1,\bullet,2,3,\bullet,\bullet]$. It is easy to obtain: $N_{inc}(\mathbf{a}_1, \mathbf{a}_2) = 14$, $\tau_x(\mathbf{a}_1, \mathbf{a}_2) = 0.1556$, $NS(\mathbf{a}_1, \mathbf{a}_2) = 0.4222$, $\hat{\tau}_x(\mathbf{a}_1, \mathbf{a}_2) = 1$, and $\widehat{NS}(\mathbf{a}_1, \mathbf{a}_2) = 0.0$.*

Example 4. *Let $\mathbf{a}_1 = [1,2,3,4,\bullet,\bullet]$ and $\mathbf{a}_2 = [1,\bullet,2,3,4,\bullet]$. It is easy to obtain: $N_{inc}(\mathbf{a}_1, \mathbf{a}_2) = 12$, $\tau_x(\mathbf{a}_1, \mathbf{a}_2) = 0.2$, $NS(\mathbf{a}_1, \mathbf{a}_2) = 0.4$, $\hat{\tau}_x(\mathbf{a}_1, \mathbf{a}_2) = 1$, and $\widehat{NS}(\mathbf{a}_1, \mathbf{a}_2) = 0.0$.*

The next example shows the results when a ranking is compared to itself and its reverse ranking for the case of complete rankings (note that $\tau_x = \hat{\tau}_x$ since $\bar{n} = n$).

Example 5. Let $\mathbf{a}_1 = [1,2,3,4,5,6]$ and $\mathbf{a}_2 = [6,5,4,3,2,1]$. Then

	$\mathbf{a}_1 \to \mathbf{a}_1$	$\mathbf{a}_1 \to \mathbf{a}_2$
N_{inc}	0	0
τ_x	1.0	-1.0
NS	0.0	1.0
$\widehat{\tau_x}$	1.0	-1.0
$\widehat{NS}$	0.0	1.0

The next example shows that τ_x does not take its limit values when the rankings are incomplete and that $\widehat{\tau_x}$ is not defined when there are no elements in common in both rankings.

Example 6. Let $\mathbf{a}_1 = [1,2,3,\bullet,\bullet,\bullet]$, $\mathbf{a}_2 = [\bullet,\bullet,\bullet,3,2,1]$, $\mathbf{a}_3 = [1,2,3,4,\bullet,\bullet]$, and $\mathbf{a}_4 = [\bullet,\bullet,4,3,2,1]$, Then

	$\mathbf{a}_1 \to \mathbf{a}_1$	$\mathbf{a}_1 \to \mathbf{a}_2$	$\mathbf{a}_3 \to \mathbf{a}_4$
N_{inc}	12	15	14
τ_x	0.2	0.0	-0.0667
NS	0.4	0.5	0.5333
$\widehat{\tau_x}$	1.0	not defined	-1.0
$\widehat{NS}$	0.0	not defined	1.0

Our main practical result in this paper is *the definition of a measure to deal not only with two rankings $\mathbf{a}_1$ and $\mathbf{a}_2$, as we have seen so far, but with a series of incomplete rankings with ties* $\{\mathbf{a}_1, \mathbf{a}_2, \ldots \mathbf{a}_m\}$ *in which,* in practical situations, some kind of time evolution is presented (e.g., a sport ranking during a session where there may be ties or inclusion/elimination of teams, charts of songs ordered on a daily/weekly basis, etc.). In order to define this measure, it will be useful to recall some concepts defined for complete rankings.

5. Treatment of More Than Two Complete Rankings. Known Results

To study the evolution of more than two rankings we will use the concept of Kendall distance defined in [10], where some weights were introduced to measure the changes when passing from one ranking to the next. After that, we will recall how to extend this definition to a series of m complete rankings, as in [13].

5.1. Kendall Distance for Complete Rankings with Penalty Parameters

We recall the definition of Kendall distance with penalty parameters p and q from [10,13].

Definition 1. *Let $\mathbf{a}$ and $\mathbf{b}$ be two complete rankings with ties of the set $N = \{1, \ldots, n\}$, and penalty parameters $p \in [0, \frac{1}{2}]$ and $q \in [0, \frac{1}{2}]$. The Kendall distance with penalty parameters p and q is defined as*

$$K^{(p,q)}(\mathbf{a},\mathbf{b}) = \sum_{\{i,j\} \in N} \bar{K}_{i,j}^{(p,q)}(\mathbf{a},\mathbf{b}) \qquad (28)$$

where $\bar{K}_{i,j}^{(p,q)}(\mathbf{a},\mathbf{b})$ is computed according to the following cases:

Case 1: If i and j are not tied in $\mathbf{a}$, nor in $\mathbf{b}$. If they cross their positions when passing from $\mathbf{a}$ to $\mathbf{b}$ then $\bar{K}_{i,j}^{(p,q)} = 1$. Otherwise, $\bar{K}_{i,j}^{(p,q)} = 0$.

Case 2: If i and j are tied in both $\mathbf{a}$ and $\mathbf{b}$. Then $\bar{K}_{i,j}^{(p,q)} = q$.

Case 3: If i and j are tied only in one ranking. Then $\bar{K}_{i,j}^{(p,q)} = p$.

Remark 6. *The penalty parameters p and q are bounded and take into account the cases where there exist tied elements in* **a**, *in* **b**, *or in both. For our purposes of measuring competitiveness, it is reasonable to assign $p = 1/2$, to represent that they are tied in one ranking, and $q = 0$ to represent that they are tied in both of them. These assignments are inspired by [10]. In particular, they proved that $p \in [0.5, 1]$ in order to get that $K^{(p,0)}$ was a metric.*

Remark 7. *Note that, by using the notation introduced in Theorem 1, it is easy to see that*

$$K_{i,j}^{(p,q)}(\mathbf{a}, \mathbf{b}) = s + p\, n_{tu} + q\, n_{tt}$$

where n_{tt} is the number of pairs $\{i,j\}$ that go from tie to tie. Therefore, by using (14) with $N_{inc} = 0$ we get

$$\tau_x(\mathbf{a}, \mathbf{b}) = 1 - \frac{4 K^{(0.5,0)}(\mathbf{a}, \mathbf{b})}{n(n-1)} \qquad (29)$$

that is, once more, a relation of the form (3). We see here another consequence of Theorem 1: it opens the possibility of defining new metrics based on putting penalties to the cases $n_{\bullet\bullet}$, $n_{\bullet*}$, etc. since it gives an explicit expression on these cases.

With the previous definitions, we can deal with the general case of the study of a series of complete rankings. We do this in the next section.

5.2. Series of Complete Rankings with Ties

In [13], it was shown how to extend Definition 1 to m complete rankings with ties in a natural way. We recall these definitions here because they will be extended in Section 6 to a series of incomplete rankings.

Definition 2. *Given m complete rankings with ties $\mathbf{a}_1, \mathbf{a}_2, \ldots \mathbf{a}_m$ of n elements, we define the evolutive Kendall distance with penalty parameters p and q as*

$$K_{ev}^{(p,q)}(\mathbf{a}_1, \mathbf{a}_2, \ldots, \mathbf{a}_m) = \sum_{i=1}^{m-1} K^{(p,q)}(\mathbf{a}_i, \mathbf{a}_{i+1}). \qquad (30)$$

When handling m rankings it is natural to include a new case (see [13]) that consists of a series of ties between a crossing (see Example 7 further on). Thus it is convenient to define a new case in the definition of $K_{ev}^{(p,q)}(\mathbf{a}_1, \mathbf{a}_2, \ldots, \mathbf{a}_m)$ according to the following rule.

Definition 3. *Given m complete rankings with ties $\mathbf{a}_1, \mathbf{a}_2, \ldots \mathbf{a}_m$ of n elements, we define the* crossing after ties coefficient $\bar{K}_{i,j}^{cat}(\mathbf{a}_1, \mathbf{a}_2, \ldots, \mathbf{a}_m)$ *following the rule*

Case 4. *If there exists a maximal set of rankings $\mathbf{a}_{t_1}, \ldots, \mathbf{a}_{t_k}$ such that for each $\ell = 1, \ldots, k$ the pair $\{i,j\}$ is not tied in $\mathbf{a}_{t_\ell}$, but is tied in $\mathbf{a}_{t_\ell+1}, \mathbf{a}_{t_\ell+2}, \ldots, \mathbf{a}_{t_\ell+s}$, with $s \geq 1$, it is not tied in $\mathbf{a}_{t_\ell+s+1}$ and, moreover, $\{i,j\}$ exchange their relative positions between $\mathbf{a}_{t_\ell}$ and $\mathbf{a}_{t_\ell+s+1}$. In this case $\bar{K}_{i,j}^{cat}(\mathbf{a}_1, \mathbf{a}_2, \ldots, \mathbf{a}_m) = k$, where k is the number of rankings in the maximal set of rankings $\mathbf{a}_{t_1}, \ldots, \mathbf{a}_{t_k}$ verifying the aforementioned property.*

Example 7. *Given the rankings with ties*

r_1	r_2	r_3	r_4	r_5	r_6
1	1,2	1,2	2	1,2	1
2	3	3	1	3	2
3	4	4	3	4	3
4			4	4	4

the corresponding $\mathbf{a}_i$ *are*

$\mathbf{a}_1$	$\mathbf{a}_2$	$\mathbf{a}_3$	$\mathbf{a}_4$	$\mathbf{a}_5$	$\mathbf{a}_6$
1	1	1	2	1	1
2	1	1	1	1	2
3	2	2	2	2	3
4	3	3	3	3	4

we have that the only nonzero crossing after ties coefficient is

$$\bar{K}^{cat}_{1,2}(\mathbf{a}_1, \mathbf{a}_2, \ldots, \mathbf{a}_6) = 2$$

since we have the appearance of the two series

$\mathbf{a}_1$	$\mathbf{a}_2$	$\mathbf{a}_3$	$\mathbf{a}_4$
1	1	1	2
2	1	1	1

and

$\mathbf{a}_4$	$\mathbf{a}_5$	$\mathbf{a}_6$
2	1	1
1	1	2

that show a series of ties between a crossing of the pair $\{i = 1, j = 2\}$.

By including the cases given by Definition 3 in the sum defined in Definition 2, in [13] a *corrected evolutive distance* in the following form is defined.

Definition 4. *Given m complete rankings with ties* $\mathbf{a}_1, \mathbf{a}_2, \ldots \mathbf{a}_m$ *of n elements we define the corrected evolutive Kendall distance with penalty parameters p and q as follows:*

$$K^{(p,q)}_{cev}(\mathbf{a}_1, \ldots, \mathbf{a}_m) = K^{(p,q)}_{ev}(\mathbf{a}_1, \ldots, \mathbf{a}_m) + \sum_{\{i,j\}} \bar{K}^{cat}_{i,j}(\mathbf{a}_1, \ldots, \mathbf{a}_m), \quad (31)$$

where the summation is over the pairs $\{i, j\}$ *that verify Case 4 in Definition 3.*

Following the same argument as in [13], it is easy to show that

$$\max[K^{(0.5,0)}_{cev}(\mathbf{a}_1, \ldots, \mathbf{a}_m)] = \frac{1}{2}(m-1)n(n-1) \quad (32)$$

Now, in analogy with (3) and (14), the Kendall's evolutive coefficient τ_{ev} for a series of m complete rankings with ties can be defined as

$$\tau_{ev}(\mathbf{a}_1, \mathbf{a}_2, \ldots \mathbf{a}_m) = 1 - \frac{4K^{(0.5,0)}_{cev}(\mathbf{a}_1, \ldots, \mathbf{a}_m)}{(m-1)n(n-1)} \quad \in [-1, 1] \quad (33)$$

With these previous definitions we can present the new coefficients for incomplete rankings with ties.

6. New Coefficients for Series of Incomplete Rankings with Ties

Given a series $\{\mathbf{a}_1, \mathbf{a}_2, \ldots, \mathbf{a}_m\}$ of incomplete rankings with ties, for each pair of rankings $\mathbf{a}_i$ and $\mathbf{a}_j$ we can use Definitions 1–4 straightforwardly to also apply for a series of incomplete rankings by assuming that there is no penalty for the case of absent elements (regarding Definitions 1 and 2) and that these absent elements (denoted by '•') do not contribute to either ties or to *crossings after ties* (regarding Definitions 3 and 4). That is, those definitions are applied as they are, ignoring the effect of the absent elements.

Keeping this in mind and, in analogy with (14), given a series of m incomplete rankings we could include the effect of the incomplete cases by defining

$$\tau_{ev}^* = 1 - \frac{2d_{evol}(\mathbf{a}_1, \mathbf{a}_2, \ldots, \mathbf{a}_m)}{\max(d_{evol})} \tag{34}$$

with

$$d_{evol}(\mathbf{a}_1, \mathbf{a}_2, \ldots, \mathbf{a}_m) = 2K_{cev}^{(p=0.5, q=0)}(\mathbf{a}_1, \ldots, \mathbf{a}_m) + \sum_{i=1}^{m-1} N_{inc}(\mathbf{a}_i, \mathbf{a}_{i+1})$$

where $N_{inc}(\mathbf{a}_i, \mathbf{a}_{i+1})$ is the number of incomplete cases when passing from ranking $\mathbf{a}_i$ to ranking $\mathbf{a}_{i+1}$. Note that the explicit form of $N_{inc}(\mathbf{a}_i, \mathbf{a}_{i+1})$ for each pair of consecutive rankings is given by (11) in Theorem 1 and Corollary 1. The value of $\max(d_{evol})$ depends on $N_{inc}(\mathbf{a}_i, \mathbf{a}_{i+1})$. We have seen in Remark 5 that the definition of τ_x corresponds to take $d_{max}(\mathbf{a}, \mathbf{b})$ as the value corresponding to $N_{inc} = 0$ (and that is the reason why τ_x is not well normalized). We can translate here the same reasoning and formalize it in the next definition.

Definition 5. *Given m incomplete rankings with ties $\mathbf{a}_1, \mathbf{a}_2, \ldots \mathbf{a}_m$ of n elements we define the corrected evolutive Kendall's τ coefficient for the series with penalty parameters $p = 0.5$ and $q = 0$ as follows:*

$$\tau_{ev}^{\bullet} = 1 - \frac{4K_{cev}^{(0.5,0)}(\mathbf{a}_1, \ldots, \mathbf{a}_m) + 2\sum_{i=1}^{m-1} N_{inc}(\mathbf{a}_i, \mathbf{a}_{i+1})}{(m-1)n(n-1)} \tag{35}$$

where $K_{cev}^{(0.5,0)}(\mathbf{a}_1, \ldots, \mathbf{a}_m)$ is given by Definition 4, and $N_{inc}(\mathbf{a}_i, \mathbf{a}_{i+1})$ is given by (11).

Here we have the same drawback as we showed for τ_x in Remark 5: $\tau_{ev}^{\bullet}$ is not properly normalized and it cannot get the values ± 1 if any $N_{inc}(\mathbf{a}_i, \mathbf{a}_{i+1}) \neq 0$. Therefore, in analogy with (26), we introduce a new coefficient in the following definition.

Definition 6. *Given m incomplete rankings with ties $\mathbf{a}_1, \mathbf{a}_2, \ldots \mathbf{a}_m$ of n elements, such that $\overline{n}_{i,i+1} > 1$, for all $i = 1, 2, \ldots, m-1$, we define the scaled corrected evolutive Kendall's τ coefficient for the series with penalty parameters $p = 0.5$ and $q = 0$ as follows:*

$$\widehat{\tau}_{ev}^{\bullet} = 1 - \frac{2K_{cev}^{(0.5,0)}(\mathbf{a}_1, \ldots, \mathbf{a}_m)}{\max(K_{cev}^{(0.5,0)}(\mathbf{a}_1, \ldots, \mathbf{a}_m))} \tag{36}$$

where $K_{cev}^{(0.5,0)}(\mathbf{a}_1, \ldots, \mathbf{a}_m)$ is given by Definition 4 and with

$$\max[K_{cev}^{(0.5,0)}(\mathbf{a}_1, \ldots, \mathbf{a}_m)] = \frac{1}{2} \sum_{i=1}^{m-1} \overline{n}_{i,i+1}(\overline{n}_{i,i+1} - 1) \tag{37}$$

where $\overline{n}_{i,i+1}$ denotes the common ranked elements between $\mathbf{a}_i$ and $\mathbf{a}_{i+1}$.

Note that we need that, for some i, $\overline{n}_{i,i+1} \neq 0$.

Remark 8. *In the limit case of m complete rankings with ties, note that Equation (37) collapses to Equation (32). Note also that $\hat{\tau}_{ev}^{\bullet}$ is affected by the crossings, the pass from tie to untie (or viceversa) and the long crossings (crossings after ties given by $\bar{K}_{i,j}^{cat}(\mathbf{a}_1, \mathbf{a}_2, \ldots, \mathbf{a}_m)$, given by Definition 3), due to the term $2K_{cev}^{(p=0.5,q=0)}(\mathbf{a}_1, \ldots, \mathbf{a}_m)$. The effect of the elements that are out of the rankings appear explicitly by the term $\bar{n}_{i,i+1}$ that does not take into account the position in $\mathbf{a}_i$ nor in $\mathbf{a}_{i+1}$. $\hat{\tau}_{ev}^{\bullet}$ is well normalized, that is $\hat{\tau}_{ev}^{\bullet} \in [-1, 1]$.*

Example 8. *Let $n = 6$. Given the series of incomplete rankings with ties $\mathbf{a}_1 = [1, 2, 3, 4, 5, 6]$, $\mathbf{a}_2 = [1, 2, 3, \bullet, \bullet, \bullet]$, and $\mathbf{a}_3 = [1, 2, \bullet, \bullet, \bullet, \bullet]$, an easy computation shows $K_{cev}(\mathbf{a}_1, \mathbf{a}_2, \mathbf{a}_3) = 0$ and thus $\hat{\tau}_{ev}^{\bullet} = 1$. Note that $\tau_{ev}^{\bullet} = 0.1333$.*

Example 9. *Let $n = 6$. Given the series of incomplete rankings with ties $\mathbf{a}_1 = [1, 2, 3, 4, 5, 6]$, $\mathbf{a}_2 = [3, 2, 1, \bullet, \bullet, \bullet]$, and $\mathbf{a}_3 = [1, 2, \bullet, \bullet, \bullet, \bullet]$, it is easy to obtain that $K_{cev}(\mathbf{a}_1, \mathbf{a}_2, \mathbf{a}_3) = 4 = \max(K_{cev})$ and thus $\hat{\tau}_{ev}^{\bullet} = -1$. Note that $\tau_{ev}^{\bullet} = -0.1333$.*

As we have seen in the above definitions, the importance of Theorem 1 and Corollary 1 consists of giving the explicit formula for $N_{inc}(\mathbf{a}_i, \mathbf{a}_{i+1})$ to allow for the computation of the coefficient $\hat{\tau}_{ev}^{\bullet}$ for the series of m incomplete rankings with ties. Note that $\hat{\tau}_{ev}^{\bullet} \in [-1, 1]$. For the particular case when the rankings are complete, we have $N_{inc}(\mathbf{a}_i, \mathbf{a}_{i+1}) = 0$ for all the pairs of consecutive rankings and $\bar{n}_{i,i+1} = n$, for $i = 1, 2, \ldots, m-1$, and therefore Equation (36) reduces to the complete case given by Equation (33), that is, $\hat{\tau}_{ev}^{\bullet}$ collapses to τ_{ev}.

Another contribution of Theorem 1 and Definition 6 is that they are useful to describe the behavior of the series of m rankings in terms of a competitivity graph. We can define a weighted graph for each one of the interactions between the elements when passing from $\mathbf{a}_i$ to $\mathbf{a}_{i+1}$: crossings, passing from tie to untie (or vice-versa), and crossing after ties. Moreover, for each kind of graph, we can add the contributions of all the pairs of consecutive rankings to obtain a *projected graph* for any interaction (crossings, passing from tie to untie (or vice-versa), and crossing after ties). The procedure is the following: First, we construct an undirected graph for each pair of rankings $\mathbf{a}_k, \mathbf{a}_{k+1}$ by identifying each element i as a node and defining an edge between i and j by the rule: there is an edge connecting $\{i, j\}$ with weight $\bar{K}_{i,j}^{(p,q)}(\mathbf{a}_k, \mathbf{a}_{k+1})$ when this weight is nonzero. By adding the $m-1$ pairs of undirected graphs we obtain a projected graph with a total sum of weights $K_{cev}^{(p=0.5,q=0)}(\mathbf{a}_1, \ldots, \mathbf{a}_m)$. By adding the *crossing after ties* term to the projected graph we have all the ingredients appearing on Definition 6. We show this procedure by using the next example with $m = 6$ and $n = 8$.

Example 10. *Given the series of incomplete rankings with ties*

$\mathbf{r}_1$	$\mathbf{r}_2$	$\mathbf{r}_3$	$\mathbf{r}_4$	$\mathbf{r}_5$	$\mathbf{r}_6$
5	2	4	6	2	1
7	1	8	1,4	1,4	5
3	8	3	3	6,7	8
8	3	2,6	8	5	3
1,4	5,7	5,7	2	3	4
	4	1	7	8	

the corresponding $\mathbf{a}_i$ are

	a₁	a₂	a₃	a₄	a₅	a₆
	5	2	6	2	2	1
	•	1	4	5	1	•
	3	4	3	3	5	4
	5	6	1	2	2	5
	1	5	5	•	4	2
	•	•	4	1	3	•
	2	5	5	6	3	•
	4	3	2	4	6	3

In this example we have $n = 8$, and an easy computation leads to the parameters shown in Table 4. For each pair of consecutive rankings it is easy to compute the parameters defined in Theorem 1: $n_{\bullet\bullet}$, $n_{\bullet*}$, $n_{*\bullet}$, n_{**}, s, n_{tu}, and n_{tt}. Then, by using Equation (10) in Theorem 1 we can obtain, for any pair of rankings, the value N_{inc}. $\bar{n}$ is the number of common elements, given by (9). The coefficient $\tau^{\bullet}_{ev}$ is given by (35), and the coefficient $\widehat{\tau}^{\bullet}_{ev}$ is given by (36). In analogy with (4) we can define the corresponding normalized mean strengths given by

$$NS^{\bullet} = \frac{(1 - \tau^{\bullet}_{ev})}{2} \qquad (38)$$

and

$$\widehat{NS}^{\bullet} = \frac{(1 - \widehat{\tau}^{\bullet}_{ev})}{2} \qquad (39)$$

Finally, in Table 4 we include the coefficients τ_x and $\widehat{\tau}_x$ given by (7) and (8), respectively. These last coefficients are included to show that our new coefficients $\tau^{\bullet}_{ev}$ and $\widehat{\tau}^{\bullet}_{ev}$ reduce to them when only a pair of rankings are considered.

Table 4. Parameters for pairs of consecutive rankings. Example 10.

	a₁ → a₂	a₂ → a₃	a₃ → a₄	a₄ → a₅	a₅ → a₆
$n_{\bullet\bullet}$	1	0	0	0	0
$n_{\bullet*}$	1	1	0	1	0
$n_{*\bullet}$	0	0	1	0	3
n_{**}	6	7	7	7	5
s	9	12	8	9	4
n_{tu}	2	0	2	1	1
n_{tt}	0	0	0	0	0
N_{inc}	13	7	7	7	18
$\bar{n}$	6	7	7	7	5
$\widehat{\tau}^{\bullet}_{ev}$	−0.3333	−0.1429	0.1429	0.0952	0.1000
$\widehat{NS}^{\bullet}$	0.6667	0.5714	0.4268	0.4524	0.4500
$\tau^{\bullet}_{ev}$	−0.1786	−0.1071	0.1071	0.0714	0.0357
$NS^{\bullet}$	0.5893	0.5536	0.4464	0.4643	0.4821
τ_x	−0.1786	−0.1071	0.1071	0.0714	0.0357
$\widehat{\tau}_x$	−0.3333	−0.1429	0.1429	0.0952	0.1000

To compute our new coefficients $\tau^{\bullet}_{ev}$ and $\widehat{\tau}^{\bullet}_{ev}$ for the whole series of rankings a_1 to a_6 we need some previous parameters. First, we need the value

$$\sum_{i=1}^{5} N_{inc}(a_i, a_{i+1}) = 52$$

To compute $K^{(p=0.5,q=0)}_{cev}(a_1, \ldots, a_6)$, given by (31), we need to know, previously, the value of the *crossing after ties* coefficients $\bar{K}^{cat}_{i,j}(a_1, \ldots, a_6)$, given by Definition 3. Note that the unique long crossing occurs for the pair $\{1, 4\}$: the elements tagged as 1 and 4 are such that 4 is above 1 in r_3, both elements are tied in rankings r_4 and r_5, and, finally, 4 is below 1 in ranking r_6. Note, for example, that the pair $\{5, 7\}$

does not accomplish the conditions of crossing after ties. Therefore the only term that contributes to $\sum_{\{i,j\}} \bar{K}_{i,j}^{cat}$ is $\bar{K}_{1,4}^{cat}(\mathbf{a}_1, \ldots, \mathbf{a}_6) = 1$.

With respect to $K_{ev}^{(p=0.5, q=0)}(\mathbf{a}_1, \ldots, \mathbf{a}_6)$, given by (30), we need to compute the terms $\bar{K}_{i,j}^{(p,q)}(\mathbf{a}_i, \mathbf{a}_{i+1})$, given by (28), for any pair of consecutive rankings. A detailed computation shows that, in this example, we have 42 *crossings* and 6 cases of *tie to untie or viceversa*. The precise pairs of elements that contribute to these cases are shown in the corresponding projected weighted graphs in Figure 1. The *crossing after ties* case is represented in Figure 2.

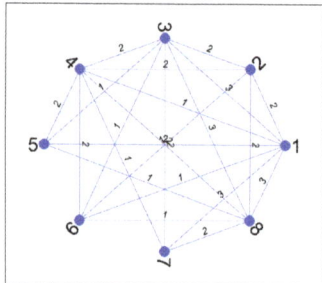
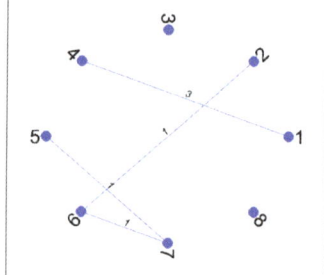

Figure 1. Projected weighted graphs representing the pairs of elements that contribute to *crossings* (**left panel**) and the pairs corresponding to the case *tie to untie or viceversa* (**right panel**), occurring in Example 10.

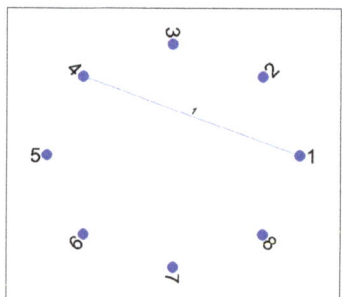

Figure 2. Projected weighted graph representing the *crossing after ties* cases occurred in Example 10.

Therefore we have all the ingredients to compute $K_{cev}^{(p=0.5, q=0)}$. That is

$$
\begin{aligned}
K_{cev}^{(p=0.5, q=0)}(\mathbf{a}_1, \ldots, \mathbf{a}_6) &= K_{ev}^{(p=0.5, q=0)}(\mathbf{a}_1, \ldots, \mathbf{a}_6) + \sum_{\{i,j\}} \bar{K}_{i,j}^{cat}(\mathbf{a}_1, \ldots, \mathbf{a}_6) \\
&= \sum_{i=1}^{5} K^{(p,q)}(\mathbf{a}_i, \mathbf{a}_{i+1}) + \sum_{\{i,j\}} \bar{K}_{i,j}^{cat}(\mathbf{a}_1, \ldots, \mathbf{a}_6)
\end{aligned}
$$

and, by Remark (7), we know that

$$K_{i,j}^{(p,q)}(\mathbf{a}, \mathbf{b}) = s + p\, n_{tu} + q\, n_{tt}$$

Therefore, we have

$$K_{cev}^{(p=0.5,q=0)}(\mathbf{a}_1,\ldots,\mathbf{a}_6) = (9+12+8+9+4) + 0.5(2+0+2+1+1) + \bar{K}_{1,4}^{cat}(\mathbf{a}_1,\ldots,\mathbf{a}_6)$$
$$= 42 + 3 + 1 = 46.$$

By using (35), we obtain

$$\tau_{ev}^\bullet = 1 - \frac{4\cdot 46 + 2\cdot 52}{(6-1)8\cdot 7} = 1 - 1.0286 = -0.0286$$

that corresponds to an equivalent normalized mean strength

$$NS^\bullet = \frac{(1-\tau_{ev}^\bullet)}{2} = 0.5143$$

Finally, regarding $\widehat{\tau}_{ev}^\bullet$, we have

$$\widehat{\tau}_{ev}^\bullet = 1 - \frac{2\cdot 46}{\frac{1}{2}(6\cdot 5 + 7\cdot 6 + 7\cdot 6 + 7\cdot 6 + 5\cdot 4)} = 1 - \frac{4\cdot 46}{176} = -0.0455$$

that corresponds to

$$\widehat{NS}^\bullet = 0.5227.$$

All in all, we conclude that $\widehat{\tau}_{ev}^\bullet$ is a proper coefficient for the evaluation of m incomplete rankings with ties and can be considered as a natural extension of the coefficient $\widehat{\tau}_x$ presented in [12]. In the next section we apply the new coefficients $\tau_{ev}^\bullet$ and $\widehat{\tau}_{ev}^\bullet$ to real rankings appearing on Spotify charts.

7. Results

Spotify is one of the major music streaming services worldwide, with 299 million monthly active users, as of July 2020 [28]. The company Spotify Technology S.A. has been listed on the New York Stock Exchange since 2018. As of September 2020, the company offers a catalog of 60 million tracks and operates in 92 countries from Albania to Vietnam [29]. Spotify divides the monthly active users into four regions [30]: Europe (35%), North America (26%), Latin America (22%) and rest of the world (17%). The app is available on several devices, such as computers, smartphones, tablets, wearable devices, etc. The users can choose between a free service (called *Freemium* or *Ad-Supported*) or a *Premium* service. In any case, the user can listen by streaming any song of the catalog (that is, the user does not own the song's digital file, but can listen to it). It is accepted that music streaming services have transformed the entire music market—see [31]—and they have evolved very fast, changing their services and capabilities. For example, Spotify has signed some partnerships with Microsoft [32], Sony [33] and Facebook [34] among other big companies. There exists a large amount of literature about Spotify, but it is mainly focused on Economics and Music. To the best of our knowledge, a small number of papers are devoted to the mathematical aspects of the rankings produced by Spotify. Among these papers, we have [35,36]. A paper that studies the relationship between personality and type of music is [37]. See [38] for more details about Spotify.

Like other services on the Internet, Spotify provides some chart lists (song rankings) based on the platform's number of streamings. To this kind of rankings belongs the Top 200 (see [39–41]), that is one of the topics of our study. Another ranking that we are interested in is called *Viral 50* which is an evolution of the original Social 50 ranking (see [42–44]) that incorporated in the song chart the effect of the social sharing of a track by Spotify users. This sharing included platforms such as Facebook and Twitter. It is not completely clear for us how this rank is computed, but it aims to gather *fresh* songs that acquire high impact on social networks by new release promotions, special apparitions on tv-shows, music festivals, tours, etc. (see [45] for an example of how a viral song transformed into a Top 100 song in 2013).

Due to the situation caused by the COVID-19 pandemic, the live music business reflected some drawbacks, such as festivals being cancelled worldwide, a reduction in public-performance licensing, and other related factors—see [46]. As an example, Warner Music Group Corp showed a total revenue fall of 1.7% in the first quarter of 2020 compared to the first quarter of 2019 [47]. Spotify also reported some impact on their business, but in the first quarter of 2020, it seemed that the consumption recovered and monthly active users increased faster in the first quarter of 2020 than in the same period of 2019 [30]. Some perturbations in Spotify streaming were also reported by the music analytic company Chartmetric that observed a change in the type of consumption of Spotify streamings by music genre in the period between 3 March 2020 to 9 April 2020, concluding that it seemed that it had been a *pandemic-induced lifestyle change* [48].

With regard to the Top 50 viral, it is reasonable to think that the fact that many artists (such as Lady Gaga, Alicia Keys, and Cardi B. [46]) have postponed big releases may have decreased the movements in these charts.

7.1. Method to Convert Spotify Lists into Incomplete Rankings

Both Spotify Top 200 and Viral 50 lists can be treated as incomplete rankings since some elements (songs) quit the list and some others that appear on the list (new songs). Let us call any of these rankings as Top k rankings. In order to handle these Top k rankings, our methodology consists of the following steps:

1. Select a set of m lists $\{v_1, v_2, \ldots, v_m\}$ with k entries in each v_i.
2. Denote as n the number of different songs that appear on these m lists. We tag these songs from 1 to n, following the order they first appear, reading the lists from the first to the last one, and each list from top to bottom. Denote t_i the tagged version of v_i, for $i = 1, 2, \ldots, m$, including all the n songs.
3. Denote r_1 a vector with entries from 1 to n. The first k values correspond to the elements in v_1.
4. Construct the rankings r_i for $2 = 1, \ldots m$, in the following form:

    (a) The first k entries of r_i are copied from t_i;
    (b) The rest of the entries form a vector s_i and come from the the elements that quit from t_{i-1} plus the elements that, being in s_{i-1}, are not included in t_i.

    These $n - k$ elements preserve their relative order. This order is not important since these elements are not included in the Top k ranking t_i.
5. From each t_i, we construct the corresponding incomplete ranking a_i given by (5).

Example 11. *Let us consider three Top 4 lists* (v_1, v_2, v_3) *and construct the corresponding three rankings* (a_1, a_2, a_3). *Here we have* $m = 3$ *and* $k = 4$.

$$
\begin{vmatrix} v_1 & v_2 & v_3 \\ A & B & F \\ B & C & C \\ C & E & B \\ D & A & E \end{vmatrix}
\quad \rightarrow \quad
\begin{matrix} t_1 & t_2 & t_3 \\ & & \\ & & \\ & & \\ s_1 & s_2 & s_3 \end{matrix}
\quad
\begin{Bmatrix} r_1 & r_2 & r_3 \\ 1 & 2 & 6 \\ 2 & 3 & 3 \\ 3 & 5 & 2 \\ 4 & 1 & 5 \\ 5 & 4 & 1 \\ 6 & 6 & 4 \end{Bmatrix}
\quad \rightarrow \quad
\begin{matrix} a_1 & a_2 & a_3 \\ 1 & 4 & \bullet \\ 2 & 1 & 3 \\ 3 & 2 & 2 \\ 4 & \bullet & \bullet \\ \bullet & 3 & 4 \\ \bullet & \bullet & 1 \end{matrix}
$$

We have denoted as s_i *the elements beyond the* k *position in each ranking* r_i. *The rankings* a_i *are constructed looking at* r_i *from positions 1 to 4. Since the elements that do no belong to* t_i *are in* s_i, *we tagged them as* •.

7.2. Comparison of Two Series of Top 200 Rankings

From the site [49] we downloaded the series of Top 200 (Global) rankings corresponding to the following time intervals:

- 2019 Series: 18 weekly rankings ranging from 28 December 2018 to 3 May 2019.
- 2020 Series: 18 weekly rankings ranging from 27 December 2019 to 1 May 2020.

The term *Global* means that the charts were produced from streaming on Spotify from all over the world. By using the methodology explained in the previous section, we convert the 18 downloaded rankings to a series of incomplete rankings (with no ties) $a_1, \ldots, a_{18}$, and we compute our parameters. This is repeated for each considered year. The results are shown in Table 5.

Table 5. Parameters for two series of incomplete rankings obtained from Spotify Top 200 lists.

	2019 Series	2020 Series
n	474	556
N_{inc}	1.6×10^6	2.4×10^6
$<\overline{n}_{i,i+1}>$	182	175
$\tau_{ev}^\bullet$	0.1256	0.0836
$NS^\bullet$	0.4372	0.4582
$\widehat{\tau}_{ev}^\bullet$	0.8540	0.8421
$\widehat{NS}^\bullet$	0.0730	0.0789

In Table 5 we have denoted by $<\overline{n}_{i,i+1}>$ the average of $\{\overline{n}_{i,i+1}\}$ for $i = 1, 2, \ldots, 17$, that is the mean number of common elements from each pair of consecutive rankings. We see that the number of songs involved in the 2019 series is $n = 474$, which is lower than the 2020 series number. This fact could indicate that there was more *activity* in the 2020 series since more new songs appeared than the previous year. By extension, we can also conclude that the *activity* on Spotify of the users was higher in the 2020 series.

The same tendency is observed by looking at N_{inc} and $\overline{n}_{i,i+1}$. Our coefficients $NS^\bullet$ and $\widehat{NS}^\bullet$ corroborate this intuition since they take higher values in the 2020 Series than in the 2019 Series. Analogously, by looking at $\tau_{ev}^\bullet$ and $\widehat{\tau}_{ev}^\bullet$, we see a decrease when comparing the 2019 Series with the 2020 Series. Recall that the coefficients $NS^\bullet$ and $\widehat{NS}^\bullet$ introduced in this paper offer a measure of the *movements* in the rankings, since they take into account the number of crossings and, in this case, that we do not have ties, due to the effect of absent elements.

In the same manner, as we did in Example 10, we can construct the projected graph corresponding to the crossings for each series. We show these graphs in Figure 3, that have been plotted with MATLAB by using the option "subspace".

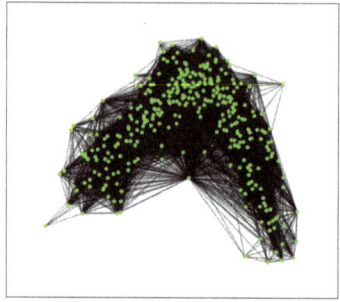 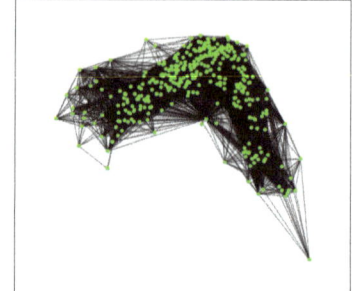

Figure 3. Graph based on crossings corresponding to the giant connected component of Top 200 2019 Series (**left panel**, 360 nodes, 16,115 edges) and Top 200 2020 Series (**right panel**, 374 nodes, 16,564 edges).

7.3. Comparison of Two Series of Viral-50 Rankings

From the site [50], we downloaded the series of Viral 50 (Global) weekly rankings corresponding to the following periods:

- 2019 Series: 18 weekly rankings ranging from 3 January 2019 to 2 May 2019.
- 2020 Series: 18 weekly rankings ranging from 2 January 2020 to 30 April 2020.

For each considered year, we convert the 18 downloaded rankings to a series of incomplete rankings (with no ties) $a_1, \ldots, a_{18}$, and we computed again the aforementioned parameters. The results are shown in Table 6.

Table 6. Parameters for two series of incomplete rankings obtained from Spotify Viral 50 lists.

	2019 Series	2020 Series
n	315	300
N_{inc}	8.3×10^5	7.5×10^5
$<\bar{n}_{i,i+1}>$	33.6	35
$\tau_{ev}^\bullet$	0.0067	0.0093
$NS^\bullet$	0.4966	0.4954
$\widehat{\tau}_{ev}^\bullet$	0.6037	0.6922
$\widehat{NS}^\bullet$	0.1982	0.1539

The number of songs involved in the 2019 series is $n = 315$, that is greater than the number involved in the 2020 series. This fact could indicate that there was less viral *activity* in the 2020 series since fewer new songs appeared than the previous year. The same tendency is observed at N_{inc}. This intuition is corroborated by our coefficients. $NS^\bullet$ and $\widehat{NS}^\bullet$ since they take lower values in the 2020 series than in the 2019 series. We also see an increase in $\tau_{ev}^\bullet$ and $\widehat{\tau}_{ev}^\bullet$ when comparing the 2019 series with the 2020 series.

If we compare these results with those obtained in the previous section, we conclude that Spotify's viral activity was negatively affected by the Pandemic. This may seem reasonable since many events that produce sharing in Social Networks, such as shows, new releases, and performances, were postponed during these months, as we have already discussed. We again plot the projected graph corresponding to the crossings for each series in Figure 4.

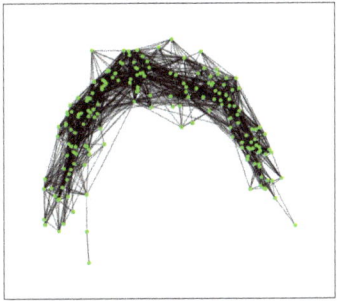
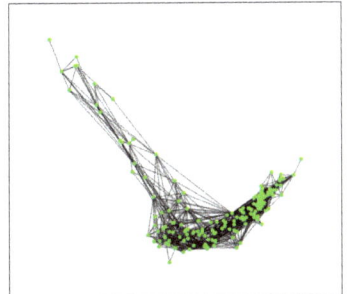

Figure 4. Graph based on crossings corresponding to the giant connected component of Viral-50 2019 Series (**left panel**, 185 nodes, 1685 edges) and Viral-50 2020 Series (**right panel**, 186 nodes, 1447 edges).

7.4. Comparison of a Series of Top 200 and a Series of Viral 50 Rankings

Given that our coefficients $\tau_{ev}^{\bullet}$, $\hat{\tau}_{ev}^{\bullet}$, $NS^{\bullet}$, and $\widehat{NS}^{\bullet}$ are normalized, we can compare series of rankings of different type. Looking at Tables 5 and 6, we conclude (e.g., looking at $\widehat{NS}^{\bullet}$) that the Viral-50 rankings present more activity than the Top 200 rankings. For example in the 2019 series the value of $\widehat{NS}^{\bullet}$ is 0.1982 for the Viral-50 rankings, and only 0.0730 for the Top 200 rankings. This conclusion seems reasonable, taking into account that the Viral-50 rankings are constructed by looking at the behaviours of songs that may rapidly change, since they are viral phenomena.

7.5. Comparison of the Evolution of Two Series of Incomplete Ranking with Ties

Spotify charts Top 200 do not present ties, but we can construct incomplete rankings with ties if we take into account the Top 200 ranking and the rest of the songs that appear in the whole studied interval. In detail, to obtain a series of incomplete rankings with ties from a Top 200 series on Spotify, we will consider the whole list of tracks along with the m rankings and focus on what happens in positions greater than 200. Using the terminology used in Example 11 we consider the elements that appear on the rankings, denoted as $\mathbf{s}_1, \mathbf{s}_2, \ldots$. In this ranking we consider the following:

(i) All the tracks in $\mathbf{s}_1$ are tied. That is $\mathbf{a}_1 = [\bullet_{1,200} \; \mathbf{1}^{n-200}]$ where $\bullet_{1,200}$ is a row vector of 200 entries of the type $\bullet$, and $\mathbf{1}^{n-200}$ is the row vector of all-ones, with $n - 200$ entries, being n the total number of different tracks in the m rankings.

(ii) For $i = 2, 3 \ldots m$, we consider that in $\mathbf{s}_i$ we have (at most) two *buckets* of tied elements. In one bucket we have the elements (if any) that come from $\mathbf{t}_{i-1}$. In the other bucket, we consider the rest of the elements of $\mathbf{s}_i$

The next example with a series of $m = 7$ Top 4 charts illustrates this methodology.

Example 12. *Let us consider the series of seven Top 4 tracks* $\mathbf{v}_i$ *with* $n = 10$ *elements* $\{A, B, \ldots, J\}$ *given by the rankings*

$$\begin{array}{ccccccc} A & A & F & G & G & G & J \\ B & B & E & H & H & I & C \\ C & E & A & C & C & B & A \\ D & F & C & E & E & A & H \end{array}$$

from these rankings we construct the rankings t_i *and* s_i *to obtain the rankings in the form*

$$
\begin{array}{c}
t_i \\
\\
\\
\\
\hline
\\
\\
s_i \\
\\
\\
\\
\end{array}
\quad
\begin{array}{ccccccc}
1 & 1 & 6 & 7 & 7 & 7 & 10 \\
2 & 2 & 5 & 8 & 8 & 9 & 3 \\
3 & 5 & 1 & 3 & 3 & 2 & 1 \\
4 & 6 & 3 & 5 & 5 & 1 & 8 \\
\hline
5 & 3 & 2 & 6 & 6 & 8 & 7 \\
6 & 4 & 4 & 1 & 1 & 3 & 9 \\
7 & 7 & 7 & 2 & 2 & 5 & 2 \\
8 & 8 & 8 & 4 & 4 & 6 & 5 \\
9 & 9 & 9 & 9 & 9 & 4 & 6 \\
10 & 10 & 10 & 10 & 10 & 10 & 4 \\
\end{array}
$$

Now, we consider the rankings s_i as a series of incomplete rankings with ties with the convention explained above and we compute the corresponding a_i vectors to obtain the rankings

a_1	a_2	a_3	a_4	a_5	a_6	a_7
•	•	•	1	1	•	•
•	•	1	2	1	•	1
•	1	•	•	•	1	•
•	1	2	2	1	2	2
1	•	•	•	•	1	2
1	•	•	1	1	2	2
1	2	2	•	•	•	1
1	2	2	•	•	1	•
1	2	2	2	1	•	1
1	2	2	2	1	2	•

Note that, since there are at most two buckets, the entries of a_i belong to the set $\{1, 2, \bullet\}$. Note also that in s_5 there is only one bucket.

By using this methodology, we have converted the series of rankings studied in Section 7.2 to the corresponding series a_i with ties. The parameters obtained are shown in Table 7.

Table 7. Series of incomplete rankings with ties obtained from Spotify Top 200 charts.

	2019 Series	2020 Series
n	474	556
N_{inc}	1.4×10^6	1.7×10^6
$<\bar{n}_{i,i+1}>$	256	331
$\tau_{ev}^{\bullet}$	0.2577	0.3108
$NS^{\bullet}$	0.3712	0.3446
$\widehat{\tau}_{ev}^{\bullet}$	0.8848	0.8757
$\widehat{NS}^{\bullet}$	0.0576	0.0621

If we look at n, N_{inc}, $<\bar{n}_{i,i+1}>$, and $\widehat{NS}^{\bullet}$ in Table 7, we conclude that there has been more activity in the 2020 Series than in the 2019 Series. However, by looking at $NS^{\bullet}$ (and $\tau_{ev}^{\bullet}$), the conclusion seems to be the reverse. Here we see, therefore, that $\tau_{ev}^{\bullet}$ and $\widehat{\tau}_{ev}^{\bullet}$ can present different tendencies. This is related to the form in which they are normalized, as we have commented in Remark 5 and in Section 6. These results provide an example of how the transformation from $\tau_{ev}^{\bullet}$ to $\widehat{\tau}_{ev}^{\bullet}$ is not linear, since $\tau_{ev}^{\bullet}$ increases from 2019 to 2020 but $\widehat{\tau}_{ev}^{\bullet}$ decreases in the same period.

In Figure 5, we show the plot of the giant component corresponding to the projected graph showing the interactions of the form *tie to untie or viceversa*. That is, there is a link between elements (nodes) i and j when the pair $\{i,j\}$ goes from tie to untie (or vice versa) in any pair of consecutive rankings $\mathbf{a}_i$ and $\mathbf{a}_{i+1}$. We see many more interactions of this type in the 2020 series than in the 2019 series.

Figure 5. Graph based on crossings of the type *from tie to untie or vice versa* corresponding to the giant connected component of 2019 Series (**left**, 377 nodes, 49,915 edges) and 2020 Series (**right**, 457 nodes, 82,051 edges). We also have 97 isolated nodes in the 2019 Series and 99 in the 2020 Series.

In Figure 6, we show the plot of the giant component corresponding to the projected graph showing the interactions of the form *tie to tie*, that is, there is a link between elements (nodes) i and j when the pair $\{i,j\}$ goes from tie to tie in any pair of consecutive rankings $\mathbf{a}_i$ and $\mathbf{a}_{i+1}$. We also see many more interactions of this type in the 2020 series than in the 2019 series.

Therefore, and taking into account the values of Table 7, we can conclude (for this artificial model of incomplete ranking with ties) that there was more activity in the 2020 series than in the 2019 series.

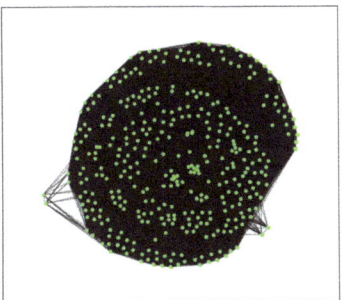 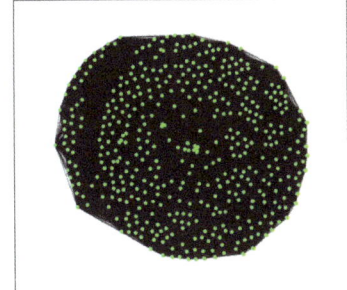

Figure 6. Graph based on crossings of the type *from tie to tie* corresponding to the giant connected component of 2019 Series (**left**, 382 nodes, 65,269 edges) and 2020 Series (**right**, 462 nodes, 101,101 edges).

We have shown the application of the new coefficients introduced in this work, as long as the utility of the visualizations based on the projected graph plots of the (evolutive) competitive graph associated to a series of incomplete rankings with or without ties.

8. Conclusions

We present the main conclusions of our work:

- We provide a theoretical result that allows for understanding, in terms of the type of interactions between pairs of elements in a series of incomplete rankings with ties, two recently introduced coefficients, given in [4,12].
- We have defined two new coefficients to characterize a series of incomplete rankings with ties in terms of the interactions mentioned above.
- We have presented a methodology to treat Spotify charts (both Top 200 and Viral 50) as a series of incomplete rankings. This methodology allows us to obtain conclusions about the *movements* in the lists and, therefore, on the *activity* of the users of the app.
- We have obtained an artificial series of incomplete rankings with ties based on Spotify Top 200 lists, to apply our coefficients and show the applicability of the method.
- The main theoretical result (Theorem 1) may serve to define new coefficients by giving weight to the interactions between pairs of elements when going from one ranking to the next one. The applications can be of interest in other fields (neuroscience, sports, bioinformatics, etc.).

Author Contributions: All authors contributed equally to this paper and have read and agreed to the published version of the manuscript.

Funding: This research was funded by the Spanish Government, Ministerio de Economía y Competividad, grant number MTM2016-75963-P.

Acknowledgments: We thank the four anonymous reviewers for their constructive comments, which helped us to improve the readability of the manuscript.

Conflicts of Interest: The funders had no role in the design of the study; in the collection, analyses, or interpretation of data; in the writing of the manuscript, or in the decision to publish the results.

References

1. Langville, A.N.; Carl, D.; Meyer, C.D. *Who's #1: The Science of Rating and Ranking*; Princeton University Press: Princeton, NJ, USA, 2012.
2. Kemeny, J.G.; Snell, J.L. Preference rankings. An axiomatic approach. In *Mathematical Models in the Social Sciences*, 2nd ed.; The MIT Press: Cambridge, MA, USA, 1973; pp. 9–23.
3. Diaconis, P.; Graham, R.L. Spearman's Footrule as a Measure of Disarray. *J. R. Stat. Soc. B Met.* **1977**, *39*, 262–268. [CrossRef]
4. Moreno-Centeno, E.; Escobedo, A.R. Axiomatic aggregation of incomplete rankings. *IIE Trans.* **2016**, *48*, 475–488. [CrossRef]
5. Criado, R.; García, E.; Pedroche, F.; Romance, M. A new method for comparing rankings through complex networks: Model and analysis of competitiveness of major European soccer leagues. *Chaos* **2013**, *23*, 043114. [CrossRef] [PubMed]
6. Fortune 500. Available online: https://fortune.com/fortune500/ (accessed on 24 September 2020).
7. Academic Ranking of World Universities ARWU 2020. Available online: http://www.shanghairanking.com/ARWU2020.html (accessed on 24 September 2020).
8. CWTS Leiden Ranking 2020. Available online: https://www.leidenranking.com/ranking/2020/list (accessed on 24 September 2020).
9. Billborad. The Hot 100. Available online: https://www.billboard.com/charts/hot-100 (accessed on 24 September 2020).
10. Fagin, R.; Kumar, R.; Mahdian, M.; Sivakumar, D.; Vee, E. Comparing Partial Rankings. *SIAM J. Discrete Math.* **2006**, *20*, 628–648. [CrossRef]
11. Cook, W.D.; Kress, M.; Seiford, L.M. An axiomatic approach to distance on partial orderings. *Rairo-Rech. Oper.* **1986**, *20*, 115–122. [CrossRef]
12. Yoo, Y.; Escobedo, A.R.; Skolfield, J.K. A new correlation coefficient for comparing and aggregating non-strict and incomplete rankings. *Eur. J. Oper. Res.* **2020**, *285*, 1025–1041. [CrossRef]

13. Pedroche, F.; Criado, R.; García, E.; Romance, M.; Sánchez, V.E. Comparing series of rankings with ties by using complex networks: An analysis of the Spanish stock market (IBEX-35 index). *Netw. Heterog. Media* **2015**, *10*, 101–125. [CrossRef]
14. Criado, R.; García, E.; Pedroche, F.; Romance, M. On graphs associated to sets of rankings. *J. Comput. Appl. Math.* **2016**, *291*, 497–508. [CrossRef]
15. Kendall, M.G. A New Measure of Rank Correlation. *Biometrika* **1938**, *30*, 81–89. [CrossRef]
16. Kendall, M.G. *Rank Correlation Methods*, 4th ed.; Griffin: London, UK, 1970.
17. Kendall, M.G.; Babington-Smith, B. The Problem of m Rankings. *Ann. Math. Stat.* **1939**, *10*, 275–287. [CrossRef]
18. Bogart, K.P. Preference structures I: Distances between transitive preference relations. *J. Math. Sociol.* **1973**, *3*, 1. [CrossRef]
19. Bogart, K.P. Preference Structures II: Distances Between Asymmetric Relations. *SIAM J. Appl. Math.* **1975**, *29*, 254–262. [CrossRef]
20. Cicirello, V.A. Kendall Tau Sequence Distance: Extending Kendall Tau from Ranks to Sequences. *arXiv* **2019**, arXiv:1905.02752v3.
21. Armstrong, R.A. Should Pearson's correlation coefficient be avoided? *Ophthal. Physl. Opt.* **2019**, *39*, 316–327. [CrossRef]
22. Redman, W. An O(n) method of calculating Kendall correlations of spike trains. *PLoS ONE* **2019**, *14*, e0212190. [CrossRef] [PubMed]
23. Pihur, V.; Datta, S.; Datta, S. RankAggreg, an R package for weighted rank aggregation. *BMC Bioinform.* **2009**, *10*, doi:10.1186/1471-2105-10-62. [CrossRef]
24. Pnueli, A.; Lempel, A.; Even, S. Transitive orientation of graphs and identification of permutation graphs. *Can. J. Math.* **1971**, *23*, 160–175. [CrossRef]
25. Gervacio, S.; Rapanut, T.; Ramos, P. Characterization and construction of permutation graphs. *Open J. Discrete Math.* **2013**, *3*, 33–38. [CrossRef]
26. Golumbic, M.; Rotem, D.; Urrutia, J. Comparability graphs and intersection graphs. *Discrete Math.* **1983**, *43*, 37–46. [CrossRef]
27. Emond, E.J.; Mason, D.W. A New Rank Correlation Coefficient with Application to the Consensus Ranking Problem. *J. Multi-Crit. Decis. Anal.* **2002**, *11*, 17–28. [CrossRef]
28. Spotify Reports Second Quarter 2020 Earnings. 29 July 2020. Available online: https://newsroom.spotify.com/2020-07-29/spotify-reports-second-quarter-2020-earnings (accessed on 24 September 2020).
29. Spotify. Company info. Available online: https://newsroom.spotify.com/company-info/ (accessed on 24 September 2020).
30. Bussines Wire. 29 April 2020. Available online: https://www.businesswire.com/news/home/20200429005216/en/ (accessed on 24 September 2020).
31. Swanson, K. A Case Study on Spotify: Exploring Perceptions of the Music Streaming Service. *J. Music Entertain. Ind. Educ. Assoc.* **2013**, *13*, 207–230. [CrossRef]
32. Warren, T. Microsoft Retires Groove Music Service, Partners with Spotify. *The Verge. Vox Media*, 2 October 2017. Available online: https://www.theverge.com/2017/10/2/16401898/microsoft-groove-music-pass-discontinued-spotify-partner (accessed on 24 September 2020).
33. Lempel, E. Spotify Launches on PlayStation Music Today. (30 March 2015) Sony. Available online: https://blog.playstation.com/2015/03/30/spotify-launches-on-playstation-music-today/ (accessed on 24 September 2020).
34. Perez, S. You Can Now Share Music from Spotify to Facebook Stories. *Techcrunch.com*, 31 August 2019. Available online: https://techcrunch.com/2019/08/30/you-can-now-share-music-from-spotify-to-facebook-stories (accessed on 24 September 2020).
35. Mähler, R.; Vonderau, P. Studying Ad Targeting with Digital Methods: the Case of Spotify. *Cult. Unbound.* **2017**, *9*, 212–221. Available online: https://cultureunbound.ep.liu.se/article/view/1820 (accessed on 24 September 2020). [CrossRef]
36. Van den Hoven, J. Analyzing Spotify Data. Exploring the Possibilities of User Data from a Scientific and Business Perspective. (Supervised by Sandjai Bhulai). Report from Vrije Universiteit Amsterdam. August 2015. Available online: https://www.math.vu.nl/~sbhulai/papers/paper-vandenhoven.pdf (accessed on 24 September 2020).

37. Greenberg, D.V.; Kosinski, M.; Stillwell, D.V.; Monteiro, B.L.; Levitin, D.J.; Rentfrow, P.J. The Song Is You: Preferences for Musical Attribute Dimensions Reflect Personality. *Soc. Psychol. Pers. Sci.* **2016**, *7*, 597–605. doi:10.1177/1948550616641473. [CrossRef]
38. Eriksson, M.; Fleischer, R.; Johansson, A.; Snickars, P.; Vonderau, P. *Spotify Teardown. Inside the Black Box of Streaming Music*; The MIT Press: Cambridge, MA, USA, 2019.
39. Spotify Charts Regional. Available online: https://spotifycharts.com/regional (accessed on 24 September 2020).
40. Harris, M.; Liu, B.; Park, C.; Ramireddy, R.; Ren, G.; Ren, M.; Yu, S.; Daw, A.; Pender, J. Analyzing the Spotify Top 200 Through a Point Process Lens. *arXiv* **2019**, arXiv:1910.01445v1.
41. Aguiar, L.; Waldfogel, J. *Platforms, Promotion, and Product Discovery: Evidence from Spotify Playlists*; JRC Digital Economy Working Paper. No. 2018-04; European Commission, Joint Research Centre (JRC): Seville, Spain, 2018.
42. Lawler, R. Spotify Charts Launch Globally, Showcase 50 Most Listened to and Most Viral Tracks Weekly. *Engadget*, 21 May 2013. Available online: https://www.engadget.com/2013-05-21-spotify-charts-launch.html (accessed on 24 September 2020).
43. Spotify says its Viral-50 chart reaches the parts other charts don't. *Music Ally Blog*, 15 July 2014. Available online: https://musically.com/2014/07/15/spotify-says-its-viral-50-chart-reaches-the-parts-other-charts-dont/ (accessed on 24 September 2020).
44. Stassen, M. Spotify Reveals New Viral 50 Chart. *MusicWeek*, 15 July 2014. Available online: https://www.musicweek.com/news/read/spotify-launches-the-viral-50-chart/059027 (accessed on 24 September 2020).
45. Bertoni, S. How Spotify Made Lorde A Pop Superstar. *Forbes*, 26 November 2013.
46. Ingham, T. Record Companies Aren't Safe From the Coronavirus Economic Fallout. *Rolling Stone*, 31 March 2020.
47. Warner Music Group Corp. Reports Results for Fiscal Second Quarter Ended 31 March 2020. Available online: https://www.wmg.com/news/warner-music-group-corp-reports-results-fiscal-second-quarter-ended-march-31-2020-34751 (accessed on 24 September 2020).
48. Joven, J.; Rosenborg, R.A.; Seekhao, N.; Yuen, M. COVID-19's Effect on the Global Music Business, Part 1: Genre. Available online: https://blog.chartmetric.com/covid-19-effect-on-the-global-music-business-part-1-genre/ (accessed on 23 April 2020).
49. Spotify. Top 200. Available online: https://spotifycharts.com/regional/global/weekly (accessed on 24 September 2020).
50. Spotify Charts. Available online: https://spotifycharts.com/viral/ (accessed on 24 September 2020).

Publisher's Note: MDPI stays neutral with regard to jurisdictional claims in published maps and institutional affiliations.

© 2020 by the authors. Licensee MDPI, Basel, Switzerland. This article is an open access article distributed under the terms and conditions of the Creative Commons Attribution (CC BY) license (http://creativecommons.org/licenses/by/4.0/).

Article

Analysis of Generalized Multistep Collocation Solutions for Oscillatory Volterra Integral Equations

Hao Chen, Ling Liu and Junjie Ma *

School of Mathematics and Statistics, Guizhou University, Guiyang 550025, China; sdch0807@163.com (H.C.); lingliu95@126.com (L.L.)
* Correspondence: jjma@gzu.edu.cn

Received: 9 October 2020; Accepted: 4 November 2020; Published: 10 November 2020

Abstract: In this work, we introduce a class of generalized multistep collocation methods for solving oscillatory Volterra integral equations, and study two kinds of convergence analysis. The error estimate with respect to the stepsize is given based on the interpolation remainder, and the nonclassical convergence analysis with respect to oscillation is developed by investigating the asymptotic property of highly oscillatory integrals. Besides, the linear stability is analyzed with the help of generalized Schur polynomials. Several numerical tests are given to show that the numerical results coincide with our theoretical estimates.

Keywords: collocation; volterra integral equation; highly oscillatory; convergence

1. Introduction

In many practical problems, such as epidemic diffusion, population dynamics and reaction processes, one may usually come across a class of Volterra integral equations (VIEs) (see [1] and references therein). Noting that most VIEs cannot be solved in closed forms, many researchers have made contributions to the numerical approaches to VIEs.

Particularly, the study of numerical solutions to VIEs with highly oscillatory Fourier or Bessel kernels has attracted much attention during the past decade. In [2], Xiang and Brunner first investigated Filon collocation approximations to highly oscillatory VIEs by employing the asymptotic property of oscillatory integrals. They found that errors of Filon collocation solutions decayed fast as the frequency increased. The third author presented an optimal convergence order for the direct Filon collocation solution to the first kind of oscillatory VIE arising in acoustic scattering in [3]. The convergence behavior of such kinds of numerical approaches was able to be revealed with the help of the detailed study of the remainder for the error function. Besides, it is noted that numerical analysis with respect to the frequency, which is usually done by solving error equations and extending van der Corput lemma (see [4] p. 333), is able to detect the ability of the numerical method to solve highly oscillatory VIEs. With these techniques in mind, several authors made great contributions to numerical solutions to highly oscillatory VIEs. For example, Galerkin and collocation solutions for VIEs with highly oscillatory trigonometric kernels were investigated in [5,6], highly oscillatory VIEs with weakly singular kernels were studied in [7], the Hermite-type Filon collocation method was presented in [8], and Clenshaw–Curtis–Filon qudrature for Cauchy singular integral equations was investigated in [9].

In this work, we consider the numerical computation of the following second-kind oscillatory VIE:

$$u(t) = f(t) + \int_0^t K(t,s)e^{i\omega g(t,s)}u(s)ds, \ t \in [0,T], \tag{1}$$

where $K(t,s)$, $g(t,s)$ and $f(t)$ are sufficiently smooth, $u(t)$ is unknown, and ω denotes the oscillation parameter. When $\omega = 0$, Equation (1) reduces to the classical VIEs. In the case of $\omega \gg 1$, the kernel in Equation (1) is highly oscillatory, and special quadrature rules should be employed in practical computation.

In the remaining part, we are restricted to the following problems. In the forthcoming section, we first develop a class of generalized multistep collocation methods ($GMC_{k_1,k_2}M$) for Equation (1) with non-oscillatory kernels, that is, $\omega = 0$. Then, classical convergence analysis and linear stability analysis are implemented. In the third section, we study the numerical solution to VIE (1) when the kernel changes rapidly, that is, $\omega \gg 1$, and present the frequency-explicit convergence analysis. Some concluding remarks are given in Section 4.

2. $GMC_{k_1,k_2}M$ in the Case of $\omega = 0$

Frequently-used approaches for VIEs include collocation methods [10], the spectral collocation method [11,12], the spectral Galerkin method [13,14], the Nyström method [15,16], and so on. Among these numerical formulae, the collocation-based approach is one of the most important tools. In general, the collocation solution is obtained by making the polynomial or piecewise polynomial satisfy the collocation equation. For one-step collocation methods, one can find detailed analysis in [10]. To increase the convergence rate without adding collocation points, Conte and Paternoster studied multistep collocation solutions with the help of employing approximations to numerical solutions in computed steps in [17]. However, multistep methods usually tend to be unstable. Fazeli et al. further investigated the stability of multistep collocation methods in [18], and found some super implicit collocation solutions with wide stability regions. On the other hand, inspired by the study of boundary value methodology for solving ODE (see [19]), several authors made contributions to boundary value solutions to Volterra functional equations [20–22]. Based on interpolation outside the current subinterval and approximated end values, the third author and Xiang devised CBVM for second-kind VIEs in [22]. Furthermore, the third author extended said kind of methodology to VIEs with weakly singular kernels by employing the fractional polynomial interplant in [23], and the block CBVM for the first-kind VIE was investigated in [24]. In this section we first investigate the construction of $GMC_{k_1,k_2}M$ with the help of local polynomial interpolation. Then, the convergence and linear stability analysis of $GMC_{k_1,k_2}M$ are considered.

2.1. Discretization of VIE

Let the interval $[0, T]$ be divided uniformly, that is,

$$X_h = \{t_j : t_j = jh, \, j = 0, 1, \cdots, N = T/h\}.$$

Then define local basic functions

$$\phi_j^{k_1,k_2}(s) = \prod_{i=-k_1, i \neq j}^{k_2+1} \frac{s-i}{j-i}, \, j = -k_1, \cdots, k_2+1, \tag{2}$$

For the first k_1 subintervals, that is, for any $t \in [t_0, t_{k_1}]$, the collocation polynomial is represented by

$$u_h(t_{k_1} + sh) = \sum_{i=-k_1}^{k_2+1} y_{k_1+i} \phi_i^{k_1,k_2}(s), \, s \in (-k_1, 0], \tag{3}$$

For $k_1 \leq n \leq N - k_2 - 1$, $u_h(t)$ over the interval $[t_n, t_{n+1}]$ is rewritten as

$$u_h(t_n + sh) = \sum_{i=-k_1}^{k_2+1} y_{n+i} \phi_i^{k_1,k_2}(s), \, s \in (0, 1], \tag{4}$$

In the last subinterval $[t_{N-k_2}, t_N]$, we rewrite $u_h(t)$ as

$$u_h(t_{N-k_2-1} + sh) = \sum_{i=-k_1}^{k_2+1} y_{N-k_2-1+i} \phi_i^{k_1,k_2}(s), \quad s \in (1, k_2+1]. \tag{5}$$

Finally, the collocation equation follows:

$$u_h(t_n) = f(t_n) + \int_0^{t_n} K(t_n, s) u_h(s) ds, \quad t_n \in X_h. \tag{6}$$

A direct calculation leads to

$$y_n - f(t_n) = \begin{cases} h \int_{-k_1}^{n-k_1} K(t_n, t_{k_1} + sh) \left(\sum_{i=-k_1}^{k_2+1} y_{k_1+i} \phi_i^{k_1,k_2}(s) \right) ds, & n = 1, \cdots, k_1, \\ h \int_{-k_1}^{0} K(t_n, t_{k_1} + sh) \left(\sum_{i=-k_1}^{k_2+1} y_{k_1+i} \phi_i^{k_1,k_2}(s) \right) ds \\ \quad + h \sum_{j=k_1+1}^{n} \int_0^1 K(t_n, t_{j-1} + sh) \left(\sum_{i=-k_1}^{k_2+1} y_{j-1+i} \phi_i^{k_1,k_2}(s) \right) ds, & n = k_1+1, \cdots, N-k_2, \\ h \int_{-k_1}^{0} K(t_n, t_{k_1} + shsh) \left(\sum_{i=-k_1}^{k_2+1} y_{k_1+i} \phi_i^{k_1,k_2}(s) \right) ds \\ \quad + h \sum_{j=k_1+1}^{N-k_2} \int_0^1 K(t_n, t_{j-1} + sh) \left(\sum_{i=-k_1}^{k_2+1} y_{j+i-1} \phi_i^{k_1,k_2}(s) \right) ds \\ \quad + h \int_1^{n-N+k_2+1} K(t_n, t_{N-k_2-1} + sh) \sum_{i=-k_1}^{k_2+1} y_{N-k_2-1+i} \phi_i^{k_1,k_2}(s) ds, & n = N-k_2+1, \cdots, N. \end{cases} \tag{7}$$

Denoting

$$\text{MOM}_{a,c,i}^{b,d} = \int_a^b K(t_c, t_d + sh) \phi_i^{k_1,k_2}(s) ds, \tag{8}$$

we have for $k = -k_1, \cdots, k_2+1$,

$$A_k^{initial} = (a_{i,j}^{(k)}) = \begin{cases} \text{MOM}_{-k_1,n,j-k_1-1}^{0,k_1}, & i \leq N, j = k+k_1+1, \\ 0, & \text{others}, \end{cases}$$

$$A_k^{main} = (b_{i,j}^{(k)}) = \begin{cases} \text{MOM}_{0,n,k}^{1,j-1}, & k_1 < i \leq N, i+k \leq j \leq i+k+\min\{0, n-N+k_2+1\}, \\ 0, & \text{others}, \end{cases}$$

$$A_k^{end} = (c_{i,j}^{(k)}) = \begin{cases} \text{MOM}_{i,n,j-N+k_2}^{n-N+k_2+1,N-k_2-1}, & i > N-k_2, j = N-k_2+k, \\ 0, & \text{others}. \end{cases}$$

Now we are able to rewrite Equation (6) in the closed form:

$$(\mathbf{I} - h\mathbf{A}(1:N, 2:N+1))\mathbf{Y} = \mathbf{F} + hy_0 \mathbf{A}(1:N, 1), \tag{9}$$

where $\mathbf{I}$ denotes the identity matrix, $\mathbf{A} = \sum_{k=-k_1}^{k_2+1} A_k^{initial} + \sum_{k=-k_1}^{k_2+1} A_k^{main} + \sum_{k=-k_1}^{k_2+1} A_k^{end}$, $\mathbf{Y} = [y_1, y_2, \cdots, y_N]^T$, and $\mathbf{F} = [f(t_1), f(t_2), \cdots, f(t_N)]^T$. By employing proper numerical integration approaches such as Clenshaw–Curtis quadrature and applying iterative solvers to Equation (9), we are able to obtain the collocation solution at the grid.

2.2. Convergence Analysis with Respect to Stepsize

Now we turn to studying the convergence behavior of the piecewise collocation polynomial computed by Equation (9). Firstly, we revisit some helpful results from approximation theory.

Lemma 1 ([10] p. 43). *Consider the following assumption.*

- *Defining abscissa $a \leq \xi_1 < ... < \xi_m \leq b$, we obtain the error between $f(x)$ and the Lagrange interpolation polynomial of degree $m-1$ with respect to the given points $\{\xi_j\}$.*

$$\varepsilon_m(f;x) = f(x) - \sum_{j=1}^{m} L_j(x) f(\xi_j), \; x \in [a,b],$$

where $L_j(x)$ denotes Lagrange basis.
- *Letting $1 \leq d \leq m$, we suppose $f(x)$ belongs to the space $C^d[a,b]$.*

Then we can represent the error function $\varepsilon_m(f;x)$ as follows.

$$\varepsilon_m(f;x) = \int_a^b \kappa_d(x,t) f^{(d)}(t) dt, \; x \in [a,b]. \tag{10}$$

Here the kernel function $\kappa_d(x,t)$ can be obtained by

$$\kappa_d(x,t) := \frac{1}{(d-1)!} \left\{ (x-t)_+^{d-1} - \sum_{j=1}^{m} L_j(x) (\xi_j - t)_+^{d-1} \right\},$$

and

$$(x-t)_+^p := \begin{cases} 0, & x < t, \\ (x-t)^p, & x \geq t. \end{cases}$$

Lemma 2 ([10] p. 81). *Suppose that there a sequence $\{k_i\}$ with $k_i \geq 0$ and another sequence $\{\varepsilon_i\}$ with $\varepsilon_0 \leq \rho_0$. Moreover, $\{k_i\}$ and $\{\varepsilon_i\}$ satisfy*

$$\varepsilon_n \leq \rho_0 + \sum_{i=0}^{n-1} q_i + \sum_{i=0}^{n-1} k_i \varepsilon_i, \; n \geq 1,$$

with $\rho_0 \geq 0, q_i \geq 0, i \geq 0$. Then

$$\varepsilon_n \leq \left(\rho_0 + \sum_{i=0}^{n-1} q_i \right) e^{\sum_{i=0}^{n-1} k_i}, \; n \geq 1.$$

Existing studies show that we cannot compute collocation boundary value solutions by recurrences. All numerical values should be computed simultaneously through solving linear systems. Note that the element of $h\mathbf{A}(1:N, 2:N+1)$ is bounded by

$$h(k_1 + k_2 + 1)\bar{K} \left\| \sum_{i=-k_1}^{k_2+1} \phi_i^{k_1,k_2}(t) \right\|_\infty \leq h\bar{K} 2^{k_1+k_2+4},$$

where $\bar{K}$ denotes the maximum of the kernel function $K(t,s)$, and the above inequality is derived from the Lesbegue constant of the polynomial interpolant (see [25]). We obtain $h\mathbf{A}(1:N,2:N+1) < 1$ whenever $h < (\bar{K} 2^{k_1+k_2+4})^{-1}$, which enables us to compute $\det(\mathbf{I} - h\mathbf{A}(1:N,2:N+1)) \neq 0$ by Gaussian elimination, as is done in [22]. Therefore, the well-posedess of the solution computed by $GMC_{k_1,k_2}M$ is guaranteed. It is noted that when we encounter stiff problems, the maximum $\bar{K}$ may be large, which implies we have to apply a particularly small stepsize h and restricts the application of the collocation method. However, due to the compactness of Volterra integral operator, the spectrum of $h\mathbf{A}(1:N, 2:N+1)$ will be found in the neighborhood of 0 with a tolerance stepsize, and the multistep collocation method is feasible in practical uses. In Figure 1, we show the discretized spectrum of $h\mathbf{A}(1:N, 2:N+1)$ by considering the kernel function $K(t,s) = 50e^{i\omega(t-s)}$ with the maximum 50. It can be seen that eigenvalues are bounded by the unit circle when the stepsize decreases to $1/64$, which guarantees the solvability of the linear system.

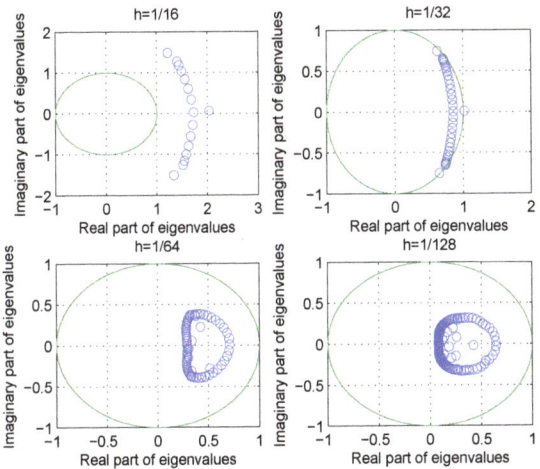

Figure 1. The spectrum of $h\mathbf{A}(1:N, 2:N+1)$ for various stepsizes h.

Furthermore, we arrive at the following theorem by employing Lemmas 1 and 2.

Theorem 3. *Suppose that $K(t,s), f(t)$ in VIE (1) are sufficiently smooth, that is, $K \in C^{k_1+k_2+2}(D)$ and $f \in C^{k_1+k_2+2}(I)$. Furthermore, let $u_h(t)$ denote the collocation polynomial computed by $GMC_{k_1,k_2}M$ with a stepsize h. Then the collocation error $e_h(t) = u(t) - u_h(t)$ in the collocation grid is bounded by*

$$\max_{t \in X_h} |e(t)| \leq Ch^{k_1+k_2+2}, \tag{11}$$

where the constant C is independent of the stepsize h but depends on T.

Proof. Note that by

$$u(t) = f(t) + \int_0^t K(t,s)u(s)ds, \ t \in [0,T],$$

and

$$u_h(t) = f(t) + \int_0^t K(t,s)u_h(s)ds, \ t \in X_h,$$

we obtain the collocation error function $e_h(t)$ satisfying

$$e_h(t) = \int_0^t K(t,s)e_h(s)ds, \ t \in X_h. \tag{12}$$

Let

$$R_n^{k_1,k_2}(v) := \int_0^{k_1+k_2+1} \kappa_{k_1}^{k_2}(v,z) u^{(k_1+k_2+2)}(t_n + zh)dz,$$

where
$$\kappa_{k_1}^{k_2}(v,z) := \frac{1}{(k_1+k_2+1)!}\left((v-z)_+^{k_1+k_2+1} - \sum_{j=-k_1}^{k_2+1}\phi_j^{k_1,k_2}(v)(j-z)_+\right).$$

For $t \in [0, t_{k_1}]$, we have
$$e_h(t) = e_h(t_{k_1} + sh) = \sum_{i=-k_1}^{k_2+1} v_{k_1+i}\phi_i^{k_1,k_2}(s) + h^{k_1+k_2+2}R_0^{k_1,k_2}(s), \quad s \in [-k_1, 0],$$

where $v_n := e_h(t_n)$. For $t \in [t_j, t_{j+1}]$, $j = k_1, \cdots, N-k_2-1$, we have
$$e_h(t) = e_h(t_j + sh) = \sum_{i=-k_1}^{k_2+1} v_{j+i}\phi_i^{k_1,k_2}(s) + h^{k_1+k_2+2}R_j^{k_1,k_2}(s), \quad s \in [0, 1].$$

For $t \in [t_{N-k_2}, t_N]$, we have
$$e_h(t) = e_h(t_{N-k_2-1} + sh) = \sum_{i=-k_1}^{k_2+1} v_{N-k_2+i}\phi_i^{k_1,k_2}(s) + h^{k_1+k_2+2}R_{N-k_2}^{k_1,k_2}(s), \quad s \in [1, k_2+1].$$

Furthermore, by letting
$$\text{RES}_{a,c,i}^{b,d} = \int_a^b K(t_c, t_d + sh)R_i^{k_1,k_2}(s)ds, \tag{13}$$

we obtain
$$v_n := \begin{cases} h\sum_{i=-k_1}^{k_2+1} v_{i+k_1}\text{MOM}_{k_1,n,0}^{n-k_1,k_1} + h^{k_1+k_2+3}\text{RES}_{k_1,n,0}^{n-k_1,k_1}, & n = 1, \cdots, k_1, \\ h\sum_{i=-k_1}^{k_2+1} v_{i+k_1}\text{MOM}_{-k_1,n,i}^{0,k_1} + h^{k_1+k_2+3}\text{RES}_{-k_1,n,0}^{0,k_1} \\ \quad + h\sum_{l=k_1}^{n-1}\sum_{i=-k_1}^{k_2+1} v_{l+i}\text{MOM}_{0,n,i}^{1,l} + h^{k_1+k_2+3}\sum_{l=k_1}^{n-1}\text{RES}_{0,n,n-1}^{1,l} & n = k_1+1, \cdots, N-k_2, \\ h\sum_{i=-k_1}^{k_2+1} v_{i+k_1}\text{MOM}_{-k_1,n,i}^{0,k_1} + h^{k_1+k_2+3}\text{RES}_{0,0,0}^{k_1,0} \\ \quad + h\sum_{l=k_1}^{N-k_2-1}\sum_{i=-k_1}^{k_2+1} v_{l+i}\text{MOM}_{0,n,i}^{1,l} + h^{k_1+k_2+3}\sum_{l=k_1}^{N-k_2-1}\text{RES}_{0,n,l}^{1,l} \\ \quad + h\sum_{i=-k_1}^{k_2+1} v_{i+N-k_2}\text{MOM}_{1,n,i}^{n-N+k_2+1,n-k_2-1} \\ \quad + h^{k_1+k_2+3}\text{RES}_{1,n,N-k_2-1}^{n-N+k_2+1,N-k_2-1} & n = N-k_2+1, \cdots, N. \end{cases}$$

Suppose that $\mathrm{MOM}_{a,c,i}^{b,d}$ and $\mathrm{RES}_{a,c,i}^{b,d}$ are bounded by the constant B. It is easily noted from Equations (8) and (13) that B does not depend on the stepsize. A direct calculation leads to

$$|v_n| \leq \begin{cases} hB \sum_{i=-k_1, i \neq n-k_1}^{k_2+1} |v_{i+k_1}| + h|v_n| + h^{k_1+k_2+3}B, & n = 1, \cdots, k_1, \\ hB \sum_{i=-k_1, i \neq n-k_1}^{k_2+1} |v_{i+k_1}| + h^{k_1+k_2+3}B + h(k_1+k_2+2)|v_n| \\ +hB \sum_{l=k_1}^{n-1} \left(\sum_{i=-k_1, i \neq n-l}^{k_2+1} |v_{l+i}| + h^{k_1+k_2+3}B \right), & n = k_1+1, \cdots, N-k_2, \\ hB \sum_{i=-k_1, i \neq n-k_1}^{k_2+1} |v_{i+k_1}| + h^{k_1+k_2+3}B + h(k_1+k_2+2)|v_n| \\ +hB \sum_{l=k_1}^{N-k_2-1} \left(\sum_{i=-k_1, i \neq n-l}^{k_2+1} |v_{l+i}| + h^{k_1+k_2+3}B \right) \\ +hB \sum_{i=-k_1, i \neq n-N+k_2+1}^{k_2+1} |v_{i+N-k_2}| + h|v_n| + h^{k_1+k_2+3}B, & n = N-k_2+1, \cdots, N. \end{cases}$$

Hence, we have

$$(1 - h(k_1+k_2+2))|v_n| \leq h^{k_1+k_2+2}B + h(k_1+k_2+2)B \sum_{i=n+1}^{n+k_2} |v_i| + h(k_1+k_2+2)B \sum_{i=1}^{n-1} |v_i|.$$

Since $1 - h(k_1+k_2+2) \approx 1$ for sufficiently small stepsize h, we obtain

$$|v_n| \leq h^{k_1+k_2+2}\tilde{B} + hk_2(k_1+k_2+2)\tilde{B}\|e_h\|_\infty + h(k_1+k_2+2)\tilde{B} \sum_{i=1}^{n-1} |v_i|.$$

According to Lemma 2, we have

$$\|e_h\|_\infty \leq e^{(k_1+k_2+2)\tilde{B}} \tilde{B} h^{k_1+k_2+2} + he^{(k_1+k_2+2)\tilde{B}}(k_1+k_2+2)k_2\tilde{B}\|e_h\|_\infty,$$

or equivalently,

$$\|e_h\|_\infty \leq \frac{e^{(k_1+k_2+2)\tilde{B}} \tilde{B}}{1 - he^{(k_1+k_2+2)\tilde{B}}(k_1+k_2+2)k_2\tilde{B}} h^{k_1+k_2+2}$$

for sufficiently small stepsize h. □

Example 1. *Let us solve VIE with* $GMC_{k_1,k_2}M$

$$u(t) = e^t + \int_0^t 2\cos(t-s)u(s)ds, t \in [0,2] \tag{14}$$

with the exact solution $u(t) = (1+t)^2 e^t$.

In this example, we test the performance of $GMC_{k_1,k_2}M$. We mainly focus on two terms of data, the maximum of error functions (INAE), and the convergence order. Computed results are shown in Tables 1–3.

Table 1. Collocation error and convergence order of $GMC_{k_1,k_2}M$ for Example 1.

	$GMC_{1,2}M$ Error	Order	$GMC_{1,3}M$ Error	Order
$N = 8$	4.10×10^{-3}	–	1.20×10^{-3}	–
$N = 16$	4.13×10^{-4}	3.31	2.72×10^{-5}	5.46
$N = 32$	1.59×10^{-5}	4.70	6.76×10^{-7}	5.33
$N = 64$	5.34×10^{-7}	4.90	1.22×10^{-8}	5.79
$N = 128$	1.72×10^{-8}	4.96	2.03×10^{-10}	5.91
$N = 256$	5.45×10^{-10}	4.98	3.27×10^{-12}	5.96
Referenced Order		5.00		6.00

Table 2. Collocation error and convergence order of $GMC_{k_1,k_2}M$ for Example 1.

	$GMC_{2,1}M$ Error	Order	$GMC_{2,3}M$ Error	Order
$N = 8$	1.57×10^{-2}	–	9.39×10^{-4}	–
$N = 16$	4.77×10^{-4}	5.04	3.91×10^{-6}	7.91
$N = 32$	1.61×10^{-5}	4.89	2.62×10^{-8}	7.22
$N = 64$	5.32×10^{-7}	4.92	1.97×10^{-10}	7.05
$N = 128$	1.71×10^{-8}	4.96	1.51×10^{-12}	7.03
Referenced Order		5.00		7.00

Table 3. Collocation error and convergence order of $GMC_{k_1,k_2}M$ for Example 1.

	$GMC_{3,1}M$ Error	Order	$GMC_{3,2}M$ Error	Order
$N = 16$	1.58×10^{-5}	–	4.57×10^{-7}	–
$N = 32$	5.54×10^{-7}	4.83	1.51×10^{-8}	4.92
$N = 64$	1.12×10^{-8}	5.63	1.60×10^{-10}	6.57
$N = 128$	1.95×10^{-10}	5.85	1.38×10^{-12}	6.86
Referenced Order		6.00		7.00

It can be seen from these tables that as the quantity of nodes increases, absolute errors decay fast, and as k_1 and k_2 get bigger, the convergence order enlarges. Besides, numerical results illustrate that $GMC_{k_1,k_2}M$ achieves the expected order of the estimate given in Theorem 3.

Remark 1. *When numerical solutions of evolution equations are considered, Courant proposes that the combination of a consistent and stable numerical approach led to its convergence, which contributes to the foundation of classical numerical analysis theory of numerical studies on differential equations. On the other hand, the above convergence analysis is based on a fixed integration interval $[0, T]$, which differs from the convergence analysis for evolution problems where we usually consider the case of $T \to \infty$. In addition, it should be noted that the convergence result in Theorem 3 does not guarantee a feasible approximation in practical computation for long-time integration, especially when we are met with stiff problems. Therefore, we give linear stability analysis of the presented collocation method in the forthcoming subsection.*

2.3. Linear Stability Analysis

For a long-time integration problem, round-off errors may dramatically affect the numerical solution. In this subsection, we analyze the collocation solution's linear stability originating from the study of numerical solutions of ordinary differential equations, where one usually considers the test equation

$$y'(t) = \lambda y(t), \text{ Re}(\lambda) < 0.$$

Particularly, Brugnano and Trigiante investigated multistep methods for solving differential problems with the above scalar equation in [19]. For the general linear multistep formula

$$\sum_{j=0}^{k} \alpha_j y_{n+j} - h\lambda \sum_{j=0}^{k} \beta_j y_{n+j} = 0,$$

we can introduce two polynomials

$$\rho(z) = \sum_{j=0}^{k} \alpha_j z^j, \ \sigma(z) = \sum_{j=0}^{k} \beta_j z^j,$$

and define the associated characteristic polynomial $\pi(z,q) = \rho(z) - q\sigma(z)$ with $q = h\lambda$. When $\pi(z,q)$ is a Schur polynomial for fixed q, the method is absolutely stable at q. For the moment the definition of the region of absolute stability is

$$\mathbb{D} := \{q \in \mathbb{C} : \pi(z,q) \text{ is a Schur polynomial}\}$$

If $\mathbb{C}^- \subseteq \mathbb{D}$, the method is said to be A−stable.

Since both of discretization of ODE and VIE result in difference equations, we can investigate the generalized multistep collocation method with the help of stability studies of ODE. Consider the following test equation:

$$u(t) = 1 + \lambda \int_0^t u(s)ds, \ t \in [0,T], \text{ Re}(\lambda) < 0. \tag{15}$$

We turn to study the linear stability of the collocation solution by investigating Equation (15). By applying $GMC_{k_1,k_2}M$ we have

$$y_j = 1 + \lambda \int_0^{jh} u_h(s)ds, j = k_1 + 1, ..., N - k_2. \tag{16}$$

Next, noting the difference between y_j and y_{j-1} in Equation (15) leads to

$$y_j - y_{j-1} = h\lambda \sum_{i=-k_1}^{k_2+1} y_{j-1+i} \int_0^1 \phi_i^k(s)ds, \ j = k_1 + 1, ..., N - k_2. \tag{17}$$

Then the characteristic polynomial is defined by

$$\pi^{k_1,k_2}(z,q) = z^{k_1+1} - z^{k_1} - q \sum_{i=0}^{k_1+k_2+1} z^i \int_0^1 \phi_{i-k_1}^{k_1,k_2}(s)ds = \rho(z) - q\sigma(z). \tag{18}$$

Before investigating the linear stability region, we introduce some helpful definitions and theorems in the version of $GMC_{k_1,k_2}M$.

Definition 4 ([19]). *For any complex number $q := h\lambda$, if the collocation solution u_h to Equation (15) computed by $GMC_{k_1,k_2}M$ goes to 0 as T goes ∞ for fixed stepsize, then $GMC_{k_1,k_2}M$ is said to be absolutely stable at q.*

Definition 5 ([19]). *For any $z \in \mathbb{S}$, if $GMC_{k_1,k_2}M$ is absolutely stable at z, then the set $\mathbb{S}$ is said to be the linear stability region of $GMC_{k_1,k_2}M$. Particularly, if the left part of the complex plane is contained in $\mathbb{S}$, then $GMC_{k_1,k_2}M$ is said to be A−stable.*

Theorem 6 ([19]). *For any complex number q, if roots of Equation* (18) *satisfy*

$$|z_1^k| \leq \cdots \leq |z_{k_1}^k| < 1 < |z_{k_1+1}^k| \leq \cdots \leq |z_{k_1+k_2+1}^k|, \tag{19}$$

then $GMC_{k_1,k_2}M$ *is stable at q.*

By a direct calculation, we find that roots of $\pi^{k_1,k_2}(z,q)$ do not satisfy the condition given in Theorem 6 in the case of $k_1 = k_2$. Hence, the region of stability cannot be shown. In Figures 2 and 3, we list the boundary locus corresponding to various multistep collocation methods with $k_1 \neq k_2$, where the boundary Γ is defined by

$$\Gamma := \{z \in \mathbb{C}, z = \frac{\rho(e^{i\theta})}{\sigma(e^{i\theta})}, 0 \leq \theta < 2\pi\}.$$

It can be seen that these trajectories are Jordan curves, which implies Γ is the boundary of corresponding absolute stability region. The stability region in Figure 2 is the part outside the boundary curves, while that in Figure 3 is the inside part. Therefore, we can conclude that $GMC_{k_1,k_2}M$ has wide stability region in the case of $k_2 > k_1$. In addition, the boundary trajectories of $GMC_{k1,k2}M$ and $GMC_{k2,k1}M$ are symmetric with respect to virtual axis.

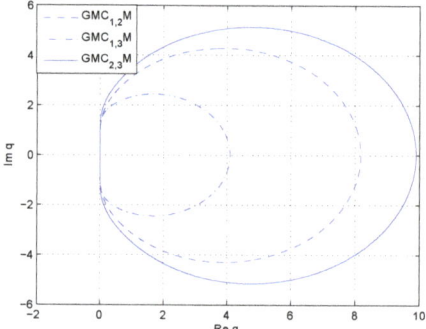

Figure 2. Linear stability region for $GMC_{1,2}M$, $GMC_{1,3}M$, $GMC_{2,3}M$.

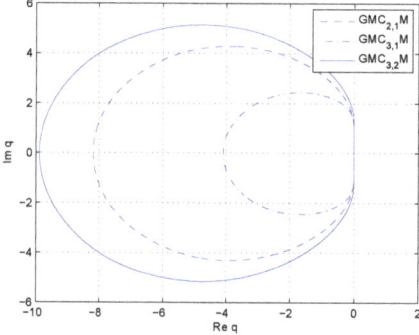

Figure 3. Linear stability region for $GMC_{2,1}M$, $GMC_{3,1}M$, $GMC_{3,2}M$.

3. $\text{GMC}_{k_1,k_2}\text{M}$ in the Case of $\omega \gg 1$

When the oscillation parameter $\omega \gg 1$ in Equation (1), classical quadrature usually results in time-consuming algorithms. Hence, we first give an efficient numerical approach for moments in Equation (6) in this section. Then the frequency-explicit convergence analysis is presented.

3.1. Fast Calculation of Moments

Numerical studies of highly oscillatory integrals (HOIs) have been intensively focused on in the past few decades. High-order algorithms, such as Filon-type quadrature [26], Levin quadrature [27], and the numerical steepest decent method [28], have been proposed. In this subsection, we consider a composite quadrature rule based on Xiang's modified Filon-type quadrature developed in [29].

Consider the computation of

$$\text{M}_{\omega,n}^{a,b} := \int_a^b K(t_n,s) e^{i\omega g(t_n,s)} \phi(s) ds, \quad n = 1, 2, \cdots, N. \tag{20}$$

When the phase has no stationary points, that is, $g'(t_n, s) \neq 0$ for any $s \in [a,b]$, let $\{c_k\}_{k=0}^v$ be the equispaced nodes on the interval $[a,b]$, that is, $c_k = a + \frac{k}{v}(b-a)$ for $k = 0, \cdots, v$. In addition, let $\{m_k\}_{k=0}^v$ denote a set of positive integers associated with nodes $\{c_k\}_{k=0}^v$, which helps represent Hermite interplant later. Furthermore, define the function

$$\sigma_k(s) = \begin{cases} \dfrac{K(t_n,s)\phi(s)}{g'(t_n,s)}, & k = 1, \\ \dfrac{\sigma_{k-1}'(s)}{g'(t_n,s)}, & k \geq 2. \end{cases}$$

Then we can find a polynomial $p(s) = \sum_{q=0}^{\hat{N}} a_q s^q$ with $\hat{N} = \sum_{k=0}^{v} m_k - 1$ satisfying

$$\begin{cases} p(g(c_0)) = \sigma_1(c_0) \\ \cdots \\ p^{(m_0-1)}(g(c_0)) = \sigma_{m_0-1}(c_0) \\ p(g(c_1)) = \sigma_1(c_1) \\ \cdots \\ p^{(m_1-1)}(g(c_1)) = \sigma_{m_1-1}(c_1) \\ \cdots \\ p(g(c_v)) = \sigma_1(c_v) \\ \cdots \\ p^{(m_v-1)}(g(c_v)) = \sigma_{m_v-1}(c_v) \end{cases}$$

With the coefficients a_q by solving the above linear system, we can approximate $\text{M}_{\omega,n}^{a,b}$ by

$$\int_{g(t_n,a)}^{g(t_n,b)} p(s) e^{i\omega s} ds = \sum_{q=0}^{\hat{N}} a_q \int_{g(t_n,a)}^{g(t_n,b)} s^q e^{i\omega s} ds,$$

where $\int_{g(t_n,a)}^{g(t_n,b)} s^q e^{i\omega s} ds$ can be calculated by incomplete Gamma function.

In the case of $g'(t_n, s) = 0$ for some $s \in [a,b]$, suppose $s = a$ without loss of generality. Then we insert the grid points

$$a, a + \frac{2^0}{\omega}, a + \frac{2}{\omega}, a + \frac{2^2}{\omega}, \cdots, a + \frac{2^m}{\omega}, b,$$

where m is the maximum integer less than $\log_2 \omega(b-a)$. Integration with Xiang's Filon quadrature in each subintervals results in the composite Filon quadrature. It is noted that the integral over the first interval is non-oscillatory and we can employ classical quadrature such as Gauss or Clenshaw–Curtis instead to avoid the stationary problem.

3.2. Convergence Analysis with Respect to the Frequency

Collocation methods with high-order quadrature usually lead to a class of fascinating algorithms, which are able to provide high-precision collocation solutions in the case of high frequency. In this subsection, we consider the general oscillator and investigate the convergence analysis for multistep collocation solutions, where the convergence order is represented by the frequency parameter ω.

Firstly, let us restrict ourselves to considering the following set of functions.

Definition 7. *Given any bivariate function $g(t,s)$ defined on $[0,T] \times [0,T]$, suppose that $g(t,s)$ has several stationary points $\xi_1, \cdots, \xi_{n_t}$ over $[0,T]$ for any fixed t, and*

$$\begin{cases} g'(t,\xi_1) = \cdots = g^{(r_1)}(t,\xi_1) = 0, g^{(r_1+1)}(t,\xi_1) \neq 0 \\ g'(t,\xi_2) = \cdots = g^{(r_2)}(t,\xi_2) = 0, g^{(r_2+1)}(t,\xi_2) \neq 0, \\ \cdots \\ g'(t,\xi_{n_t}) = \cdots = g^{(r_N)}(t,\xi_{n_t}) = 0, g^{(r_{n_t}+1)}(t,\xi_{n_t}) \neq 0. \end{cases}$$

Let $\rho(t) = \max\limits_{i=1,\cdots,n_t} \{r_i\}$ and $r = \sup\limits_{t \in [0,T]} \{\rho(t)\}$. Then $g(t,s)$ is said to be in $\mathcal{A}(r)$.

Secondly, we give a slight extension of the classical van der Corput Lemma (see [4] p. 333).

Lemma 8. *Suppose that $g(t,s) \in \mathcal{A}(r)$. Moreover, suppose $\phi(s) \in C^1(a,b)$ and $\phi'(s)$ is integrable. We can conclude that*

$$\left| \int_a^b \phi(s) e^{i\omega g(t_n,s)} ds \right| \leq C \omega^{-1/(r+1)}. \, n = 1, 2, \cdots, N.$$

Here the constant C is independent of ω.

Finally, we are able to develop the convergence behavior of collocation polynomials computed by $GMC_{k_1,k_2}M$ in the highly oscillatory case.

Theorem 9. *Assume both of $g(t,s) \in \mathcal{A}(r)$ and f are sufficiently smooth. Then the numerical solutions derived from $GMC_{k_1,k_2}M$ for VIE (1) satisfy*

$$\max_{t \in I_h} \{|u(t) - u_h(t)|\} = O(\omega^{-1/(r+1)}), \omega \to \infty. \tag{21}$$

Proof. To begin with, we explore the boundedness of the solution $u(t)$ to Equation (1) and its derivative. By applying Picard iteration, we can rewrite $u(t)$ as

$$u = f + \sum_{j=1}^{\infty} (\mathbf{K}^j f). \tag{22}$$

Here $\mathbf{K}$ denotes the integral operator

$$(\mathbf{K}\phi)(t) := \int_0^t K(t,s) e^{i\omega g(t,s)} \phi(s) ds$$

According to Lemma 8, we get that $u(t)$ is bounded as $\omega \to \infty$. On the other hand, the derivative can be rewritten by a direct calculation

$$u'(t) = f'(t) + \sum_{j=1}^{\infty} \left(K(t,t) f(t) e^{i\omega g(t,t)} (\mathbf{K}^{j-1} f)(t) + i\omega \int_0^t K(t,s) f(s) g'(t,s) e^{i\omega g(t,s)} (\mathbf{K}^{j-1} f)(s) ds \right)$$
$$= f'(t) + \sum_{j=1}^{\infty} \left(\mathcal{I}_j + \mathcal{II}_j \right),$$

where

$$\mathcal{I}_j := K(t,t) f(t) e^{i\omega g(t,t)} (\mathbf{K}^{j-1} f)(t), \quad \mathcal{II}_j := i\omega \int_0^t K(t,s) f(s) g'(t,s) e^{i\omega g(t,s)} (\mathbf{K}^{j-1} f)(s) ds.$$

By letting $\omega \to \infty$, $\mathcal{I}_j$ is bounded due to Lemma 8, and $\mathcal{II}_j$ is bounded by noting that $g'(t,s)$ vanishes at $s = 0$.

When noting that the collocation error function $e_h(t)$ defined in the previous section satisfies

$$e_h(t) = \int_0^t K(t,s) e^{i\omega g(t,s)} e_h(s) ds, \ t \in X_h, \tag{23}$$

we obtain

$$(\mathbf{I} - h\mathbf{A}(1:N, 2:N+1)) \mathbf{e}_h = \mathbf{R}, \tag{24}$$

where

$$\mathbf{e}_h = \begin{pmatrix} e_h(t_1) \\ e_h(t_2) \\ \cdots \\ e_h(t_N) \end{pmatrix}, \mathbf{R} = \begin{pmatrix} h^{k_1+k_2+3} \mathrm{RES}_{0,0,0}^{1,0} \\ h^{k_1+k_2+3} \mathrm{RES}_{0,0,0}^{2,0} \\ \cdots \\ h^{k_1+k_2+3} \mathrm{RES}_{0,0,0}^{k_1,0} + h^{k_1+k+2+3} \sum_{l=k_1}^{N-k_2-1} \mathrm{RES}_{0,N,N}^{1,l} + h^{k_1+k_2+3} \mathrm{RES}_{0,n,N-k_2-1}^{k_2+1,N-k_2-1} \end{pmatrix}$$

Since both of $u(t)$ and $u_h(t)$ are bounded as $\omega \to \infty$, employing Lemma 8 implies

$$|\mathrm{MOM}_{a,c,i}^{b,d}| \leq C\omega^{-1/(r+1)}, |\mathrm{RES}_{a,c,i}^{b,d}| \leq C\omega^{-1/(r+1)}.$$

Hence for fixed stepsize h, $\mathbf{I} - h\mathbf{A}(1:N, 2:N+1)$ is invertible for sufficiently large ω, and we can represent $\mathbf{e}_h$ by

$$\mathbf{e}_h = (\mathbf{I} - h\mathbf{A}(1:N, 2:N+1))^{-1} \mathbf{R}.$$

By noting that maximum of $\mathbf{R}$ goes to 0 with a speed of $O(\omega^{-1/(r+1)})$ as ω goes to ∞, we obtain the estimate (21). □

In the following example, we test the convergence rate of $GCM_{1,2}M$ in the case of high frequency.

Example 2. *In this example, we solve the following VIE with $GCM_{1,2}M$,*

$$u(t) + \int_0^t e^{i\omega(t-s)} u(s) ds = e^t, t \in [0,1]. \tag{25}$$

The exact solution is $u(t) = \left(\int_0^t (-ce^s) e^{-cs} ds + 1 \right) e^{ct}, c = i\omega - 1.$

In Figure 4, we plot the scaled infinite norm of absolute error according to the corresponding order by letting $N = 32$, and ω varies from 50 to 1000. The left part shows the infinite norm of the error and the right part shows the absolute error scaled by corresponding rates. It can be seen that the increase of the frequency parameter ω makes the absolute error get smaller. This indicates as the kernel becomes more highly oscillatory, computed approximation becomes more accurate. Considering the right part of Figure 4, we find that when the frequency parameter ω reaches 150, the curve turns to a horizontal straight line, which is in agreement with the estimate given in Theorem 9.

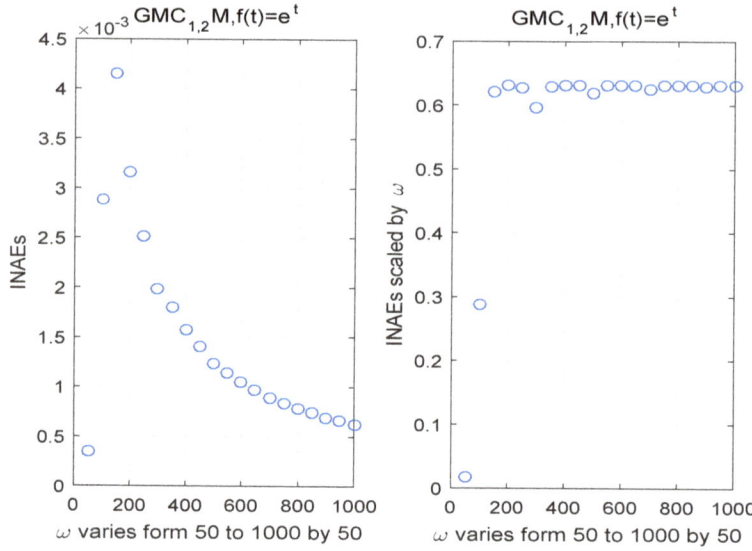

Figure 4. $GMC_{k_1,k_2}M$ for the highly oscillatory problem.

4. Final Remark

For VIEs with oscillatory and non-oscillatory kernels, we have investigated the generalized multistep collocation solution to VIE (1). Detailed convergence properties with respect to the stepsize and oscillation are presented. Noting that the new approach coupled with mild composite oscillatory quadrature rules is able to produce high-order approximation as the frequency goes to infinity, we could expect it is valuable to conduct further studies in related highly oscillatory problems, such as oscillatory Riemann–Hilbert problems, spectral calculation of oscillatory Fredholm operators, and so on.

Author Contributions: H.C. and J.M. conceived and designed the experiments; H.C. and L.L. performed the experiments; H.C. and J.M. analyzed the data; J.M.contributed reagents/materials/analysis tools; H.C. and J.M. wrote the paper. All authors have read and agreed to the published version of the manuscript.

Funding: This work was supported by National Natural Science Foundation of China (number 11901133) and the Science and Technology Foundation of Guizhou Province (number QKHJC[2020]1Y014).

Conflicts of Interest: The authors declare no conflict of interest.

Abbreviations

The following abbreviations are used in this manuscript:

VIE	Volterra integral equation
$GMC_{k_1,k_2}M$	generalized multistep collocation method
ODE	ordinary differential equation
CBVM	collocation boundary value method
HOI	highly oscillatory integral

References

1. Brunner, H. *Volterra Integral Equations: An Introduction to Theory and Applications*; Cambridge University Press: Cambridge, UK, 2017.
2. Xiang, S.; Brunner, H. Efficient methods for Volterra integral equations with highly oscillatory Bessel kernels. *BIT Numer. Math.* **2013**, *53*, 241–263. [CrossRef]
3. Ma, J.; Xiang, S.; Kang, H. On the convergence rates of Filon methods for the solution of a Volterra integral equation with a highly oscillatory Bessel kernel. *Appl. Math.* **2013**, *26*, 699–705. [CrossRef]
4. E Stein. *Harmonic Analysis: Real-Variable Methods, Orthogonality, and Oscillatory Integrals*; Princeton University Press: Princeton, NJ, USA, 1993.
5. Xiang, S.; He, K. On the implementation of discontinuous Galerkin methods for Volterra integral equations with highly oscillatory Bessel kernels. *Appl. Math. Comput.* **2013**, *219*, 4884–4891. [CrossRef]
6. Ma, J.; Fang, C.; Xiang, S. Modified asymptotic orders of the direct Filon method for a class of Volterra integral equations. *J. Comput. Appl. Math.* **2015**, *281*, 120–125. [CrossRef]
7. Ma, J.; Kang, H. Frequency-explicit convergence analysis of collocation methods for highly oscillatory Volterra integral equations with weak singularities. *Appl. Numer. Math.* **2020**, *151*, 1–12. [CrossRef]
8. Fang, C.; He, G.; Xiang, S. Hermite-Type collocation methods to solve Volterra integral equations with highly oscillatory Bessel kernels. *Symmetry* **2019**, *11*, 168. [CrossRef]
9. Saira, S.X.; Liu, G. Numerical solution of the Cauchy-type singular integral equation with a highly oscillatory kernel function. *Mathematics* **2019**, *7*, 872. [CrossRef]
10. H. Brunner. *Collocation Methods for Volterra Integral and Related Functional Equations*; Cambridge University Press: Cambridge, UK, 2004.
11. Shen, J.; Tang, T.; Wang, L. *Spectral Methods: Algorithms, Analysis and Applications*; Springer: Berlin/Heidelberg, Germany, 2011.
12. Li, X.; Tang, T. Convergence analysis of Jacobi spectral collocation methods for Abel-Volterra integral equations of second kind. *Front. Math. China* **2012**, *7*, 69–84. [CrossRef]
13. Xie, Z.; Li, X.; Tang, T. Convergence analysis of spectral Galerkin methods for Volterra type integral equations. *J. Sci. Comput.* **2012**, *53*, 414–434. [CrossRef]
14. Cai, H.; Qi, J. A Legendre-Galerkin method for solving general Volterra functional integral equations. *Numer. Algorithms* **2016**, *73*, 1159–1180. [CrossRef]
15. Berrut, J.P.; Hosseini, S.A.; Klein, G. The linear barycentric rational quadrature method for Volterra integral equations. *SIAM J. Sci. Comput.* **2014**, *36*, A105–A123. [CrossRef]
16. Li, M.; Huang, C. The linear barycentric rational quadrature method for auto-convolution Volterra integral equations. *J. Sci. Comput.* **2019**, *78*, 549–564. [CrossRef]
17. Conte, D.; Paternoster, B. Multistep collocation methods for Volterra integral equations. *Appl. Numer. Math.* **2009**, *59*, 1721–1736. [CrossRef]
18. Fazeli, S.; Hojjati, G. Numerical solution of Volterra integro-differential equations by superimplicit multistep collocation methods. *Numer. Algorithms* **2015**, *68*, 741–768. [CrossRef]
19. Brugnano, L.; Trigiante, D. *Solving Differential Problems by Multistep Initial and Boundary Value Methods*; Cordon and Breach Science Publishers: Concord, ON, Canada, 1998.
20. Chen, H.; Zhang, C. Boundary value methods for Volterra integral and integro-differential equations. *Appl. Math. Comput.* **2011**, *218*, 2619–2630. [CrossRef]
21. Li, C.; Zhang, C. Block boundary value methods applied to functional differential equations with piecewise continuous arguments. *Appl. Numer. Math.* **2017**, *115*, 214–224. [CrossRef]
22. Ma, J.; Xiang, S. A collocation boundary value method for linear Volterra integral equations. *J. Sci. Comput.* **2017**, *71*, 1–20. [CrossRef]
23. Ma, J.; Liu, H. Fractional collocation boundary value methods for the second kind Volterra equations with weakly singular kernels. *Numer. Algorithms* **2020**, *84*, 743–760. [CrossRef]
24. Liu, L.; Ma, J. Block collocation boundary value solutions of the first-kind Volterra integral equations. *Numer. Algorithms* **2020**, doi:10.1007/s11075-020-00917-6. [CrossRef]
25. Trefethen, L.N.; Weideman, J.A.C. Two results on polynomial interpolation in equally spaced points. *J. Approx. Theory* **1991**, *65*, 247–260. [CrossRef]

26. Iserles, A.; Nørsett, S.P. Efficient quadrature of highly oscillatory integrals using derivatives. *Proc. R. Soc. Math. Phys. Eng. Sci.* **2005**, *461*, 1383–1399. [CrossRef]
27. Levin, D. Procedures for computing one- and two-dimensional integrals of functions with rapid irregular oscillations. *Math. Comput.* **1982**, *38*, 531–538. [CrossRef]
28. Milovanovic, G.V. Numerical calculation of integrals involving oscillatory and singular kernels and some applications of quadratures. *Comput. Math. Appl.* **1998**, *36*, 19–39. [CrossRef]
29. Xiang, S. Efficient Filon-type methods for $\int_a^b f(x)e^{i\omega g(x)}dx$. *Numer. Math.* **2007**, *105*, 633–658. [CrossRef]

Publisher's Note: MDPI stays neutral with regard to jurisdictional claims in published maps and institutional affiliations.

© 2020 by the authors. Licensee MDPI, Basel, Switzerland. This article is an open access article distributed under the terms and conditions of the Creative Commons Attribution (CC BY) license (http://creativecommons.org/licenses/by/4.0/).

Article

Relating Hydraulic Conductivity Curve to Soil-Water Retention Curve Using a Fractal Model

Carlos Fuentes [1], Carlos Chávez [2,*] and Fernando Brambila [3]

[1] Mexican Institute of Water Technology, Paseo Cuauhnáhuac Núm. 8532, Jiutepec 62550, Mexico; cfuentes@tlaloc.imta.mx
[2] Water Research Center, Department of Irrigation and Drainage Engineering, Autonomous University of Queretaro, Cerro de las Campanas SN, Col. Las Campanas, Queretaro 76010, Mexico
[3] Facultad de Ciencias, Universidad Nacional Autónoma de México, Ciudad Universitaria, Mexico City 04510, Mexico; fernando.brambila@ciencias.unam.mx
* Correspondence: chagcarlos@uaq.mx; Tel.: +52-442-192-1200 (ext. 6036)

Received: 26 October 2020; Accepted: 8 December 2020; Published: 10 December 2020

Abstract: In the study of water transference in soil according to Darcy law, the knowledge of hydrodynamic characteristics, formed by the water retention curve $\theta(\psi)$, and the hydraulic conductivity curve $K(\psi)$ are of great importance. The first one relates the water volumetric content (θ) with the water-soil pressure (ψ); the second one, the hydraulic conductivity (K) with the water-soil pressure. The objective of this work is to establish relationships between both curves using concepts of probability theory and fractal geometry in order to reduce the number of unknown functions. The introduction of four definitions used at the literature of the pore effective radius that is involve in the general model has permitted to establish four new specials models to predict the relative hydraulic conductivity. Some additional considerations related to the definitions of flow effective area and the tortuosity factor have allow us to deduce four classical models that are extensively used in different studies. In particular, we have given some interpretations of its empirical parameters in the fractal geometry context. The resulting functions for hydrodynamic characteristics can be utilized in many studies of water movement in the soil.

Keywords: areal porosity; volumetric porosity; fractal area-volume relationship; tortuosity factor; joint probability

1. Introduction

Darcy's law [1] establishes that the water flow in porous media is proportional to the hydraulic gradient; the proportionality coefficient is denoted hydraulic conductivity (K). The law, discovered in the context of water flow in saturated soils, has since been generalized to flow in unsaturated soil [2]. In saturated soils, conductivity is independent of water pressure, whereas in unsaturated soils it is a highly nonlinear function of pressure (ψ), or volumetric water content (θ) [3–8].

The saturated hydraulic conductivity, denoted K_S, is at most a function of spatial coordinates. In unsaturated soils the hydraulic conductivity is a function of water pressure $K(\psi)$ as well as spatial coordinates. In such cases the soil-water retention curve $\theta(\psi)$ is needed to relate volumetric water content to soil-water pressure. The two curves $\theta(\psi)$ and $K(\psi)$ are known as the soil hydrodynamic characteristics and are important to the study of mass and transfers such as infiltration, drainage and evaporation, and groundwater recharge [2,9–12].

The aim of the present work is to establish relationships between the soil-water retention curve and hydraulic conductivity curve, using concepts of probability theory and fractal geometry in order to reduce the amount of unknown functions in the unsaturated soil zone. Soil here is considered as a set of Lebesgue measure different than zero. This does not consider sets which porosity is a unity, such as Menger sponge, which may not represent natural soil.

2. Materials and Methods

2.1. A General Model of Hydraulic Conductivity

A conceptual model for the hydraulic conductivity based on Poiseuille law of water flow in capillary tubes has been proposed at the literature [13–18]. The model has the general form:

$$K = fC_f \int_\Omega (R/T)^2 d\omega \tag{1}$$

where Ω represents the water flow area divided by total area of the exposed face [13]; $f = \rho_w g/\eta$ is the fluidity, ρ_w is water density, η is the dynamical viscosity coefficient, g is the gravitational acceleration; R is the pore radius; the non-dimensional coefficient (C_f) take in account the irregular shape of the pore perimeter, for a circular pore $C_f = 1/8$; if R is taken as the hydraulic radius then C_f is called Koseny coefficient [19] with $C_f = 1/2$ for a circle; T is the tortuosity factor defined as $T = dz_f/dz \geq 1$, where z is the rectilinear path of water particles following macroscopic direction of the movement and z_f is the actual path of water particles [20]; ω is the water flow effective area relative to total area of soil or partial effective areal porosity.

The effective area definition is established by Fuentes et al. [14–18] from the probabilistic idea of Childs and Collis–George [21] and fractal geometry concepts. After a perpendicular cut to macroscopic trajectory of water we obtain two faces, which are located at z and z + dz positions (Figure 1); the radii of pores of the z-face are denoted by r and those of the z + dz-face are denoted by ρ. A water particle in a pore of z-face can continue its trajectory by the same pore or by other pore of equal or different radius. The introduction of the joint probability of the two faces at intermediate point z + 1/2 dz allows the modeling of these possibilities.

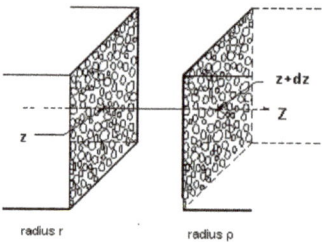

Figure 1. A cross section in the perpendicular macroscopic direction of the water flow. In the z-face the different pore radius are represented by r and in the z + dz-face the different pore radius are represented by ρ, where dz is the pore size order.

We consider a completely saturated soil. Childs and Collis–George [21] assume that the pore size distribution f(r) is the same at both faces and that $d\theta(r) = f(r)dr$ is the water content in the pore interval containing r, $(r - 1/2\, dr, r + 1/2\, dr)$, and $d\theta(\rho) = f(\rho)d\rho$ is the water content in the pore interval containing ρ, $(\rho - 1/2\, d\rho, \rho + 1/2\, d\rho)$, at the other face. The completely random joint probability of all the pores represented by these intervals is equal to the product of these probabilities, provided that dz is of the order of a pore size and less than a characteristic particle size. The product of $d\theta(r)$ and $d\theta(\rho)$ represents the flow effective area $d\omega(r,\rho) = d\theta(r)d\theta(\rho) = f(r)dr\, f(\rho)d\rho$, which integration over whole pore domain gives the total effective flow area $\mu = \phi\phi = \phi^2$, where μ represents also a total effective areal porosity and ϕ the total volumetric porosity. In a parallel capillary system, a water particle moves always in the same capillary and the effective area is equal to volumetric porosity $d\omega(r) = d\theta(r) = f(r)dr$, which integration gives $\mu = \phi$ [22].

In probabilistic terms, the Purcell model represents a complete correlation between the two faces, whereas the Childs and Collis–George model represents a complete decorrelation, these models represent the possible extreme behaviors. Mualem and Dagan [13] show that the Purcell model can be formally deduced from the Childs and Collis–George model if the flow effective area is defined as $d\omega(r, \rho) = f(r)dr\, \delta(\rho - r)\, d\rho$, where δ indicates the Dirac delta.

An intermediate approach between the Purcell and Childs and Collis–George models is proposed by Millington and Quirk [23]. These authors suppose that, if the areal porosity at each face is ϕ^s, then the effective areal porosity at intermediate point is $\mu = \phi^s \phi^s$ where $s = 1/2$ represent the Purcell model and $s = 1$ the Childs and Collis–George model. Further, since $\phi^{2s} \leq \phi$, Millington and Quirk proposed to add the larger solid area $(1 - \phi)^s$ to ϕ^{2s}, in order to obtain the total area, that is $(1 - \phi)^s + \phi^{2s} = 1$. In most studies about soil structure, it is studied as a fractal object [16,17,24–28]. In this context, we'll offer an interpretation of the relationship $(1 - \phi)^s + \phi^{2s} = 1$.

In Euclidean geometry we have $V \propto L^3$ and $A \propto L^2$, where V, A, and L are the volume, area and length in an object, respectively; for example in a sphere $V = 4/3\pi r^3$, $A = 4\pi r^2$, $L = r$; since $L \propto V^{1/3}$ we have $A \propto V^{2/3}$. Using the Mandelbrot [29] area–volume relationship in fractal geometry we have $A \propto V^{D/E}$, where D is the particle surface fractal dimension or the particle-pore interface fractal dimension and $E = 3$ is the Euclidean dimension of the space where the object is embedded.

In a cross section, the total area is the sum of the solid cross-sectional area and the pore cross-sectional area. If $\mu_s = 1 - \mu$ is the solid cross-sectional area relative to the total area (call it the areal solidity), and $\phi_s = 1 - \phi$ is the solid volume relative to soil volume (call it volumetric solidity), then $\mu_s = \phi_s^s$, or $1 - \mu = (1 - \phi)^s$, where $s = D/E$ is the fractal dimension relative to Euclidean dimension. Following the probabilistic idea, the areal and volumetric porosities relationship resulting is $\mu = \phi^s \phi^s = \phi^{2s}$; $s = 1/2$ corresponds to Purcell model and $s = 1$ to Childs and Collis–George model. Notice that the solid area takes the exponent s rather than 1-s; this is because $\mu_s = \phi_s^s$ is Hausdorff measure of the solid phase. The exponent 1-s is important in the calculus of the parallel body volume of a fractal [30].

The r-parallel body of a set F is defined by $P_r(F) = \{x \in \mathfrak{R}^E : |x - y| \leq r, y \in F\}$ where $\mathfrak{R}^E$ is the E-cartesian product of real numbers and represents soil (both pores and solids), F is the solids and $\mathfrak{R}^E - F$ the pores. Clearly, F is contained in $P_r(F)$. The volume of a parallel body is obtained as the product of the cover set's volume, cr^E where c is a form coefficient ($c = 1$ if all covers are parallelepipeds), and the number of covers; therefore: $\text{vol}_E(P_r) = N_r c r^E$. Considering that $N_r \propto r^{-D}$ when $r \to 0$, we obtain $\text{vol}_E(P_r) \propto r^{E-D}$. It must be noted that the body parallel volume of the solids is not the same as the porous volume; it would, however, be the same if porosity tends to unity and if F were dense in $\mathfrak{R}^E$.

Since $\mu_s + \mu = 1$, the relationship between the relative fractal dimension and the total volumetric porosity is defined implicitly by the equation:

$$(1 - \phi)^s + \phi^{2s} = 1 \tag{2}$$

where $s = D/E$. It can be shown that $\mu \leq \phi$, $s \to 1/2$ when $\phi \to 0$, and $s \to 1$ when $\phi \to 1$, in other words $1/2 < s < 1$ when $0 < \phi < 1$.

From mass additive property we deduce the relationship $\rho_t = \rho_s \phi_s + \rho_v \phi$, where ρ_t is the total density of soil, ρ_s the density of solids, and ρ_v the density of pores; if this last one is considered null, we get the classic $\phi = 1 - \rho_t/\rho_s$ formula for estimating porosity, where ρ_t becomes the total density of dry soil. The comparison with Equation (2) allows us to deduce that $\rho_s/\rho_t = \phi_s^{s-1}$ and $\rho_v/\rho_t = \phi^{2s-1}$. The first could also be written as $\rho_s/\rho_t = \phi_s^{s-1}/\phi_s$, which is the quotient of Hausdorff measure and Lebesgue measure of solids. The second one can be interpreted as the quotient of the pores Hausdorff measure and the pores parallel body volume if we write $\rho_v/\rho_t = \phi^s/\phi^{1-s}$. It can be shown that $\phi_s \leq \phi_s^{1-s}/$ and $\phi \leq \phi^{1-s}$.

Clearly the relation s(φ) defined by Equation (2) cannot by applied, for instance, to Menger sponge, where φ = 1 and s = log20/3log3 ≅ 0.9089; in other words, this relationship cannot be applied to any abstract sets where Lebesgue measure is zero ($φ_s = 0$) and s < 1. Because relation s(φ) established a one-to-one relationship between porosity and the fractal dimension of solid pore interphase, it may not work for certain types of soils. However, their implications in the modeling of hydraulic conductivity are being investigated.

For the modeling of the hydraulic conductivity of unsaturated soils we accept the classic hypothesis that the water is contained in saturated pores with radius r, where 0 < r < R, for a given water content θ(R). Consequently, the effective area of flux, or partial area porosity, is the generalization of $μ = φ^s φ^s$:

$$dω(r, ρ) = dθ^s(r)dθ^s(ρ) \tag{3}$$

As concerns the tortuosity factor, it was demonstrated in Fuentes et al. [17] that R is the pore radius measured perpendicularly to actual trajectory of the water particles (z_f) and R_s is its projection in the macroscopic direction (z), consequently $R_s/R = dz/dz_f = 1/T$. According to the probabilistic idea we have $R_s \propto R^s R^s = R^{2s}$, which we can write as an equality $R_s/R_{so} = (R/R_o)^{2s}$ with $R_{so} = R_o/T_o$, where R_o is a reference radius and T_o the associated tortuosity factor. The pore radius-tortuosity factor relationship resulting is:

$$T(R) = T_o(R_o/R)^δ \tag{4}$$

where 0 < δ = 2s − 1 < 1.

The general hydraulic conductivity model results of the introduction of the Equations (2) and (4) in the Equation (1):

$$K = f \frac{C_f}{T_o^2 R_o^{2(2s-1)}} \int_Ω R^{4s}(r, ρ) dθ^s(r) dθ^s(ρ) \tag{5}$$

The saturated hydraulic conductivity (K_s) is obtained from Equation (5), replace Ω by the total pore domain $Ω_T$.

2.2. Classical Models of Hydraulic Conductivity

Classic models reported in literature may be deduced from the proposed general model, provided that certain hypotheses are introduced. From Equation (3) we can deduce $dω(r, ρ) = s^2[θ(r)]^{s-1}[θ(ρ)]^{s-1} dθ(r) dθ(ρ)$, with 0 < r < R and 0 < ρ < R. Assuming that the multiplicative function of the θ(r) and θ(ρ) differentials can be replaced by a medium value, which clearly depends on a superior limit, we have:

$$dω(r, ρ; R) = [θ(R)]^{2s-2} dθ(r) dθ(ρ) \tag{6}$$

where the s^2 term has been eliminated to satisfy:

$$\int_0^R \int_0^R dω(r, ρ; R) = ω = θ^{2s} \tag{7}$$

Likewise, tortuosity will depend only on the major radius, that of θ(R). We know that for a small R θ(R) = φ(R/R_o)^λ, where R_o is the radius of a reference pore and λ > 0 is an index of pores (Brooks and Corey, 1964). From Equation (4) we obtain that:

$$T(R) = T_o \left[\frac{φ}{θ(R)} \right]^γ \text{ with } γ = \frac{δ}{λ} \tag{8}$$

In several articles [24,25] has been suggested that partial volumetric porosity is proportional to the volume of the parallel body, which means that θ(R) ∝ R^{E-D}, where the fractal dimension D is estimated from soil-water retention curve, a result that's valid when porosity tends to unity. If that's the case, we get:

$$\lambda = E - D \tag{9}$$

Considering Equations (6) and (8), Equation (1) becomes:

$$K = f\frac{C_f}{T_o^2}\phi^{2s-2}\left[\frac{\theta}{\phi}\right]^p \int_\Omega [R(r,\rho)]^2 d\theta(r)d\theta(\rho) \text{ with } p = (2s-2) + (2\gamma) \tag{10}$$

In the power p, the first addend represents the global effects of the correlation among pores, whereas the second represents the global effects caused by the tortuosity of flow trajectories.

For each relation between the effective radius R(r,ρ) and the radii r and ρ we can deduce a special model of the hydraulic conductivity. We will obtain four models corresponding to four relations R(r,ρ) utilized in models of the hydraulic conductivity reported at the literature: (i) 'small pore' model R(r,ρ) = min(r,ρ) used by Childs and Collis–George [21]; (ii) 'geometric pore' model R(r,ρ) = $\sqrt{r\rho}$ used by Mualem [31]; (iii) 'neutral pore' model R(r,ρ) = either r or ρ corresponding to Burdine (1953) model; and iv) 'large pore' model R(r,ρ) = max(r,ρ) used by Fuentes [14]. Following the general indications of Brutsaert [32] for the integration of the Equation (10), the resulting special models are:

Small pore model: $R(r,\rho) = \min(r,\rho)$.

$$K(\theta) = f\frac{2C_f}{T_o^2}\phi^{2s-2}\left[\frac{\theta}{\phi}\right]^p \int_0^\theta [\theta - \vartheta] r^2 d\vartheta \tag{11}$$

This is the model of Childs and Collis–George [21], with a correction factor $[\theta/\phi]^p$.

Geometric pore model: $R(r,\rho) = \sqrt{r\rho}$.

$$K(\theta) = f\frac{C_f}{T_o^2}\phi^{2s-2}\left[\frac{\theta}{\phi}\right]^p \left[\int_0^\theta r d\vartheta\right]^2 \tag{12}$$

This model presents the structure of Mualem's model [31] with p = 1/2.

Neutral pore model: $R(r,\rho)$ = either r or ρ.

$$K(\theta) = f\frac{C_f}{T_o^2}\phi^{2s-1}\left[\frac{\theta}{\phi}\right]^{p+1} \int_0^\theta r^2 d\vartheta \tag{13}$$

This model presents the structure of Burdine's model [33] with p = 1.

Large pore model: $R(r,\rho) = \max(r,\rho)$.

$$K(\theta) = f\frac{2C_f}{T_o^2}\phi^{2s-2}\left[\frac{\theta}{\phi}\right]^p \int_0^\theta r^2 \vartheta d\vartheta \tag{14}$$

Childs and Collis–George [21] suppose that the resistance to the flow is determined by the small pore. Whereas Fuentes [14] proposes that, to deduce the opposite behavior, the conductance is determined by the pore of greater size [17]. Mualem [31] gives a little more weight to the large pore by proposing the geometric mean. The Burdine model is deduced when the pores sizes have the same weight.

The classic models are reported for relative hydraulic conductivity [$K_r(\theta)$] and they are in function of the retention curve $\psi(\theta)$. These are obtained from Equations (11)–(14) with the rule $K_r(\theta) = K(\theta)/K(\theta_s)$, and the introduction of the Young–Laplace–Jurin law for the capillary rise phenomena:

$$\psi = -\frac{\ell_L^2}{R}\cos(\alpha_c) \tag{15}$$

where the scale or capillary number (ℓ_L) is defined by $\ell_L = \sqrt{2\sigma/\rho_w g}$ [34], σ is the interfacial tension, $\ell_L \cong 0.386$ cm at 20 °C; α_c is the contact angle formed between the air–water interface and the solid particles, assumed generally constant and equal to zero.

Classic models also consider the residual water content θ_r, defined by Brooks and Corey [35] as $K(\theta_r) = 0$. This can be incorporated in the precedent models replacing θ by the effective water content $\theta_{ef} = \theta - \theta_r$, and ϕ by the effective volumetric porosity $\phi_{ef} = \phi - \theta_r$. The exponent s must be calculated by replacing ϕ with effective porosity (ϕ_{ef}) in Equation (3); θ_r is added to solid particles, i.e., $\phi_{sef} = 1 - \phi_{ef} = \phi_s + \theta_r$. In the classical conductivity models, the porosity ϕ is replaced by the volumetric water content to natural saturation θ_s, when entrapped air is considered.

The power p that appears in models (11)–(14) has been considered as an empiric parameter. This power ($p = p_1 + p_2$) is the result of the effects of correlation between pores, $[\theta(R)]^{p_1}$ with $p_1 = 2s - 2$, and tortuosity $T^2(R) = T_0^2[\phi/\theta(R)]^{p_2}$ with $p_2 = 2(2s-1)/\lambda$. To know the order of magnitude of the power p_2 we assume that the particle surface fractal dimension is roughly equal to the fractal dimension estimated from the soil-water retention curve, i.e., from Equation (9) $\lambda \cong 3(1-s)$, hence $p_2 \cong 2(2s-1)/3(1-s)$. In consequence, the value of p may be estimated from porosity, some of which are shown in Table 1.

Table 1. Predicted values of the exponent p of the classical hydraulic conductivity models, which results from the effect of the pore correlation (p_1) and the tortuosity factor (p_2), for some values of the total volumetric porosity.

ϕ	S = D/3	p_1	p_2	$p = p_1 + p_2$
0	1/2	−1	0	−1
0.3671	2/3	−2/3	2/3	0
1/2	0.6942	−0.6115	0.8470	0.2355
0.6180	0.7202	−0.5596	1.0494	0.4898
1	1	0	∞	∞

The approximate value of $p \approx 1/2$ was obtained by Mualem [31] from the calibration of Equation (12) over the experimental data of 45 soils reported in different works, with total volumetric porosity in the range of $0.4 < \phi < 0.7$. According to Table 1, these soils may be represented by a soil with an average porosity of roughly 0.6.

3. Results and Discussion

3.1. Some New Models of Hydraulic Conductivity

We may obtain new hydraulic conductivity models from the general model established by Equation (5), without the introductions of those hypotheses established in Equations (6) and (8) to deduce the classic conductivity models, which may be restricted.

Each of the R(r,ρ) relationships mentioned above corresponds to a specific model of hydraulic conductivity. And again, following Brutsaert [32] for the integration of the Equation (5), the resulting special models are:

Small pore model: $R(r, \rho) = \min(r, \rho)$

$$K(R) = f \frac{2C_f}{T_o^2 R_o^{2(2s-1)}} \int_0^R [\theta^s(R) - \theta^s(r)] \, r^{4s} d\theta^s(r) \tag{16}$$

Geometric pore model: $R(r, \rho) = \sqrt{r\rho}$.

$$K(R) = f \frac{C_f}{T_o^2 R_o^{2(2s-1)}} \left[\int_0^R r^{2s} d\theta^s(r) \right]^2 \tag{17}$$

Neutral pore model: $R(r, \rho) = r$ or $R(r, \rho) = \rho$.

$$K(R) = f \frac{C_f}{T_o^2 R_o^{2(2s-1)}} \theta^s(R) \left[\int_0^R r^{4s} d\theta^s(r) \right] \tag{18}$$

Large pore model: $R(r, \rho) = \max(r, \rho)$.

$$K(R) = f \frac{2C_f}{T_o^2 R_o^{2(2s-1)}} \int_0^R r^{4s} \theta^s(r) d\theta^s(r) \tag{19}$$

Note that the Equations (17) and (18) can be generalized assuming $R(r, \rho) = r^\alpha \rho^{1-\alpha}$, where $0 \leq \alpha \leq 1$; the Equation (17) follow with $\alpha = 1/2$ and the Equation (18) with $\alpha = 0$ or $\alpha = 1$.

To obtain specific functions from new special models is necessary to provide the function $\theta(R)$. This can be obtained from the soil-water retention curve $\theta(\psi)$, relating the soil-water content to soil-water pressure, and the Laplace law, defined by the Equation (15).

From Equations (16)–(19) we can obtain the corresponding models to calculate the relative hydraulic conductivity from the retention curve:

$$\frac{K(\Theta)}{K_s} = \int_0^\Theta (\Theta^s - \vartheta^s) \frac{\vartheta^{s-1}}{|\psi(\vartheta)|^{4s}} d\vartheta \bigg/ \int_0^1 (1 - \vartheta^s) \frac{\vartheta^{s-1}}{|\psi(\vartheta)|^{4s}} d\vartheta \tag{20}$$

$$\frac{K(\Theta)}{K_s} = \left[\int_0^\Theta \frac{\vartheta^{s-1}}{|\psi(\vartheta)|^{2s}} d\vartheta \bigg/ \int_0^1 \frac{\vartheta^{s-1}}{|\psi(\vartheta)|^{2s}} d\vartheta \right]^2 \tag{21}$$

$$\frac{K(\Theta)}{K_s} = \Theta^s \left[\int_0^\Theta \frac{\vartheta^{s-1}}{|\psi(\vartheta)|^{4s}} d\vartheta \bigg/ \int_0^1 \frac{\vartheta^{s-1}}{|\psi(\vartheta)|^{4s}} d\vartheta \right] \tag{22}$$

$$\frac{K(\Theta)}{K_s} = \int_0^\Theta \frac{\vartheta^{2s-1}}{|\psi(\vartheta)|^{4s}} d\vartheta \bigg/ \int_0^1 \frac{\vartheta^{2s-1}}{|\psi(\vartheta)|^{4s}} d\vartheta \tag{23}$$

where $\Theta = (\theta - \theta_r)/(\theta_s - \theta_r)$ is an effective degree of saturation.

3.2. Applications

3.2.1. Brooks and Corey Equation

The equation proposed by Brooks and Corey [35] to represent the soil-water retention curve is:

$$\Theta = (\psi_{cr}/\psi)^\lambda \tag{24}$$

if $\psi < \psi_{cr}$, and $\Theta = 1$ if $\psi_{cr} \leq \psi$; where ψ_{cr} is a critical pressure and $\lambda > 0$ is an index of the pore distribution. The assumption $\lambda = E - D$ [24,25] is not used here.

The introduction of the Equation (24) in the Equations (20)–(23) gives the same expression for the relative hydraulic conductivity:

$$K(\Theta)/K_s = \Theta^{2s(2/\lambda+1)} \tag{25}$$

The saturated hydraulic conductivity is given by:

$$K_s = fC_f(\theta_s - \theta_r)^{2s}(R_o/T_o)^2 \Lambda \tag{26}$$

where we have defined $R_o = \lambda_L(\lambda_L/|\psi_{cr}|)$. The factor Λ is different for each model: small pore $\Lambda_s = 1/[2(2/\lambda+1/2)(2/\lambda+1)]$; geometric pore $\Lambda_g = 1/(2/\lambda+1)^2$; neutral pore $\Lambda_N = 1/[2(2/\lambda+1/2)]$; and large pore $\Lambda_L = 1/(2/\lambda+1)$. The following inequalities are satisfied $\Lambda_s < \Lambda_g < \Lambda_N < \Lambda_L$, the equalities are given at extremes ($\lambda \to 0, \infty$). When $\lambda \to 0$ we have: $\Lambda_s = \lambda^2/8 < \Lambda_g = \lambda^2/4 < \Lambda_N = \lambda/4 < \Lambda_L = \lambda/2$.

The corresponding saturated hydraulic conductivity value satisfies the inequalities: $K_{ss} < K_{sg} < K_{sN} < K_{sL}$.

3.2.2. Generalized Power Function

One of the larger groups of models used to represent the soil–water retention curve is the following power function [36]:

$$\psi = \psi_d \Theta^{-1/\lambda}\left(1 - \Theta^{1/m}\right)^{1/n} \tag{27}$$

where ψ_d is a pressure scale; $m > 0$, $n > 0$ and $\lambda > 0$ are three form parameters.

In Equation (27), we can note when $\Theta \to 0$ we obtain the Brooks and Corey [35] equation, and when $\lambda = mn$ we obtain the van Genuchten equation [37]:

$$\Theta(\psi) = \left[1 + (\psi/\psi_d)^n\right]^{-m} \tag{28}$$

Introducing the Equation (27) in the Equations (20)–(23) we obtain, respectively:

$$\frac{K(\Theta)}{K_s} = \frac{\Theta^s B_I(\Theta^{1/m}; 4sm/\lambda + sm, 1 - 4s/n) - B_I(\Theta^{1/m}; 4sm/\lambda + 2sm, 1 - 4s/n)}{B(4sm/\lambda + sm, 1 - 4s/n) - B(4sm/\lambda + 2sm, 1 - 4s/n)} \tag{29}$$

$$K(\Theta)/K_s = \left[\beta_I(\Theta^{1/m}; 2sm/\lambda + sm, 1 - 2s/n)\right]^2 \tag{30}$$

$$K(\Theta)/K_s = \Theta^s \beta_I(\Theta^{1/m}; 4sm/\lambda + sm, 1 - 4s/n) \tag{31}$$

$$K(\Theta)/K_s = \beta_I(\Theta^{1/m}; 4sm/\lambda + 2sm, 1 - 4s/n) \tag{32}$$

where $\beta_I(x; p, q) = B_I(x; p, q)/B(p, q)$, $B_I(x; p, q)$ is the incomplete beta function of variable x and parameters $p > 0$ and $q > 0$ and $B(p,q) = B_I(1;p,q)$ is the complete beta function.

We can obtain closed-form equations accepting the van Genuchten [37] idea consisting in to assign integral values to parameter p of the beta function and specially p = 1; this conduces to impose relationships between the form parameters of the soil–water retention curve. This idea is only applicable to models defined by Equations (30)–(32) because the unicity of these relationships. From Equation (29) we can obtain only closed-form equations of the first or second integral of the numerator, the result is an incomplete closed-form formula (semi closed-form) of the conductivity.

Small pore model:

$$\frac{K(\Theta)}{K_s} = \frac{\Theta^s B_I(\Theta^{1/m}; 1, 1-4s/n) - B_I(\Theta^{1/m}; 1+sm, 1-4s/n)}{B(1, 1-4s/n) - B(1+sm, 1-4s/n)}, \quad \lambda = \frac{4sm}{1-sm} \tag{33}$$

$$\frac{K(\Theta)}{K_s} = \frac{\Theta^s B_I(\Theta^{1/m}; 1-sm, 1-4s/n) - B_I(\Theta^{1/m}; 1, 1-4s/n)}{B(1-sm, 1-4s/n) - B(1, 1-4s/n)}, \quad \lambda = \frac{4sm}{1-2sm} \tag{34}$$

with $n > 4s$ and $B_I(\Theta^{1/m}; 1, 1-4s/n) = (1-4s/n)^{-1}\left[1 - \left(1 - \Theta^{1/m}\right)^{1-4s/n}\right]$.

Geometric pore model:

$$K(\Theta)/K_s = \left[1 - \left(1 - \Theta^{1/m}\right)^{1-2s/n}\right]^2, \quad \lambda = \frac{2sm}{1-sm} \tag{35}$$

with $n > 2s$.

Neutral pore model:

$$K(\Theta)/K_s = \Theta^s\left[1 - \left(1 - \Theta^{1/m}\right)^{1-4s/n}\right], \quad \lambda = \frac{4sm}{1-sm} \tag{36}$$

with $n > 4s$.

Large pore model:

$$K(\Theta)/K_s = 1 - \left(1 - \Theta^{1/m}\right)^{1-4s/n}, \quad \lambda = \frac{4sm}{1-2sm} \tag{37}$$

with $n > 4s$.

We can note that soil–water retention curves induced by Equations (33) and (34) are equals to those induces by Equations (36) and (37), respectively.

The use of models (33)–(37) reduces the form parameters number of the soil–water retention curve defined by Equation (27): the three independent parameters {m, n, λ} are reduced to two parameters {m, n}.

The form parameters can even be reduced to one. If we assume $\lambda = mn$ in the Equation (27) we obtain van Genuchten equation, Equation (28), which makes the function $\Theta(\psi)$ explicit where the form parameters {m, n} are still independent. If we accept the relationships between λ and m used to obtain Equations (33)–(37), the Equation (27) will have only one form parameter (m). The corresponding models of the conductivity associated to van Genuchten equation with a form parameter are the following:

Small pore model:

$$\frac{K(\Theta)}{K_s} = \frac{\Theta^s B_I(\Theta^{1/m}; 1, sm) - B_I(\Theta^{1/m}; 1+sm, sm)}{B(1, sm) - B(1+sm, sm)}, \quad 0 < sm = 1 - 4s/n < 1 \tag{38}$$

$$\frac{K(\Theta)}{K_s} = \frac{\Theta^s B_I(\Theta^{1/m}; 1-sm, sm) - B_I(\Theta^{1/m}; 1, sm)}{B(1-sm, sm) - B(1, sm)}, \quad 0 < 2sm = 1 - 4s/n < 1 \tag{39}$$

Geometric pore model:

$$K(\Theta)/K_s = \left[1 - \left(1 - \Theta^{1/m}\right)^{sm}\right]^2, \quad 0 < sm = 1 - 2s/n < 1 \tag{40}$$

Neutral pore model:

$$K(\Theta)/K_s = \Theta^s\left[1 - \left(1 - \Theta^{1/m}\right)^{sm}\right], \quad 0 < sm = 1 - 4s/n < 1 \qquad (41)$$

Large pore model:

$$K(\Theta)/K_s = 1 - \left(1 - \Theta^{1/m}\right)^{2sm}, \quad 0 < 2sm = 1 - 4s/n < 1 \qquad (42)$$

A first evaluation of the predictive capacity of the relative hydraulic conductivity models defined by Equations (38)–(42), on fifty soils of the GRIZZLY database reported by Haverkamp et al. [38], was presented by Fuentes et al. [18] with acceptable results.

The classical version of the models defined by Equations (21) and (22) corresponds to the Mualem [31] and Burdine [33] models that are defined in Equations (12) and (13). The use of these models has been proved by different authors [37]. We will show the capacity of prediction of the model that corresponds to the hypothesis of the large pore defined in Equation (23) and in particular the close-form equation of the relative hydraulic conductivity defined in Equation (42).

We use three of the five soils that van Genuchten [37] analyzes, holding the values θ_r and θ_s that the author use and the parameter s is estimated from θ_s. In Table 2 we present some properties of the three soils. The parameters ψ_d and m obtained by least squares method corresponding to three different models are presented in Table 3.

Table 2. Some physical properties of the three analyzed soils.

Soil Name	θ_s (cm³/cm³)	θ_r (cm³/cm³)	K_s (cm/day)	s
Hygiene sandstone	0.250	0.153	108.0	0.642
Touchet Silt Loam G.E.3	0.469	0.190	303.0	0.688
Silt Loam G.E.3	0.396	0.131	4.96	0.673

Table 3. Parameters values of the soil water retention curve of the three analyzed soils corresponding to three different models, Equations (40)–(42).

Soil Name	Geometric Pore		Neutral Pore		Large Pore	
	$-\psi_d$ (cm)	m	$-\psi_d$ (cm)	m	$-\psi_d$ (cm)	m
Hygiene sandstone	146.71	1.3176	142.23	1.1020	129.61	0.6000
Touchet Silt Loam G.E.3	213.94	1.1896	205.17	0.9554	185.86	0.5329
Silt Loam G.E.3	253.28	0.5421	176.75	0.2687	165.63	0.2197

Figures 2–4 present the fitted soil water retention curves, and the predicted relative hydraulic conductivity curves by the geometric pore, neutral pore and large pore models for the three studied soils. We can observe that the predictions are good enough in these three soils. In addition, in Figure 5 we illustrate the prediction capability of the small pore model using the relationships between m and n provides by the neutral pore and large pore models.

The analysis performed on the classical models and the comparison between experimental and predicted relative hydraulic conductivity with the four special new models allows us to show that the different models may be used to estimate relative hydraulic conductivity.

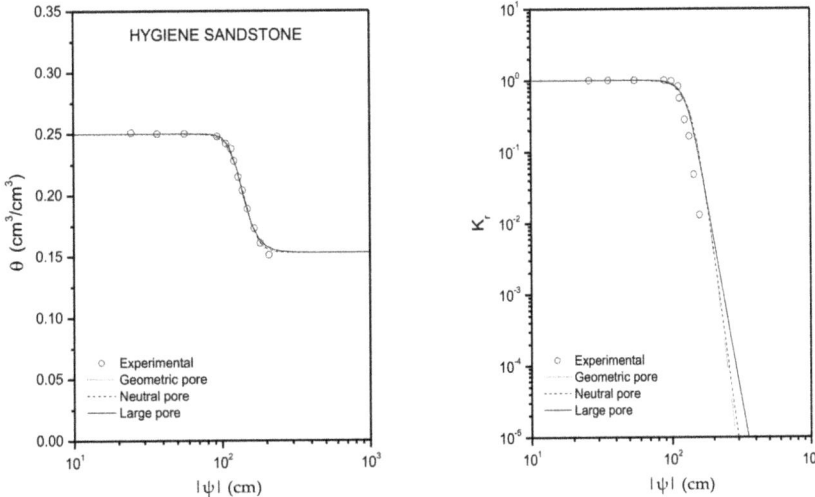

Figure 2. Observed and calculated curves of the soil hydrodynamic characteristics of hygiene sandstone, Equations (40)–(42).

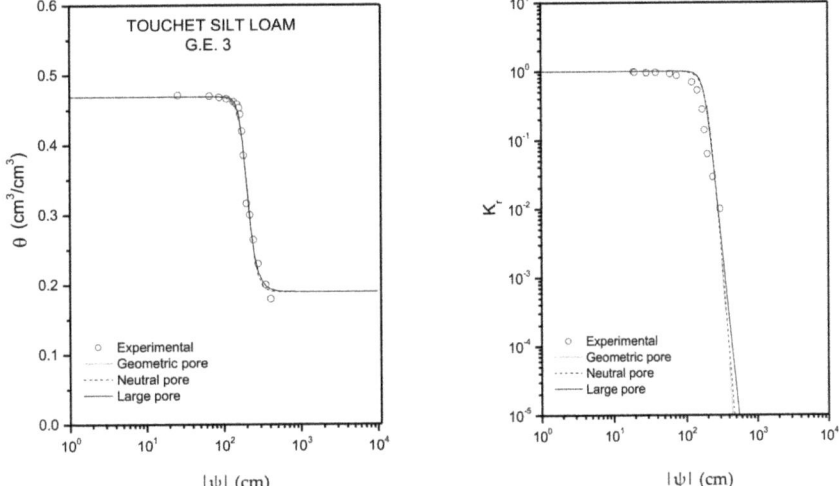

Figure 3. Observed and calculated curves of the soil hydrodynamic characteristics of Touchet silt loam G.E.3, Equations (40)–(42).

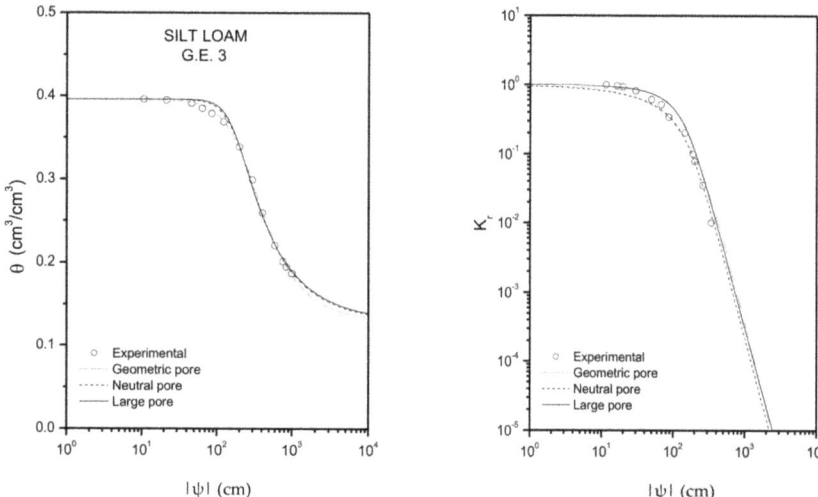

Figure 4. Observed and calculated curves of the soil hydrodynamic characteristics of silt loam G.E.3, Equations (40)–(42).

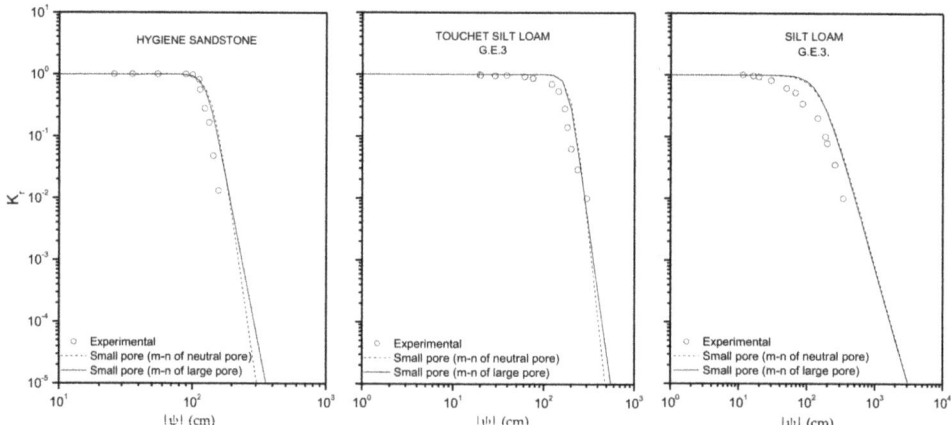

Figure 5. Observed and calculated curves of the relative hydraulic conductivity with small pore model, Equations (38) and (39).

4. Conclusions

The proposed hydraulic conductivity model has been deduced using classic probabilistic theory and fractal geometry concepts in order to approach the flow effective area and the tortuosity of flow trajectories for each pore radius. The introduction of four definitions used at the literature of the pore effective radius that is involved in the general model has permitted to establish four new models to predict the relative hydraulic conductivity.

Some additional considerations related to the definitions of flow effective area and the tortuosity factor have allow us to deduce four classical models that are extensively used in different studies. In particular, we have given some interpretations of its empirical parameters in the fractal geometry context. The power function proposed by Brooks and Corey (1964) to represent the water retention curve leads to a power function to represent the relative hydraulic conductivity. The difference between the new models consists in the prediction of the relative hydraulic conductivity.

We have studied a general power function with three form parameters to represent the water retention curve reported by Braddock et al. [36], which presents as a special case the equation of Brooks and Corey [35] with a form parameter and the van Genuchten equation [37] with two form parameters. With this function we have gotten close-form equations of hydraulic conductivity with two and one form parameter. The general power function of the soil water retention curve and the close-form equations of hydraulic conductivity, obtained through special fractal models, may be used for the study of mass and energy transferences through soil, as infiltration, drainage, evaporation and ground water recharge.

Author Contributions: Formal analysis, C.F., C.C. and F.B.; investigation, C.F., C.C. and F.B.; software, C.F., C.C. and F.B.; C.F., C.C. and F.B.; writing—original draft, C.F. and F.B.; writing—review and editing, C.C. All authors have read and agreed to the published version of the manuscript.

Funding: This research received no external funding.

Conflicts of Interest: The authors declare no conflict of interest.

References

1. Darcy, H. Dètermination des lois d'ècoulement de l'eau à travers le sable. In *Les Fontaines Publiques de la Ville de Dijon*; Dalmont, V., Ed.; Victor Dalmont: Paris, France, 1856; pp. 590–594.
2. Buckingham, E. *Studies on the Movement of Soil Moisture*; Bulletin 38; U.S. Department of Agriculture Bureau of Soils: Washington, DC, USA, 1907.
3. Fuentes, S.; Trejo-Alonso, J.; Quevedo, A.; Fuentes, C.; Chávez, C. Modeling Soil Water Redistribution under Gravity Irrigation with the Richards Equation. *Mathematics* **2020**, *8*, 1581. [CrossRef]
4. Fu, Q.; Hou, R.; Li, T.; Li, Y.; Liu, D.; Li, M. A new infiltration model for simulating soil water movement in canal irrigation under laboratory conditions. *Agric. Water Manag.* **2019**, *213*, 433–444.
5. Mahallati, S.; Pazira, E.; Abbasi, F.; Babazadeh, H. Estimation of soil water retention curve using fractal dimension. *J. Appl. Sci. Environ. Manag.* **2018**, *22*, 173–178. [CrossRef]
6. Hossein, B.; Golnaz, E.Z. Estimation of the soil water retention curve using penetration resistance curve models. *Comput. Electron. Agric.* **2018**, *144*, 329–343.
7. Zhang, J.; Wang, Z.; Luo, X. Parameter Estimation for Soil Water Retention Curve Using the Salp Swarm Algorithm. *Water* **2018**, *10*, 815. [CrossRef]
8. Wang, L.; Huang, C.; Huang, L. Parameter Estimation of the Soil Water Retention Curve Model with Jaya Algorithm. *Comput. Electron. Agric.* **2018**, *151*, 349–353. [CrossRef]
9. Fuentes, C.; Chávez, C.; Quevedo, A.; Trejo-Alonso, J.; Fuentes, S. Modeling of Artificial Groundwater Recharge by Wells: A Model Stratified Porous Medium. *Mathematics* **2020**, *8*, 1764. [CrossRef]
10. Fuentes, C.; Chávez, C. Analytic Representation of the Optimal Flow for Gravity Irrigation. *Water* **2020**, *12*, 2710. [CrossRef]
11. Ket, P.; Oeurng, C.; Degré, A. Estimating Soil Water Retention Curve by Inverse Modelling from Combination of in Situ Dynamic Soil Water Content and Soil Potential Data. *Soil Syst.* **2018**, *2*, 55. [CrossRef]
12. Baiamonte, G. Analytical solution of the Richards equation under gravity-driven infiltration and constant rainfall intensity. *J. Hydrol. Eng.* **2020**, *25*, 04020031. [CrossRef]
13. Mualem, Y.; Dagan, G. Hydraulic conductivity of soils: Unified approach to the statistical models. *Soil Sci. Soc. Am. J.* **1978**, *42*, 392–395. [CrossRef]
14. Fuentes, C. Approche Fractale des Transferts Hydriques dans les sols non Saturès. Ph.D. Thesis, Université Joseph Fourier de Grenoble, Grenoble, France, 1992; p. 267.
15. Fuentes, C.; Vauclin, M.; Parlange, J.-Y.; Haverkamp, R. A note on the soil-water conductivity of a fractal soil. *Transp. Porous Media* **1996**, *23*, 31–36. [CrossRef]
16. Fuentes, C.; Vauclin, M.; Parlange, J.-Y.; Haverkamp, R. Soil-water conductivity of a fractal soil. In *Fractals in Soil Science*; Baveye, P.H., Parlange, J.-Y., Stewart, B.A., Eds.; CRC Press: Boca Raton, FL, USA, 1998; pp. 333–340.
17. Fuentes, C.; Brambila, F.; Vauclin, M.; Parlange, J.-Y.; Haverkamp, R. Modelación fractal de la conductividad hidráulica de los suelos no saturados. *Ing. Hidraul. Mex.* **2001**, *16*, 119–137.

18. Fuentes, C.; Antonino, A.C.D.; Sepúlveda, J.; Zataráin, F.; De León, B. Predicción de la conductividad hidráulica relativa con modelos fractales. *Ing. Hidraul. Mex.* **2003**, *18*, 31–40.
19. Bear, J. *Dynamics of Fluids in Porous Media*; Dover Publications, Inc.: New York, NY, USA, 1972; p. 764.
20. Dullien, F.A.L. *Porous Media, Fluid Transport and Pore Structure*; Academic Press: New York, NY, USA, 1979; p. 574.
21. Childs, E.C.; Collis-George, N. The permeability of porous materials. *Proc. R. Soc. Ser. A* **1950**, *201*, 392–405.
22. Purcell, W.R. Capillary pressures- their measurement using mercury and the calculation of permeability thereform. *Pet. Trans. Am. Inst. Min. Metall. Eng.* **1949**, *186*, 39–48.
23. Millington, R.J.; Quirk, J.P. Permeability of porous solids. *Trans. Faraday Soc.* **1961**, *57*, 1200–1206. [CrossRef]
24. Rieu, M.; Sposito, G. Fractal fragmentation, soil porosity, and soil-water properties: I. Theory. *Soil Sci. Soc. Am. J.* **1991**, *55*, 1231–1238. [CrossRef]
25. Rieu, M.; Sposito, G. Fractal fragmentation, soil porosity, and soil-water properties: II. Applications. *Soil Sci. Soc. Am. J.* **1991**, *55*, 1239–1244. [CrossRef]
26. Oleschko, K.; Fuentes, C.; Brambila, F.; Álvarez, R. Linear fractal analysis of three mexican soils in different management systems. *Soil Technol.* **1997**, *10*, 185–206. [CrossRef]
27. Huang, G.H.; Zhang, R.D.; Huang, Q.Z. Modeling Soil Water Retention Curve with a Fractal Method. *Pedosphere* **2006**, *16*, 137–146. [CrossRef]
28. Ding, D.; Zhao, Y.; Feng, H. A user-friendly modified pore-solid fractal model. *Sci. Rep.* **2016**, *6*, 39029. [CrossRef] [PubMed]
29. Mandelbrot, B.B. *The Fractal Geometry of Nature*; Freeman: San Francisco, CA, USA, 1983; p. 460.
30. Falconer, K. *Fractal Geometry, Mathematical Foundations and Applications*; John Wiley & Sons: England, UK, 1990; p. 288.
31. Mualem, Y. A new model for predicting the hydraulic conductivity of unsaturated porous media. *Water Resour. Res.* **1976**, *12*, 513–522. [CrossRef]
32. Brutsaert, W. Some methods of calculating unsaturated permeability. *Trans. ASAE* **1967**, *10*, 400–404. [CrossRef]
33. Burdine, N.T. Relative permeability calculation from size distribution data. *Trans. AIME* **1953**, *198*, 71–78. [CrossRef]
34. Landau, L.; Lifchitz, E. *Physique Théorique. Tome 6: Mécanique des Fluids*, 2nd ed.; Editions Mir: Moscow, Russia, 1989; p. 748.
35. Brooks, R.H.; Corey, A.T. Hydraulic properties of porous media. In *Hydrology Papers*; Colorado State University: Colorado, CO, USA, 1964; Volume 3.
36. Braddock, R.D.; Parlange, J.-Y.; Lee, H. Application of a soil water hysteresis model to simple water retention curves. *Trans. Porous Media* **2001**, *44*, 407–420. [CrossRef]
37. Van Genuchten, M.T. A closed-form equation for predicting the hydraulic conductivity of unsaturated soils. *Soil Sci. Soc. Am. J.* **1980**, *44*, 892–898. [CrossRef]
38. Haverkamp, R.; Zammit, C.; Bouraoui, F.; Rajkai, K.; Arrúe, J.L.; Heckmann, N. *GRIZZLY, Grenoble Catalogue of Soils: Survey of Soil Field Data and Description of Particle-Size, Soil Water Retention and Hydraulic Conductivity Functions*; Laboratoire d'Etude des Transferts en Hydrologie et Environnement (LTHE): Grenoble, France, 1998.

Publisher's Note: MDPI stays neutral with regard to jurisdictional claims in published maps and institutional affiliations.

© 2020 by the authors. Licensee MDPI, Basel, Switzerland. This article is an open access article distributed under the terms and conditions of the Creative Commons Attribution (CC BY) license (http://creativecommons.org/licenses/by/4.0/).

Article

Developing and Applying a Selection Model for Corrugated Box Precision Printing Machine Suppliers

Chin-Tsai Lin and Cheng-Yu Chiang *

Department of Business Administration, Ming Chuan University, 250 Zhong Shan N. Rd., Sec. 5, Taipei 111, Taiwan; ctlin@mail.mcu.edu.tw
* Correspondence: 04119050@me.mcu.edu.tw; Tel.: +886-04-2206-1660-630

Abstract: Corrugated box printing machines are precision equipment produced by markedly few manufacturers. They involve high investment cost and risk. Having a corrugated box precision printing machine (CBPPM) supplier with a good reputation enables a corrugated box manufacturer to maintain its competitive advantage. Accordingly, establishing an effective CBPPM supplier selection model is crucial for corrugated box manufacturers. This study established a two-stage CBPPM supplier selection model. The first stage involved the use of a modified Delphi method to construct a supplier selection hierarchy with five criteria and 14 subcriteria. In the second stage, an analytic network process was employed to calculate the weights of criteria and subcriteria and to determine the optimal supplier. According to the results, the five criteria in the model, in descending order of importance, are quality, commitment, cost, service attitude, and reputation. This model can provide insights for corrugated box manufacturers formulating their CBPPM supplier selection strategy.

Keywords: corrugated box printing machine; modified Delphi method; analytic network process (ANP); supplier

Citation: Lin, C.-T.; Chiang, C.-Y. Developing and Applying a Selection Model for Corrugated Box Precision Printing Machine Suppliers. *Mathematics* 2021, 9, 68. https://doi.org/10.3390/math9010068

Received: 20 November 2020
Accepted: 24 December 2020
Published: 30 December 2020

Publisher's Note: MDPI stays neutral with regard to jurisdictional clai-ms in published maps and institutio-nal affiliations.

Copyright: © 2020 by the authors. Licensee MDPI, Basel, Switzerland. This article is an open access article distributed under the terms and conditions of the Creative Commons Attribution (CC BY) license (https://creativecommons.org/licenses/by/4.0/).

1. Introduction

The global e-commerce market is rapidly developing, with exponential growth in online and TV shopping as well as demand for global shipping. Because most products purchased online or through TV shopping channels (e-commerce) are packaged using corrugated boxes for shipping, the development of e-commerce has contributed to the growth of the corrugated box industry. According to Smithers Pira [1], the global packaging market attained a value of US$917 billion in 2019. Research and Markets (2019) revealed that the corrugated box market reached a value of US$184.377 billion in 2019. Corrugated boxes have become the most adopted packaging materials in the packaging industry. With the continuous and rapid development of the e-commerce market, corrugated boxes, as the main packaging products, will inevitably grow rapidly accompanied with the development of the packaging industry, thus driving the rapid growth of the corrugated box precision printing machine equipment industry. For Tsao (2011) [2], the corrugated box precision printing machine is accompanied by the development of the corrugated box packaging industry. The main manufacturers of the corrugated box precision printing machine industry are currently concentrated in Europe, the United States, Japan, South Korea, Taiwan and China. Manufacturers with advanced production technology in Japan and Taiwan in the Asian region, mainly in the high-tech field, provide the best marketing and after-sales service system in the corrugated box printing machine manufacturers [3].

In the booming Internet and TV shopping consumption era, these consumer packaging have gradually become a visible part of people's lives. The increasing variety of consumer products and complexity of shipping methods have contributed to the importance of corrugated boxes as a packaging material. The demand for corrugated boxes is rapidly growing worldwide, contributing to the development of the corrugated box precision printing machine (CBPPM) industry. The sales value of the global CBPPM industry grew from US$5.499 billion in 2014 to US$7.312 in 2019, and the growth trend is expected to continue (Figure 1). As people's standard of living improves, they expect better appearance and quality of paper boxes rather than just basic paper box packaging. These expectations are closely related to the development of the CBPPM industry and spur market demands for corrugated boxes and for corrugated box precision printing machines.

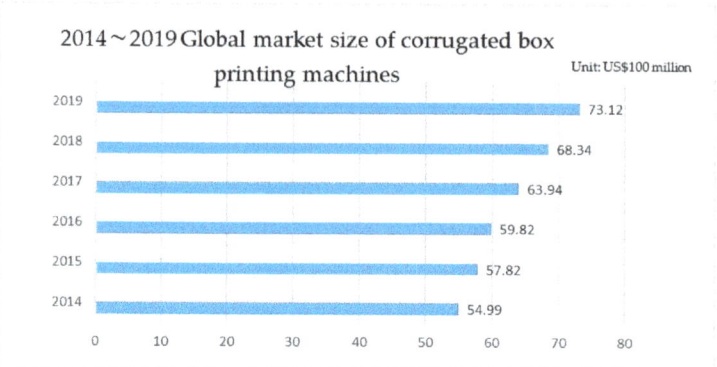

Figure 1. Global market size of corrugated box printing machines [4].

Despite such a large business opportunity in the global CBPPM industry, global manufacturers face challenges such as high investment cost, long research and development (R&D) periods, consumer demand for customization, and high risk. Accordingly, the establishment of a decision-making model for the selection of CBPPM suppliers has become critical for corrugated box manufacturers to maintain their competitive advantage.

The selection of suppliers is a crucial but complex decision-making problem, and its ultimate goal is to find sustainable suppliers with the best potential of providing raw materials and components within a cost budget. According to Ptak and Schragenhiem [5], disruptions in the procurement process can undermine productivity, leading to serious consequences such as bad reputation caused by late delivery or the loss of customers. Therefore, suppliers play a crucial role in procurement activities. In the competitive global environment, most businesses have revisited their procurement strategy and established partnerships with their key suppliers. Despite its recognized importance, cost reduction is not the only critical factor. This study can provide insights into decision making strategies and sustainable operations that can be adopted by the CBPPM industry.

2. Literature Review

2.1. Supplier Selection

The selection of suppliers is critical. Having an appropriate supplier enables a company to offer competitive prices, deliver the correct quantity of products on time, produce high-quality products, and enhance its corporate image and reputation. Labib [6] considered product quality and delivery to be of greater importance than cost. Tam and Tummala [7] argued that the selection criteria for telecommunication service suppliers include quality, cost, problem-solving skills, expertise, delivery time, the ability to satisfy consumer needs, experience, and reputation. Liao and Kao [8] evaluated suppliers with the following criteria: depth of relationship, quality, shipping ability, guaranteed standard, and experience. Basnet [9] suggested that for both local and international businesses, quality, the ability to deliver on time, and performance are the most critical elements in supply chain management.

An excellent supplier satisfies a company's demands for raw materials, products, quality, and services. A company cannot find a high-quality and cost-efficient supplier without having a plan. A critical competency of a procurement specialist is to, by using a rigorous and systematic method, find, evaluate, and select the most suitable supplier for a company [10]. Hsu [11] proposed the following evaluation methods for supplier selection: (a) benchmarking, (b) categorical method, (c) weighted-point method, (d) cost-ratio method, and (e) unit total cost. Considering conflicts among supplier selection indicators, Shirouyehzad [12] employed a strengths–weaknesses–opportunities–threats analysis to evaluate suppliers qualitatively and quantitatively; Shirouyehzad used the technique for order performance by "similarity to the ideal solution" to determine the weights of indicators and adopted a linear planning method to allocate orders. Supplier evaluation methods fall into three major categories: qualitative analyses, quantitative analyses, and methods combining qualitative and quantitative analyses.

Chin [13] defined suppliers as business entities that provide products or services to a buyer and charge the buyer with remuneration in return; such provision encompasses raw materials, equipment, tools, and other resources. The management of suppliers involves active attitudes gradually established in the process of communication and interaction with the suppliers [14]. Shima Aghai [15] proposed a fuzzy multiobjective planning model that incorporates a wide range of factors, namely qualitative, quantitative, risk, and volume discount factors, in supplier selection; this model can be used to select suppliers and optimize supply volume. Supplier selection largely determines subsequent endeavors of establishing buyer–supplier partnerships and increasing supplier capabilities through supplier development programs [16]. The importance of this process for companies is reflected in the final price of products. The price of raw materials, as the main part of the product, is crucial [17,18]. Supplier selection is among the key tasks of supply management [19]. Accordingly, this study constructed a supplier selection model suitable for CBPPM suppliers to help companies maintain competitive advantage.

2.2. Analytic Network Process

The analytic network process (ANP) involves using pairwise comparisons to reveal the relative importance of decision-making features at each level on a 1–9 ratio scale. Establishing a pairwise comparison matrix, calculating the eigenvalue and eigenvector, and conducting a consistency test can avoid evaluation accuracy being undermined by the decision maker's adoption of multiple criteria. The levels are then aggregated to yield a priority vector of the relative importance of alternatives; subsequently, the optimal alternative is determined according to their relative weights as indicated in the vector. ANP, whose theory and application were introduced by Saaty [20], is derived from the analytical hierarchy process (AHP) and is aimed at solving problems involving dependence and feedback among elements in decision making. Overall, the ANP is a mathematical theory capable of solving dependence and feedback problems systematically.

The ANP comprises four steps: (1) constructing a hierarchical structure of the problem, (2) establishing the pairwise comparison matrix and calculating the eigenvector, (3) obtaining the supermatrices and weights, and (4) determining the optimal alternative.

Step 1: Constructing a Hierarchical Structure of the Problem

Determine the decision-making problem and construct a hierarchical structure for the problem; and describe the problem in detail and divide it into a hierarchical network.

Saaty [21] divided the ANP into two parts. The first part involves evaluating the network relationships between criteria and subcriteria; these relationships affect the relationships within a system. The second part is constituted by the network relationships between elements and clusters. According to a network system can be divided into various clusters to form a complex network structure. Figure 2a,b conceptualizes the AHP and ANP, respectively. Saaty [20] presented the interdependent relationships between clusters and elements in a diagram and used arrows to indicate relationships and interaction between them. For example, Figure 2b depicts interdependent elements.

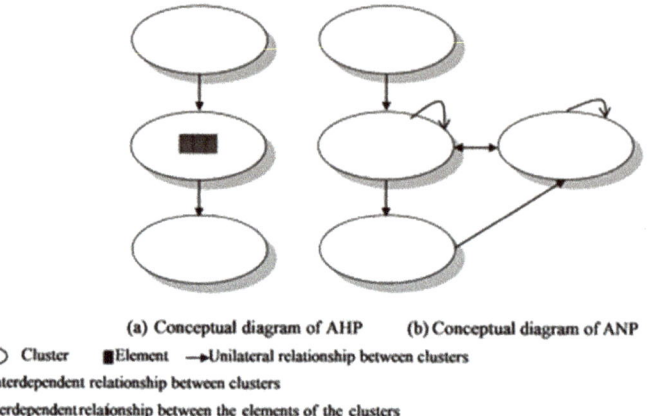

(a) Conceptual diagram of AHP   (b) Conceptual diagram of ANP

○ Cluster   ▪Element   →Unilateral relationship between clusters
↔Interdependent relationship between clusters
↶ Interdependent relationship between the elements of the clusters

Figure 2. Conceptual diagrams of the analytical hierarchy process (AHP) and analytic network process (ANP) [22].

Step 2: Establishing the Pairwise Comparison Matrix and Calculating the Eigenvector

Saaty [20] recommended the use of the 1–9 ratio scale in pairwise comparison. In ANP pairwise comparisons, the limiting influence of each criterion is calculated to establish the supermatrices.

The pairwise comparison matrix (A) is formed by experts making pairwise comparisons between criteria. Through a hierarchical analysis, the eigenvector (W) of the maximum eigenvalue (λmax) is obtained to satisfy the equation $A \times W = \lambda\text{max} \times W$. Then, λmax can be used to calculate the consistency index (CI); a satisfactory consistency level is indicated by CI $\leq$ 0.1. According to Saaty [23], a CI < 0.1 suggests the judgments made by experts are consistent. CI and consistency ratio are calculated using the following equations:

$$CI = \frac{\lambda_{max} - n}{n - 1} \quad (1)$$

$$CR = \frac{CI}{RI} \quad (2)$$

RI = random index.

Step 3: Obtaining the Supermatrices and Weights

The supermatrices comprise an unweighted supermatrix, a weighted supermatrix, and a limiting supermatrix, which can be used to obtain the weights of criteria and subcriteria.

A supermatrix is composed of various submatrices, and each ratio scale in the submatrices represents the influence of elements in a cluster on elements in other clusters (i.e., outer dependence) or on other elements in the same cluster (i.e., inner interdependence). Finally, the criteria and subcriteria of all dimensions are listed (respectively) at the left and top of a matrix to form a complete supermatrix, as shown in (3).

Because an unweighted supermatrix (W) may not be column-stochastic (i.e., each column does not sum to (1), it must be converted using the following process. No conversion is needed if the dimension column is stomatic (sum = 1). For nonstochastic columns, relative importance is applied on the submatrix of criteria columns to obtain the weighted supermatrix (W′). Subsequently, the supermatrix is subject to a limiting process, namely raising W′ to the power of 2k + 1 (k is an arbitrarily large number) until the interdependent relationships converge, to obtain the relative weights of criteria [20].

$$W = \begin{array}{c} C_1 \\ \\ C_2 \\ \\ C_n \end{array} \begin{array}{c} e_{11} \\ e_{12} \\ \vdots \\ e_{1m_1} \\ e_{21} \\ e_{22} \\ \vdots \\ e_{2m_2} \\ e_{n_1} \\ e_{n_2} \\ \vdots \\ e_{nm_n} \end{array} \overset{\begin{array}{cccccc} & C_1 & & C_2 & & C_n \\ e_{11} \cdots e_{1m_1} & e_{21} \cdots e_{2m_2} & \cdots & e_{n_1} \cdots e_{nm_n} \end{array}}{\begin{bmatrix} W_{11} & W_{12} & \cdots & W_{1n} \\ W_{12} & W_{22} & \cdots & W_{2n} \\ \vdots & \vdots & \ddots & \vdots \\ W_{n1} & W_{n2} & \cdots & W_{nn} \end{bmatrix}} \quad (3)$$

As an example, the following is the supermatrix (W_h) of a three-level hierarchical structure [20]:

$$W_h = \begin{bmatrix} 0 & 0 & 0 \\ W_{21} & 0 & 0 \\ 0 & W_{32} & I \end{bmatrix} \quad (4)$$

where W_{21} is the eigenvector of criteria under the decision-making goal, W_{32} is the eigenvector of the pairwise comparison matrix between alternatives under each criterion, and I is the identity matrix; a 0 indicates the relationship between identical or two independent elements or criteria without interdependences.

For interdependent criteria, a network structure must be used in place of a hierarchical structure. Accordingly, the supermatrix is updated to W_n in (5), where W_{22} represents the interdependence of the criteria [20].

$$W_n = \begin{bmatrix} 0 & 0 & 0 \\ W_{21} & W_{22} & 0 \\ 0 & W_{32} & W_{33} \end{bmatrix} \quad (5)$$

This study employed the *ANP* to obtain the weights of elements and weights. Therefore, W_n must be modified as W'_n, as presented in (6).

In (6), W_{22} and W_{33} respectively represent the interdependence weights of the elements and criteria.

$$W'_n = \begin{bmatrix} 0 & 0 & 0 \\ W_{21} & W_{22} & 0 \\ 0 & W_{32} & W_{33} \end{bmatrix} \quad (6)$$

The exponent of the matrix reaches an extremum where the matrix converges, thus the extremum holds constant. To achieve matrix convergence, the weighted supermatrix is raised to the power of $2k + 1$, where $k \to \infty$, as in (7). This yields a new matrix, the limiting supermatrix (W_{ANP}; [20]), and the finalized weights of criteria and subcriteria can then be obtained.

$$W_{ANP} = \lim_{k \to \infty} (W'_n)^{2k+1} \quad (7)$$

Step 4: Determining the Optimal Alternative

According to the limiting supermatrix W_{ANP} in (7), the weights can be obtained through multiple matrix calculations. These weights are then used as the basis for arranging the priority of alternatives.

3. Proposed Model

This study established a two-stage CBPPM supplier selection model. In Stage 1, a modified Delphi method and content validity ratio were used to determine the criteria and subcriteria for supplier selection as well as the interdependence between criteria and subcriteria. In Stage 2, the ANP was used to calculate the weights of criteria and subcriteria.

The two-stage supplier selection model is as follows [24–26]:

3.1. Stage 1: Establish a Hierarchical Network

This stage involves the use of the modified Delphi method comprising four steps, as follows [27,28]:

1. Step 1: Define the criteria.
2. Step 2: Convene an expert panel.
3. Step 3: Conduct a questionnaire survey on the panel.
4. Step 4: Determine the standard of consistency within the panel.

3.2. Stage 2: Select the Optimal Supplier with the ANP

This stage involves the four steps of ANP, as follows [18]:

1. Step 1: Establish the pairwise comparison matrix.
2. Step 2: Calculate the eigenvalue and eigenvector.
3. Step 3: Form the supermatrix and obtain the weights.
4. Step 4: Select the optimal procurement alternative.

4. Results and Discussion

SUNRISE, established in 1996 with a capital of NT$150 million, is a CBPPM manufacturer that sells machines mostly to paper box manufacturers in Taiwan, China, Europe, Southeast Asia, Middle America, and the Middle East. With a revenue of US$36 million in 2019, it is now the largest CBPPM manufacturer in Asia and the second largest in the world. It has thus become the hidden champion of the industry in the Taiwanese market, with patents in various countries. The high-capacity fixed-type CBPPM is its most precise, expensive, and sold machine. This CBPPM (Figure 3), which can print more than 300 color corrugated boxes per minute, contributes nearly 35% of the company's revenue (http://www.sunrisemachinery.com) [29].

Figure 3. High-capacity fixed-type corrugated box precision printing machine (CBPPM) (http://www.sunrisemachinery.com).

The CBPPM industry is relatively closed compared with other industries in Taiwan. Despite the enormous business opportunity in CBPPM manufacturing, no more than 30 CBPPM manufacturers exist in Taiwan. The R&D of CBPPMs involve an extremely high cost and a 3–5-year period (or longer). The R&D and sales expenses for a CBPPM total more than US$3 million. Although a new CBPPM has a product life cycle of more than 10 years on average, its high investment cost and slow return on investment discourage new investors. The industry also has high entry barriers because it involves (1) complex and specialized technologies, (2) specialized assembly technicians who require extensive training, and (3) a high level of working capital. A monthly working capital of more than US$120 million is required for the warehousing of components alone. Accordingly, procurement plays a critical role in the operation of a CBPPM manufacturer, which must establish a collaborative supply chain management system that integrates upstream and downstream suppliers well to shift from the red ocean strategy—which focuses on competition and price cuts—toward the blue ocean strategy, manufacturing products of high value at low cost [30].

This study adopted SUNRISE as an example and optimized its supplier selection process for the five firms that supply the most electronic control components to it. The optimization was conducted using the modified Delphi method and ANP to verify the feasibility of the study's proposed supplier selection model based on these two methods.

This study used the five firms as alternatives to conduct a supplier selection process as follows.

4.1. Stage 1: Establish a Hierarchical Network

Step 1: Define the Criteria

Six key members of SUNRISE (board director, director of plant operations, chief R&D officer, chief procurement officer, junior procurement officer, and procurement specialist) were invited to determine 11 criteria and 64 subcriteria for the supplier selection (Table 1).

Table 1. Criteria and subcriteria determined by six key members of SUNRISE.

Criteria	Definition	Subcriteria
Organization management	Effective process of realizing organizational goals through interaction, coordination, collaboration, and task delegation among all organization members, facilitated by establishing organizational structure, job roles or titles, and clear responsibilities and liabilities	(1) Emergency response (2) Employer–employee relationship (3) Government policy (4) Competitor behavior (5) Competitive analysis of the industry
Financial position	Management of asset purchases (investment), capital loans (financing), operation cash flows (working capital), and profit allocation given the overall goals	(1) Financial stability (2) Property risk management (3) Activity ratio (4) Investment in derivatives
Quality	Whether the product or service conforms to or surpasses the client's expectation	(1) Continuous improvement (2) Product reliability (3) Quality records (4) Solving quality problems (5) Quality management system for substandard products (6) Repair and compensation claims
Delivery	(1) The period between when an order is placed and its delivery by the supplier (2) Delivery = time spent in administrative procedure + procurement + production + shipping + inspection + other operations	(1) Stable supply of orders (2) Commitment to the delivery of orders (3) Accuracy and reliability of supply (4) On-time delivery (5) Ability to deliver orders at short notice (6) Ability to manage inventory
Commitment	A contract made with mutual agreement of all parties	(1) Commitment to orders (2) Stable supply (3) Accuracy and reliability (4) Speed of delivery (5) Commitment to the delivery time

Table 1. Cont.

Criteria	Definition	Subcriteria
Cost	All costs incurred during a company's acquisition of products or services and all expenses, which are the cost invested by a company in its business activities to make profit	(1) Procurement cost (2) Reflects real-time prices (3) Transportation cost (4) Price competitiveness (5) Ability to negotiate prices (6) Controlling price with volume (7) Discounts for cash payment
Production capacity	The maximum volume of products produced or raw materials processed by all fixed assets in a company within the contract period and under the given technological conditions	(1) Product stability (2) Production capacity and output value (3) Productivity (4) Expected sales and production capacity (5) Contracting or outsourcing
Technical capability	The level of understanding of and familiarity with a certain activity, particularly interaction with others, in relation to a method, process, program, or technique	(1) Ability to continuously improve (2) Ability to innovate techniques (3) Ability to provide technical support (4) Ability to change designs (5) Core technical skills
Service attitude	An activity or a benefit that is provided by one party to another, is intangible, and does not involve change of rights in remuneration	(1) Continuously reporting back to client (2) Attitude (3) Ability to manage customer complaints (4) Ability to supply spare parts (5) Negotiation with suppliers (6) Ability to conduct training (7) Maintenance of product safety (8) After-sales repair (9) After-sales services
Reputation	The sum of a company's value-creation capabilities generated from its acquisition of recognition by society and then of resources, opportunities, and support	(1) Integrity (2) Value of business reputation (3) Business competitiveness (4) Enhancement of corporate value (5) Improvement of profit (6) Corporate social responsibility (7) Profit increase (8) Financial robustness
Environmental protection product management	Manufacturing, use, and processing of products, conforming to environmental requirements, causing no or very little harm to the environment and conducive to resource circulation and product repurposing	(1) RoHS Regulations on Banned Substances in Components (2) RoHS monitoring and documentation on inbound materials (3) RoHS training (4) Provision of guarantee and a third-party report

Step 2: Convene an Expert Panel

According to Murry and Hammons [28], the appropriate size of an expert panel is more than 10 members, but an excessively large panel (with more than 30 members) can complicate the research work and create difficulty for the panel to reach a conclusion. On this basis and in consideration of feasibility and available research resources, the present study determined that the expert panel size be 23 members from the industry, government, and academia (Table 2).

Table 2. Composition of the expert panel.

Expert Category	Place of Employment	Number of People	Percentage (%)
Industry	Manufacturers specializing in the design, production, and sale of CBPPMs (each with over 20 years of experience in selling and manufacturing CBPPMs [note 1])	12	52.17
Government	Bureau of Foreign Trade (Ministry of Economic Affairs); Industrial Development Bureau (Ministry of Economic Affairs); Taiwan External Trade Development Council; National Taiwan Bureau of Taipei (Ministry of Finance)	5	21.74
Academia	Five from academic institutions; one from The Global Logistics & Commerce Council of Taiwan	6	26.09
	Total	23	100

Six manufacturers were interviewed; four completed a questionnaire.

Step 3: Conduct a Questionnaire Survey on the Panel

The first survey was administered to 23 experts who expressed their willingness to participate through mail; 20 questionnaire responses were returned for a response rate of 86.96%. Subsequently, a second survey was administered to the 20 experts (i.e., excluding the three who did not return a response) along with statistical charts for the first survey. In the second survey, 20 questionnaires were distributed, and all were returned. This study employed a two-round modified Delphi method, repeating the administration of the survey until consensus was established (Table 3).

Table 3. Comparison of survey response rates in the two-round modified Delphi method.

Category	First Round	Second Round
Number of copies distributed	23	20
Number of responses	20	20
Response rate (%)	86.96	100

Step 4: Determine the Standard of Consistency within the Panel

After a preliminary version of the questionnaire was created, the modified Delphi method was used to verify its content. A total of 20 experts from industry, government, and academia were recruited to determine the validity of the items. The experts were asked to rate each item on a 5-point Likert scale (1 = *Very dissatisfied* to 5 = *Very satisfied*) according to its appropriateness and relevance to the research topic as well as to determine the importance of each criterion and subcriterion. The content validity ratio (CVR) formula proposed by Lawshe [31] was employed to calculate the level of agreement among the experts. The ratings were used to calculate the CVR; in this study, a rating = 5 was determined to be the standard. Specifically, the CVR for each expert was calculated by dividing the number of items rated as 5 points by the total number of items. The CVRs for the 20 experts ranged between 0.7 and 1.00. This indicated the content validity of the questionnaire, with an average CVR ≥ 0.7 [32].

After deleting criteria and subcriteria with a CVR < 0.7, five criteria and 14 subcriteria remained. The five criteria were quality, commitment, cost, service attitude, and reputation. The 14 subcriteria were product reliability, quality management system for substandard products, commitment to orders, stable supply, accuracy and reliability, on-time delivery, price reduction, price competitiveness, attitude, ability to manage customer complaints, negotiation with suppliers, after-sales services, integrity, and profit increase. On the basis of the experts' input, the criteria were interdependent (Figure 4); inner interdependence was present between subcriteria (Table 4). According to these results, the CBPPM supplier selection hierarchical network was established (Figure 5).

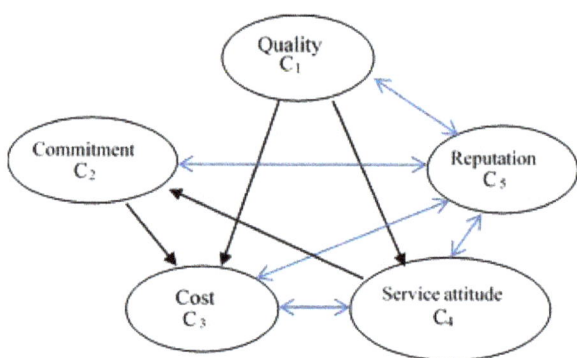

Figure 4. Relationships among CBPPM supplier selection criteria.

Table 4. Relationships among CBPPM supplier selection subcriteria.

Criteria	Subcriteria	Relationship
Quality (C_1)	Product reliability (C_{11})	Interdependent with C_{21}, C_{22}, C_{23}, C_{24}, C_{31}, and C_{32}
	Quality management system for substandard products (C_{12})	Interdependent with C_{21}, C_{22}, C_{23}, C_{24}, C_{31}, and C_{32}
Commitment (C_2)	Commitment to orders (C_{21})	Interdependent with C_{11}, C_{12}, C_{31}, and C_{32}
		Unilaterally dominant over C_{51} and C_{52}
	Stable supply (C_{22})	Interdependent with C_{11}, C_{12}, C_{31}, and C_{32}
		Unilaterally dominant over C_{51} and C_{52}
	Accuracy and reliability (C_{23})	Interdependent with C_{11}, C_{12}, C_{31}, and C_{32}
		Unilaterally dominant over C_{43} and C_{44}
	On-time delivery (C_{24})	Interdependent with C_{11}, C_{12}, C_{31}, and C_{32}
		Unilaterally dominant over C_{41}, C_{42}, and C_{43}
Cost (C_3)	Price reduction (C_{31})	Interdependent with C_{11}, C_{12}, C_{21}, C_{22}, C_{23}, C_{24}, C_{43}, and C_{52}
	Price competitiveness (C_{32})	Interdependent with C_{11}, C_{12}, C_{21}, C_{22}, C_{23}, C_{24}, C_{42}, and C_{51}
Service attitude (C_4)	Attitude (C_{41})	Unilaterally dominant over C_{11}, C_{12}, C_{31}, C_{32}, and C_{51}
	Ability to manage customer Complaints (C_{42})	Interdependent with C_{32} and C_{51}
		Unilaterally dominant over C_{11}, C_{12}, C_{31}, and C_{52}
	Negotiation with suppliers (C_{43})	Interdependent with C_{31} and C_{51}
		Unilaterally dominant over C_{11}, C_{12}, and C_{32}
	After-sales services (C_{44})	Unilaterally dominant over C_{11}, C_{12}, C_{31}, C_{32}, and C_{51}
Reputation (C_5)	Integrity (C_{51})	Interdependent with C_{32}, C_{43}, and C_{52}
		Unilaterally dominant over C_{11}, C_{12}, C_{31}, and C_{42}
	Profit increase (C_{52})	Interdependent with C_{31} and C_{51}
		Unilaterally dominant over C_{11}, C_{12}, and C_{32}

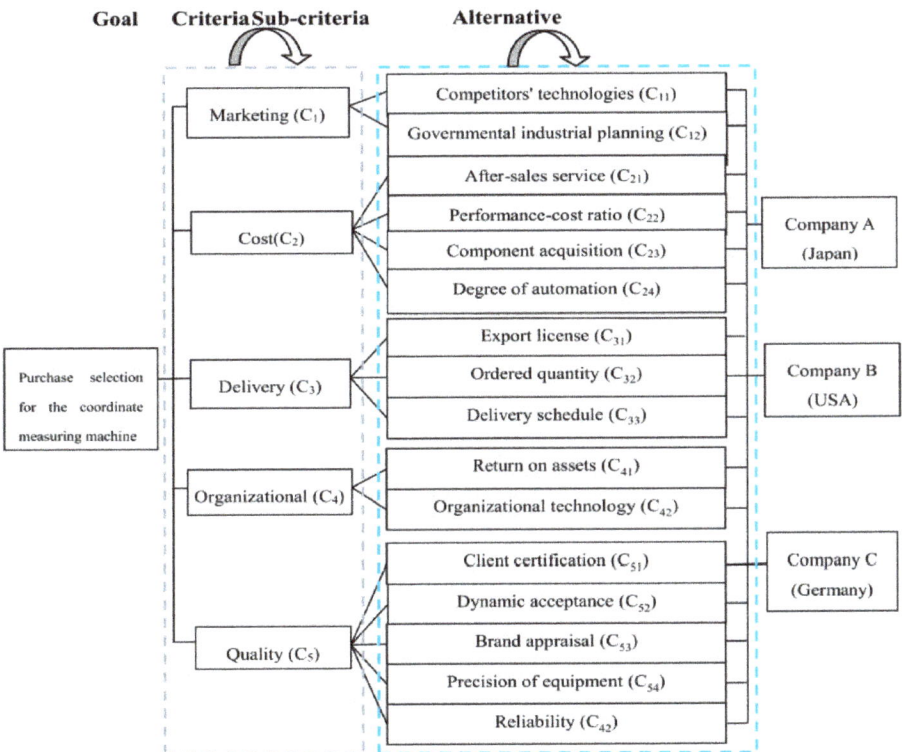

Figure 5. Hierarchical network for CBPPM supplier selection.

4.2. Stage 2: Select the Optimal Supplier with the ANP

Step 1: Establish the Pairwise Comparison Matrix

A panel of 20 experts was convened to determine the relative importance of each criterion in the ANP questionnaire. Table 5 depicts the resulting pairwise comparison matrix W_{21}.

Table 5. Pairwise comparison matrix W_{21} and the eigenvector.

Criteria	Quality (C_1)	Commitment (C_2)	Cost (C_3)	Service Attitude (C_4)	Reputation (C_5)	Eigenvector
Quality C_1	1	2	5	5	3	0.4206
Commitment C_2	0.5000	1	4	3	3	0.2827
Cost C_3	0.2000	0.2500	1	3	0.5000	0.0985
Service attitude C_4	0.2000	0.3333	0.3333	1	0.5000	0.0655
Reputation C_5	0.3333	0.3333	2	2	1	0.1327

(1) $\lambda \max = 5.2394$; (2) CI = 0.0598 and consistency ratio (CR) = $0.0534 \leq 0.1$.

Step 2: Calculate the Eigenvalue and Eigenvector

Super Decisions software was used to calculate the maximum eigenvalue $\lambda_{\max} = 5.2394$ and corresponding eigenvector x = (0.4206, 0.2827, 0.0985, 0.0655, 0.1327, rightmost column of Table 5) for the pairwise comparison matrix.

Equations (1) and (2) were used to obtain CI = 0.0598 and CR = 0.0534, both of which indicate satisfactory consistency.

Step 3: Form the Supermatrix and Obtain the Weights

After calculating the weights for W_{21}, the eigenvector matrix W_{32} was formed. For example, the pairwise comparison matrix and eigenvector for subcriteria C_{11} and C_{12} under criterion C_1 are presented in Table 6. Table 7 compiles W_{21} and W_{32}.

Table 6. Pairwise comparison matrix and eigenvector for subcriteria under criterion C_1.

Subcriteria Under C_1	Product Reliability (C_{11})	Quality Management System for Substandard Products (C_{12})	Eigenventor
Product reliability (C_{11})	1	4	0.8000
Quality management system for substandard products (C_{12})	0.2500	1	0.2000

(1) λmax = 2.0000; (2) CI = 0 and CR = 0 $\leq$ 0.1.

Table 7. Weights for criteria and subcriteria.

Criteria	Criteria Weight (W_{21})	Subcriteria	Subcriteria Weight (W_{32})
Quality (C_1)	0.4206	Product reliability (C_{11})	0.8000
		Quality management system for substandard products (C_{12})	0.2000
Commitment (C_2)	0.2827	Commitment to orders (C_{21})	0.0493
		Stable supply (C_{22})	0.2075
		Accuracy and reliability (C_{23})	0.2701
		On-time delivery (C_{24})	0.4731
Cost (C_3)	0.0985	Price reduction (C_{31})	0.1111
		Price competitiveness (C_{32})	0.8889
Service attitude (C_4)	0.0655	Attitude (C_{41})	0.6642
		Ability to manage customer complaints (C_{42})	0.0903
		Negotiation with suppliers (C_{43})	0.0957
		After-sales services (C_{44})	0.1498
Reputation (C_5)	0.1327	Integrity (C_{51})	0.7500
		Profit increase (C_{52})	0.2500

Matrix W_{22} represents the pairwise comparison results for the five criteria with the presence of inner interdependence. The eigenvector matrix formed with the eigenvectors is shown in (8). Matrix W_{33} is the eigenvector matrix representing the pairwise comparison results for the 14 subcriteria with the presence of inner interdependence, as presented in (10).

An unweighted supermatrix is formed by combining matrices W_{21}, W_{22}, W_{32}, and W_{33}, as expressed in (9). Table 8 reveals the details of the supermatrix. In this study, matrices W_{22} and W_{32} are each assigned a weight of 0.5 to obtain the weighted supermatrix (Table 9).

Table 10 illustrates the limiting supermatrix, and Equation (11) provides the weights of all subcriteria (W_{ANP}) [33].

Table 8. Unweighted supermatrix.

Unweighted Super Matrix	Goal	(C_1)	(C_2)	(C_3)	(C_4)	(C_5)	(C_{11})	(C_{12})	(C_{21})	(C_{22})	(C_{23})	(C_{24})	(C_{31})	(C_{32})	(C_{41})	(C_{42})	(C_{43})	(C_{44})	(C_{51})	(C_{52})
Goal	0	0	0	0	0	0	0	0	0	0	0	0	0	0	0	0	0	0	0	0
(C_1)	0.4206	0.4296	0	0	0	0.1779	0	0	0	0	0	0	0	0	0	0	0	0	0	0
(C_2)	0.2827	0	0.4806	0	0	0.1982	0	0	0	0	0	0	0	0	0	0	0	0	0	0
(C_3)	0.0985	0.0784	0.1140	0.1655	0.1093	0.1402	0	0	0	0	0	0	0	0	0	0	0	0	0	0
(C_4)	0.0655	0.0820	0	0.6098	0.5725	0.2043	0	0	0	0	0	0	0	0	0	0	0	0	0	0
(C_5)	0.1327	0.4100	0.4054	0.2247	0.2090	0.2793	0	0	0	0	0	0	0	0	0	0	0	0	0	0
(C_{11})	0	0.8	0	0	0	0	0.2808	0	0.2936	0.4116	0.1736	0.0479	0.2522	0.2176	0.3302	0.3702	0.3495	0.2489	0.0561	0.0466
(C_{12})	0	0.2	0	0	0	0	0	0.4158	0.1587	0.0454	0.0645	0.0520	0.0442	0.0942	0.2869	0.2201	0.2409	0.0986	0.0346	0.0440
(C_{21})	0	0	0.25	0	0	0	0.0292	0.0430	0.0497	0	0	0	0.1451	0.0725	0	0	0	0	0	0
(C_{22})	0	0	0.25	0	0	0	0.1200	0.0459	0	0.1593	0.0341	0	0.1937	0.1582	0	0	0	0	0	0
(C_{23})	0	0	0.25	0	0	0	0.3515	0.0853	0	0	0.1691	0	0.1535	0.1617	0	0	0	0	0	0
(C_{24})	0	0	0.25	0	0	0	0.1004	0.1071	0	0	0.0545	0.1278	0.0921	0.1608	0	0	0	0	0	0
(C_{31})	0	0	0	0.5	0	0	0.0586	0.1324	0.2257	0.0340	0.0447	0.0591	0.0257	0	0.0437	0.0253	0.0286	0.0339	0.0466	0.1211
(C_{32})	0	0	0	0.5	0	0	0.0595	0.1706	0.1849	0.0533	0.0436	0.0513	0	0.0263	0.0437	0.0317	0.0534	0.0427	0.0581	0.2903
(C_{41})	0	0	0	0	0.25	0	0	0	0	0.1261	0	0.1474	0	0	0.1143	0.0869	0	0	0.1663	0
(C_{42})	0	0	0	0	0.25	0	0	0	0	0	0.1702	0.3575	0.0339	0.0420	0	0	0.0800	0	0.2076	0
(C_{43})	0	0	0	0	0.25	0	0	0	0	0	0.1860	0.1570	0	0	0	0	0	0.2475	0	0
(C_{44})	0	0	0	0	0.25	0	0	0	0	0	0.2299	0	0	0	0	0	0	0.3284	0	0
(C_{51})	0	0	0	0	0	0.5	0	0	0.0624	0	0	0	0.0666	0	0.1812	0.1449	0.2476	0	0.3273	0.2687
(C_{52})	0	0	0	0	0	0.5	0	0	0.0250	0	0	0	0.0596	0	0	0.1210	0	0	0.1033	0.2294

Table 9. Weighted supermatrix.

Weighted Super Matrix	Goal	(C_1)	(C_2)	(C_3)	(C_4)	(C_5)	(C_{11})	(C_{12})	(C_{21})	(C_{22})	(C_{23})	(C_{24})	(C_{31})	(C_{32})	(C_{41})	(C_{42})	(C_{43})	(C_{44})	(C_{51})	(C_{52})
Goal	0	0	0	0	0	0	0	0	0	0	0	0	0	0	0	0	0	0	0	0
(C_1)	0.4206	0.2148	0	0	0	0	0	0	0	0	0	0	0	0	0	0	0	0	0	0
(C_2)	0.2827	0	0.2403	0	0	0.0889	0	0	0	0	0	0	0	0	0	0	0	0	0	0
(C_3)	0.0985	0.0392	0.0570	0.0828	0.0546	0.0991	0	0	0	0	0	0	0	0	0	0	0	0	0	0
(C_4)	0.0655	0.0410	0	0.3049	0.2862	0.0701	0	0	0	0	0	0	0	0	0	0	0	0	0	0
(C_5)	0.1327	0.2050	0.2027	0.1123	0.1045	0.1022	0	0	0	0	0	0	0	0	0	0	0	0	0	0
						0.1397														
(C_{11})	0	0.4	0	0	0	0	0.2808	0	0.2936	0.4116	0.1736	0.0479	0.2522	0.2176	0.3302	0.3702	0.3495	0.2489	0.0561	0.0466
(C_{12})	0	0.1	0	0	0	0	0	0.4158	0.1587	0.0454	0.0645	0.0520	0.0442	0.0942	0.2869	0.2201	0.2409	0.0986	0.0346	0.0440
(C_{21})	0	0	0.1250	0	0	0	0.0292	0.0430	0.0497	0	0	0	0.1451	0.0725	0	0	0	0	0	0
(C_{22})	0	0	0.1250	0	0	0	0.1200	0.0459	0	0.1593	0.0341	0	0.1937	0.1582	0	0	0	0	0	0
(C_{23})	0	0	0.1250	0	0	0	0.3515	0.0853	0	0	0.1691	0.1278	0.1535	0.1617	0	0	0	0	0	0
(C_{24})	0	0	0.1250	0	0	0	0.1004	0.1071	0	0	0.0545	0.0591	0.0921	0.1608	0	0	0	0	0	0
(C_{31})	0	0	0	0.25	0	0	0.0586	0.1324	0.2257	0.0340	0.0447	0.0513	0.0257	0	0.0437	0.0253	0.0286	0.0339	0.0466	0.1211
(C_{32})	0	0	0	0.25	0	0	0.0595	0.1706	0.1849	0.0533	0.0436	0.1474	0	0.0263	0.0437	0.0317	0.0534	0.0427	0.0581	0.2903
(C_{41})	0	0	0	0	0.1250	0	0	0	0	0.1261	0	0.3575	0	0.0420	0.1143	0.0869	0	0	0.1663	0
(C_{42})	0	0	0	0	0.1250	0	0	0	0	0.1702	0.1860	0.1570	0.0339	0	0	0	0.0800	0	0	0
(C_{43})	0	0	0	0	0.1250	0	0	0	0	0	0.2299	0	0	0	0	0	0	0.2475	0.2076	0
(C_{44})	0	0	0	0	0.1250	0	0	0	0	0	0	0	0	0	0	0	0	0	0	0
(C_{51})	0	0	0	0	0	0.25	0	0	0.0624	0	0	0	0	0.0666	0.1812	0.1449	0.2476	0.3284	0.3273	0.2687
(C_{52})	0	0	0	0	0	0.25	0	0	0.0250	0	0	0	0.0596	0	0	0.1210	0	0	0.1033	0.2294

Table 10. Limiting supermatrix.

Limit Super Matrix	Goal	(C_1)	(C_2)	(C_3)	(C_4)	(C_5)	(C_{11})	(C_{12})	(C_{21})	(C_{22})	(C_{23})	(C_{24})	(C_{31})	(C_{32})	(C_{41})	(C_{42})	(C_{43})	(C_{44})	(C_{51})	(C_{52})
Goal	0	0	0	0	0	0	0	0	0	0	0	0	0	0	0	0	0	0	0	0
(C_1)	0	0	0	0	0	0	0	0	0	0	0	0	0	0	0	0	0	0	0	0
(C_2)	0	0	0	0	0	0	0	0	0	0	0	0	0	0	0	0	0	0	0	0
(C_3)	0	0	0	0	0	0	0	0	0	0	0	0	0	0	0	0	0	0	0	0
(C_4)	0	0	0	0	0	0	0	0	0	0	0	0	0	0	0	0	0	0	0	0
(C_5)	0	0	0	0	0	0	0	0	0	0	0	0	0	0	0	0	0	0	0	0
(C_{11})	0.2094	0.2094	0.2094	0.2094	0.2094	0.2094	0.2094	0.2094	0.2094	0.2094	0.2094	0.2094	0.2094	0.2094	0.2094	0.2094	0.2094	0.2094	0.2094	0.2094
(C_{12})	0.1139	0.1139	0.1139	0.1139	0.1139	0.1139	0.1139	0.1139	0.1139	0.1139	0.1139	0.1139	0.1139	0.1139	0.1139	0.1139	0.1139	0.1139	0.1139	0.1139
(C_{21})	0.0257	0.0257	0.0257	0.0257	0.0257	0.0257	0.0257	0.0257	0.0257	0.0257	0.0257	0.0257	0.0257	0.0257	0.0257	0.0257	0.0257	0.0257	0.0257	0.0257
(C_{22})	0.0675	0.0675	0.0675	0.0675	0.0675	0.0675	0.0675	0.0675	0.0675	0.0675	0.0675	0.0675	0.0675	0.0675	0.0675	0.0675	0.0675	0.0675	0.0675	0.0675
(C_{23})	0.1244	0.1244	0.1244	0.1244	0.1244	0.1244	0.1244	0.1244	0.1244	0.1244	0.1244	0.1244	0.1244	0.1244	0.1244	0.1244	0.1244	0.1244	0.1244	0.1244
(C_{24})	0.0647	0.0647	0.0647	0.0647	0.0647	0.0647	0.0647	0.0647	0.0647	0.0647	0.0647	0.0647	0.0647	0.0647	0.0647	0.0647	0.0647	0.0647	0.0647	0.0647
(C_{31})	0.0578	0.0578	0.0578	0.0578	0.0578	0.0578	0.0578	0.0578	0.0578	0.0578	0.0578	0.0578	0.0578	0.0578	0.0578	0.0578	0.0578	0.0578	0.0578	0.0578
(C_{32})	0.0692	0.0692	0.0692	0.0692	0.0692	0.0692	0.0692	0.0692	0.0692	0.0692	0.0692	0.0692	0.0692	0.0692	0.0692	0.0692	0.0692	0.0692	0.0692	0.0692
(C_{41})	0.0204	0.0204	0.0204	0.0204	0.0204	0.0204	0.0204	0.0204	0.0204	0.0204	0.0204	0.0204	0.0204	0.0204	0.0204	0.0204	0.0204	0.0204	0.0204	0.0204
(C_{42})	0.0424	0.0424	0.0424	0.0424	0.0424	0.0424	0.0424	0.0424	0.0424	0.0424	0.0424	0.0424	0.0424	0.0424	0.0424	0.0424	0.0424	0.0424	0.0424	0.0424
(C_{43})	0.0680	0.0680	0.0680	0.0680	0.0680	0.0680	0.0680	0.0680	0.0680	0.0680	0.0680	0.0680	0.0680	0.0680	0.0680	0.0680	0.0680	0.0680	0.0680	0.0680
(C_{44})	0.0380	0.0380	0.0380	0.0380	0.0380	0.0380	0.0380	0.0380	0.0380	0.0380	0.0380	0.0380	0.0380	0.0380	0.0380	0.0380	0.0380	0.0380	0.0380	0.0380
(C_{51})	0.0763	0.0763	0.0763	0.0763	0.0763	0.0763	0.0763	0.0763	0.0763	0.0763	0.0763	0.0763	0.0763	0.0763	0.0763	0.0763	0.0763	0.0763	0.0763	0.0763
(C_{52})	0.0222	0.0222	0.0222	0.0222	0.0222	0.0222	0.0222	0.0222	0.0222	0.0222	0.0222	0.0222	0.0222	0.0222	0.0222	0.0222	0.0222	0.0222	0.0222	0.0222

$$W_{22} = \begin{bmatrix} & C_1 & C_2 & C_3 & C_4 & C_5 \\ C_1 & 0.4296 & 0 & 0 & 0 & 0.1779 \\ C_2 & 0 & 0.4806 & 0 & 0.1093 & 0.1982 \\ C_3 & 0.0784 & 0.1140 & 0.1655 & 0.1093 & 0.1402 \\ C_4 & 0.0820 & 0 & 0.6098 & 0.5725 & 0.2043 \\ C_5 & 0.4100 & 0.4054 & 0.2247 & 0.2090 & 0.2793 \end{bmatrix} \quad (8)$$

$$W_n = \begin{bmatrix} 0 & 0 & 0 \\ W_{21} & W_{22} & 0 \\ 0 & W_{32} & W_{33} \end{bmatrix} \quad (9)$$

$W_{33} =$

	C_{11}	C_{12}	C_{21}	C_{22}	C_{23}	C_{24}	C_{31}	C_{32}	C_{41}	C_{42}	C_{43}	C_{44}	C_{51}	C_{52}
C_{11}	0.2808	0	0.2936	0.4116	0.1736	0.0479	0.2522	0.2176	0.3302	0.3702	0.3495	0.2489	0.0561	0.0466
C_{12}	0	0.4158	0.1587	0.0454	0.0645	0.0520	0.0442	0.0942	0.2869	0.2201	0.2409	0.0986	0.0346	0.0440
C_{21}	0.0292	0.0430	0.0497	0	0	0	0.1451	0.0725	0	0	0	0	0	0
C_{22}	0.1200	0.0459	0	0.1593	0.0341	0	0.1937	0.1582	0	0	0	0	0	0
C_{23}	0.3515	0.0853	0	0	0.1691	0	0.1535	0.1617	0	0	0	0	0	0
C_{24}	0.1004	0.1071	0	0	0.0545	0.1278	0.0921	0.1608	0	0	0	0	0	0
C_{31}	0.0586	0.1324	0.2257	0.0340	0.0447	0.0591	0.0257	0	0.0437	0.0253	0.0286	0.0339	0.0466	0.1211
C_{32}	0.0595	0.1706	0.1849	0.0533	0.0436	0.0513	0	0.0263	0.0437	0.0317	0.0534	0.0427	0.0581	0.2903
C_{41}	0	0	0	0.1261	0	0.1474	0	0	0.1143	0	0	0	0	0
C_{42}	0	0	0	0	0	0.3575	0	0.0420	0	0.0869	0	0	0.1663	0
C_{43}	0	0	0	0.1702	0.1860	0.1570	0.0339	0	0	0	0.0800	0	0.2076	0
C_{44}	0	0	0	0	0.2299	0	0	0	0	0	0	0.2475	0	0
C_{51}	0	0	0.0624	0	0	0	0	0.0666	0.1812	0.1449	0.2476	0.3284	0.3273	0.2687
C_{52}	0	0	0.0250	0	0	0	0.0596	0	0	0.1210	0	0	0.1033	0.2294

(10)

$$W_{ANP} = \begin{array}{r|c} & Goal \\ (C_{11}) & 0.2094 \\ (C_{12}) & 0.1139 \\ (C_{21}) & 0.0257 \\ (C_{22}) & 0.0675 \\ (C_{23}) & 0.1244 \\ (C_{24}) & 0.0647 \\ (C_{31}) & 0.0578 \\ (C_{32}) & 0.0692 \\ (C_{41}) & 0.0204 \\ (C_{42}) & 0.0424 \\ (C_{43}) & 0.0680 \\ (C_{44}) & 0.0380 \\ (C_{51}) & 0.0763 \\ (C_{52}) & 0.0222 \end{array} \quad (11)$$

Table 11 lists the weights of the five alternatives calculated according to the pairwise comparison matrix for subcriterion C_{12} (quality management system for substandard products).

Table 11. Weights of the five alternatives calculated using the pairwise comparison matrix for subcriterion C_{12}.

(C_{12})	Company A	Company B	Company C	Company D	Company E	Weight
Company A	1	3	0.5	0.5	0.3333	0.1470
Company B	0.3333	1	1	0.3333	0.3333	0.1024
Company C	2	1	1	1	1	0.2154
Company D	2	3	1	1	1	0.2538
Company E	3	3	1	1	1	0.2814

(1) λmax = 5.3250; (2) CI = 0.0598 and CR = 0.0534 ≤ 0.1.

Table 12 presents the weights of the five alternatives calculated using the pairwise comparison matrices of all subcriteria.

Table 12. Eigenvectors of five alternatives under each criterion.

Alternatives	(C_{11})	(C_{12})	(C_{21})	(C_{22})	(C_{23})	(C_{24})	(C_{31})	(C_{32})	(C_{41})	(C_{42})	(C_{43})	(C_{44})	(C_{51})	(C_{52})
Company A	0.3274	0.1470	0.2346	0.2183	0.2086	0.2909	0.1033	0.0957	0.0787	0.1421	0.0566	0.0718	0.1688	0.0877
Company B	0.1299	0.1024	0.1660	0.0986	0.2630	0.1470	0.1818	0.1599	0.1814	0.1459	0.1308	0.1388	0.1601	0.1498
Company C	0.1331	0.2154	0.1978	0.0888	0.0947	0.1062	0.1377	0.1733	0.1814	0.0878	0.1218	0.0988	0.1217	0.2148
Company D	0.2383	0.2538	0.1694	0.2730	0.2881	0.2479	0.2256	0.2428	0.3149	0.4103	0.5491	0.2862	0.2435	0.2739
Company E	0.1714	0.2814	0.2321	0.3213	0.1457	0.2080	0.3515	0.3283	0.2435	0.2139	0.1417	0.4045	0.3059	0.2739

Step 4: Select the Optimal Procurement Alternative

On the basis of (11) and Table 12, the priority vector of the five alternatives is obtained, as shown in (12).

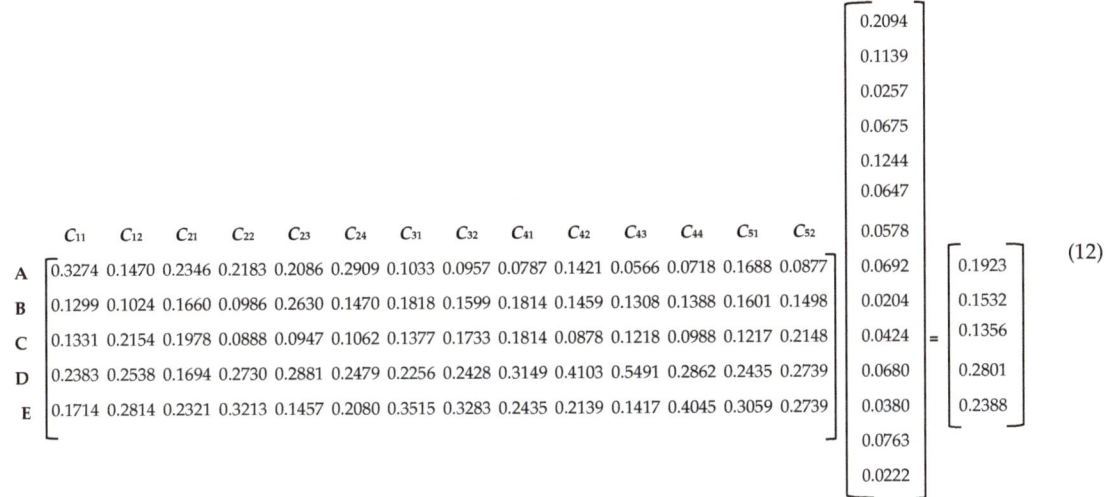

(12)

According to (12), in the supplier selection process, the companies were ranked as follows: Company D (0.2801), Company E (0.2388), Company A (0.1923), Company B (0.1532), and Company C (0.1356; Figure 6).

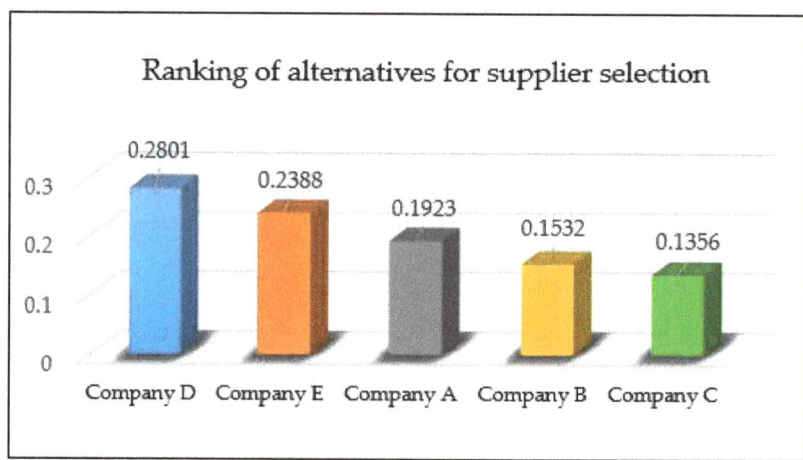

Figure 6. Ranking of alternatives for supplier selection.

5. Conclusions

The continuous, rapid development of the e-commerce market has contributed to fast growth in the packaging industry and is allowing the CBPPM industry to thrive. Accordingly, the establishment of an appropriate CBPPM supplier selection decision-making model has become critical for corrugated box manufacturers to maintain their competitive advantage.

According to the empirical results, Labib [6] considered product quality and delivery to be of greater importance than cost. Basnet [9] suggested that for both local and international businesses, quality, the ability to deliver on time, and performance are the most critical elements in supply chain management. Tam and Tummala [7] argued that the selection criteria for telecommunication service suppliers include quality, cost, problem-solving skills, expertise, delivery time, the ability to satisfy consumer needs, experience, and reputation. Four of the criteria obtained in this study are quality, commitment, cost and reputation, which is the same as Tam and Tummala [7], the new model is of high practical value and enables enterprises to consider and evaluate alternative solutions from multiple perspectives, thus facilitating sustainable operation and development. A quality-oriented company has an influence from higher managers to the employees of different functional departments. It not only can prevent the problems facing the products but also improve the current situation continuously. A corporate culture of having quality as the first priority has always been regarded as one of the main elements of a successful implementation of total quality management.

Through a rigorous research design, this study formed a panel of experts to build a CBPPM supplier selection model, offering insight for the corrugated box industry. The following conclusions were drawn:

1. This study convened a panel comprising 20 experts and scholars from the industry, government, and academia and employed the ANP to determine the weights of criteria as follows: quality (0.4206), commitment (0.2827), reputation (0.1327), cost (0.0985), and service attitude (0.0655).
2. The three subcriteria assigned the most weight were product reliability (0.2094), accuracy and reliability (0.1244), and quality management system for substandard

products (0.1139), two of which belonged to the "quality" criteria. A CBPPM manufacturer can face an enormous loss if it manufactures products using substandard components. Scrupulous product inspection and component control practices guarantee a long, useful life and high quality of machines, which in turn enhances the manufacturer's commitment, service attitude, and reputation.

3. The proposed CBPPM supplier selection model was verified to be feasible. Additionally, the robustness of this two-stage model was tested using the ranking of alternative suppliers. The ranking remained the same according to a sensitivity analysis of the five suppliers, which indicates the robustness of the model and its suitability for adoption by companies for supplier selection.

4. A corporate culture emphasizing quality is commonly considered a main factor for successful total quality management [34,35]. Quality orientation is the extent to which companies emphasize quality, their attitude toward quality, and the effort they make to enhance quality. The establishment of a quality-oriented philosophy within a company creates a top-down drive for quality problem prevention and continuous improvement among company members at all levels. Forza and Filippini [36] researched total quality management practice in companies and observed that companies improved the consistency of product quality as well as customer satisfaction through emphasis on quality, maintained raw material quality by strengthening connections with suppliers, obtained improvement plans initiated by employees through employee education and training, and elevated the overall process control by enhancing communication with suppliers and employees.

Author Contributions: Conceptualization, C.-T.L. and C.-Y.C.; methodology, Analytic Network Process (ANP); software, SuperDecisions; validation, C.-T.L. and C.-Y.C.; formal analysis, C.-T.L.; investigation, C.-Y.C.; resources, C.-T.L.; data curation, C.-Y.C.; writing—original draft preparation, C.-Y.C.; writing—review and editing, C.-Y.C.; visualization, NA; supervision, C.-T.L.; project administration, C.-Y.C.; funding acquisition, NA. All authors have read and agreed to the published version of the manuscript.

Funding: This research received no external funding.

Data Availability Statement: We fully adhere to its Code of Conduct and to its Best Practice Guidelines.

Conflicts of Interest: The authors declare no conflict of interest.

References

1. Smithers Pira. Available online: https://www.smithers.com/home (accessed on 27 May 2020).
2. Tsao, C.-P. *Talking about the Development of Corrugated Box Precision Printing Machine*; China Packaging Federation: Beijing, China, 2011; Volume 31, pp. 59–61.
3. China Packaging Federation. *Development Status of Domestic and Foreign Corrugated Box Printing Machine Industry*; China Packaging Federation: Beijing, China, 2011; Volume 31, pp. 31–36.
4. China Packaging Federation. Available online: http://www.cpta.org.cn/ (accessed on 28 May 2020).
5. Ptak, C.A.; Schragenheim, E. *ERP: Tools, Techniques, and Applications for Integrating the Supply Chain*; CRC Press: Boca Raton, FL, USA, 2003; pp. 9–10.
6. Labib, A.W. A supplier selection model: A comparison of fuzzy logic and the analytic hierarchy process. *Int. J. Prod. Res.* **2011**, *49*, 6287–6299. [CrossRef]
7. Tam, M.C.Y.; Tummala, V.M.R. An application of the AHP in vendor selection of a telecommunications system. *Omega* **2001**, *29*, 171–182. [CrossRef]
8. Liao, C.N.; Kao, H.P. An integrated fuzzy TOPSIS and MCGP approach to supplier selection in supply chain management. *Expert Syst. Appl.* **2011**, *38*, 10803–10811. [CrossRef]
9. Basnet, C.; Corner, F.; Wisner, F.; Tan, K.C. Banchmarking supply chain management practice in New Zealand. *Supply Chain Manag.* **2003**, *8*, 57–64. [CrossRef]
10. Hsu, C.P. *Procurement and Supply Management*; BEST-WISE Book: Taipei, Taiwan, 2015.
11. Hsu, C.P. *Purchasing and Supply Management*, 5th ed.; BEST-WISE Book: Taipei, Taiwan, 2017.
12. Shirouyehzad, H.; Pourjavad, E. Evaluating manufacturing systems by fuzzy ANP: A case study. *Appl. Manag. Sci.* **2014**, *6*, 65–83.
13. Chin, C.Y. *Supplier Management Manual*; Constitutional Enterprise Management (Group) Company: Taipei Taiwan, 2013.
14. Pipaç, M.D. Interdependence between management, attitude and organizational behavior. *Qual. Access* **2015**, *16*, 99–104.

15. Aghai, S.; Mollaverdi, N.; Sabbagh, M. A fuzzy multi-objective programming model for supplier selection with volume discount and risk criteria. *Int. J. Adv. Manufact. Technol.* **2014**, *71*, 1483–1492. [CrossRef]
16. Yawar, S.A.; Seuring, S. Management of social issues in supply chains: A literature review exploring social issues, actions and performance outcomes. *J. Bus. Ethics* **2017**, *141*, 621–643. [CrossRef]
17. Ramanathan, R. Supplier selection problem: Integrating DEA with the approaches of total cost of ownership and AHP. *Supply Chain Manag.* **2007**, *12*, 258–261. [CrossRef]
18. Bai, C.; Sarkis, J. Integrating sustainability into supplier selection with grey system and rough set methodologies. *Int. J. Prod. Econ.* **2010**, *124*, 252–264. [CrossRef]
19. Zhong, L.; Yao, L. An ELECTRE I-based multi-criteria group decision making method with interval type-2 fuzzy numbers and its application to supplier selection. *Appl. Soft Comp.* **2017**, *57*, 556–576. [CrossRef]
20. Saaty, T.L. *Decision Making with Dependence and Feedback: The Analytic Network Process*, 1st ed.; RWS Publications: Pittsburgh, PA, USA, 1996.
21. Saaty, T.L. *RANK, Normalization and Idealization in the Analytic Hierarchy Process*; 7th ISAHP: Bali, Indonesia, 2003.
22. Momoh, J.A.; Zhu, J.Z. Application of AHP/ANP to unit commitment in the deregulated power industry. In Proceedings of the SMC'98 Conference Proceedings, 1998 IEEE International Conference on Systems, Man, and Cybernetics (Cat. No.98CH36218), San Diego, CA, USA, 14 October 1998; Volume 1, pp. 817–822.
23. Saaty, T.L. *Analytic Hierarchy Process: Planning, Priority Setting, Resource Allocation*, 1st ed.; McGraw-Hill: New York, NY, USA, 1980.
24. Lin, C.T.; Wu, C.S. Selecting marketing strategy for private hotels in Taiwan using the analytic hierarchy process. *Serv. Ind. J.* **2008**, *28*, 1077–1091. [CrossRef]
25. Lin, C.T.; Tsai, M.C. Location choice for direct foreign investment in new hospitals in China by using ANP and TOPSIS. *Qual. Quant.* **2010**, *44*, 375–390. [CrossRef]
26. Lin, C.T.; Hung, K.P.; Hu, S.H. A decision-making model for evaluating and selecting suppliers for the sustainable operation and development of enterprises in the aerospace industry. *Sustainability* **2018**, *10*, 735.
27. Linstone, H.A.; Turoff, M. *The Delphi Method: Techniques and Applications*; Addison-Wesley Publishing Company: Reading, PA, USA, 1975.
28. Murry, J.W.; Hommons, J.O. Delphi: A versatile methodology for conducting qualitative research. *Rev. High. Educ.* **1995**, *18*, 423–436. [CrossRef]
29. Sunrise Pacific Co. Available online: http://www.sunrisemachinery.com/ (accessed on 23 June 2018).
30. Chou, W.J. Supply Chain Integration Based on Fuzzy Multi-criteria Group Decision Methods. Master's Thesis, Department of Information Management, Chaoyang University of Technology, Taichung City, Taiwan, 2006.
31. Lawshe, C.H. A quantitative approach to content validity. *Person. Psychol.* **1975**, *28*, 563–575. [CrossRef]
32. Waltz, C.W.; Strickland, O.L. *Lenz ER: Measurement in Nursing Research*; F.A. Davis Co.: Philadelphia, PA, USA, 1991.
33. Chang, S.S. *Fuzzy Multi-Criteria Decision Making for Evaluation Method*, 2nd ed.; Wu-Nan Book Inc.: Taipei, Taiwan, 2016; pp. 547–576.
34. Saraph, J.V.; Sebastian, R.J. Developing a quality culture. *Qual. Progr.* **1993**, *26*, 73–78.
35. Westbrook, J.D. Organizational culture and its relationship to TQM. *Ind. Manag.* **1993**, *35*, 1–3.
36. Forza, C.; Filippini, R. TQM impact on quality conformance and customer satisfaction: A causal model. *Int. J. Prod. Econ.* **1998**, *55*, 1–20. [CrossRef]

Article

Convergence and Stability of a Parametric Class of Iterative Schemes for Solving Nonlinear Systems

Alicia Cordero, Eva G. Villalba , Juan R. Torregrosa * and Paula Triguero-Navarro

Multidisciplinary Institute of Mathematics, Universitat Politènica de València, 46022 València, Spain; acordero@mat.upv.es (A.C.); egarvil@posgrado.upv.es (E.G.V.); ptrinav@posgrado.upv.es (P.T.-N.)
* Correspondence: jrtorre@mat.upv.es

Abstract: A new parametric class of iterative schemes for solving nonlinear systems is designed. The third- or fourth-order convergence, depending on the values of the parameter being proven. The analysis of the dynamical behavior of this class in the context of scalar nonlinear equations is presented. This study gives us important information about the stability and reliability of the members of the family. The numerical results obtained by applying different elements of the family for solving the Hammerstein integral equation and the Fisher's equation confirm the theoretical results.

Keywords: nonlinear system; iterative method; divided difference operator; stability; parameter plane; dynamical plane

Citation: Cordero, A.; Villalba, E.G.; Torregrosa, J.R.; Triguero-Navarro, P. Convergence and Stability of a Parametric Class of Iterative Schemes for Solving Nonlinear Systems. *Mathematics* **2021**, *9*, 86. https://dx.doi.org/10.3390/math9010086

Received: 1 December 2020
Accepted: 24 December 2020
Published: 3 January 2021

Publisher's Note: MDPI stays neutral with regard to jurisdictional claims in published maps and institutional affiliations.

Copyright: © 2021 by the authors. Licensee MDPI, Basel, Switzerland. This article is an open access article distributed under the terms and conditions of the Creative Commons Attribution (CC BY) license (https://creativecommons.org/licenses/by/4.0/).

1. Design of a Parametric Family of Iterative Methods

The need to find a solution $\bar{x}$ of equations or systems of nonlinear equations of the form $F(x) = 0$, where $F : D \subseteq \mathbb{R}^n \to \mathbb{R}^n$, $n \geq 1$, is present in many problems of applied mathematics as a basis for solving other more complex ones. In general, it is not possible to find the exact solution to this type of equations, so iterative methods are required in order to approximate the desired solution.

The essence of these methods is to find, through an iterative process and, from an initial approximation $x^{(0)}$ close to a solution $\bar{x}$, a sequence $\{x^{(k)}\}$ of approximations such that, under different requirements, $\lim_{k \to \infty} x^{(k)} = \bar{x}$.

It is well known that one of the most used iterative methods, due to its simplicity and efficiency, is Newton's scheme, whose iterative expression is

$$x^{(k+1)} = x^{(k)} - [F'(x^{(k)})]^{-1} F(x^{(k)}), \quad k = 0, 1, 2, \ldots \tag{1}$$

where $F'(x^{(k)})$ denotes the derivative or the Jacobian matrix of function F evaluated in the kth iteration $x^{(k)}$. In addition, this method has great importance in the study of iterative methods because it presents quadratic convergence under certain conditions and has great accessibility, that is, the region of initial estimates $x^{(0)}$ for which the method converges is wide, at least for polynomials or polynomial systems.

Based on Newton-type methods and by using different procedures, many iterative schemes for solving $F(x) = 0$ have been presented in the last years. Refs. [1,2] compile many of the methods recently designed to solve this type of problem. These books give us good overviews about this area of research.

In this paper, we use a convex combination of the methods presented by Chun et al. in [3] and Maheswari in [4]. As the mentioned schemes are designed for nonlinear equations and they have as the first step Newton's method, we use the following algebraic manipulation in order to extend the mentioned schemes to nonlinear systems:

$$\frac{f(y^{(k)})}{f(x^{(k)})} = \frac{f(y^{(k)})}{(x^{(k)}-y^{(k)})f'(x^{(k)})} = \frac{f(y^{(k)}) - f(x^{(k)}) + f(x^{(k)})}{(x^{(k)}-y^{(k)})f'(x^{(k)})} = -\frac{[x^{(k)},y^{(k)};f]}{f'(x^{(k)})} + 1.$$

Therefore, the parametric family of iterative methods for solving nonlinear systems that we propose has the following iterative expression:

$$\begin{cases} y^{(k)} = x^{(k)} - F'(x^{(k)})^{-1}F(x^{(k)}), \\ H(x^{(k)},y^{(k)},\gamma) = I + \frac{\gamma}{2}I + (1-\gamma)B_k^{-1} - (1-\gamma)B_k(2I - B_k) - \frac{\gamma}{2}F'(x^{(k)})^{-1}F'(y^{(k)}) \\ x^{(k+1)} = x^{(k)} - H(x^{(k)},y^{(k)},\gamma)F'(x^{(k)})^{-1}F(x^{(k)}) \\ \text{for } k = 0,1,2,\ldots, \end{cases} \quad (2)$$

where $x^{(0)}$ is the initial estimation, $B_k = F'(x^{(k)})^{-1}P^{(k)}$ and $P^{(k)} = [x^{(k)},y^{(k)};F]$ is the divided difference operator defined as

$$[x,y;F](x-y) = F(x) - F(y), \quad x,y \in \mathbb{R}^n.$$

The rest of the paper is organized as follows: Section 2 is devoted to analyze the convergence of family (2) in terms of the values of parameter γ. In Section 3, we study the dynamical behavior of the class on quadratic polynomials in the context of scalar equations. This study allows for selecting the members that are more stable in the family. In the numerical section, (Section 4), we apply the proposed class on different examples such as the Hammerstein integral equation and the Fisher's equation in order to confirm the theoretical results obtained in Sections 2 and 3. We finish the work with some conclusions and the references used in it.

2. Convergence Analysis

Let us consider function $F : D \subseteq \mathbb{R}^n \to \mathbb{R}^n$, differentiable in the convex set $D \subset \mathbb{R}^n$ which contains a solution $\bar{x}$ of the nonlinear equation $F(x) = 0$. From the Genochi–Hermite formula (see [5]) of the divided difference operator

$$[x+h,x;F] = \int_0^1 F'(x+th)dt \quad (3)$$

and by performing the Taylor's expansion of $F'(x+th)$ on the point x and integrating, we obtain the following development:

$$[x+h,x;F] = F'(x) + \frac{1}{2}F''(x)h + \frac{1}{6}F'''(x)h^2 + O(h^3), \quad (4)$$

which we will use in the proof of the following result, when the order of convergence of family is established.

Theorem 1. *Let $F : D \subseteq \mathbb{R}^n \longrightarrow \mathbb{R}^n$ be a sufficiently Fréchet differentiable function in a convex neighborhood D of $\bar{x}$, being $F(\bar{x}) = 0$. We suppose the Jacobian matrix $F'(x)$ is continuous and non-singular in $\bar{x}$. Then, taking an initial estimate $x^{(0)}$ close enough to $\bar{x}$, the sequence of iterates $\{x^{(k)}\}$ generated with family (2) converges to $\bar{x}$ with the following error equation:*

$$e_{k+1} = \frac{\gamma}{2}(C_3 + 4C_2^2)e_k^3 + (\gamma C_4 + (4-13\gamma)C_2^3 + 3\gamma C_2 C_3 + (-1+\frac{5}{2}\gamma)C_3 C_2)e_k^4 + O(e_k^5), \quad (5)$$

where $C_j = \frac{1}{j!} F'(\alpha)^{-1} F^{(j)}(\alpha) \in L_j(\mathbb{R}^n, \mathbb{R}^n)$, $L_j(\mathbb{R}^n, \mathbb{R}^n)$ being the set of j-linear functions of bounded functions, $j = 2, 3, \ldots$ and $e_k = x^{(k)} - \bar{x}$. In the particular case in which $\gamma = 0$, the error equation is

$$e_{k+1} = (4C_2^3 - C_3 C_2) e_k^4 + O(e_k^5), \tag{6}$$

and so the method has an order of convergence four.

Proof. We consider the Taylor's expansion of $F(x^{(k)})$ around $\bar{x}$:

$$F(x^{(k)}) = \Gamma\left(e_k + C_2 e_k^2 + C_3 e_k^3 + C_4 e_k^4 + C_5 e_k^5 + O(e_k^6)\right), \tag{7}$$

where $\Gamma = F'(\bar{x})$, $e_k = x^{(k)} - \bar{x}$ and $C_j = \frac{F'(\bar{x})^{-1} F^{(j)}(\bar{x})}{j!} \in L_j(\mathbb{R}^n, \mathbb{R}^n), j = 2, 3, \ldots$

In a similar way, the derivatives of $F(x^{(k)})$ around $\bar{x}$ take the form:

$$\begin{aligned}
F'(x^{(k)}) &= \Gamma\left[I + 2C_2 e_k + 3C_3 e_k^2 + 4C_4 e_k^3 + 5C_5 e_k^4\right] + O(e_k^5), \\
F''(x^{(k)}) &= \Gamma\left[2C_2 + 6C_3 e_k + 12C_4 e_k^2 + 20C_5 e_k^3\right] + O(e_k^4), \\
F'''(x^{(k)}) &= \Gamma\left[6C_3 + 24C_4 e_k + 60C_5 e_k^2\right] + O(e_k^3).
\end{aligned} \tag{8}$$

From the development of $F'(x^{(k)})$ around $\bar{x}$, we calculate the inverse

$$F'(x^{(k)})^{-1} = \left[I + X_2 e_k + X_3 e_k^2 + X_4 e_k^3 + X_5 e_k^4\right] \Gamma^{-1} + O(e_k^5), \tag{9}$$

with X_2, X_3, X_4 and X_5 satisfying $\left[F'(x^{(k)})\right]^{-1} F'(x^{(k)}) = I$.

Therefore,

- $X_2 = -2C_2$,
- $X_3 = 4C_2^2 - 3C_3$,
- $X_4 = -8C_2^3 + 6C_2 C_3 + 6C_3 C_2 - 4C_4$,
- $X_5 = 16C_2^4 + 9C_3^2 + 8C_2 C_4 + 8C_4 C_2 - 12C_2^2 C_3 - 12C_2 C_3 C_2 - 12C_3 C_2^2 - 5C_5$.

Applying (7) and (9), we obtain

$$F'(x^{(k)})^{-1} F(x^{(k)}) = e_k - C_2 e_k^2 + (-2C_3 + 2C_2^2) e_k^3 + (-3C_4 + 4C_2 C_3 + 3C_3 C_2 - 4C_2^3) e_k^4 + O(e_k^5). \tag{10}$$

Then, we obtain the error equation of the first step of the parametric family (2):

$$\begin{aligned}
y^{(k)} - \bar{x} &= x^{(k)} - \bar{x} - F'(x^{(k)})^{-1} F(x^{(k)}) = \\
&= C_2 e_k^2 + (2C_3 - 2C_2^2) e_k^3 + (3C_4 - 4C_2 C_3 - 3C_3 C_2 + 4C_2^3) e_k^4 + O(e_k^5).
\end{aligned} \tag{11}$$

Substituting this expression in the Taylor expansion of $F(y^{(k)})$ around $\bar{x}$, we get:

$$F(y^{(k)}) = \Gamma\left[C_2 e_k^2 + (2C_3 - 2C_2^2) e_k^3 + (3C_4 - 4C_2 C_3 - 3C_3 C_2 + 5C_2^3) e_k^4\right] + O(e_k^5). \tag{12}$$

Furthermore,

$$F'(y^{(k)}) = \Gamma[I + 2C_2^2 e_k^2 + (4C_2 C_3 - 4C_2^3) e_k^3 + (6C_2 C_4 - 8C_2^2 C_3 - 6C_2 C_3 C_2 + 8C_2^4 + 3C_3 C_2^2) e_k^4] + O(e_k^5). \tag{13}$$

Multiplying expressions (9) and (13), we obtain:

$$\begin{aligned}
F'(x^{(k)})^{-1} F'(y^{(k)}) = \ & I - 2C_2 e_k + (-3C_3 + 6C_2^2) e_k^2 + (-4C_4 + 10C_2 C_3 + 6C_3 C_2 - 16C_2^3) e_k^3 + \\
& + (-5C_5 + 14C_2 C_4 + 9C_3^2 - 28C_2^2 C_3 + 8C_4 C_2 - 18C_2 C_3 C_2 - 15C_3 C_2^2 + 40C_2^4) e_k^4 + \\
& + O(e_k^5).
\end{aligned} \tag{14}$$

To obtain the development of the divided difference operator of (2), we use the Taylor series expansion of (4), considering in this case $x + h = y$ and, so, $h = y - x = -F'(x^{(k)})^{-1}F(x^{(k)})$. Therefore, substituting (8) and (10) in (4), we obtain

$$[x^{(k)}, y^{(k)}; F] = \Gamma[I + C_2 e_k + (C_3 + C_2^2)e_k^2 + (C_4 + C_3 C_2 + 2C_2 C_3 - 2C_2^3)e_k^3 + \\ + (C_5 + C_4 C_2 + 2C_3^2 - C_3 C_2^2 + 3C_2 C_4 - 4C_2^2 C_3 - 3C_2 C_3 C_2 + 4C_2^4)e_k^4] + O(e_k^5), \quad (15)$$

To calculate the inverse of this operator, we search

$$[x^{(k)}, y^{(k)}; F]^{-1} = \left[I + Y_2 e_k + Y_3 e_k^2 + Y_4 e_k^3 + Y_5 e_k^4\right]\Gamma^{-1} + O(e_k^5), \quad (16)$$

with Y_2, Y_3, Y_4 and Y_5 satisfying $[x^{(k)}, y^{(k)}; F]^{-1}[x^{(k)}, y^{(k)}; F] = I$.
Thus,

- $Y_2 = -C_2$,
- $Y_3 = -C_3$,
- $Y_4 = -C_4 - C_2 C_3 + 3C_2^3$,
- $Y_5 = -C_5 - 2C_2 C_4 - 9C_2^4 - C_3^2 + 2C_3 C_2^2 + 6C_2^2 C_3 + 5C_2 C_3 C_2$.

Now, using (9) and (15), we obtain B_k,

$$B_k = C_2^2 e_k^2 + (2C_2 C_3 + 2C_3 C_2 - 6C_2^3)e_k^3 + \\ + (3C_2 C_4 - 10C_2 C_3 C_2 - 12C_2^2 C_3 + 25C_2^4 + 4C_3^2 - 10C_3 C_2^2 + 3C_4 C_2)e_k^4 + O(e_k^5), \quad (17)$$

and using (8) and (16), we calculate B_K^{-1},

$$B_k^{-1} = I + C_2 e_k + (2C_3 - 2C_2^2)e_k^2 + (3C_4 - 4C_2 C_3 - 2C_3 C_2 + 3C_2^3)e_k^3 + \\ + (4C_5 - 6C_2 C_4 - 2C_4 C_2 - 4C_3^2 - 3C_2^4 + 2C_3 C_2^2 + 3C_2 C_3 C_2 + 6C_2^2 C_3)e_k^4 + O(e_k^5). \quad (18)$$

Substituting the expressions (10), (14), (17), and (18) in the scheme (2), we get the error equation of the parametric family

$$e_{k+1} = x^{(k+1)} - \bar{x} = \tfrac{\gamma}{2}(C_3 + 4C_2^2)e_k^3 + (\gamma C_4 + (4 - 13\gamma)C_2^3 + 3\gamma C_2 C_3 + (-1 + \tfrac{5}{2}\gamma)C_3 C_2)e_k^4 + O(e_k^5). \quad (19)$$

Finally, from the error equation, we conclude that the parametric family (2) has order 3 for all $\gamma \neq 0$ and order 4 for $\gamma = 0$, being in this last case the error equation

$$e_{k+1} = (4C_2^3 - C_3 C_2)e_k^4 + O(e_k^5). \quad (20)$$

□

In the next section, we analyze the dynamical behavior of the parametric family (2) on quadratic scalar polynomials.

3. Complex Dynamics

The dynamical analysis of (2) is performed throughout this section in terms of complex analysis. The order of convergence is not the only important criterion to study when evaluating an iterative scheme. The validity of a method also depends on other aspects such as knowing how it behaves based on the initial estimates that are taken, that is, how wide the set of initial estimations is for which the method is convergent. For this reason, it is necessary to introduce several tools that allow for a more exhaustive study.

The analysis of the dynamics of a method is becoming one of the most investigated parts within the study of iterative methods since it allows for classifying the different iterative schemes, not only from the point of view of their speed of convergence, but also analyzing its behavior based on the initial estimate taken (see, for example, [6–13]). This study allows for visualizing graphically the set of initial approximations that converge to a given root or to points that are not roots of the equation. In addition, it provides important information about the stability and reliability of the iterative method.

In this paper, we focus on studying the complex dynamic of the parametric family (2) on quadratic polynomials of the form $p(z) = (z-a)(z-b)$, where $a, b \in \mathbb{C}$. For this study, we need to present the result called the Scaling Theorem, since it allows us to conjugate the dynamical behavior of one operator with the behavior associated with another, conjugated through an affine application, that is, our operator has the same stability on all quadratic polynomials. This result will be of great use to us since we can apply the Möbius transformation on the operator $R_{p,\gamma}$ associated with our parametric family acting on $p(z)$, assuming that the conclusions obtained will be of general application for any quadratic polynomial used.

Theorem 2 (Scaling Theorem for family (2)). *Let $f(z)$ be an analytic function in the Riemann sphere $\hat{\mathbb{C}}$ and let $T(z) = \alpha z + \beta$ be an affine transformation with $\alpha \neq 0$. We consider $g(z) = \lambda (f \circ T)(z)$, $\lambda \neq 0$. Let $R_{f,\gamma}$ and $R_{g,\gamma}$ be the fixed point operators of the family (2) associated with the functions f and g, respectively, that is to say,*

$$R_{f,\gamma}(z) = z + \left[-\frac{\gamma}{2}\left(3 - \frac{f'(y)}{f'(z)}\right) + (1-\gamma)\left(\frac{1}{\frac{f(y)}{f(z)} - 1} - \left(\frac{f(y)}{f(z)}\right)^2\right)\right] \frac{f(z)}{f'(z)}, \quad (21)$$

$$R_{g,\gamma}(z) = z + \left[-\frac{\gamma}{2}\left(3 - \frac{g'(y)}{g'(z)}\right) + (1-\gamma)\left(\frac{1}{\frac{g(y)}{g(z)} - 1} - \left(\frac{g(y)}{g(z)}\right)^2\right)\right] \frac{g(z)}{g'(z)}, \quad (22)$$

where $y = z - \frac{f(z)}{f'(z)}$ and $z \in \mathbb{C}$. Then, $R_{f,\gamma}$ is analytically conjugated to $R_{g,\gamma}$ through T, that is to say,

$$(T \circ R_{g,\gamma} \circ T^{-1})(z) = R_{f,\gamma}(z).$$

Proof. Taking into account that $T(x-y) = T(x) - T(y) + \beta$, $T(x+y) = T(x) + T(y) - \beta$ and $g'(z) = \alpha \lambda f'(T(z))$, so

$$(T \circ R_{g,\gamma} \circ T^{-1})(z) = T(R_{g,\gamma}(T^{-1})(z)) =$$
$$= T\left(T^{-1}(z) + \left[-\frac{\gamma}{2}\left(3 - \frac{g'(T^{-1}(y))}{g'(T^{-1}(z))}\right) + (1-\gamma)\left(\frac{1}{\frac{g(T^{-1}(y))}{g(T^{-1}(z))} - 1} - \left(\frac{g(T^{-1}(y))}{g(T^{-1}(z))}\right)^2\right)\right] \frac{g(T^{-1}(z))}{g'(T^{-1}(z))} \right),$$

where $y = z - \frac{g(z)}{g'(z)}$, $T(T^{-1}(z)) = z$ and

$$T\left(T^{-1}(y)\right) = T\left(T^{-1}(z) - \frac{g(T^{-1}(z))}{g'(T^{-1}(z))}\right) = T\left(T^{-1}(z) - \frac{f(z)}{\alpha f'(z)}\right) = z - T\left(\frac{f(z)}{\alpha f'(z)}\right) + \beta = z - \frac{f(z)}{f'(z)} = y.$$

Therefore, substituting these equalities and simplifying, we have

$$(T \circ R_{g,\gamma} \circ T^{-1})(z) =$$
$$= T\left(T^{-1}(z) + \left[-\frac{\gamma}{2}\left(3 - \frac{f'(y)}{f'(z)}\right) + (1-\gamma)\left(\frac{1}{\frac{f(y)}{f(z)} - 1} - \left(\frac{f(y)}{f(z)}\right)^2\right)\right] \frac{f(z)}{\alpha f'(z)} \right)$$
$$= z + T\left(-\frac{\gamma}{2}\left(3 - \frac{f'(y)}{f'(z)}\right) + (1-\gamma)\left(\frac{1}{\frac{f(y)}{f(z)} - 1} - \left(\frac{f(y)}{f(z)}\right)^2\right) \frac{f(z)}{\alpha f'(z)} \right) - \beta$$
$$= z + T\left[-\frac{\gamma}{2}\left(3 - \frac{f'(y)}{f'(z)}\right) + (1-\gamma)\left(\frac{1}{\frac{f(y)}{f(z)} - 1} - \left(\frac{f(y)}{f(z)}\right)^2\right)\right] \frac{f(z)}{\alpha f'(z)},$$

then $(T \circ R_{g,\gamma} \circ T^{-1})(z) = R_{f,\gamma}(z)$, that is to say, $R_{f,\gamma}$ and $R_{g,\gamma}$ are analytically conjugated by $T(z)$. □

Now, we can apply the Möbius transformation on the operator associated with the parametric family (2) in order to obtain an operator that does not depend on the constants a and b and, thus, be able to study the dynamical behavior of this family for any quadratic polynomial. The Möbius transformation, in this case, is $h(z) = \frac{z-a}{z-b}$ and has the following properties:

(i) $h(\infty) = 1$ **(ii)** $h(a) = 0$ **(iii)** $h(b) = \infty$.

The fixed-point rational operator of family (2) on $p(z)$ has the expression

$$O_\gamma(z) = (h \circ R_{p,\gamma} \circ h^{-1})(z) = \frac{z^3(2\gamma z^2 + 3\gamma z + 2\gamma + z^5 + 5z^4 + 10z^3 + 9z^2 + 4z)}{2\gamma z^5 + 3\gamma z^4 + 2\gamma z^3 + 4z^4 + 9z^3 + 10z^2 + 5z + 1}. \quad (23)$$

We can also deduce from (23) that the order of the methods for quadratic polynomials is 3 when $\gamma \neq 0$ and 4 when $\gamma = 0$.

3.1. Fixed Points

The orbit of a point $z \in \mathbb{C}$ is defined (see, for example, [14,15]) as the set of the successive applications of the rational operator, i.e.,

$$\{z, O_\gamma(z), O_\gamma^2(z), \ldots\}.$$

The performance of the orbit of z is deduced attending to its asymptotic behavior. A point x^T is said to be T-periodic if $O_\gamma^T(z) = z$ and $O_\gamma^t(z) \neq z$, for $t < T$. For $T = 1$, this point is a fixed point.

Therefore, a fixed point is one that is kept invariant by the operator O_γ, that is, it is one that satisfies the equation $O_\gamma(z) = z$. All the roots of the quadratic polynomial are, of course, fixed points of the O_γ operator. However, it may happen that fixed points appear that do not correspond to any root; we call these points strange fixed points. These points are not desirable from a numerical point of view because when an initial estimate is taken that is in the neighborhood of a strange fixed point, there is a possibility that the numerical method will converge to it, that is, to a point that is not a solution of the equation. Strange fixed points often appear when iterative methods are analyzed and their presence can show the instability of the method.

Fixed points can be classified according to the behavior of the derivative operator on them; thus, a fixed point z^* can be:

- Repulsor, if $|O'_\gamma(z^*)| > 1$;
- Parabolic, if $|O'_\gamma(z^*)| = 1$;
- Attracting, if $|O'_\gamma(z^*)| < 1$;
- Superattracting, if $|O'_\gamma(z^*)| = 0$.

Moreover, the basin of attraction $\mathcal{A}(z^*)$ of an attracting fixed point z^* is the set of initial guesses whose orbits tend to z^*. Therefore, the set of points whose orbit tends to an attracting fixed point defines the Fatou set $\mathcal{F}(O_\gamma)$, while its complement is the Julia set $\mathcal{J}(O_\gamma)$.

In what follows, we study what are the fixed points of operator O_γ and their character depending on the value of parameter γ. The proof of the following result is straightforward, as it only needs to solve the equation $O_\gamma(z) = z$.

Proposition 1. *By analyzing the equation $O_\gamma(z) = z$, one obtains the following statements:*

(i) $z = 0$ and $z = \infty$ are superattracting fixed points for each value of γ.

(ii) $z = 1$ is a strange fixed point when $\gamma \neq -\dfrac{29}{7}$.

(iii) the roots of polynomial

$$k(t) = 1 + 6t + (16 - 2\gamma)t^2 + (21 - 3\gamma)t^3 + (16 - 2\gamma)t^4 + 6t^5 + t^6, \quad (24)$$

which we denote by $Ex_i(\gamma)$, where $i = 1, 2, \ldots, 6$, are also strange fixed points for each value of γ.

We need the expression of the differentiated operator to analyze the stability of the fixed points and to obtain the critical points:

$$O'_\gamma(z) = \frac{z^2(z+1)^4 \left(\gamma \left(6z^6 + 8z^5 + 7z^4 + 7z^2 + 8z + 6 \right) + z \left(16z^4 + 41z^3 + 60z^2 + 41z + 16 \right) \right)}{\left(2\gamma z^5 + (3\gamma + 4)z^4 + (2\gamma + 9)z^3 + 10z^2 + 5z + 1 \right)^2},$$

It is clear that 0 and ∞ are always superattracting fixed points because they come from the roots of the polynomial, and the order of the iterative methods is higher than 2, but the stability of the other fixed points can change depending on the values of parameter γ.

Proposition 2. *The character of the strange fixed point $z = 1$ is as follows:*

(a) *If $\gamma = -\frac{29}{7}$, then $z = 1$ is not a strange fixed point.*

(b) *If $Re(\gamma) < -\frac{125}{7}$ or $Re(\gamma) > \frac{67}{7}$, then $z = 1$ is an attracting point.*

(c) *If $Re(\gamma) \in \left[-\frac{125}{7}, \frac{67}{7} \right]$ and $Im(\gamma)^2 + \left(Re(\gamma) + \frac{29}{7} \right)^2 > \frac{9216}{49}$, then $z = 1$ is an attracting point.*

(d) *$z = 1$ cannot be a superattracting point.*

(e) *If $Re(\gamma) \in \left[-\frac{125}{7}, \frac{67}{7} \right]$ and $Im(\gamma)^2 + \left(Re(\gamma) + \frac{29}{7} \right)^2 = \frac{9216}{49}$, then $z = 1$ is a parabolic point.*

(f) *In another case, $z = 1$ is the repulsor.*

Proof. We obtain that

$$|O'_\gamma(1)| = \left| \frac{96}{7\gamma + 29} \right|.$$

It is not difficult to check that $|O'_\gamma(1)|$ cannot be 0, so $z = 1$ cannot be a superattractor, and, when $\gamma = -\frac{29}{7}$, $z = 1$ is not a fixed point.

Now, we are going to study when $z = 1$ is an attracting point. It is easy to check that $|O'_\gamma(1)| < 1$ is equivalent to $96^2 < |29 + 7\gamma|^2$. Rewriting the last expression, we obtain the following inequality:

$$8375 < 406 Re(\gamma) + 49 Re(\gamma)^2 + 49 Im(\gamma)^2.$$

Let us see when this inequality is verified. When $8375 - 406 Re(\gamma) - 49 Re(\gamma)^2 < 0$, that is, $\left(Re(\gamma) - \frac{67}{7} \right) \left(Re(\gamma) + \frac{125}{7} \right) > 0$, $z = 1$ is an attracting point, so we obtain that $z = 1$ is an attracting point when $Re(\gamma) > \frac{67}{7}$ or $Re(\gamma) < -\frac{125}{7}$. When we have $Re(\gamma) \in \left[-\frac{125}{7}, \frac{67}{7} \right]$, we need $Im(\gamma)$ to satisfy $8375 < 406 Re(\gamma) + 49 Re(\gamma)^2 + 49 Im(\gamma)^2$, for $z = 1$ being a superattractor.

We are going to study when $z = 1$ is a parabolic point. $z = 1$ will be a parabolic point when $8375 - 406 Re(\gamma) - 49 Re(\gamma)^2 = 49 Im(\gamma)^2$, that is, $z = 1$ is a parabolic point when $Re(\gamma) \in \left[-\frac{125}{7}, \frac{67}{7} \right]$ and $49 Im(\gamma)^2 = -Re(\gamma)^2 - 406 Re(\gamma) + 8375$. □

Now, we establish the stability of the strange fixed points that are roots of the polynomial (24). To do this, we calculate these roots noting that this polynomial is a sixth degree symmetric polynomial, that is, it is a polynomial that can be reduced to a third degree one, and that satisfies the following properties:

- $t = 0$ is not the root;
- if $t = \alpha$ is the root, $t = \frac{1}{\alpha}$ is also the root.

Performing the reduction of (24), we obtain:

$$1 + 6t + (16 - 2\gamma)t^2 + (21 - 3\gamma)t^3 + (16 - 2\gamma)t^4 + 6t^5 + t^6 = 0$$
$$\leftrightarrow (\frac{1}{t^3} + t^3) + 6(\frac{1}{t^2} + t^2) + (16 - 2\gamma)(\frac{1}{t} + t) + 21 - 3\gamma = 0$$
$$\leftrightarrow z^3 + 6z^2 + (13 - 2\gamma)z + 9 - 3\gamma = 0,$$

where $z = \frac{1}{t} + t$, $z^2 - 2 = \frac{1}{t^2} + t^2$ and $z^3 - 3z = \frac{1}{t^3} + t^3$. Now, we calculate the roots of this polynomial and obtain:

$$z_1(\gamma) = \frac{\sqrt[3]{\frac{2}{3}}(2\gamma - 1)}{\sqrt[3]{-9\gamma + \sqrt{3\gamma((75 - 32\gamma)\gamma - 78)} + 93} + 9} + \frac{\sqrt[3]{-9\gamma + \sqrt{3\gamma((75 - 32\gamma)\gamma - 78)} + 93} + 9}{\sqrt[3]{2}3^{2/3}} - 2,$$

$$z_2(\gamma) = \frac{\sqrt[3]{-\frac{2}{3}}(1 - 2\gamma)}{\sqrt[3]{-9\gamma + \sqrt{3\gamma((75 - 32\gamma)\gamma - 78)} + 93} + 9} + \frac{(-1)^{2/3}\sqrt[3]{-9\gamma + \sqrt{3\gamma((75 - 32\gamma)\gamma - 78)} + 93} + 9}{\sqrt[3]{2}3^{2/3}} - 2,$$

$$z_3(\gamma) = \frac{(-1)^{2/3}\sqrt[3]{\frac{2}{3}}(2\gamma - 1)}{\sqrt[3]{-9\gamma + \sqrt{3\gamma((75 - 32\gamma)\gamma - 78)} + 93} + 9} - \frac{\sqrt[3]{-\frac{1}{2}}\sqrt[3]{-9\gamma + \sqrt{3\gamma((75 - 32\gamma)\gamma - 78)} + 93} + 9}{3^{2/3}} - 2.$$

To calculate the roots of polynomial (24) from the $z_i(\gamma)$, $i = 1, 2, 3$, we undo the variable change since $t = \frac{z_i(\gamma) \pm \sqrt{z_i(\gamma)^2 - 4}}{2}$. Therefore, we obtain the roots of the sixth degree polynomial, which are conjugated two by two

$$Ex_1(\gamma) = \frac{z_1(\gamma) + \sqrt{z_1(\gamma)^2 - 4}}{2}, \quad Ex_2(\gamma) = \frac{z_1(\gamma) - \sqrt{z_1(\gamma)^2 - 4}}{2},$$
$$Ex_3(\gamma) = \frac{z_2(\gamma) + \sqrt{z_2(\gamma)^2 - 4}}{2}, \quad Ex_4(\gamma) = \frac{z_2(\gamma) - \sqrt{z_2(\gamma)^2 - 4}}{2},$$
$$Ex_5(\gamma) = \frac{z_3(\gamma) + \sqrt{z_3(\gamma)^2 - 4}}{2}, \quad Ex_6(\gamma) = \frac{z_3(\gamma) - \sqrt{z_3(\gamma)^2 - 4}}{2}.$$

Now, we study when the roots of the polynomial (24) are superattractors. For them, we solve $|O'_\gamma(Ex_i(\gamma))| = 0$ for all $i = 1, \ldots, 6$, and we get the following relevant values of γ:

- $\gamma_1 = 0.8114608325277108$,
- $\gamma_2 = 5.5908453191613585$,
- $\gamma_3 = 0.7671008924094337 + 0.7784254153980097i$,
- $\gamma_4 = 0.7671008924094337 - 0.7784254153980097i$.

Next, we are going to study the character of the fixed points by analyzing those values of γ close to the values of the parameter for which some $Ex_i(\gamma)$ is a supertractor. To do this, we study how $|O'_\gamma(Ex_i(\gamma))|$ behaves near the four previous values, and we obtain regions where some of the roots will be attractors. These regions are represented in Figure 1.

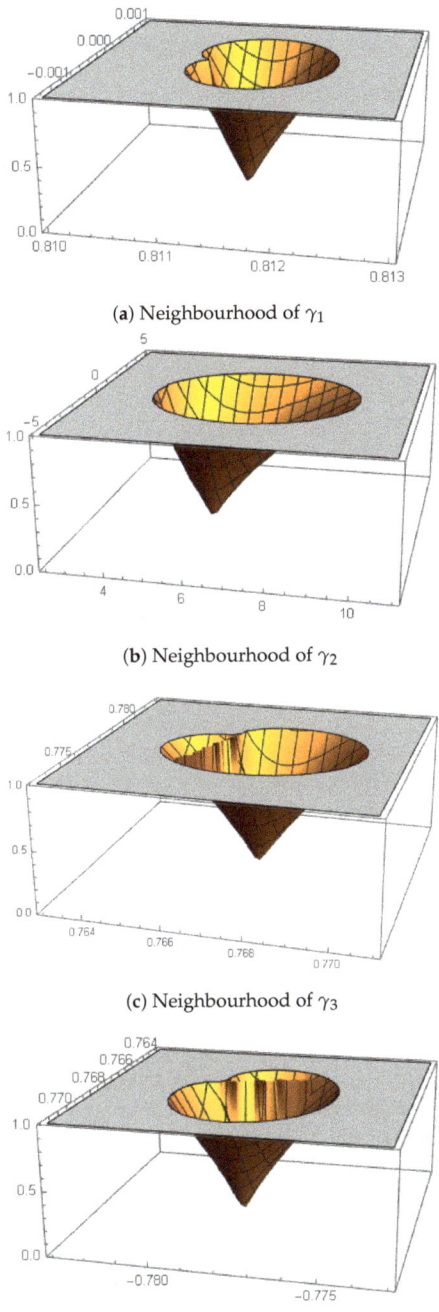

(a) Neighbourhood of γ_1

(b) Neighbourhood of γ_2

(c) Neighbourhood of γ_3

(d) Neighbourhood of γ_4

Figure 1. Character of the roots of polynomial $k(t)$: (**a**) γ_1, (**b**) γ_2, (**c**) γ_3, (**d**) γ_4.

3.2. Critical Points

The relevance of knowing that the free critical points (that is, critical points different from the roots of the polynomial) is based on this known fact: each invariant Fatou component contains, at least, one critical point. Operator $O_\gamma(z)$ has as critical points $z = 0$, $z = -1$, $z = \infty$, and the roots of the polynomial

$$q(t) = 6\gamma + (16 + 8\gamma)t + (41 + 7\gamma)t^2 + 60t^3 + (41 + 7\gamma)t^4 + (16 + 8\gamma)t^5 + 6\gamma t^6,$$

which we denote by $Zx_i(\gamma)$, where $i = 1, \ldots, 6$.

Let us remark that $z = -1$ is a preimage of the fixed point $z = 1$. We can see that $q(t)$ is a symmetric polynomial, so we can obtain the roots of $q(t)$ obtaining roots of a polynomial of degree 3. The polynomial reduced of $q(t)$ is the following one that we obtain analogously to the polynomial (24):

$$\hat{q}(t) = 6\gamma t^3 + (16 + 8\gamma)t^2 + (41 - 11\gamma)t + 28 - 16\gamma.$$

In order to calculate the roots z of $q(t)$, we need to obtain the roots of $\hat{q}(t)$ and apply the following expression to them $\frac{z \pm \sqrt{z^2 - 4}}{2}$. Thus, the roots of $q(t)$ are conjugated.

Now, we are going to study the asymptotic behavior of the critical points to establish if there are different convergence basins than those generated by the roots. For the free critical point -1, we have $O_\gamma(-1) = 1$, who is a strange fixed point, so the parameter plane associated with this critical point is not significative, since we know the stability of $z = 1$.

The other free critical points are roots of a polynomial that depends on γ; for that, we draw the parameter planes. As we have that the roots are conjugated, we will only draw three planes. We use as an initial estimate a free critical point that depends on γ. We establish a mesh in the complex plane of 500×500 points. Each point of the mesh corresponds to a parameter value. In each of them, the rational function is iterated to obtain the orbit of the critical point as a function of γ. If that orbit converges to $z = 0$ or to $z = \infty$ in less than 40 iterations, that point of the mesh is painted red; otherwise, the point appears in black.

As we can see, there are many values of the parameter γ that would result in a method in which the free critical points converge to one of the two roots. As it is observed in Figure 2, they are located in the red area on the right side of the plane. Moreover, some black areas can be identified as the regions of stability of those fixed points that can be attracting, such as Figure 1b, whose stability region appears in black on the right side of Figure 2c.

Now, we select some stable (in red in parameter planes) and unstable values of γ (in black) in order to show their performance.

In the case of dynamical planes, the value of the parameter γ is fixed. Each point in the complex plane is considered as a starting point of the iterative scheme, and it is painted in different colors depending on the point that it has converged to. In this case, we paint in blue points what converged to ∞, and in orange points what converged to 0. These dynamical planes have been generated with a mesh of 500×500 points and a maximum of 40 iterations per point. We mark strange fixed points with white circles, the fixed point $z = 0$ with a white star, and free critical points with white squares (again, the routines used appear in [6]).

One value of the parameter that would be an interesting value is $\gamma = 0$ because it is the only one that obtains order 4. In that case, we obtain the dynamical plane that we can see in Figure 3a. In this case, two free critical points are in each basin of attraction, and the strange fixed points are in the boundary of both basins of attraction, so they are repulsive. In that case, the method is stable, and, as we can see, almost every point converges to 0 or ∞ (Let us notice that, in practice, any initial estimation taken in the Julia set will converge to 0 or to ∞, due to the rounding error).

Other value for the parameter that we study is $\gamma = 1$, Figure 3b. As we can see, this dynamical plane is similar to that of $\gamma = 0$, but, in this case, we obtain less free critical points and less strange fixed points, due to the simplification of the rational function for this value of γ.

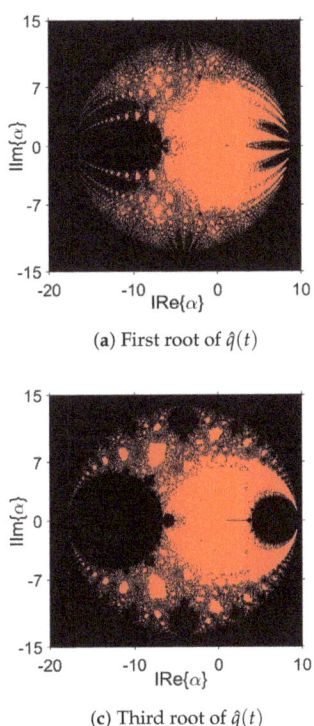

(a) First root of $\hat{q}(t)$

(b) Second root of $\hat{q}(t)$

(c) Third root of $\hat{q}(t)$

Figure 2. Parameter planes of $O_\gamma(z)$.

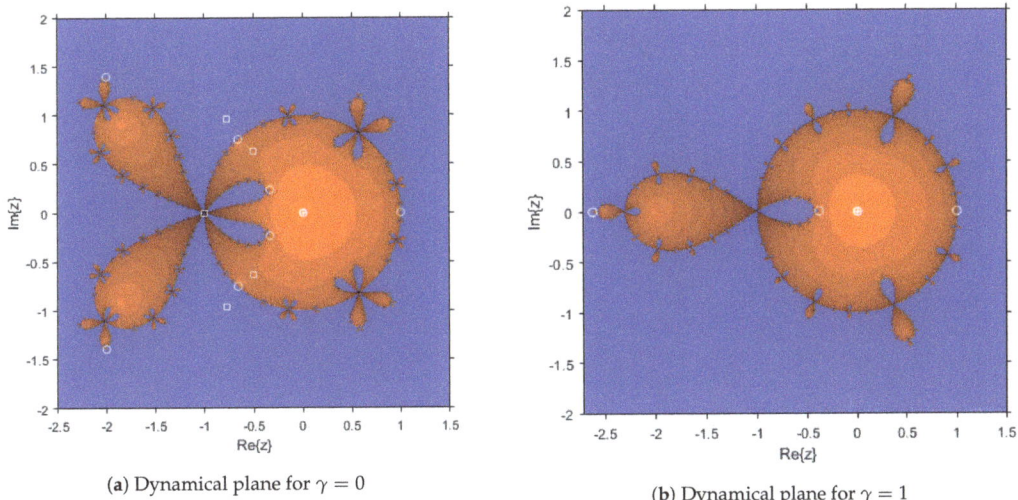

(a) Dynamical plane for $\gamma = 0$

(b) Dynamical plane for $\gamma = 1$

Figure 3. Dynamical planes of $\gamma = 0$ and $\gamma = 1$.

Carrying out numerous experiments, we have realized that the simplest dynamics is that of the methods with parameter $\gamma = 0$ and $\gamma = 1$. Next, we will see other dynamical planes associated with other values of the parameter γ. Some of these planes do not have a bad dynamics, although it is not as simple as the previous ones. This is the case of $\gamma = 2$, Figure 4b, or the case of $\gamma = 2i$, Figure 4a.

However, values such as $\gamma = -10 + i$, $\gamma = -5$ or $\gamma = -\frac{29}{7}$ present a dynamical plane with the same number of basins of attraction but with more complicated performance. We can see some of these dynamical planes in Figures 5a,b and 6a. There are also parameter values for which the number of basins of attraction increases, for example, $\gamma = 5$ (Figure 6b). These cases should be avoided since our method may not converge to the roots and may end up converging to other points.

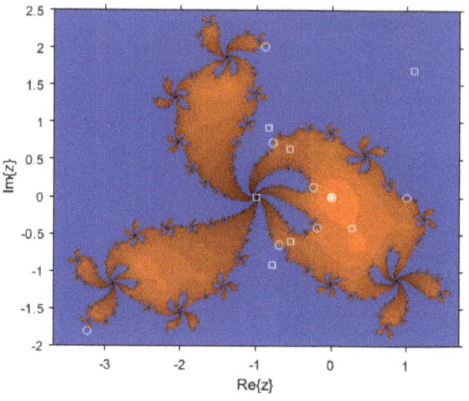

(**a**) Dynamical plane for $\gamma = 2i$

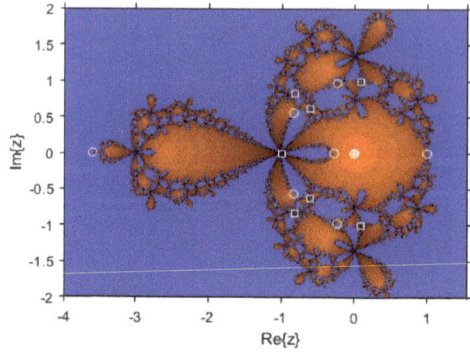

(**b**) Dynamical plane for $\gamma = 2$

Figure 4. Dynamical planes of $\gamma = 2i$ and $\gamma = 2$.

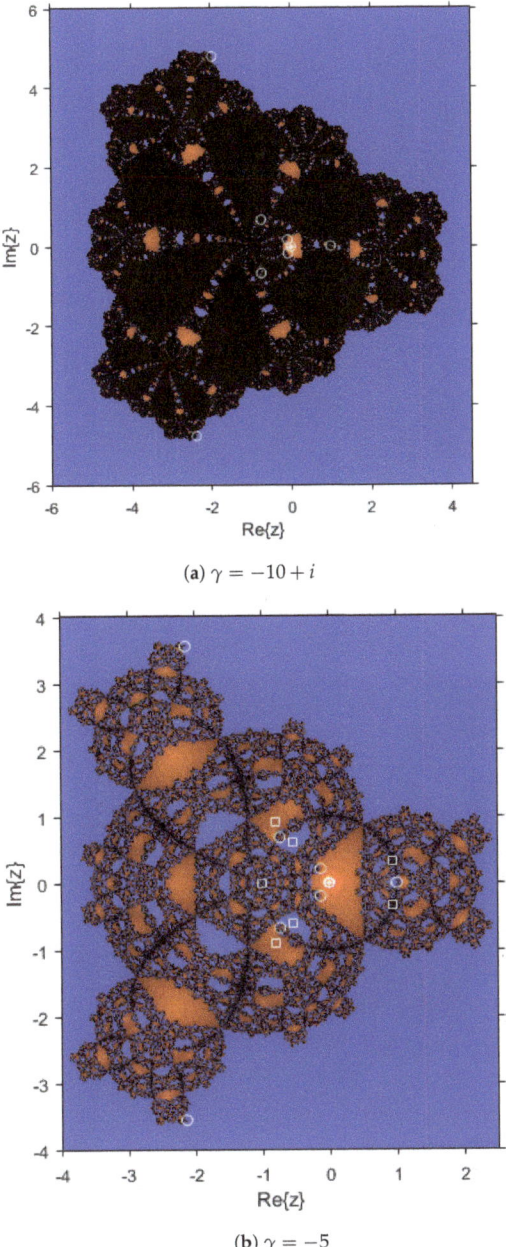

(a) $\gamma = -10 + i$

(b) $\gamma = -5$

Figure 5. Dynamical planes of $\gamma = -10 + i$ and $\gamma = -5$.

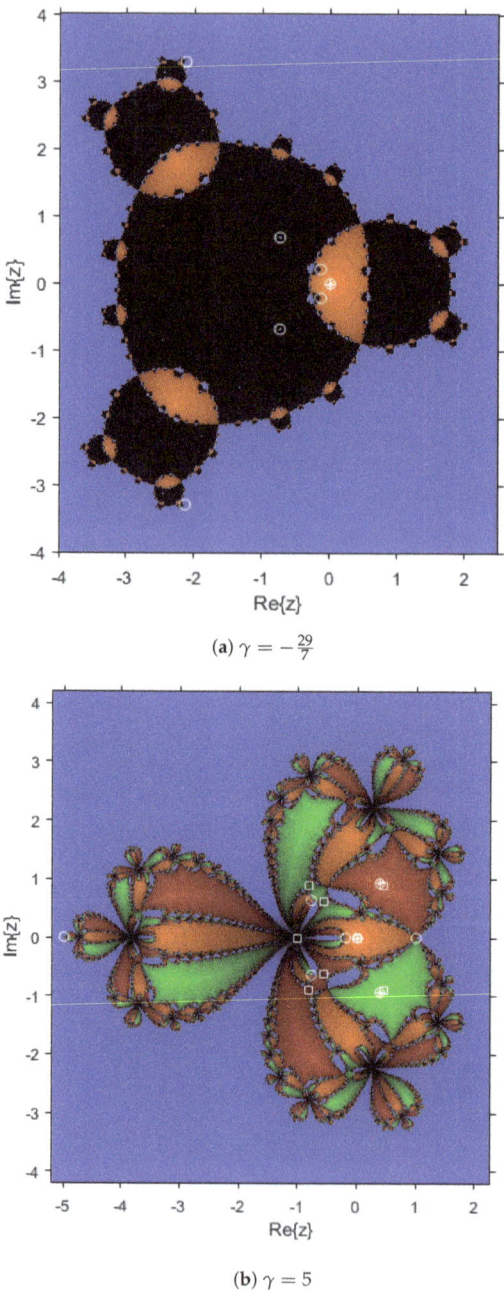

(a) $\gamma = -\frac{29}{7}$

(b) $\gamma = 5$

Figure 6. Dynamical planes of $\gamma = -\frac{29}{7}$ and $\gamma = 5$.

4. Numerical Experiments

In this section, we compare different iterative methods of the parametric family (2), solving two classical problems of applied mathematics: the Hammerstein integral equation and the Fisher partial derivative equation. We are going to use elements for the proposed class for which we have studied the dynamical plane because we want to verify that, although some of them have complicated dynamics, they can be methods that give good numerical results.

For the computational calculations, Matlab R2020b with variable precision arithmetics with 1000 digits of mantissa is used. From an initial estimation $x^{(0)}$, the different algorithms calculate iterations until the stoping criterium $\|x^{(k+1)} - x^{(k)}\| < tol$ is satisfied.

For the different examples and algorithms, we compare the approximation obtained, the norm of the function in the last iterate, the norm of the distance between the last two approximations, the number of iterations needed to satisfy the required tolerance, the computational time and the approximate computational convergence order (ACOC), defined by Cordero and Torregrosa in [16], which has the following expression:

$$p \approx ACOC = \frac{\ln(\|x^{(k+1)} - x^{(k)}\|_2 / \|x^{(k)} - x^{(k-1)}\|_2)}{\ln(\|x^{(k)} - x^{(k-1)}\|_2 / \|x^{(k-1)} - x^{(k-2)}\|_2)}.$$

4.1. Hammerstein Equation

In this example, we consider the well-known Hammerstein integral equation (see [5]), which is given as follows:

$$x(s) = 1 + \frac{1}{5} \int_0^1 F(s,t) x(t)^3 dt, \qquad (25)$$

where $x \in \mathbb{C}[0,1]$, $s, t \in [0,1]$ and the kernel F is

$$F(s,t) = \begin{cases} (1-s)t & t \leq s, \\ s(1-t) & s \leq t. \end{cases}$$

We transform the above equation into a finite-dimensional nonlinear problem by using the Gauss–Legendre quadrature formula given as $\int_0^1 f(t) dt \approx \sum_{j=1}^{7} \omega_j f(t_j)$, where the nodes t_j and the weights ω_j are determined for $n = 7$ by the Gauss–Legendre quadrature formula. In this case, the nodes and the weights are in Table 1.

Table 1. Weights and nodes of the Gauss–Legendre quadrature.

i	Weight ω_i	Abscissa t_i
1	0.0647424831	0.0254460438
2	0.1398526957	0.1292344072
3	0.1909150252	0.2970774243
4	0.2089799185	0.5
5	0.1909150252	0.7029225757
6	0.1398526955	0.8707655928
7	0.0647424831	0.9745539561

By denoting the approximations of $x(t_i)$ by x_i ($i = 1, \ldots, 7$), one gets the system of nonlinear equations:

$$5x_i - 5 - \sum_{j=1}^{7} a_{ij} x_j^3 = 0,$$

where $i = 1, \ldots, 7$ and

$$a_{ij} = \begin{cases} \omega_j t_j (1 - t_i) & j \leq i, \\ \omega_j t_i (1 - t_j) & i < j. \end{cases}$$

Starting from an initial approximation $x^{(0)} = (-1, \ldots, -1)^T$ and with a tolerance of $tol = 10^{-15}$, we run the parametric family for different values of the parameter γ. The numerical results are shown in Table 2.

Table 2. Hammerstein results for different parameters.

Parameter γ	$v\|\|F(x^{(k+1)})\|\|_2$	$\|\|x^{(k+1)} - x^{(k)}\|\|_2$	Iteration	ACOC	Time
0	5.40317×10^{-46}	1.82600×10^{-184}	4	3.99753	38.0469
1	1.1060×10^{-20}	7.36657×10^{-63}	4	2.85884	33.8594
$-10+i$	4.02251×10^{-45}	3.70484×10^{-135}	6	2.98801	84.8594
$-29/7$	1.73829×10^{-32}	1.18363×10^{-97}	5	2.98095	44.0781
-5	8.18771×10^{-29}	1.48807×10^{-86}	5	2.97987	46.2500
5	6.98712×10^{-28}	9.02414×10^{-84}	5	2.97222	36.3281
$2i$	5.87285×10^{-47}	2.22194×10^{-141}	5	2.98606	35.3281
2	5.36968×10^{-17}	8.93118×10^{-52}	4	2.93508	25.8750

In all cases, we obtain as an approximation of the solution of Equation (25) the following vector $x^{(k+1)} = (1.0026875, 1.0122945, 1.0229605, 1.0275616, 1.0229605, 1.0122945, 1.0026875)^T$.

In the case of the Hammerstein integral equation, we see that the numerical results of the parametric family (2) for different values of γ are quite similar. The main difference observed between the methods is that the ACOC for $\gamma = 0$ is 4, and, for the rest of the methods, it is approximately 3. On the other hand, we note that the method with $\gamma = -10 + i$ needs to perform a larger number of iterations than the rest of the methods to satisfy the required tolerance, so the time it takes to approximate the solution is also longer. Finally, taking into account the columns that measure the error of the approximation, that is, $\|F(x^{(k+1)})\|_2$ and $\|x^{(k+1)} - x^{(k)}\|_2$, we see that iterative methods that get lower errors are those associated with the parameters $\gamma = 0$ and $\gamma = 2$. These results confirm the information obtained in the dynamical section.

4.2. Fisher Equation

In this second example, we are going to study the equation proposed in [17] by Fisher to model the diffusion process in population dynamics. The analytical expression of this partial derivative equation is as follows:

$$u_t(x,t) = Du_{xx}(x,t) + ru(x,t)\left(1 - \frac{u(x,t)}{p}\right), \quad x \in [a,b], \ t \geq 0, \tag{26}$$

where $D \leq 0$ is the diffusion constant, r is the level of growth of the species, and p is the carrying capacity.

In this case, we will study the Fisher equation for the values $p = 1, r = 1$, and $D = 1$ in the spatial interval $[0,1]$ and with the initial condition $u(x,0) = \text{sech}^2(\pi x)$ and null boundary conditions.

We transform the problem we just described in a set of nonlinear systems by applying an implicit method of finite differences, providing the estimated solution in the instant t_k from the estimated one in t_{k-1}. We denote the spatial step by $h = \dfrac{1}{n_x}$ and the temporal step by $k = \dfrac{T_{max}}{n_t}$, where T_{max} is the final instant and n_x and n_t are the number of subintervals in x and t, respectively. Therefore, we define a mesh of the domain $[0,1] \times [0, T_{max}]$, composed of points (x_i, t_j), as follows:

$$x_i = 0 + ih, \quad i = 0, \ldots, n_x, \quad t_j = 0 + jk, \quad j = 0, \ldots, n_t.$$

Our objective is to approximate the solution of problem (26) in these points of the mesh, solving as many nonlinear systems as there are temporary nodes t_j in the mesh. For this, we use the following finite differences:

$$u_t(x,t) \approx \frac{u(x,t) - u(x,t-k)}{k}$$

$$u_{xx}(x,t) \approx \frac{u(x+h,t) - 2u(x,t) + u(x-h,t)}{h^2}.$$

We observe that, for the time step, we use first order backward divided differences and for the spatial step they are second order centered divided differences.

By denoting $u_{i,j}$ as the approximation of the solution at (x_i, t_j), and, by replacing it in the Cauchy problem, we get the system

$$ku_{i+1,j} + (kh^2 - 2k - h^2)u_{i,j} - kh^2 u_{i,j}^2 + ku_{i-1,j} = -h^2 u_{i,j-1},$$

for $i = 1, 2, \ldots, n_x - 1$ and $j = 1, 2, \ldots, n_t$. The unknowns of this system are $u_{1,j}, u_{2,j}, \ldots, u_{n_x-1,j}$, that is, the approximations of the solution in each spatial node for the fixed instant t_j.

In this example, we are going to work with the parameters $T_{max} = 10$, $n_x = 10$ and $n_t = 50$. As we have said, it is necessary to solve as many systems as there are temporary nodes t_j; for each of these systems, we use the parametric family (2) to approximate its solution. Thus, starting from an initial approximation $u_{i,0} = \text{sech}^2(\pi x_i)$, $i = 0, \ldots, n_x$, with a tolerance of 10^{-6}, we execute the parametric family for different values of γ so that we get Table 3.

Table 3. Fisher results for different parameters.

Parameter γ	$\|F(x^{(k)})\|_2$	$\|x^{(k+1)} - x^{(k)}\|_2$	Iteration	ACOC	Time
0	1.00166×10^{-8}	1.12488×10^{-35}	3	4.21099	213.4219
1	1.9199×10^{-16}	5.88036×10^{-50}	4	2.99609	248.7344
$-10+i$	8.08037×10^{-9}	4.65282×10^{-26}	5	3.01506	352.6563
$-29/7$	1.8002×10^{-7}	2.00583×10^{-22}	4	2.86978	247.9844
-5	1.89574×10^{-19}	2.9985×10^{-58}	5	2.99569	267.2969
5	2.4177×10^{-17}	6.2774×10^{-52}	5	2.99654	275.7344
$2i$	2.27659×10^{-11}	1.96645×10^{-34}	4	2.97846	252.8438
2	9.67264×10^{-12}	1.50906×10^{-35}	4	3.00948	231.2188

In all cases, we obtain as an approximation of the solution of problem (26) the following vector $x^{(k+1)} = (0, 4.32639, 0.708718, 0.853425, 0.918847, 0.93729, 0.918847, 0.853425, 0.708718, 0.432639, 0)^T$.

In this case, it can seen that the results are very similar, although there are subtle differences. For example, the method when $\gamma = 0$ uses a smaller number of iterations than the rest to satisfy the required tolerance, although this does not make it much faster than the rest of the methods since the difference in time is seconds. On the other hand, if we look at the time column, we can see that there is a method that stands out for its slowness; this is the case of $\gamma = -10 + i$. Again, we note that the ACOC of the methods roughly match the theoretical predictions made throughout the article. Observing the columns of the errors, we find similar results as well and that, in this case, having a higher tolerance than in the first example, no great differences are observed in these results.

5. Conclusions

A parametric family of iterative methods for solving nonlinear systems is presented. The dynamical analysis of the class on quadratic polynomials is done in order to select the members of the family with better stability properties. We prove that there exist a wide set of real and complex values of the parameter for which the corresponding methods are stable. That is, the set of initial estimations converging to the roots is very wide. In particular, we have stated that those procedures with $\gamma = 0$, $\gamma = 1$, and $\gamma = 2$ are especially stable, although some other ones can also show similar dynamical properties. Two numerical examples related to Hammerstein's equation and Fisher's equation allow us to confirm the theoretical results corresponding to the convergence and the stability of the proposed class.

Author Contributions: The individual contributions of the authors are as follows: conceptualization, J.R.T.; writing—original draft preparation, E.G.V. and P.T.-N.; validation, A.C.; numerical experiments, E.G.V. and P.T.-N. All authors have read and agreed to the published version of the manuscript

Funding: This research was supported by Ministerio de Ciencia, Innovación y Universidades PGC2018-095896-BC22 (MCIU/AEI/FEDER, UE).

Institutional Review Board Statement: Not applicable.

Informed Consent Statement: Not applicable.

Data Availability Statement: Not applicable.

Acknowledgments: The authors would like to thank the anonymous reviewers for their useful comments that have improved the final version of this manuscript.

Conflicts of Interest: The authors declare that there is no conflict of interest regarding the publication of this paper.

References

1. Petković, M.S.; Neta, B.; Petković, L.D.; Džunić, J. *Multipoint Methods for Solving Nonlinear Equations*; Elsevier: Amsterdam, The Netherlands, 2013.
2. Amat, S.; Busquier, S. (Eds). *Advances in Iterative Methods for Nonlinear Equations*; Springer: Cham, Switzerland, 2016.
3. Chun, C.; Kim, Y. Several New Third-Order Iterative Methods for Solving Nonlinear Equations. *Acta Appl. Math.* **2010**, *109*, 1053–1063. [CrossRef]
4. Maheshwari, A.K. A fourth order iterative method for solving nonlinear equation. *Appl. Math. Comput.* **2009**, *211*, 383–391. . [CrossRef]
5. Ortega, J.M.; Rheinboldt, W.C. *Iterative Solution of Nonlinear Equations in Several Variables*; Academic Press: Cambridge, MA, USA, 1970.
6. Chicharro, F.I.; Cordero, A.; Torregrosa, J.R. Drawing dynamical and parameters planes of iterative families and methods. *Sci. World J.* **2013**, 780153. [CrossRef] [PubMed]
7. Hernández-Verón, M.A.; Magre nán, À.; Rubio, M.J. Dynamics and local convergence of a family of derivative-free iterative processes, *J. Comput. Appl. Math.* **2019**, *354*, 414–430.
8. Chicharro, F.I.; Cordero, A.; Garrido, N.; Torregrosa, J.R. Generating root-finder iterative methods of second order: convergence and stability. *Axioms* **2019**, *8*, 55. [CrossRef]
9. Lee, M.Y.; Kim, Y.I.; Neta, B. A generic family of optimal sixteenth-order multiple-root finders and their dynamics underlying purely imaginary extraneous fixed points. *Mathematics* **2019**, *7*, 562. [CrossRef]
10. Chicharro, F.I.; Cordero, A.; Garrido, N.; Torregrosa, J.R. Generalized high-order classes for solving nonlinear systems and their applications. *Mathematics* **2019**, *7*, 1194. [CrossRef]
11. Chicharro, F.I.; Cordero, A.; Garrido, N.; Torregrosa, J.R. Wide stability in a new family of optimal fourth-order iterative methods. *Comput. Math. Methods* **2019**, *1*, e1023. [CrossRef]
12. Sharma, D.; Parhi, S.K. Local Convergence and Complex Dynamics of a Uni-parametric Family of Iterative Schemes. *Int. J. Appl. Comput. Math.* **2020**, *6*, 1–16. [CrossRef]
13. Behl, R.; Bhalla, S.; Magreñán, Á.A.; Kumar, S. An efficient high order iterative scheme for large nonlinear systems with dynamics, *Comput. Appl. Math.* **2020**, 113249. [CrossRef]
14. Blanchard, P. Complex analitic dynamics on the Riemann splere. *Bull. Am. Math. Soc.* **1984**, *11*, 85–141. [CrossRef]
15. Devaney, R.L. *An Introduction to Chaotic Dynamical Systems*; Addison-Wesley: Bostom, MA, USA, 1989.
16. Cordero, A.; Torregrosa, J.R. Variants of Newton's method using fifth-order quadrature formulas. *Appl. Math. Comput.* **2007**, *190*, 686–698. [CrossRef]
17. Fisher, R.A. The wave of advance of advantageous genes. *Ann. Eugen.* **1937**, *7*, 353–429. [CrossRef]

Article

Quadrature Integration Techniques for Random Hyperbolic PDE Problems

Rafael Company [1],*, Vera N. Egorova [2] and Lucas Jódar [1]

[1] Instituto de Matemática Multidisciplinar, Universitat Politècnica de València, Camino de Vera s/n, 46022 Valencia, Spain; ljodar@imm.upv.es

[2] Departamento de Matemática Aplicada y Ciencias de la Computación, Universidad de Cantabria, Avenida de los Castros s/n, 39005 Santander, Spain; vera.egorova@unican.es

* Correspondence: rcompany@imm.upv.es

Abstract: In this paper, we consider random hyperbolic partial differential equation (PDE) problems following the mean square approach and Laplace transform technique. Randomness requires not only the computation of the approximating stochastic processes, but also its statistical moments. Hence, appropriate numerical methods should allow for the efficient computation of the expectation and variance. Here, we analyse different numerical methods around the inverse Laplace transform and its evaluation by using several integration techniques, including midpoint quadrature rule, Gauss–Laguerre quadrature and its extensions, and the Talbot algorithm. Simulations, numerical convergence, and computational process time with experiments are shown.

Keywords: random hyperbolic model; random laplace transform; numerical integration; monte carlo method; numerical simulation; talbot algorithm

Citation: Company, R.; Egorova, V.N.; Jódar, L. Quadrature Integration Techniques for Random Hyperbolic PDE Problems. *Mathematics* 2021, 9, 160. https://doi.org/10.3390/math9020160

Received: 7 October 2020
Accepted: 12 January 2021
Published: 14 January 2021

Publisher's Note: MDPI stays neutral with regard to jurisdictional clai-ms in published maps and institutio-nal affiliations.

Copyright: © 2021 by the authors. Licensee MDPI, Basel, Switzerland. This article is an open access article distributed under the terms and conditions of the Creative Commons Attribution (CC BY) license (https://creativecommons.org/licenses/by/4.0/).

1. Introduction

Random hyperbolic partial differential equations (PDEs) are mathematical models that describe wave phenomena with applications in various fields: fluid mechanics [1,2], electromagnetic radiation [3], geosciences [4], and many others. The theory of hyperbolic problems has been well developed based on the assumption that parameters of the model, such as coefficients or initial values are exactly known, which is not available in the real world, where error measurement and the unavailability of the measurement occur. It causes the increasing interest for the random models, which can estimate the impact of the uncertainty to the predicted solution.

The solution is found numerically due to the complexity of random models. Following the mean square approach [5], we can extend existing numerical methods for deterministic problems to the random case by applying the Monte Carlo method [6,7] in order to approximate the statistical moments of the solution. Nevertheless, iterative numerical methods require the storage of the preliminary results and huge number of repetitions, which leads to the the necessity of enormous computational resources and makes them not appropriated to deal with random models. Thus, it becomes urgent to search for an accurate and fast numerical algorithm. Integral transform is a good alternative, as it allows us to construct the solution at one fixed point, not necessarily in the whole domain as it occurs in the case of the finite difference methods, as it is shown in the literature [8].

Integral transform methods convert the original random PDE to an ordinary differential equation (ODE), which can be solved analytically, in some cases, or numerically. Once obtained the solution of the random ODE, the inverse transform is applied in order o restore the solution of the original problem. This inverse transform can be done by the definition, i.e., integrating over the infinite domain, or by using some numerical techniques [9]. There are several widely used methods: Fourier Series, Stehfest approach [10], and Talbot inverse algorithm [11]. Because the inverse Laplace transform is ill-posed problem, the regularization property of the numerical algorithm is necessary. In this sense, the Talbot inverse becomes the best option, since it guarantees the regularization property, while other numerical inversion schemes fail in dealing with noisy data [12].

In this work, we construct a numerical solution for random hyperbolic PDE models, not only by constructing the approximating stochastic process solution, but also while computing its expectation and variance. Thinking of practical applications, we deal with random models where the uncertainty is described by stochastic processes (s.p.'s) having a finite degree of randomness ([5], p. 37); this means that the involved s.p.'s take the form

$$g(x) = G(x, V_1, V_2, \ldots, V_m),\tag{1}$$

where V_i, $1 \leq i \leq m$, are mutually independent random variables (r.v.'s).

We propose an analytic-numerical approach that is based on random integral transform technique combined with various numerical integration methods, such as midpoint rule, Gauss-quadratures, and Talbot inverse [11]. The Monte Carlo method is used for the evaluations of the integrands involving the solution of random ordinary differential problems and also for the computation of the expectation and variance of the approximating stochastic process solution. The oscillatory nature of the appearing integrands deserves careful attention, because not all of the quadrature rules are advisable [13–15].

The proposed analytical–numerical approach for solving random hyperbolic PDE problems considered in this paper includes known state of the art of numerical integration methods, which are compared between themselves in terms of accuracy and computational time: the midpoint quadrature rule, the Talbot algorithm for Laplace inverse, the Gauss–Laguerre quadrature, the Exponential-Fitting Gauss-Laguerre quadrature, and the adaptative quadrature. This comparison is provided to highlight the advantages and drawbacks of each method. Moreover, this complex approach is compared with standard finite-difference methods for solving the random hyperbolic PDE problem. In all cases, the Monte Carlo simulations are used in order to calculate the statistical moments of the random solution process.

The rest of the paper is organized, as follows. In Section 2, the random hyperbolic PDE problem is formulated and the random Laplace transform method is briefly described. Section 3 proposes numerical integration methods for Laplace inverse, while Section 4 gives an algorithm for Monte Carlo simulations. All of the proposed methods are compared by the series of numerical tests in Section 5. Section 6 discusses the results.

All of the numerical tests have been carried out by MatLAB, version R2020a, for Windows 10 Home (64-bit), Intel(R) Core(TM) i5-8265U CPU, 1.60 GHz.

2. Preliminaries and Integral Transform for Random Hyperbolic PDE

This section begins by recalling previous results and definitions [8,16]. Let us consider a complete probability space $(\Omega, \mathcal{F}, \mathbb{R})$ and the set L_p with the p-norm of a real-values random variable $Y \in L_p(\Omega)$, as defined by

$$\|Y\|_p = (\mathbb{E}[|Y|^p])^{1/p}, \quad p \geq 1,\tag{2}$$

where the expectation $\mathbb{E}[|Y|^p] < \infty$, and $L_p(\Omega)$ is a Banach space [17]. By using definition (2), the integrability, continuity, and differentiability of a function $Y(t) \in L_p(\Omega)$ can be defined straightforwardly.

Note that, if $p = 2$, then it is a mean square (m.s.) case. Let $\mathcal{C}$ be the class of all m.s. locally integrated two-stochastic processes (s.p.'s) $h(t)$ defined in $\mathbb{R}$ such that $h(t) = 0$, for all negative arguments and the two-norm satisfies

$$\exists c \geq 0, M > 0: \quad \|h(t)\|_2 \leq M \exp(ct), \quad \forall t \geq 0.\tag{3}$$

Subsequently, for $h(t) \in \mathcal{C}$, the m.s. integral

$$H(s) = \mathcal{L}[h(t)](s) = \int_0^\infty h(t) \exp(-st) dt, \tag{4}$$

where s is a complex number with real part $Re(s) > c_0 \geq 0$, and it is called the random Laplace transform of 2-s.p. $h(t)$. The constant c_0 is chosen, such that $Re(s) > c_0$ specifies the region where $H(s)$ is analytic and it has some form of singularity on the line $Re(s) = c_0$ [9]. If $H(s)$ is known, then the random inverse transform for $t > 0$ is defined, as follows

$$h(t) = \frac{1}{2\pi i} \int_{\alpha-i\infty}^{\alpha+i\infty} H(s) \exp(st) ds, \tag{5}$$

where i stands for the imaginary unit and $\alpha > c_0$ [16].

For the purposes of present study, we recall some of the important properties of the random Laplace transform (4): if s.p. $h(t)$ is twice m.s. differentiable and $h'(t)$, $h''(t)$ belong to $\mathcal{C}$, then

$$\mathcal{L}[h'(t)](s) = sH(s) - h(0+), \qquad \mathcal{L}[h''(t)](s) = s^2 H(s) - sh(0+) - h'(0+). \tag{6}$$

In this paper, we consider a one-dimensional random hyperbolic PDE modelling the s.p. of the vibrating string motion $u(x,t)$, depending on the spatial variable x and time t,

$$u_{tt}(x,t)(\xi) = a(x)(\xi) u_{xx}(x,t)(\xi) + b(x)(\xi) u_x(x,t)(\xi) + c(\xi) u(x,t)(\xi), \quad x \in [0, L], \ t > 0, \ \xi \in \Omega, \tag{7}$$

$$u(x,0)(\xi) = f_0(x)(\xi), \qquad u_t(x,0)(\xi) = f_1(x)(\xi), \tag{8}$$
$$u(0,t)(\xi) = g_0(t)(\xi), \qquad u(L,t)(\xi) = g_1(t)(\xi), \tag{9}$$

where $a(x)(\xi) > 0$, $b(x)(\xi)$ are m.s.-continuous stochastic processes with a finite degree of randomness and absolutely integrable with respect to the spatial variable in $\mathbb{R}$; $c(\xi)$ is a random variable (r.v.). The s.p.'s $f_0(x)(\xi)$, $f_1(x)(\xi)$, $g_0(t)(\xi)$, and $g_1(t)(\xi)$ are functions depending on a finite number of r.v. that represent random initial and boundary conditions with a finite degree of randomness.

The random hyperbolic partial differential equation (PDE) (7) is solved using an analytic-numerical method that is based on Laplace transform combined with an appropriate numerical integration technique. In this paper, we consider various quadratures for inverse Laplace transform.

Following the ideas of [8,18], let us define the random Laplace transform with respect to the temporal variable, as

$$U(x,s)(\xi) = \mathcal{L}[u(x,t)(\xi)]. \tag{10}$$

Because $u(x,t)(\xi)$ is a twice m.s. differentiable s.p., one gets

$$\mathcal{L}[u_{tt}(x,t)(\xi)] = s^2 U(x,s)(\xi) - su(x,0)(\xi) - u_t(x,0)(\xi) = s^2 U(x,s)(\xi) - sf_0(x)(\xi) - f_1(x)(\xi). \tag{11}$$

Subsequently, (7) is transformed to the following random non-homogeneous ordinary differential equation (ODE) with respect to the spatial variable

$$a(x)(\xi) U_{xx}(x,s)(\xi) + b(x)(\xi) U_x(x,s)(\xi) + (c - s^2) U(x,s)(\xi) = -[sf_0(x)(\xi) + f_1(x)(\xi)], \tag{12}$$

for $x \in [0, L]$, $\xi \in \Omega$.

Assuming $a(x)(\xi) > 0$ for each event $\xi \in \Omega$, one gets

$$U_{xx}(x,s)(\xi) + \frac{b(x)(\xi)}{a(x)(\xi)} U_x(x,s)(\xi) + \frac{c - s^2}{a(x)(\xi)} U(x,s)(\xi) = -\frac{sf_0(x)(\xi) + f_1(x)(\xi)}{a(x)(\xi)}. \tag{13}$$

Equation (13) is a linear second order ODE with respect to the spatial variable, which can be analytically solved in some cases, or numerically in other cases. Because the boundary conditions (9) for the PDE are functions on t, the boundary conditions for (13) are the corresponding Laplace transforms of (9):

$$U(0,s)(\xi) = \mathcal{L}[g_0(t)(\xi)], \quad U(L,s)(\xi) = \mathcal{L}[g_1(t)(\xi)]. \tag{14}$$

Once obtaining the solution $U(x,s)(\xi)$, a real-valued $u(x,t)(\xi)$ is restored by while using random inverse Laplace transform that is given by (5). Taking advantage of the relationship between the inverse Laplace transform and Fourier cosine integrals, see [9], the following formula is used

$$u(x,t)(\xi) = \frac{2e^{\alpha t}}{\pi} \int_0^\infty Re[U(x, \alpha + iw)(\xi)] \cos(wt) dw, \quad \xi \in \Omega, \tag{15}$$

where $Re[\cdot]$ stands for the real part of a complex number. Note that the integrand appearing in (15) has an oscillatory kernel that deserves special care for the numerical integration.

3. Numerical Integration Methods

This section describes briefly acknowledged integration methods for the integrals of the type (15).

THe numerical solution of Equation (7) is constructed in the domain $\Delta = [0; L] \times [0; T]$ for each fixed event ξ. Let us introduce a uniform grid $\{x_j, t^n\}$, such that

$$x_j = jh, \ h = \frac{L}{N_x}, \ j = 0, \ldots, N_x; \quad t^n = nk, \ k = \frac{T}{N_t}, \ n = 0, \ldots, N_t. \tag{16}$$

At each node (x_j, t^n), the numerical solution is defined by $u_j^n(\xi)$ for each realization of ξ and it is obtained by approximating the integral (15). Hence, at every fixed (x_j, t^n), the following function is defined

$$f_{j,n}(w) = f_{j,n}(w, \xi) = Re[U(x_j, \alpha + iw)(\xi)] \cos(wt^n), \tag{17}$$

where $U(x_j, \alpha + iw)(\xi)$ is the numerical solution of ODE (13) at the point x_j for fixed value of $s = \alpha + iw$. Now, we briefly describe all of the considered methods for numerical integration.

3.1. Midpoint Quadrature Rule

The midpoint quadrature rule is a method of approximation of integral (15) based on the Riemann sums, the simplest case of Newton–Cotes open formulas, for truncated domain $[0, R]$. In the general case, the midpoint quadrature rule is written, as follows

$$\int_0^\infty f(w) dw \approx \int_0^R f(w) dw = \sum_{k=0}^N f(w_{k+1/2}) h_{MP} + O(h^2), \tag{18}$$

where $w_{k+1/2} = \left(k + \frac{1}{2}\right) h_{MP}$, $h_{MP} = \frac{R}{N}$, $k = 0, \ldots, N-1$.

It is well known that the main advantage of this method is its simplicity of implementation and the consideration of all the information regarding the integrand, which makes it applicable for a wider class of integrand functions [14]. However, the high accuracy of the quadrature requires large enough value of N, leading to the increasing computational cost. In the case of improper integral (in the infinite domain), the method can also be sensitive to the choice of R.

3.2. Gauss-Laguerre Quadrature

The novelty of Gauss quadratures is to choose nodes where the integrand is evaluated in order to minimize the error of approximation. It is a good alternative to Newton–Cotes formulas, especially when the evaluation of function itself requires a lot of computational resources, because good accuracy can be reached with a small number, four or five, of nodes if the integrand is well conditioned. This is not the case when the integrand is of oscillatory type [19].

The improper integral is approximated by Gauss–Laguerre (GL) quadrature of N_{GL} nodes by the following sum, see [20],

$$\int_0^\infty f(w)dw \approx \sum_{k=1}^{N_{GL}} \gamma_k f(w_k) e^{w_k}, \tag{19}$$

where w_k is the k-th root of Laguerre polynomial $L_{N_{GL}}(w)$, γ_k is the weight of the quadrature given by

$$\gamma_k = \frac{w_k}{(N_{GL}+1)^2 \left[L_{N_{GL}+1}(w_k)\right]^2}, \quad k = 1, \ldots, N_{GL}. \tag{20}$$

3.3. Exponentially-Fitted Gauss-Laguerre Quadrature

Exponential fitting is an approach that is used in numerical differentiation, interpolation, and integration for improving the accuracy of the methods. Because integrand in (15) is oscillating, Exponentially-fitted Gauss–Laguerre quadrature (EF-GL), as proposed in [21], could be a good option. For EF-GL, nodes and weights depend on integrand and cannot be defined a priori. The computation of these N_{GL} pairs of nodes and weights is based on the solution of a nonlinear system of N_{GL} equations, which leads to additional computational cost. In [21], the numerical algorithm is described in details. Further, in Section 5, we compare the accuracy and computational time of GL and EF-GL quadrature rules.

3.4. Talbot Inverse

The method of Talbot for the Laplace inversion problem [11] is based on numerical contour integration. Instead of formula (15), the Bromwich integral is used

$$u_j^n(\xi) = \frac{1}{2\pi i} \int_{\alpha-i\infty}^{\alpha+i\infty} e^{st^n} U(x_j, s) ds. \tag{21}$$

The contour deformation is used in order to obtain the Hankel contour and exploit the exponential factor, which makes the integral suitable for further application of a Newton–Cotes formula [22]. The Talbot inversion quadrature for N_{TI} nodes is written, as follows

$$u_j^n(\xi) = \frac{2}{5t^n} \sum_{k=0}^{N_{TI}-1} Re\left[\gamma_k U(x_j, \frac{w_k}{t^n})\right], \tag{22}$$

where w_k are the nodes and γ_k are the weights defined by

$$w_0 = \frac{2N_{TI}}{5}, \quad w_k = \frac{2\pi k}{5}\left(\cot\left(\frac{k\pi}{N_{TI}}\right) + i\right), \tag{23}$$

$$\gamma_0 = 0.5 \exp(w_0), \quad \gamma_k = \exp(w_k) \cdot \left[1 + \frac{k\pi}{N_{TI}}\left(1 + \cot^2\left(\frac{k\pi}{N_{TI}}\right)\right) - i \cot\left(\frac{k\pi}{N_{TI}}\right)\right]. \tag{24}$$

Here, the number of nodes N_{TI} should be chosen in accordance with desired accuracy: for n significant digits $N_{TI} = \lceil 1.7n \rceil$. It shows the flexibility of the method and the high degree of accuracy with fast convergence. Moreover, as stated in [12], the main advantage of the Talbot algorithm is the regularization property, which means the ability to handle noisy data. It is important for the inverse Laplace transform problem due to its ill-posedness and it becomes even more urgent in the random case dealing with perturbed initial conditions or parameters of the problem.

Summarizing, a numerical solution is constructed following the steps of Algorithm 1 for all of the described methods.

Algorithm 1: Numerical solution for deterministic string vibrating problem

---

Initialization: set the mesh $\{x_j, t^n\}$ by (16);
Set initial conditions $u(x_j, 0) = f_0(x_j), j = 0, \ldots, N_x$;
Set $\alpha > c_0$;
Set number of nodes of the quadrature N;
Set $n = 0$;
while $t^n < T$ **do**
    Increment n;
    for $j = 0, \ldots, N_x$ **do**
        Compute nodes and weigths $\{w_k, \gamma_k\}$ of the chosen quadrature
        - Midpoint rule: uniform grid with N nodes ;
        - GL quadrature: nodes w_k are the roots of the Laguerre polynomial of N-th order, $k = 1, \ldots, N$;
        - EF-GL quadrature [21]: nodes w_k and weights γ_k are found by solving nonlinear system of $2N$ equations;
        - Talbot inverse: nodes w_k and weights $\gamma_k, k = 0, \ldots, N - 1$ are defined by (23)–(24);
        Get the approximated value u_j^n:
        - Midpoint rule: integral in (15) is approximated by (18);
        - GL and EF-GL quadratures: integral in (15) is approximated by (19)–(20);
        - Talbot inverse: formula (22) ;
    end
end

---

4. Monte Carlo Method for Random Hyperbolic PDE

The coefficients of the random m.s. Equation (7) and corresponding initial and boundary conditions (9) are stochastic processes (s.p's) that are defined in a complete probability space $(\Omega, \mathcal{F}, \mathbb{P})$, i.e., s.p.'s $a(x), b(x), f_0(x), f_1(x), g_0(x)$ and $g_1(x)$ are described as continuous s.p.'s with with one-degree of randomness.

The solution of the random m.s. problem is approximated by using the the Monte Carlo approach [6,7], when the expectation $\mathbb{E}[u(x,t)]$ is approximated by the average of a sufficiently large number of realizations $\xi \in \Omega$ of the corresponding deterministic realized transformed random ordinary differential problem. The Algorithm 2 describes the steps of the numerical solution.

Algorithm 2: Numerical solution for random hyperbolic PDE problem

---

Initialization: set the mesh $\{x_j, t^n\}$ by (16);
Set number of the MC realizations N_{MC};
Generate N_{MC} random variables for s.p.'s $a(x), b(x), f_0(x), f_1(x), g_0(x)$;
Choose the method of numerical integration **for** $m = 1, \ldots N_{MC}$ **do**
    Define s.p.'s $a(x), b(x), f_0(x), f_1(x), g_0(x)$ for fixed realization;
    Run Algorithm 1 to obtain the numerical solution u_m of the deterministic problem;
    Increment m;
end
Compute $\mathbb{E}[u] = \sum_{m=1}^{N_{MC}} \frac{u_m}{N_{MC}}$;
Compute $\mathbb{E}[u^2] = \sum_{m=1}^{N_{MC}} \frac{u_m^2}{N_{MC}}$;
Compute $\sqrt{\text{Var}[u]} = \sqrt{\mathbb{E}[u^2] - (\mathbb{E}[u])^2}$

5. Numerical Results

This section deals with the comparison of the above-described methods of numerical integration and Laplace inversion for several test problems.

5.1. Deterministic PDE Problem with Constant Coefficients

We start with simple one dimensional deterministic problem with a known analytical solution in order to check the viability of the proposed numerical integration techniques. The deterministic example corresponds to one fixed event $\xi \in \Omega$. Instead of the bounded spatial domain $[0; L]$, the whole real axis $\mathbb{R}$ is considered. Thus, no boundary conditions are needed. We also assume that $a > 0$, b, and c are constants, i.e., the following wave equation is considered

$$u_{tt}(x,t) = a^2 u_{xx}(x,t) + b u_x(x,t) + c u(x,t), \quad x \in \mathbb{R}, t > 0, \tag{25}$$

subject to initial conditions $u(x,0) = f_0(x)$, $u_t(x,0) = f_1(x)$.

This problem admits an analytical solution that can be written in terms of Bessel function of the first kind, see [23], p. 574, Equation 6.1.5, as follows

- for $c - \frac{1}{4}a^{-2}b^2 = \sigma^2 > 0$:

$$\begin{aligned}u(x,t) = &\frac{1}{2}f(x+at)\exp\left(\frac{bt}{2a}\right) + \frac{1}{2}f(x-at)\exp\left(-\frac{bt}{2a}\right) \\&+ \frac{\sigma t}{2a}\exp\left(-\frac{bx}{2a^2}\right)\int_{x-at}^{x+at}\exp\left(\frac{b\xi}{2a^2}\right)\frac{I_1\left(\sigma\sqrt{t^2-(x-\xi)^2/a^2}\right)}{\sqrt{t^2-(x-\xi)^2/a^2}}f(\xi)d\xi \\&+ \frac{1}{2a}\exp\left(-\frac{bx}{2a^2}\right)\int_{x-at}^{x+at}\exp\left(\frac{b\xi}{2a^2}\right)I_0\left(\sigma\sqrt{t^2-(x-\xi)^2/a^2}\right)g(\xi)d\xi,\end{aligned} \tag{26}$$

where $I_0(z)$ and $I_1(z)$ are the modified Bessel function of the first kind;

- for $c - \frac{1}{4}a^{-2}b^2 = -\sigma^2 < 0$:

$$\begin{aligned}u(x,t) = &\frac{1}{2}f(x+at)\exp\left(\frac{bt}{2a}\right) + \frac{1}{2}f(x-at)\exp\left(-\frac{bt}{2a}\right) \\&- \frac{\sigma t}{2a}\exp\left(-\frac{bx}{2a^2}\right)\int_{x-at}^{x+at}\exp\left(\frac{b\xi}{2a^2}\right)\frac{J_1\left(\sigma\sqrt{t^2-(x-\xi)^2/a^2}\right)}{\sqrt{t^2-(x-\xi)^2/a^2}}f(\xi)d\xi \\&+ \frac{1}{2a}\exp\left(-\frac{bx}{2a^2}\right)\int_{x-at}^{x+at}\exp\left(\frac{b\xi}{2a^2}\right)J_0\left(\sigma\sqrt{t^2-(x-\xi)^2/a^2}\right)g(\xi)d\xi,\end{aligned} \tag{27}$$

where $J_0(z)$ and $J_1(z)$ are Bessel function of the first kind.

In order to test the proposed numerical integration methods we apply Laplace transform, as described in Section 2, and obtain a deterministic version of Equation (13):

$$U_{xx}(x,s) + \frac{b}{a^2}U_x(x,s) + \frac{c-s^2}{a^2}U(x,s)(\xi) = -\frac{sf_0(x) + f_1(x)}{a^2}. \tag{28}$$

Applying the non-unitary Fourier transform with angular frequency

$$\hat{U}(w,s) = \mathcal{F}[U(x,s)] = \int_{-\infty}^{\infty} U(x,s)\exp(-ixw)dx, \tag{29}$$

Equation (13) takes the following form

$$-w^2\hat{U}(w,s) + iw\frac{b}{a^2}\hat{U}(w,s) + \frac{c-s^2}{a^2}\hat{U}(w,s) = -\mathcal{F}\left[\frac{sf_0(x) + f_1(x)}{a^2}\right]. \tag{30}$$

Algebraic Equation (30) is solved directly

$$\hat{U}(w,s) = \frac{-\mathcal{F}\left[\frac{sf_0(x)+f_1(x)}{a^2}\right]}{-w^2 + iw\frac{b}{a^2} + \frac{c-s^2}{a^2}}. \tag{31}$$

Hence, the solution $U(x,s)$ of (28) can be obtained by applying inverse Fourier transform to (31).

In the next Example 1, we consider a particular case of Equation (25) with constant coefficients and trigonometric initial conditions.

Example 1. *Let us consider deterministic problem* (25) *with coefficients* $a = 2$, $b = 1$, $c = 3$, *and initial conditions* $f_0(x) = \cos(x)$ *and* $f_1(x) = \sin(x)$.

Numerical solution is constructed in the truncated domain $[0,L] \times [0,T]$, $L = 5$, $T = 1$, for discrete uniformly distributed nodes (16), $N_x = 5$, $N_t = 10$, by applying the described in previous section methods, see Algorithm 1. We set $\alpha = 1$.

Applying the inverse Fourier transform to (31), one obtains

$$U(x,s) = \frac{1}{2\pi}\left[\frac{\pi e^{-xi}(s+i)}{a^2 + bi + (s^2 - c)} + \frac{\pi e^{xi}(s-i)}{a^2 - bi + (s^2 - c)}\right], \quad s = \alpha + iw, \tag{32}$$

where i is the imaginary unit. Once the solution of ODE (28) is obtained, formula (15) is used to restore the solution of the PDE while using various numerical integration techniques.

Note that Equation (25) admits the analytical solution, as described above. Because the function $u(x,t)$ is close to zero, we compute the relative error of the discrete numerical solution at the mesh nodes in order to estimate the accuracy of the methods

$$\text{RelErr}(j,n) = \frac{|u_{\text{ref}}(x_j,t^n) - U_{\text{num}}(x_j,t^n)|}{|u_{\text{ref}}(x_j,t^n)|}, \tag{33}$$

where U_{num} is the matrix of numerical solution $U_{\text{num}} = \{u_j^n\}$, $j = 0,\ldots,N_x$, $n = 0,\ldots,N_t$, as computed by Algorithm 1; $u_{\text{ref}}(x_j,t^n)$ is the reference value at the point (x_j,t^n). In this example, as the exact solution is known, the reference value is equal to this exact solution. For other cases where the exact solution is not available, a reference value is obtained using accurate finite difference method (FDM) for solving the original PDE (7). The total computational time for the proposed methods are presented in Table 1, together with the maximum of RelErr(j,n).

The adaptative quadrature (MatLAB function `integral` [24]) has the same order of accuracy as the midpoint rule, but it requires greater computational resources. Thus, it will not be considered in further more complicated examples.

For the Talbot algorithm $M = 17$ is chosen to guarantee the accuracy up to 10 significant digits [22]. Even in that case, the method performs much faster than standard numerical integration methods for (15). Thus, the Talbot inverse method is found to be the most effective method for the deterministic case with constant coefficients.

Table 1. Comparison of various numerical methods for problem (25) with $a = 2$, $b = 1$ and $c = 3$ (Example 1).

Method	Error	CPU-Time, s
Midpoint rule ($R = 10^4$, $h_{MQ} = 0.1$)	4.4700×10^{-7}	0.59
Gauss-Laguerre ($N_{GL} = 25$)	3.5551×10^{-1}	0.05
EF-GL (five nodes)	7.9303	3.97
Talbot inverse ($M = 17$)	7.3457×10^{-11}	0.02
Adaptative quadrature	4.8559×10^{-6}	7.53

The relative errors for Midpoint rule and Talbot inverse methods are plotted in Figure 1. Because no boundary conditions are posed for the problem, the largest values of the relative errors are situated at the boundary $x = L$.

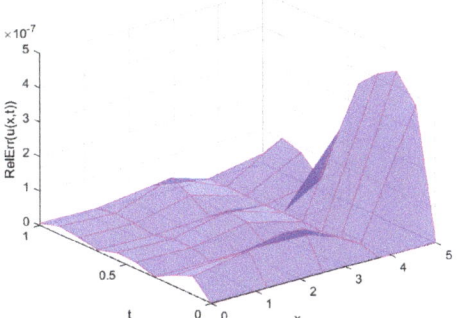 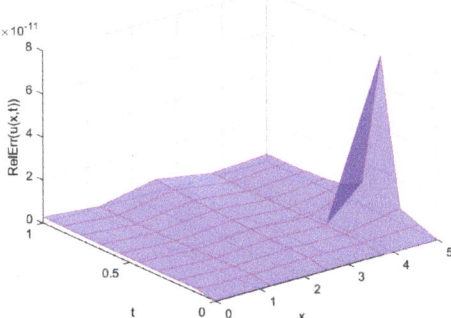

Figure 1. Distribution of the relative error among space and time for midpoint rule (**left**) and Talbot inverse (**right**) methods in Example 1.

Table 2 presents a comparison of GL and EF-GL quadratures in terms of the maximum relative error and the CPU-time varying the number of nodes N_{GL}. It is important to notice that the CPU-time may vary from simulation to simulation, thus only the order should be taken into account. In the case of GL quadrature, we find out that the computational time is similar with increasing number of nodes, while the CPU-time for EF-GL method is increasing exponentially. The convergence of the GL quadrature is shown, while taking the results shown in Table 1 into account: the error reduces significantly with an increasing number of nodes. The potential improvement of the GL method by the exponential fitting expectedly has higher computational cost, due to the solution of the nonlinear system at each point of the computational domain. However, the accuracy of the EF-Gl quadrature for this example with oscillating integrand has not been improved when comparing with the standard GL rule. Thus, it will not be considered in further more complicated examples.

Table 2. Gauss–Laguerre (GL) and Exponentially-fitted Gauss–Laguerre quadrature (EF-GL) methods results, depending on number of nodes of the quadrature for Example 1.

N_{GL}	3	5	8	15
GL Error	8.2739	7.4679	2.5601	1.4560
GL CPU-time, s	0.02	0.02	0.05	0.05
EF-GL Error	9.4937	7.9303	7.5438	3.9366×10^3
EF-GL CPU-time, s	0.44	1.84	15.84	430.89

The accuracy of the midpoint rule depends on the truncation R and step size h_{MP}. A bigger domain, as well as smaller step size, lead to an increased computational time. Figure 2 presents the plots of errors and the CPU-time for fixed step size $h_{MP} = 10^{-1}$ with respect to increasing domain. The accuracy in dependence on the step size h_{MP} is also studied. In Table 3, the maximum relative error is reported for various h_{MP} and fixed $R = 10^4$. The maximum relative error is decreasing with step size until 4.4699×10^{-7} ($h_{MP} = 1/16$); further fragmentation of the step size does not reduce the error for $R = 10^4$.

Table 3. The maximum relative error and computational time of the midpoint quadrature rule with respect to step size h_{MP} for Example 1.

h_{MP}	Error	CPU-Time, s
1/1	2.8832×10^{-1}	0.06
1/2	3.1862×10^{-2}	0.06
1/4	7.0942×10^{-5}	0.19
1/8	4.4700×10^{-7}	0.22
1/16	4.4699×10^{-7}	0.53
1/32	4.4699×10^{-7}	1.03
1/64	4.4699×10^{-7}	1.47

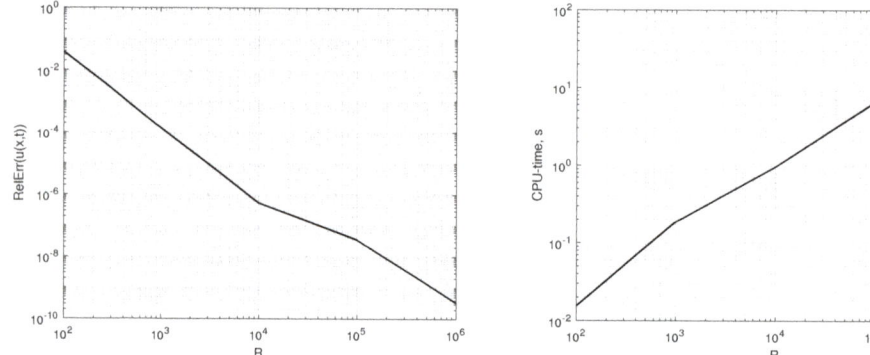

Figure 2. Error and total computational time of the midpoint rule with respect to the domain size R with fixed $h = 10^{-1}$ for Example 1.

5.2. Deterministic PDE with Non-Constant Coefficients

In the case of non-constant coefficients in (13), the analytical solution is not always available; thus, FDM is applied to construct a reference numerical solution. Note that the function $U(x,s)$ that is used in expression (17) means the value of the numerical solution of the ODE (13) at the fixed point x for fixed parameter s.

Equation (13) is discretized by the central differences on the same mesh $\{x_j\}$, $j = 0, \ldots, N_x$, as follows

$$\frac{U_{j+1} - 2U_j + U_{j-1}}{h^2} + \frac{b(x_j)}{a(x_j)} \frac{U_{j+1} - U_{j-1}}{2h} + \frac{c - s^2}{a(x_j)} U_j = -\frac{s f_0(x_j) + f_1(x_j)}{a(x_j)}, \quad j = 1, \ldots, N_x - 1, \tag{34}$$

where U_j stands for the approximated value of $U(x,s)$ at the node x_j. The values at the boundaries are found from the boundary conditions by applying the Laplace transform

$$U_0 = \mathcal{L}[g_0(t)], \quad U_{N_x} = \mathcal{L}[g_1(t)]. \tag{35}$$

Hence, the integrand (17) has to be evaluated at each fixed node of the computational grid in order to approximate integral (15), which provokes a significant augment of the CPU-time. In the next example, we increase the complexity by regarding a variable coefficients deterministic problem.

Because the analytical solution for the deterministic PDE problem in general form (7) is not available, a numerical method has to be employed to obtain the reference numerical solution. We consider an explicitly centred in time and space finite difference scheme for the mesh function $u_j^n \approx u(x_j, t^n)$:

$$\frac{u_j^{n+1} - 2u_j^n + u_j^{n-1}}{(\Delta t)^2} = a(x_j)\frac{u_{j+1}^n - 2u_j^n + u_{j-1}^n}{(\Delta x)^2} + b(x_j)\frac{u_{j+1}^n - u_{j-1}^n}{2\Delta x} + cu_j^n, \quad (36)$$

where $j = 1, \ldots, N_x$, $n = 2, \ldots, N_t$. The initial conditions (8) are used in order to obtain the solution at the first time levels t^0 and t^1. The derivative in (8) is approximated by the forward difference. Because the considered scheme is conditionally stable, the step sizes Δt and Δx are chosen to guarantee the stability. In order to obtain a good approximation, which could be considered as the reference solution, the mesh should be chosen appropriately fine.

Example 2. Let us consider a deterministic vibrating string problem (7) on rectangle $[0, L] \times [0, T]$, $L = 0.5$, $T = 0.2$. We set non-constant coefficients $a(x) = 9x + 1$, $b = -e^x$, $c = -5$, initial conditions $f_0(x) = x(x - L)$ and $f_1(x) = 0$, and boundary conditions $g_0(t) = g_1(t) = 0$.

The numerical solution is constructed by the Algorithm 1, choosing $N_x = 10$, $N_t = 5$. For the midpoint rule, $N = 100$ and $R = 100$ are used. Table 4 presents the comparison of the methods in terms of maximum relative error and computational time. The reference solution is the numerical solution that is computed by the FDM (36) in refined mesh ($N_x = 100$, $N_t = 16{,}000$), which preserves the stability of the scheme. Because an explicit method is used and no iterative procedures are needed for solving nonlinear system at each time-level, the total computational time is comparably small: 0.15 s. Figure 3 plots the reference solution.

Table 4. A comparison of various methods of numerical integration for Example 2.

Method	Error	CPU-Time, s
Midpoint quadrature	4.5628×10^{-2}	116.38
Talbot inverse	7.1305×10^{-2}	51.00
Gauss-Laguerre (9 nodes)	7.4450×10^{-1}	5.00
Gauss-Laguerre (25 nodes)	7.9014×10^{-2}	17.48

Figure 4 plots the solution at the moment $t = T$. The midpoint rule and Talbot inverse method perform more accurately than GL quadrature of nine nodes, but they require more computational time due to larger number of calls of integrand (17). However, taking 25 nodes in the GL quadrature, the accuracy has been improved significantly.

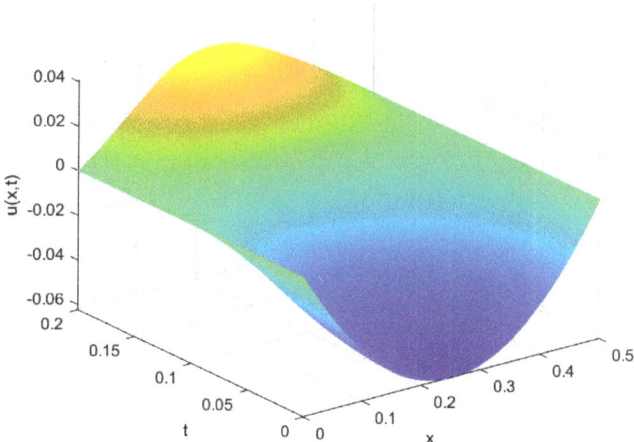

Figure 3. Reference solution for Example 2 computed by the finite difference method (FDM) (36).

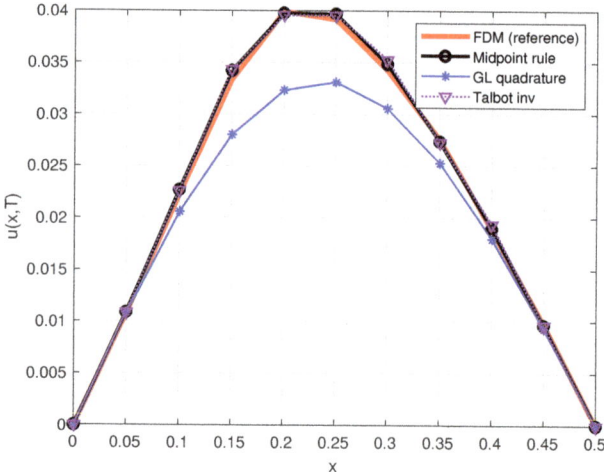

Figure 4. Numerical solution for Example 2 at the moment $t = T$ obtained by considered methods.

5.3. Random PDE with Constant Coefficients

In this subsection, we deal with random models with constant coefficient random variables. It is remarkable that, in this case, we need not only the computation of the approximation s.p. solution, but also the computation of its statistical moments.

Example 3. *We consider a random version of problem* (25), *with* $a \sim \mathcal{N}(2, 0.25)$, $b, c \sim \text{Beta}(2, 5)$. *In order to approximate the mean and variance of the solutions, the Monte Carlo method with N_{MC} simulations is used.*

Expectation and variance of the exact solution for the random hyperbolic PDE (25) are plotted in Figure 5. As in previous examples, we compare the proposed methods of integration and Laplace inverse in terms of maximum relative error and computational time. Table 5 presents the results for various N_{MC}. The CPU-time refers to the total computational time for all N_{MC} simulations. Note that, for 1000 simulations, the exact solution (26)–(27) requires 28.41 s to perform the simulations. Thus, Midpoint rule ($R = 100$, $h = 0.1$), Talbot inverse and GL quadrature require less computational time than calculation by the exact formula. As expected, the computational time is increasing with the number of simulations linearly, but errors preserve the order in most cases.

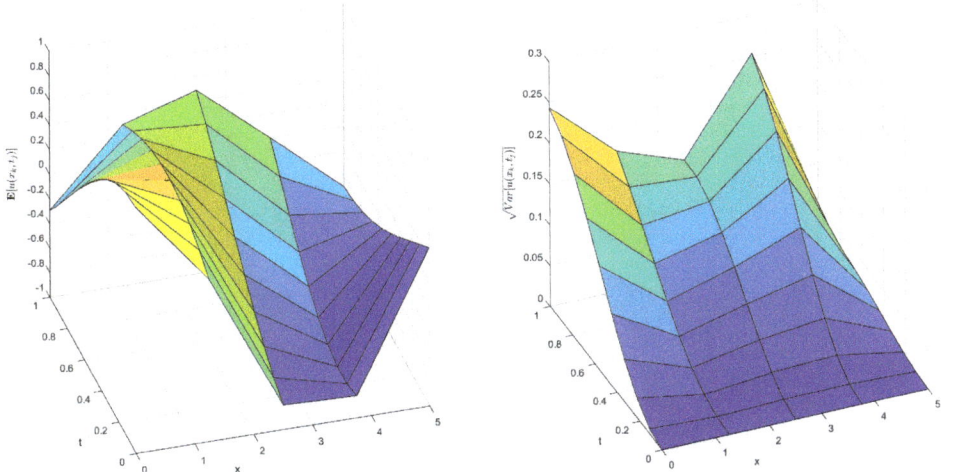

Figure 5. Expectation and variance of the exact solution for the random hyperbolic partial differential equation (PDE) (25) with $a \sim \mathcal{N}(2, 0.25)$, $b, c \sim \text{Beta}(2, 5)$, performed using the Monte-Carlo method with $N_{MC} = 10^3$ simulations.

Table 5. Comparison of various methods of numerical integration for the random hyperbolic PDE (25) with $a \sim \mathcal{N}(2, 0.25)$, $b, c \sim \text{Beta}(2, 5)$.

Method	Error of Mean	Error of Variance	CPU-Time, s
	$N_{MC} = 500$		
Midpoint rule	5.6048×10^{-2}	2.3034×10^{-2}	4.88
Talbot inverse	5.0744×10^{-2}	2.3032×10^{-2}	6.83
GL quadrature (3 nodes)	1.9009×10^{-1}	2.1436×10^{-1}	2.92
	$N_{MC} = 1000$		
Midpoint rule	4.1520×10^{-2}	2.5102×10^{-2}	7.52
Talbot inverse	4.0345×10^{-2}	2.5102×10^{-2}	12.67
GL quadrature	1.9086×10^{-1}	2.1503×10^{-1}	5.64
	$N_{MC} = 2000$		
Midpoint rule	3.7210×10^{-2}	1.2823×10^{-2}	16.88
Talbot inverse	3.1905×10^{-2}	1.2823×10^{-2}	24.86
GL quadrature	1.9001×10^{-1}	2.1443×10^{-1}	11.30
	$N_{MC} = 4000$		
Midpoint rule	4.6382×10^{-2}	9.8022×10^{-3}	32.89
Talbot inverse	4.1078×10^{-2}	9.7971×10^{-3}	51.09
GL quadrature	1.8993×10^{-1}	2.1581×10^{-1}	24.31

5.4. Random PDE with Non-Constant Coefficients

To complete the study, a random variable coefficient problem is considered.

Example 4. *The vibration of the string in* $[0, L]$ *is described by Equation* (7), *subject to the initial conditions* $f_0(x) = x(x - L)$ *and* $f_1(x) = 0$; *and boundary conditions* $g_0(t) = g_1(t) = 0$. *We set up the parameters:*

$$L = 0.5, \quad T = 0.2, \quad a(x) = \varphi x + 1, \quad \varphi \sim \mathcal{N}(9, 0.5), \quad b(x) = -e^x, \quad c \sim \text{Beta}(2, 5). \quad (37)$$

Unlike the deterministic Example 2 with non-constant coefficients where FDM provides a reference analytical solution, reference values are not available here due to the computational complexity that arises in the evaluation of the statistical moments of the approximate stochastic process when time step advances [18]. A survival reference FDM solution is taking the Monte Carlo method for an appropriate set of realizations. In this case, the number of realizations is $N_{MC} = 10^3$ and CPU-time is 16,212 s.

Figure 6 plots the numerical solution. The zero-variance at the boundaries is caused by the boundary conditions. Similar plots are obtained for the considered methods. Thus, we compare them in terms of the maximum relative error, see Table 6. As it is expected from the previous examples, the most accurate solution is obtained by the midpoint rule and Talbot inverse, although this advantage pays the price of additional computational cost.

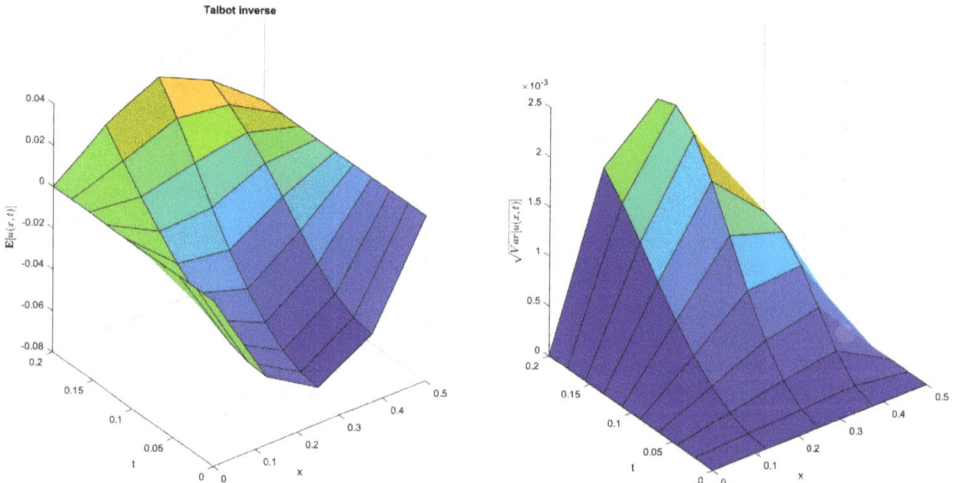

Figure 6. Expectation and variance of the numerical solution for the random hyperbolic PDE (25) with $\varphi \sim \mathcal{N}(9, 0.5)$, $b(x) = -e^x$, $c \sim \text{Beta}(2, 5)$, performed by using the Monte-Carlo method with $N_{MC} = 10^3$ simulations.

Table 6. Comparison of various methods of numerical integration for Example 4, $N_{MC} = 1000$.

Method	Error of Mean	Error of Variance	CPU-Time, s
Midpoint rule	3.8216×10^{-2}	2.2638×10^{-2}	65,385.00
Talbot inverse	3.4976×10^{-2}	2.1375×10^{-2}	28,965.54
GL quadrature (9 nodes)	1.8671×10^{-1}	2.5603×10^{-2}	7192.14

6. Conclusions

The solution of a random hyperbolic PDE problem is a challenging task that is demanded in many practical applications. Computing an expression of the approximating stochastic process makes the computation of its statistical moments available. In this paper, we propose a combination of the random Laplace transform with the numerical integration techniques for its inverse, and the Monte Carlo method for the evaluation of numerical solution of the transformed problem at a particular required point.

The Monte Carlo simulations require a fast and efficient basis numerical algorithm for solving deterministic hyperbolic PDE problem, for every fixed realization. FDM could not be an option due to the high computational cost and memory requirements. In order to avoid the numerical differentiation of the PDE, Laplace transform is applied, which results in ODE equation. In some cases, as it has been shown in present paper, the analytical solution of ODE is known; thus, we use numerical integration methods for inverse Laplace transform. If the solution of ODE is not available, then numerical techniques for boundary value problem have to be employed.

Several numerical integration methods have been considered: midpoint rule and GL-quadrature for improper integrals. However, due to the oscillatory behaviour of the integrand function GL quadrature with a small number of nodes shows comparatively poor results, while the midpoint rule is comparable with Talbot's Laplace inverse for random hyperbolic PDEs. The proposed complex analytic-numerical approach is compared with the classical explicit FDM scheme for the original random PDE problem.

Author Contributions: R.C., V.N.E. and L.J. contributed equally to this work. All authors have read and agreed to the published version of the manuscript.

Funding: This research has been funded by the Spanish Ministerio de Economía, Industria y Competitividad (MINECO), the Agencia Estatal de Investigación (AEI) and Fondo Europeo de Desarrollo Regional (FEDER UE) grant MTM2017-89664-P.

Institutional Review Board Statement: Not applicable.

Informed Consent Statement: Not applicable.

Data Availability Statement: We state that data are available to the readers.

Conflicts of Interest: The authors declare no conflict of interest.

Abbreviations

The following abbreviations are used in this manuscript:

FDM	finite difference method
GL	Gauss-Laguerre
m.s.	mean square
ODE	ordinary differential equation
PDE	partial differential equation
r.v.	random variable
s.p.	stochastic process

References

1. Pettersson, M.P.; Iaccarino, G.; Nordström, J. *Polynomial Chaos Methods for Hyperbolic Partial Differential Equations. Numerical Techniques for Fluid Dynamics Problems in the Presence of Uncertainties*; Springer International Publishing: Cham, Switzerland, 2015; p. 214.
2. Yeh, K.C.; Liu, C.H. Wave Propagation in Random Media. In *Theory of Ionospheric Waves*; Academic Press: Cambridge, MA, USA, 1972; Volume 17, pp. 308–366. [CrossRef]
3. Gibson, W.C. *The Method of Moments in Electromagnetics*; Taylor & Francis Group: Abingdon, UK, 2008.
4. Vergara, R.C. Development of Geostatistical Models Using Stochastic Partial Differential Equations. Ph.D. Thesis, Université Paris Sciences et Lettres, Paris, France, 2018.
5. Soong, T. *Random Differential Equations in Science and Engineering*, 1st ed.; Academic Press: Cambridge, MA, USA, 1973; Volume 103.
6. Kroese, D.P.; Taimre, T.; Botev, Z. *Handbook of Monte Carlo Methods*; John Wiley & Sons: New York, NY, USA, 2011.
7. Asmussen, S.; Glynn, P. *Stochastic Simulation: Algorithms and Analysis*; Springer Science & Business Media: Berlin/Heidelberg, Germany, 2007; Volume 57.
8. Casabán, M.C.; Company, R.; Jódar, L. Non-gaussian quadrature integral transform solution of parabolic models with a finite degree of randomness. *Mathematics* **2020**, *8*, 1112. [CrossRef]
9. Davies, B.; Martin, B. Numerical inversion of the laplace transform: A survey and comparison of methods. *J. Comput. Phys.* **1979**, *33*, 1–32. [CrossRef]

10. Stehfest, H. Algorithm 368: Numerical Inversion of Laplace Transforms [D5]. *Commun. ACM* **1970**, *13*, 47–49. [CrossRef]
11. Talbot, A. The Accurate Numerical Inversion of Laplace Transforms. *IMA J. Appl. Math.* **1979**, *23*, 97–120. [CrossRef]
12. Defreitas, C.L.; Kane, S.J. The noise handling properties of the Talbot algorithm for numerically inverting the Laplace transform. *J. Algorithms Comput. Technol.* **2018**, *13*, 1748301818797069. [CrossRef]
13. Iserles, A. On the numerical quadrature of highly-oscillating integrals I: Fourier transforms. *IMA J. Numer. Anal.* **2004**, *24*, 365–391. [CrossRef]
14. Davis, P. J. *Methods of Numerical Integration*; Dover Publications: Mineola, NY, USA, 2007; p. 612.
15. Casabán, M.C.; Company, R.; Egorova, V.N.; Jódar, L. Integral transform solution of random coupled parabolic partial differential models. *Math. Methods Appl. Sci.* **2020**, *43*, 8223–8236. [CrossRef]
16. Casabán, M.C.; Cortés, J.C.; Jódar, L. A random Laplace transform method for solving random mixed parabolic differential problems. *Appl. Math. Comput.* **2015**, *259*, 654–667. [CrossRef]
17. Arnold, L. *Stochastic Differential Equations Theory and Applications*; John Wiley: Hoboken, NJ, USA, 1974.
18. Casabán, M.C.; Company, R.; Jódar, L. Numerical Integral Transform Methods for Random Hyperbolic Models with a Finite Degree of Randomness. *Mathematics* **2019**, *7*, 853. [CrossRef]
19. Iserles, A.; Nørsett, S.P. On Quadrature Methods for Highly Oscillatory Integrals and Their Implementation. *BIT Numer. Math.* **2004**, *44*, 755–772. [CrossRef]
20. Shao, T.S.; Frank, T.C.; Chen, R.M. Tables of zeros and Gaussian weights of certain associated Laguerre polynomials and the related generalized Hermite polynomials. *Math. Comput.* **1964**, *18*, 598–616. [CrossRef]
21. Conte, D.; Ixaru, L.G.; Paternoster, B.; Santomauro, G. Exponentially-fitted Gauss-Laguerre quadrature rule for integrals over an unbounded interval. *J. Comput. Appl. Math.* **2014**, *255*, 725–736. [CrossRef]
22. Abate, J.; Whitt, W. A Unified Framework for Numerically Inverting Laplace Transforms. *INFORMS J. Comput.* **2006**, *18*, 408–421. [CrossRef]
23. Polyanin, A.D.; Nazaikinskii, V.E. *Handbook of Linear Partial Differential Equations for Engineers and Scientists*, 2nd ed.; CRC Press: Boca Raton, FL, USA, 2016; p. 1643.
24. Shampine, L. Vectorized adaptive quadrature in MATLAB. *J. Comput. Appl. Math.* **2008**, *211*, 131–140. [CrossRef]

Article

Approximating the Density of Random Differential Equations with Weak Nonlinearities via Perturbation Techniques

Juan-Carlos Cortés [†], Elena López-Navarro [†], José-Vicente Romero [†] and María-Dolores Roselló [*,†]

Instituto Universitario de Matemática Multidisciplinar, Universitat Politècnica de València, Camino de Vera s/n, 46022 Valencia, Spain; jccortes@imm.upv.es (J.-C.C.); ellona1@upvnet.upv.es (E.L.-N.); jvromero@imm.upv.es (J.-V.R.)
* Correspondence: drosello@imm.upv.es
† These authors contributed equally to this work.

Abstract: We combine the stochastic perturbation method with the maximum entropy principle to construct approximations of the first probability density function of the steady-state solution of a class of nonlinear oscillators subject to small perturbations in the nonlinear term and driven by a stochastic excitation. The nonlinearity depends both upon position and velocity, and the excitation is given by a stationary Gaussian stochastic process with certain additional properties. Furthermore, we approximate higher-order moments, the variance, and the correlation functions of the solution. The theoretical findings are illustrated via some numerical experiments that confirm that our approximations are reliable.

Keywords: stochastic perturbations; random nonlinear oscillator; maximum entropy principle; probability density function; stationary Gaussian noise

Citation: Cortés, J.-C.; López-Navarro, E.; Romero, J.-V.; Roselló, M.-D. Approximating the Density of Random Differential Equations with Weak Nonlinearities via Perturbation Techniques. *Mathematics* **2021**, *9*, 204. https://doi.org/10.3390/math9030204

Received: 27 December 2020
Accepted: 15 January 2021
Published: 20 January 2021

Publisher's Note: MDPI stays neutral with regard to jurisdictional claims in published maps and institutional affiliations.

Copyright: © 2021 by the authors. Licensee MDPI, Basel, Switzerland. This article is an open access article distributed under the terms and conditions of the Creative Commons Attribution (CC BY) license (https://creativecommons.org/licenses/by/4.0/).

1. Introduction and Motivation

The analysis of stochastic perturbations in nonlinear dynamical systems is a hot topic in applied mathematics [1,2] with many applications in apparently different areas such as control [3], economy [4] and especially in dealing with nonlinear vibratory systems. The study of systems subject to vibrations is encountered, for example, in Physics (in the analysis of different types of oscillators) and in Engineering (in the analysis of road vehicles, response of structures to earthquakes' excitations or to sea waves). The nature of vibrations in this type of systems is usually random because they are spawned by complex factors that are not known in a deterministic manner but statistically characterized via measurements that often contain errors and uncertainties. Although, oscillators in Physics and Engineering systems have been extensively studied in the deterministic case [5,6], and particularly, in the nonlinear case [7–9], due to the above-mentioned facts the stochastic analysis is more suitable since provides better understanding of their dynamics.

Many vibratory systems are governed by differential equations with small nonlinear terms of the following form,

$$\ddot{X}(t) + \beta \dot{X}(t) + \omega_0^2 (X(t) + \epsilon g(X(t))) = Y(t), \quad t > 0. \tag{1}$$

Here, $X(t)$ denotes the position (usually of the angle w.r.t. an origin) of the oscillatory system at the time instant t, the parameter β is given by $\beta := 2\xi\omega_0$, being ξ the damping constant and $\omega_0 > 0$ the undamped angular frequency, and finally, ϵ is a small perturbation ($|\epsilon| \ll 1$) affecting a nonlinear function of the position, $g(X(t))$. The expression $X(t) + \epsilon g(X(t))$ is referred to as the nonlinear restoring term. The right-hand side term, $Y(t)$, stands for an external source/forcing term (vibration) acting upon the system. In the setting of random vibration systems, $Y(t)$ is assumed to be a stochastic process, termed stochastic excitation, having certain characteristics that in the present study will be specified later.

Notice that the nonlinear restoring term in Equation (1) involves the parameter ϵ, which determines the magnitude of the nonlinear perturbation, whose shape is given by $g(X(t))$. When $\epsilon = 0$, Equation (1) describes a random linear oscillator. In [10], authors analyze this class of oscillators considering two cases for the stochastic source term $Y(t)$, first when is Gaussian and, secondly, when it can be represented via a Karhunen–Loève expansion. In the case that $\epsilon \neq 0$, the inclusion of the nonlinear term makes more difficult (even simply impossible) to exactly solve Equation (1). An effective method to construct reliable approximations of Equation (1) in the case that ϵ represents a small parameter is the perturbation technique [11–15]. In the stochastic setting, this method has been successfully applied to study different type of oscillators subject to random vibrations. After pioneer contributions by Crandall [16,17], the analysis of random vibration systems has attracted many researchers (see, for instance, in [15,18,19] for a full overview of this topic). In [20], approximations of quadratic and cubic nonlinear oscillators subject to white noise excitations are constructed by combining the Wiener–Hermite expansion and the homotopy perturbation technique. The aforementioned approximations correspond to the first statistical moments (mean and variance) because, as authors indicate in the introduction section, the computation of the probability density function (PDF) is usually very difficult to obtain. In [21], the authors extend the previous analysis to compute higher-order statistical moments of the oscillator response in the case the nonlinearity is only quadratic. The previous methodology is extended and algorithmically automated in [22]. In [23], the author considers the interesting scenario of an harmonic oscillator with a random mass and analyses important dynamic characteristics such as the stochastic stability and the resonance phenomena. To conduct that study, a new type of Brownian motion is introduced. The perturbation technique has also been used to approximate the first moments, mainly the mean and the variance, of some oscillators subject to small nonlinearities. The computational procedures of this method often requires amendments to the existing solution codes, so it is classified as an intrusive method. A spectral technique that allows overcoming this drawback is non-intrusive polynomial chaos expansion (PCE) in which simulations are used as black boxes and the calculation of chaos expansion coefficients for response metrics of interest is based on a set of simulation response evaluations. In the recent paper [24], authors design an interesting hybrid non-intrusive procedure that combine PCE with Chebyshev Surrogate Method to analyze a number of uncertain physical parameters and the corresponding transient responses of a rotating system.

Besides computing the first statistical moments of the response or performing a stability analysis of systems under stochastic vibrations, we must emphasize that the computation of the finite distribution (usually termed "fidis") associated to the stationary solution, and particularly of the stationary PDF, is also a major goal in the realm of vibratory systems with uncertainties. Some interesting contributions in this regard include [25,26]. In [25], the authors first present a complete overview of methods and techniques available to determine the stationary PDF of nonlinear oscillators excited by random functions. Second, nonlinear stochastic oscilators excited by a combination of Gaussian and Poisson white noises are fully analyzed. The study is based on solving the forward generalized Kolmogorov partial differential equation (PDE) using the exponential-polynomial closure method. The theoretical analysis is accompanied with several illustrative examples. In the recent contribution [26], authors propose a new method to compute a closed-form solution of stationary PDF of single-degree-of-freedom vibro-impact systems under Gaussian white noise excitation. The density is obtained by solving the Fokker–Planck–Kolmogorov PDE using the iterative method of weighted residue combined with the concepts of the circulatory and potential probability flows. Apart from obtaining the density of the solutions, it is worth to pointing out that in some recent contributions one also determines the densities of key quantities, that belong to Reliability Theory, like the first-passage time for vibro-impact systems with randomly fluctuating restoring and damping terms (see [27] and references therein).

In this paper, we address the study of random cross-nonlinear oscillators subject to small perturbations affecting the nonlinear term, g, which depend on both the position, $X(t)$, and the velocity, $\dot{X}(t)$,

$$\ddot{X}(t) + 2\zeta\omega_0\dot{X}(t) + \epsilon g(X(t), \dot{X}(t)) + \omega_0^2 X(t) = Y(t). \tag{2}$$

Here, the stochastic derivatives are understood in the mean square sense [28] (Chapter 4). In our subsequent analysis, we will consider the case that $g(X(t), \dot{X}(t)) = X^2(t)\dot{X}(t)$ and the excitation $Y(t)$ is a mean square differentiable and stationary zero-mean Gaussian stochastic process whose correlation function, $\Gamma_{YY}(\tau)$, is known. On the other hand, assuming that $Y(t)$ is a stationary and Gaussian stochastic process is a rather intuitive concept, which has been extensively used in both theoretical and practical studies [29,30]. Stationarity means that the statistical properties of the process do not vary significantly over time/space. This feature is usually met in a number of modeling problems as the surface of the sea in both spatial and time coordinates, noise in time in electric circuits under steady-state operations, homogeneous impurities in engineering materials and media, for example [28] (Chapter 3).

Now, we list the main novelties of our contribution.

- We combine mean square calculus and the stochastic perturbation method to study a class of nonlinear oscillators whose nonlinear term, g, involves both position, $X(t)$, as velocity, $\dot{X}(t)$, specifically, we consider the case $g = g(X(t), \dot{X}(t)) = X^2(t)\dot{X}(t)$. This corresponds to the most complicated case, usually termed *cross-nonlinearity*.
- The oscillator is subject to random excitations driven by a stochastic process, $Y(t)$, having the following properties: $Y(t)$ is mean square differentiable and stationary zero-mean Gaussian.
- We compute reliable approximations, not only of the mean, the variance, and the covariance (as is usually done), but also of higher moments (including the asymmetry and the kurtosis) of the steady-state of the above-described nonlinear oscillator.
- We combine the foregoing information related to higher moments and the entropy method to construct reliable approximations of the probability density function of the steady-state solution. The approximation is quite accurate as it is based on higher moments.

To the best of our knowledge, this is the first time that stochastic nonlinear oscillators with the above-described type of cross-nonlinearities is studied using our approach, i.e., combining mean square calculus and the stochastic perturbation method. In this sense, we think that our approach may be useful to extend our study to stochastic nonlinear oscillators having more general cross-nonlinearities, in particular of the form $g(X(t), \dot{X}(t)) = X^n(t)\dot{X}^m(t)$, for $n \geq 3$ and $m \geq 2$.

The paper is organized as follows. In Section 2, we introduce the auxiliary stochastic results that will be used throughout the whole paper. This section is intended to help the reader to better understand the technical aspects of the paper. Section 3 is divided into two parts. In Section 3.1, we apply the perturbation technique to construct a first-order approximation of the stationary solution stochastic process of model (2) with $g(X(t), \dot{X}(t)) = X^2(t)\dot{X}(t)$. In Section 3.2, we determine expressions for the first higher-order moments, the variance, the covariance, and the correlation of the aforementioned first-order approximation. These expressions will be given in terms of certain integrals of the correlation function of the Gaussian noise, $Y(t)$, and of the classical impulse response function to the linearized oscillator associated to Equation (2). In Section 4, we take advantage of the results given in Section 3 to construct reliable approximations of the PDF of the stationary solution using the principle of maximum entropy. In Section 5, we illustrate all theoretical findings by means of several illustrative examples. Our numerical results are compared with Monte Carlo simulations and with the application of Euler–Maruyama numerical scheme, showing full agreement. Conclusions are drawn in Section 6.

2. Stochastic Preliminaries

For the sake of completeness, in this section we will introduce some technical stochastic results that will be required throughout the paper.

Hereinafter, we will work on a complete probability space $(\Omega, \mathcal{F}, \mathbb{P})$, i.e., Ω is a sample space; $\mathcal{F}$ is a σ-algebra of sets of Ω, usually called events; and $\mathbb{P}$ is a probability measure. To simplify, we will omit the sample notation, so the input and the solution stochastic processes involved in Equation (2) will be denoted by $Y(t) \equiv \{Y(t) : t \geq 0\}$ and $X(t) \equiv \{X(t) : t \geq 0\}$, respectively, rather than $\{Y(t;\omega) : t \geq 0, \omega \in \Omega\}$ and $\{X(t;\omega) : t \geq 0, \omega \in \Omega\}$, respectively.

The following result will be applied to calculate some higher-order moments of the solution stochastic process, $X(t)$, of the random differential Equation (2), since as it shall be seen later, $X(t)$ depends on a product of the stochastic excitation, $Y(t)$, evaluated at a finite number of instants, say $t_1, t_2, \ldots, t_n$, $Y(t_i) = Y_i$, $1 \leq i \leq n$.

Proposition 1 (p. 28, [28]). *Let the random variables $Y_1, Y_2, \ldots, Y_n$ be jointly Gaussian with zero mean, $\mathbb{E}\{Y_i\} = 0, 1 \leq i \leq n$. Then, all odd order moments of these random variables vanish and for n even,*

$$\mathbb{E}\{Y_1 Y_2 \cdots Y_n\} = \sum_{m_1, m_2, \ldots, m_n} \mathbb{E}\{Y_{m_1} Y_{m_2}\} \mathbb{E}\{Y_{m_3} Y_{m_4}\} \cdots \mathbb{E}\{Y_{m_{n-1}} Y_{m_n}\}.$$

The sum above is taken over all possible combinations of $n/2$ pairs of n random variables. The number of terms in the summation is $1 \cdot 3 \cdot 5 \cdots (n-3) \cdot (n-1)$.

The two following results permit interchange the expectation operator with the mean square derivative and the mean square integral. In [28] (Equation (4.130) in Section 4.4.2), the first result is established for $n = 2$ and then it follows straightforwardly by induction.

Proposition 2. *Let $\{Y(t) : t \geq 0\}$ be a mean square differentiable stochastic process. Then,*

$$\mathbb{E}\{Y(t_1) \cdots Y(t_{n-1}) \dot{Y}(t_n)\} = \frac{\partial}{\partial t_n}(\mathbb{E}\{Y(t_1) \cdots Y(t_n)\}), \quad t_1, \ldots, t_n \geq 0,$$

provided the above expectations exists.

Proposition 3 (p. 104, [28]). *Let $\{Y(t) : -\infty \leq a \leq t \leq b \leq +\infty\}$ be a second-order stochastic process integrable in the mean square sense and $h(t)$ a Riemann integrable deterministic function on $t \in (a, b)$. Then,*

$$\mathbb{E}\left\{\int_a^b h(t) Y(t) \, dt\right\} = \int_a^b h(t) \mathbb{E}\{Y(t)\} \, dt.$$

The following is a distinctive property of Gaussian processes since they preserve Gaussianity under mean square integration.

Proposition 4 (p. 112, [28]). *Let $\{Y(t) : a \leq t \leq \infty\}$ be a Gaussian process and let $h(t)$ be a Riemann integrable deterministic function on (a, t) such that the following mean square integral,*

$$X(t) = \int_a^t h(t, \tau) Y(\tau) d\tau,$$

exists, then $\{X(t) : t \geq a\}$ is a Gaussian process.

3. Probabilistic Model Study

As it has been indicated in Section 1, in this paper we will study, from a probabilistic standpoint, the random cross-nonlinear oscillator

$$\ddot{X}(t) + 2\zeta\omega_0 \dot{X}(t) + \epsilon X^2(t) \dot{X}(t) + \omega_0^2 X(t) = Y(t). \tag{3}$$

The analysis will be divided into two steps. First, in Section 3.1 we will apply the perturbation technique to obtain an approximation, $\widehat{X}(t)$, of the stationary solution stochastic process, $X(t)$. Then, in Section 3.2 we will take advantage of $\widehat{X}(t)$ to determine reliable approximations of the main statistical functions of $X(t)$, namely, the first higher-order moments, $\mathbb{E}\{X^n(t)\}, n = 1, \ldots, 5$; the variance, $\mathbb{V}\{X(t)\}$; the covariance, $\mathbb{C}\mathrm{ov}\{X(t_1), X(t_2)\}$; and the correlation, $\Gamma_{XX}(\tau)$.

3.1. Perturbation Technique

Let us consider the Equation (3). The main idea of the stochastic perturbation technique is to consider that the solution $X(t)$ can be expanded in the powers of the small parameter ϵ ($|\epsilon| \ll 1$),

$$X(t) = X_0(t) + \epsilon X_1(t) + \epsilon^2 X_2(t) + \cdots \tag{4}$$

Replacing expression (4) into Equation (3), yields the following sequence of linear differential equations, with random inputs

$$\begin{aligned}
\epsilon^0 &: \ddot{X}_0(t) + 2\zeta\omega_0 \dot{X}_0(t) + \omega_0^2 X_0(t) &&= Y(t), \\
\epsilon^1 &: \ddot{X}_1(t) + 2\zeta\omega_0 \dot{X}_1(t) + \omega_0^2 X_1(t) &&= -X_0^2(t)\dot{X}_0(t), \\
\epsilon^2 &: \ddot{X}_2(t) + 2\zeta\omega_0 \dot{X}_2(t) + \omega_0^2 X_2(t) &&= -2X_0(t)X_1(t)\dot{X}_0(t) - X_0^2(t)\dot{X}_1, \\
\vdots & \quad \vdots && \quad \vdots
\end{aligned} \tag{5}$$

Notice that each equation can be solved in cascade. As usual, when applying the perturbation technique, we take the first-order approximation

$$\widehat{X}(t) = X_0(t) + \epsilon X_1(t). \tag{6}$$

This entails that in our subsequent development we will only need the two first equations listed in (5).

As indicated in Section 1, now we will focus on the analysis of the steady-state solution. Using the linear theory, the two first equations in (5) can be solved using the convolution integral [31]:

$$X_0(t) = \int_0^\infty h(s) Y(t-s)\, ds, \tag{7}$$

and

$$X_1(t) = \int_0^\infty h(s) \left[-X_0^2(t-s)\dot{X}_0(t-s)\right] ds, \tag{8}$$

where

$$h(t) = \begin{cases} (\omega_0^2 - \zeta^2\omega_0^2)^{-\frac{1}{2}} e^{-\zeta\omega_0 t} \sin\left[(\omega_0^2 - \zeta^2\omega_0^2)^{\frac{1}{2}} t\right] & \text{if } t > 0, \\ 0 & \text{if } t \le 0, \end{cases} \tag{9}$$

is the impulse response function for the underdamped case $\zeta^2 < 1$. This situation corresponds to the condition in which damping of an oscillator causes it to return to equilibrium with the amplitude gradually decreasing to zero (in our random setting it means that the expectation of the amplitude is null); system returns to equilibrium faster but overshoots and crosses the equilibrium position one or more times. Although, they are no treated hereinafter, two more situations are also possible, namely, critical damping and overdamping. The former corresponds to $\zeta^2 = 1$ and in that case the damping of an oscillator causes it to return as quickly as possible to its equilibrium position without oscillating back and forth about this position, while the latter corresponds to $\zeta^2 > 1$, and in this situation damping of an oscillator causes it to return to equilibrium without oscillating; oscillator moves more slowly toward equilibrium than in the critically damped system [32].

3.2. Approximation of the Main Statistical Moments

This subsection is devoted to calculate the main probabilistic information of the stationary solution stochastic process, $X(t)$, of model (3). As it has been previously pointed out, to this end, we assume that the input term $Y(t)$ is a stationary zero-mean ($\mathbb{E}\{Y(t)\} = 0$) Gaussian stochastic process whose correlation function, $\Gamma_{YY}(\tau)$, is given. We will further assume that $Y(t)$ is mean square differentiable. This additional hypothesis will be apparent later. At this point, it is convenient to recall that for any stationary stochastic process its correlation function is even, so $\Gamma_{YY}(\tau) = \Gamma_{YY}(-\tau)$, (p. 47, [28]). This property will be extensively applied throughout our subsequent developments.

To compute the mean of the approximation, we first take the expectation operator in (6),

$$\mathbb{E}\{\widehat{X}(t)\} = \mathbb{E}\{X_0(t)\} + \epsilon \mathbb{E}\{X_1(t)\}. \tag{10}$$

Therefore, we now need to determine both $\mathbb{E}\{X_0(t)\}$ and $\mathbb{E}\{X_1(t)\}$. To compute the $\mathbb{E}\{X_0(t)\}$ we again use the expectation operator in (7),

$$\mathbb{E}\{X_0(t)\} = \mathbb{E}\left\{\int_0^\infty h(s)Y(t-s)\,ds\right\} = \int_0^\infty h(s)\mathbb{E}\{Y(t-s)\}\,ds = 0, \tag{11}$$

where we have applied Proposition 3 and that $\mathbb{E}\{Y(t)\} = 0$.

Now, we deal with the computation of $\mathbb{E}\{X_1(t)\}$ in an analogous manner but using the representation of $X_1(t)$ given in (8),

$$\begin{aligned}
\mathbb{E}\{X_1(t)\} &= \mathbb{E}\left\{\int_0^\infty h(s)\left[-X_0^2(t-s)\dot{X}_0(t-s)\right]ds\right\} = \int_0^\infty h(s)\mathbb{E}\left\{\left[-X_0^2(t-s)\dot{X}_0(t-s)\right]\right\}ds \\
&= -\int_0^\infty h(s)\int_0^\infty h(s_1)\int_0^\infty h(s_2)\int_0^\infty h(s_3)\mathbb{E}\{Y(t-s-s_1)Y(t-s-s_2)\dot{Y}(t-s-s_3)\}\,ds_3\,ds_2\,ds_1\,ds \\
&= 0.
\end{aligned} \tag{12}$$

Notice that the assumption of mean square differentiability of the input process $Y(t)$ appears naturally at this stage.

Let us justify the last step in expression (12). Let us denote $u_1 = t - s - s_1$, $u_2 = t - s - s_2$ and $u_3 = t - s - s_3$, then applying Propositions 2 and 1, both with $n = 3$, one gets

$$\mathbb{E}\{Y(t-s-s_1)Y(t-s-s_2)\dot{Y}(t-s-s_3)\} = \mathbb{E}\{Y(u_1)Y(u_2)\dot{Y}(u_3)\} = \frac{\partial}{\partial u_3}\mathbb{E}\{Y(u_1)Y(u_2)Y(u_3)\} = 0.$$

Therefore, substituting (11) and (12) into (10), we obtain the expectation of the approximation is null,

$$\mathbb{E}\{\widehat{X}(t)\} = \mathbb{E}\{X_0(t)\} + \epsilon \mathbb{E}\{X_1(t)\} = 0. \tag{13}$$

From the approximation (6) and neglecting the term ϵ^2, the second-order moment for $\widehat{X}(t)$ is given by

$$\mathbb{E}\left\{\widehat{X}^2(t)\right\} = \mathbb{E}\left\{X_0^2(t)\right\} + 2\epsilon \mathbb{E}\{X_0(t)X_1(t)\}. \tag{14}$$

The first addend can be calculated using expression (7) and Fubini's theorem,

$$\begin{aligned}
\mathbb{E}\left\{X_0^2(t)\right\} &= \int_0^\infty h(s)\int_0^\infty h(s_1)\mathbb{E}\{Y(t-s)Y(t-s_1)\}\,ds_1\,ds \\
&= \int_0^\infty h(s)\int_0^\infty h(s_1)\Gamma_{YY}(s-s_1)\,ds_1\,ds.
\end{aligned} \tag{15}$$

Notice that we have used that $Y(t)$ is a stationary process, so

$$\mathbb{E}\{Y(t-s)Y(t-s_1)\} = \Gamma_{YY}(t-s_1-(t-s)) = \Gamma_{YY}(s-s_1).$$

Now, we calculate the second addend in (14). To this end, we substitute the expressions of $X_0(t)$ and $X_1(t)$ given in (7) and (8), respectively,

$$\mathbb{E}\{X_0(t)\,X_1(t)\} = \int_0^\infty h(s)\mathbb{E}\{X_0(t)[-X_0^2(t-s)\dot{X}_0(t-s)]\}\,ds$$

$$= \int_0^\infty h(s)\mathbb{E}\left\{-\int_0^\infty h(s_1)Y(t-s_1)\,ds_1 \int_0^\infty h(s_2)Y(t-s-s_2)\,ds_2 \int_0^\infty h(s_3)Y(t-s-s_3)\,ds_3 \int_0^\infty h(s_4)\dot{Y}(t-s-s_4)\,ds_4 \right\}ds$$

$$= -\int_0^\infty h(s) \int_0^\infty h(s_1) \int_0^\infty h(s_2) \int_0^\infty h(s_3) \int_0^\infty h(s_4)\mathbb{E}\{Y(t-s_1)Y(t-s-s_2)Y(t-s-s_3)\dot{Y}(t-s-s_4)\}\,ds_4\,ds_3\,ds_2\,ds_1\,ds$$

$$\stackrel{(I)}{=} -\int_0^\infty h(s) \int_0^\infty h(s_1) \int_0^\infty h(s_2) \int_0^\infty h(s_3) \int_0^\infty h(s_4)\Big(\Gamma_{YY}(s_1-s-s_2)\Gamma'_{YY}(s_3-s_4) + \Gamma_{YY}(s_1-s-s_3)\Gamma'_{YY}(s_2-s_4) \quad (16)$$

$$+ \Gamma'_{YY}(s_1-s-s_4)\Gamma_{YY}(s_2-s_3)\Big)\,ds_4\,ds_3\,ds_2\,ds_1\,ds$$

$$\stackrel{(II)}{=} -\int_0^\infty h(s) \int_0^\infty h(s_1) \int_0^\infty h(s_2) \int_0^\infty h(s_3) \int_0^\infty h(s_4)\Big(2\Gamma_{YY}(s_1-s-s_2)\Gamma'_{YY}(s_3-s_4)$$

$$+ \Gamma'_{YY}(s_1-s-s_4)\Gamma_{YY}(s_2-s_3)\Big)\,ds_4\,ds_3\,ds_2\,ds_1\,ds.$$

Observe that in the step (I) of the above expression, we have first applied Proposition 2 and second Proposition 1. Indeed, let us denote by $u_1 = t - s_1$, $u_2 = t - s - s_2$, $u_3 = t - s - s_3$ and $u_4 = t - s - s_4$, then by Proposition 2, with $n = 4$, one gets

$$\mathbb{E}\{Y(t-s_1)Y(t-s-s_2)Y(t-s-s_3)\dot{Y}(t-s-s_4)\} = \frac{\partial}{\partial u_4}\mathbb{E}\{Y(u_1)Y(u_2)Y(u_3)Y(u_4)\},$$

and now we apply Proposition 1, with $n = 4$, to the right-hand side. This yields

$$\mathbb{E}\Big\{Y(t-s_1)Y(t-s-s_2)Y(t-s-s_3)\dot{Y}(t-s-s_4)\Big\} =$$

$$= \frac{\partial}{\partial u_4}\Big(\mathbb{E}\{Y(u_1)Y(u_2)\}\mathbb{E}\{Y(u_3)Y(u_4)\} + \mathbb{E}\{Y(u_1)Y(u_3)\}\mathbb{E}\{Y(u_2)Y(u_4)\} + \mathbb{E}\{Y(u_1)Y(u_4)\}\mathbb{E}\{Y(u_2)Y(u_3)\}\Big)$$

$$= \frac{\partial}{\partial u_4}\big(\Gamma_{YY}(u_2-u_1)\Gamma_{YY}(u_4-u_3) + \Gamma_{YY}(u_3-u_1)\Gamma_{YY}(u_4-u_2) + \Gamma_{YY}(u_4-u_1)\Gamma_{YY}(u_3-u_2)\big)$$

$$= \Gamma_{YY}(u)|_{u=s_1-s-s_2}\Gamma'_{YY}(u)|_{u=s_3-s_4} + \Gamma_{YY}(u)|_{u=s_1-s-s_3}\Gamma'_{YY}(u)|_{u=s_2-s_4} + \Gamma'_{YY}(u)|_{u=s_1-s-s_4}\Gamma_{YY}(u)|_{u=s_2-s_3}.$$

In step (II) of expression (16) we have taken advantage of the symmetry of the indexes. Then, substituing (15) and (16) in (14) one gets

$$\mathbb{E}\big\{\widehat{X}^2(t)\big\} = \int_0^\infty h(s) \int_0^\infty h(s_1)\Gamma_{YY}(s-s_1)\,ds_1\,ds - 2\epsilon \left(\int_0^\infty h(s) \int_0^\infty h(s_1) \int_0^\infty h(s_2) \int_0^\infty h(s_3) \int_0^\infty h(s_4) \right.$$

$$\left. \cdot \Big(2\Gamma_{YY}(s_1-s-s_2)\Gamma'_{YY}(s_3-s_4) + \Gamma'_{YY}(s_1-s-s_4)\Gamma_{YY}(s_2-s_3)\Big)\,ds_4\,ds_3\,ds_2\,ds_1\,ds \right). \quad (17)$$

Notice that $\mathbb{E}\big\{\widehat{X}^2(t)\big\}$ does not depend on t. This is consistent with the fact that we are dealing with the stochastic analysis of the stationary solution. The same feature will hold when computing higher-order moments, $\mathbb{E}\big\{\widehat{X}^n(t)\big\}$, $n > 2$, later.

As $\mathbb{E}\big\{\widehat{X}(t)\big\}$ is null (see (13)), then the variance of the solution coincides with $\mathbb{E}\big\{\widehat{X}^2(t)\big\}$.

Now, we calculate the third-order moment of $\widehat{X}(t)$ keeping up to the first-order term of perturbation ϵ. Therefore,

$$\mathbb{E}\big\{\widehat{X}^3(t)\big\} = \mathbb{E}\big\{X_0^3(t)\big\} + 3\epsilon\mathbb{E}\big\{X_0^2(t)X_1(t)\big\}. \quad (18)$$

Reasoning analogously as we have shown before, we obtain

$$\mathbb{E}\big\{X_0^3(t)\big\} = \int_0^\infty h(s) \int_0^\infty h(s_1) \int_0^\infty h(s_2)\mathbb{E}\{Y(t-s)Y(t-s_1)Y(t-s_2)\}\,ds_2\,ds_1\,ds = 0, \quad (19)$$

where we have applied Proposition 1 in the last step.

The second addend in (18) is calculated using Propositions 1 and 2,

$$\mathbb{E}\{X_0^2(t)X_1(t)\} = \int_0^\infty h(s)\mathbb{E}\{X_0^2(t)\left[-X_0^2(t-s)\dot{X}_0(t-s)\right]\}ds$$
$$= -\int_0^\infty h(s)\int_0^\infty h(s_1)\int_0^\infty h(s_2)\int_0^\infty h(s_3)\int_0^\infty h(s_4)\int_0^\infty h(s_5)\mathbb{E}\{Y(t-s)Y(t-s_1)Y(t-s-s_3)Y(t-s-s_4) \quad (20)$$
$$\cdot \dot{Y}(t-s-s_5)\}ds_5\,ds_4\,ds_3\,ds_2\,ds_1\,ds = 0.$$

From (19) and (20), we obtain

$$\mathbb{E}\{\widehat{X}^3(t)\} = \mathbb{E}\{X_0^3(t)\} + 3\epsilon\mathbb{E}\{X_0^2(t)X_1(t)\} = 0.$$

Using again the first-order approximation of the perturbation ϵ, in general, it can be straightforwardly seen that

$$\mathbb{E}\{\widehat{X}^n(t)\} = 0, \quad n = 1, 3, 5, \ldots. \quad (21)$$

Indeed, we know that,

$$\mathbb{E}\{\widehat{X}^n(t)\} = \mathbb{E}\{X_0^n(t)\} + n\epsilon\mathbb{E}\{X_0^{n-1}(t)X_1(t)\}. \quad (22)$$

On the one hand, let us observe that applying first Fubini's theorem and Proposition 3, and second Proposition 1 for n odd, one gets

$$\mathbb{E}\{X_0^n(t)\} = \mathbb{E}\{\left(\int_0^\infty h(s)Y(t-s)\,ds\right)^n\} = \int_0^\infty h(s_1)\cdots\int_0^\infty h(s_n)\mathbb{E}\{Y(t-s_1)\cdots Y(t-s_n)\}\,ds_n\cdots ds_1 = 0.$$

On the other hand, using the same reasoning as in (20),

$$\mathbb{E}\{X_0^{n-1}(t)X_1(t)\} = \int_0^\infty h(s)\mathbb{E}\{X_0^{n-1}(t)\left[-X_0^2(t-s)\dot{X}_0(t-s)\right]\}ds = 0,$$

where first we have applied Proposition 2, in order to put the first derivative out of the expectation, and second, we have utilized that $X_0^{n-1}(t)$, $X_0^2(t-s)$ and $\dot{X}_0(t-s)$ depend upon $n-1$, 2 and 1 terms of $Y(\cdot)$, respectively, together with Proposition 1 (notice that $n+2$ is odd as n is odd).

To complete the information of statistical moments of the approximation, we also determine $\mathbb{E}\{\widehat{X}^4(t)\}$.

The fourth-order moment of $\widehat{X}(t)$, based on the first-order approximation via the perturbation method, is given by

$$\mathbb{E}\{\widehat{X}^4(t)\} = \mathbb{E}\{X_0^4(t)\} + 4\epsilon\mathbb{E}\{X_0^3(t)X_1(t)\}. \quad (23)$$

Reasoning analogously as we have shown in previous sections, we obtain for the first addend

$$\mathbb{E}\{X_0^4(t)\} = 3\int_0^\infty h(s)\int_0^\infty h(s_1)\int_0^\infty h(s_2)\int_0^\infty h(s_3)\Gamma_{YY}(s-s_1)\Gamma_{YY}(s_2-s_3)\,ds\,ds_1\,ds_2\,ds_3, \quad (24)$$

and for the second addend

$$
\begin{aligned}
\mathbb{E}\{X_0^3(t)X_1(t)\} &= -\int_0^\infty \tilde{h}(s)\int_0^\infty \tilde{h}(s_1)\int_0^\infty \tilde{h}(s_2)\int_0^\infty \tilde{h}(s_3)\int_0^\infty \tilde{h}(s_4)\int_0^\infty \tilde{h}(s_5)\int_0^\infty h(s_6)\mathbb{E}\{Y(t-s_1)\\
&\quad \cdot Y(t-s_2)Y(t-s_3)Y(t-s-s_4)Y(t-s-s_5)\dot{Y}(t-s-s_6)\}\,ds\,ds_1\,ds_2\,ds_3\,ds_4\,ds_5\,ds_6 \\
&= -\int_0^\infty \tilde{h}(s)\int_0^\infty \tilde{h}(s_1)\int_0^\infty \tilde{h}(s_2)\int_0^\infty \tilde{h}(s_3)\int_0^\infty \tilde{h}(s_4)\int_0^\infty \tilde{h}(s_5)\int_0^\infty h(s_6)\Big(6\,\Gamma'_{YY}(s_5-s_6) \\
&\quad \cdot \Gamma_{YY}(s_1-s_2)\Gamma_{YY}(s_3-s-s_4) + 3\,\Gamma'_{YY}(s_1-s-s_6)\big(2\,\Gamma_{YY}(s_2-s-s_4)\Gamma_{YY}(s_3-s-s_5) \\
&\quad + \Gamma_{YY}(s_2-s_3)\Gamma_{YY}(s_4-s_5)\big)\Big)\,ds\,ds_1\,ds_2\,ds_3\,ds_4\,ds_5\,ds_6\,.
\end{aligned}
\tag{25}
$$

Observe that in the last step of the above expression, first we have used Proposition 2, and second, Proposition 1. From this last proposition, we know that exist 15 combinations, but we can reduce the expression by the symmetry of involved indexes.

Now we deal with the approximation of the correlation function of $X(t)$ via (6), i.e., taking the first-order approximation of the perturbation expansion,

$$
\Gamma_{\widehat{X}\widehat{X}}(\tau) = \mathbb{E}\{\widehat{X}(t)\widehat{X}(t+\tau)\} = \mathbb{E}\{X_0(t)X_0(t+\tau)\} + \epsilon[\mathbb{E}\{X_0(t)X_1(t+\tau)\} + \mathbb{E}\{X_1(t)X_0(t+\tau)\}]. \tag{26}
$$

The first addend in (26) corresponds to the correlation function of $X_0(t)$. It can be expressed as

$$
\begin{aligned}
\mathbb{E}\{X_0(t)X_0(t+\tau)\} &= \int_0^\infty \int_0^\infty h(s)h(s_1)\mathbb{E}\{Y(t-s)Y(t+\tau-s_1)\}\,ds\,ds_1 \\
&= \int_0^\infty \int_0^\infty h(s)h(s_1)\Gamma_{YY}(\tau-s_1+s)\,ds_1\,ds\,.
\end{aligned}
$$

The two last addends in (26) represent the cross-correlation of $X_0(t)$ and $X_1(t)$. They are given, respectively, by

$$
\begin{aligned}
\mathbb{E}\{X_0(t)X_1(t+\tau)\} &= \int_0^\infty h(s)\mathbb{E}\Big\{X_0(t)[-X_0^2(t+\tau-s)\dot{X}_0(t+\tau-s)]\Big\}\,ds \\
&= -\int_0^\infty \tilde{h}(s)\int_0^\infty \tilde{h}(s_1)\int_0^\infty \tilde{h}(s_2)\int_0^\infty \tilde{h}(s_3)\int_0^\infty \tilde{h}(s_4)\bigg\{2\,\Gamma_{YY}(\tau-s-s_2+s_1)\Gamma'_{YY}(s_3-s_4) \\
&\quad + \Gamma'_{YY}(\tau-s-s_4+s_1)\Gamma_{YY}(s_2-s_3)\bigg\}\,ds_4\,ds_3\,ds_2\,ds_1\,ds\,.
\end{aligned}
$$

and

$$
\begin{aligned}
\mathbb{E}\{X_1(t)X_0(t+\tau)\} &= \mathbb{E}\bigg\{\int_0^\infty -h(s)X_0^2(t-s)\dot{X}_0(t-s)X_0(t+\tau)\bigg\}\,ds \\
&= -\int_0^\infty \tilde{h}(s)\int_0^\infty \tilde{h}(s_1)\int_0^\infty \tilde{h}(s_2)\int_0^\infty \tilde{h}(s_3)\int_0^\infty \tilde{h}(s_4)\bigg\{\Gamma_{YY}(s_1-s_2)\Gamma'_{YY}(\tau-s_4+s+s_3) \\
&\quad + 2\,\Gamma'_{YY}(s_1-s_3)\Gamma_{YY}(\tau-s_4+s+s_2)\bigg\}\,ds_4\,ds_3\,ds_2\,ds_1\,ds\,.
\end{aligned}
$$

Summarizing,

$$\begin{aligned}
\Gamma_{\widehat{X}\widehat{X}}(\tau) = &\int_0^\infty \int_0^\infty \overset{*}{h}(s) h(s_1) \Gamma_{YY}(\tau - s_1 + s)\, ds\, ds_1 \\
&- \epsilon \int_0^\infty \overset{*}{h}(s) \int_0^\infty h(s_1) \int_0^\infty h(s_2) \int_0^\infty h(s_3) \int_0^\infty h(s_4) \left\{ 2\Gamma_{YY}(\tau - s - s_2 + s_1) \Gamma'_{YY}(s_3 - s_4) \right. \\
&+ \Gamma'_{YY}(\tau - s - s_4 + s_1) \Gamma_{YY}(s_2 - s_3) + \Gamma_{YY}(s_1 - s_2) \Gamma'_{YY}(\tau - s_4 + s + s_3) \\
&\left. + 2\Gamma'_{YY}(s_1 - s_3) \Gamma_{YY}(\tau - s_4 + s + s_2) \right\} ds_4\, ds_3\, ds_2\, ds_1\, ds\, .
\end{aligned} \qquad (27)$$

As $\mathbb{E}\{\widehat{X}(t)\} = 0$, we observe that the covariance and correlation functions of $\widehat{X}(t)$ coincide,
$$\mathbb{C}\mathrm{ov}\{\widehat{X}(t_1), \widehat{X}(t_2)\} = \Gamma_{\widehat{X}\widehat{X}}(\tau), \quad \tau = |t_1 - t_2|.$$

4. Approximating the PDF via the Maximum Entropy Principle

So far, we have calculated approximations of the moments $\mathbb{E}\{\widehat{X}^n(t)\}$, $n = 1, \ldots, 5$ to the first-order approximation, $\widehat{X}(t)$, via the perturbation method, of the steady-state solution of the random nonlinear oscillator (3). Although this is an important information, a more ambitious goal is the approximation of the PDF, say $f_{\widehat{X}(t)}(x)$, as from it one can calculate key stochastic information as the probability that the output lies in a specific interval of interest, say $[a_1, a_2]$,
$$\mathbb{P}\{a_1 \leq \widehat{X}(t) \leq a_2\} = \int_{a_1}^{a_2} f_{\widehat{X}(t)}(x)\, dx,$$

for any arbitrary fixed time t. Furthermore, from the knowledge of the PDF one can easily compute confidence intervals at a specific confidence level $\alpha \in (0,1)$,
$$1 - \alpha = \mathbb{P}\{\mu_{\widehat{X}}(t) - k\sigma_{\widehat{X}}(t) \leq \widehat{X}(t) \leq \mu_{\widehat{X}}(t) + k\sigma_{\widehat{X}}(t)\} = \int_{\mu_{\widehat{X}}(t) - k\sigma_{\widehat{X}}(t)}^{\mu_{\widehat{X}}(t) + k\sigma_{\widehat{X}}(t)} f_{\widehat{X}(t)}(x)\, dx,$$

where $\mu_{\widehat{X}}(t) = \mathbb{E}\{\widehat{X}(t)\} = 0$ (see (13)) and $\sigma_{\widehat{X}}(t) = \sqrt{\mathbb{V}\{\widehat{X}(t)\}}$. Usually α is taken as $\alpha = 0.05$ so that 95% confidence intervals are built, and $k > 0$ must be determined numerically.

As we have calculated the approximations $\mathbb{E}\{\widehat{X}^n(t)\}$, $n = 1, \ldots, 5$, a suitable method to approximate the PDF, $f_{\widehat{X}(t)}(x)$, is the Principle of Maximum Entropy (PME), [33]. For t fixed, the PME seeks for a PDF, $f_{\widehat{X}(t)}(x)$, that maximizes the so-called Shannon's Entropy, of random variable $\widehat{X}(t)$ with support $[a, b]$, defined via the following functional,

$$\mathcal{S}\{f_{\widehat{X}(t)}(x)\} = -\int_a^b f_{\widehat{X}(t)}(x) \log(f_{\widehat{X}(t)}(x))\, dx, \qquad (28)$$

satisfying the following restrictions

$$\int_a^b f_{\widehat{X}(t)}(x)\, dx = 1, \qquad (29)$$

$$\mathbb{E}\{\widehat{X}^n(t)\} = \int_a^b x^n f_{\widehat{X}(t)}(x)\, dx = m_n, \quad n = 1, \ldots, M. \qquad (30)$$

Condition (29) guarantees $f_{\widehat{X}(t)}(x)$ is a PDF, and the M conditions given in (30) impose that the sampled moments, m_n, match the moments, $\mathbb{E}\{\widehat{X}^n(t)\}$, obtained in our setting by the stochastic perturbation method. For each t fixed, the maximization of functional (28) subject to the constrains (29)–(30) can be solved via the auxiliary Lagrange function

$$\mathcal{L}\left\{f_{\widehat{X}(t)}, \lambda_0, \ldots, \lambda_M\right\} = \mathcal{S}\left\{f_{\widehat{X}(t)}(x)\right\} + \sum_{i=0}^{M} \lambda_i \left[m_i - \int_a^b x^i f_{\widehat{X}(t)}(x)\, dx\right],$$

where $m_0 = 1$. It can be seen that the form of the PDF is given by [33]

$$f_{\widehat{X}(t)}(x) = \mathbb{1}_{[a,b]}\, e^{-\sum_{i=0}^M \lambda_i x^i},$$

where $\mathbb{1}_{[a,b]}$ denotes the characteristic function on the interval $[a,b]$.

In Section 3, we have approximated, via the stochastic perturbation technique, the moments $\mathbb{E}\{\widehat{X}^n(t)\}$ for $n = 1, 2, \ldots, 5$. Therefore, to apply the PME we will take $M = 5$ in (30). Notice that, in practice, to calculate the parameters λ_i, $i = 0, 1, \ldots, 5$, we will need to numerically solve the system of nonlinear Equations (29) and (30).

5. Numerical Examples

This section is devoted to illustrate the theoretical findings obtained in previous sections. We take the following data for the parameters of the random nonlinear oscillator (3), $\zeta = 0.05$ ($\zeta^2 < 1$) and $\omega_0 = 1$.

Example 1. *Let us consider as input excitation the trigonometric stochastic process defined by $Y(t) = \xi_1 \cos(t) + \xi_2 \sin(t)$, where $\xi_1, \xi_2 \sim N(0,1)$ are independent. Observe that $Y(t)$ satisfies the hypotheses, i.e., $\mathbb{E}\{Y(t)\} = 0$, $Y(t)$ is Gaussian, mean square differentiable with respect to t, and stationary, with its correlation being $\Gamma_{YY}(t_1, t_2) = \cos(t_1 - t_2)$ or $\Gamma_{YY}(\tau) = \cos(\tau)$. Substituting this data into Equation (3), we obtain*

$$\ddot{X}(t) + 0.1 \dot{X}(t) + \epsilon X^2(t) \dot{X}(t) + X(t) = \xi_1 \cos(t) + \xi_2 \sin(t), \quad \xi_1, \xi_2 \sim N(0,1). \tag{31}$$

Now we shall obtain approximations to the first moments, $\mathbb{E}\{\widehat{X}^i(t)\}, i = 1, \ldots, 5$, the correlation function and the variance, $\mathbb{V}\{\widehat{X}(t)\}$, of the approximate solution $\widehat{X}(t)$ of random nonlinear oscillator (31).

As we have seen in the expression (21), the moments of odd order are null, so, in this case, $\mathbb{E}\{\widehat{X}(t)\} = \mathbb{E}\{\widehat{X}^3(t)\} = \mathbb{E}\{\widehat{X}^5(t)\} = 0$. Now, we sequentially derive some bounds for the perturbation parameter ϵ using the positiveness of even order moments, i.e., $\mathbb{E}\{\widehat{X}^2(t)\} > 0$ and $\mathbb{E}\{\widehat{X}^4(t)\} > 0$. First, it is easy to check that, using expression (17), the second-order moment is given by

$$\mathbb{E}\{\widehat{X}^2(t)\} = 100 - 200000\epsilon, \tag{32}$$

so we obtain the bound $\epsilon < 0.0005$. Since $E\{\widehat{X}(t)\} = 0$, observe that the variance of the first-order approximation is given by (32). Second, using expressions (23)–(25),

$$\mathbb{E}\{\widehat{X}^4(t)\} = 30,000 - \frac{1,153,800,000,000}{6409}\epsilon. \tag{33}$$

This provides a stronger bound, $\epsilon < 0.000166641$.

Now, applying (27), we obtain the following approximation of the correlation function,

$$\Gamma_{\widehat{X}\widehat{X}}(\tau) = 100(1 - 2000\epsilon)\cos(\tau). \tag{34}$$

In Figure 1, we show the graphical representation of the correlation function, $\Gamma_{\widehat{X}\widehat{X}}(\tau)$, given in the expression (34) for different values of ϵ. We can see the higher the perturbation ϵ, the lower the variability. This graphical behavior is in full agreement with the physical interpretation of the oscillator dynamics. Indeed, let us rewrite Equation (31) as follows,

$$\ddot{X}(t) + (0.1 + \epsilon X^2(t))\dot{X}(t) + X(t) = \xi_1 \cos(t) + \xi_2 \sin(t).$$

As $\epsilon > 0$ increases, the damped coefficient $0.1 + \epsilon X^2(t)$ does, so the mechanical system reduces its oscillations. It should be noted that $\epsilon = 0.0004$ only satisfies the first bound ($\epsilon < 0.0005$); however, we can observe that the corresponding approximation preserves symmetry of correlation function. This might be due to the sample regularity of the random excitation, $Y(t)$, which is differentiable.

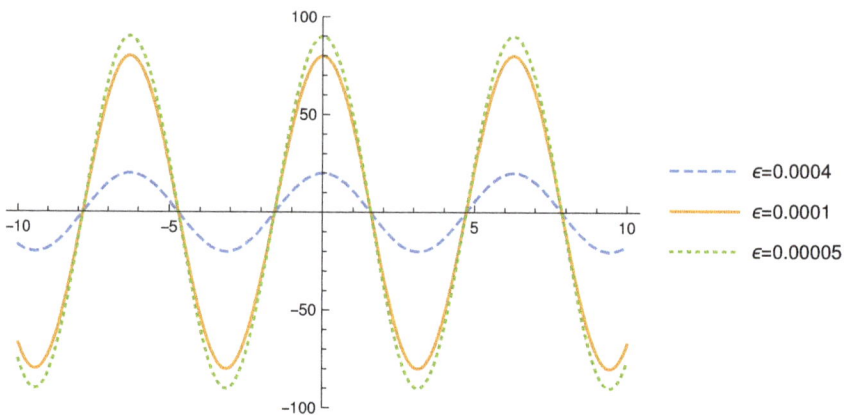

Figure 1. Correlation function $\Gamma_{\widehat{X}\widehat{X}}(\tau)$ of $X(t)$ for different values of ϵ. Example 1.

For the approximation of the PDF, $f_{\widehat{X}(t)}(x)$, we apply the results exhibited in Section 4 based on PME by taking $\epsilon = 0.00005$, which satisfies the stronger bound previously determined ($\epsilon < 0.000166641$). We first compute the approximation based on the three first moments

$$f_{\widehat{X}(t)}(x) = e^{-1-2.181+1.045 \cdot 10^{-5}x - 0.005x^2 - 4.9217 \cdot 10^{-8} x^3},$$

and, second, the approximation based on the five first moments

$$f_{\widehat{X}(t)}(x) = e^{-1-2.243+2.552 \cdot 10^{-8}x - 0.004x^2 - 2.177 \cdot 10^{-9} x^3 - 3.789 \cdot 10^{-6} x^4 + 6.754 \cdot 10^{-13} x^5}.$$

In Figure 2, we compare both graphical representations. From them, we can observe that both plots are quite similar, so giving evidence that computations are consistent.

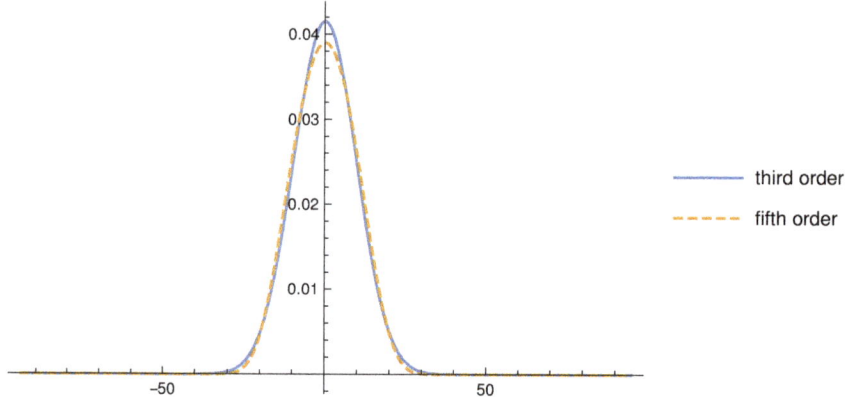

Figure 2. Approximation of PDF, $f_{\widehat{X}(t)}(x)$, using until the third and the fifth-order moment for $\epsilon = 0.00005$ via the PME. Example 1.

Finally, to check that our approximations are reliable, we compare the mean and standard deviation of the approximate solution obtained via the perturbation method against the ones calculated by Monte Carlo. The results are collected in Table 1. We can observe that both approximations agree.

Table 1. Comparison between perturbation method and Monte Carlo simulations using $\epsilon = 0.00005$. Example 1.

	Perturbation Method	Monte Carlo (1000 Simulations)	Monte Carlo (10,000 Simulations)
Mean	0	0.188808	−0.114379
Standard deviation	9.48714	9.31356	9.49534

Example 2. *To previously perform our theoretical analysis, we have required the stationary Gaussian stochastic excitation $Y(t)$ be differentiable in the mean square sense (or equivalently, its correlation function, $\Gamma_Y(\tau)$, be twice differentiable in the ordinary sense at $\tau = 0$ [28] (Chapter 4)), so having differentiable sample trajectories [34]. If we carefully revise our previous development, we can notice this is an hypothesis coming from the fact the nonlinearity cross-term depends upon $\dot{X}(t)$. In this second example, we shall show that using the general concept of differentiability, in the sense of distributions, we can still obtain good results via the perturbation techniques when the excitation is not differentiable. To this end, we have chosen, $Y(t) = \xi(t)$, a Gaussian white-noise process with zero-mean and correlation function $\Gamma_{YY}(\tau) = \frac{1}{2}W\delta(\tau)$, where $\delta(\tau)$ is the Dirac delta function and W is the noise power. This type of random noise has been extensively used in the literature since the earliest contributions [17]. Observe that $Y(t) = \xi(t)$ is a stationary zero-mean Gaussian process but is not mean square differentiable (as its correlation function, given by the Dirac delta function, is not differentiable) and, consequently, its sample trajectories are not differentiable either. In this case, Equation (3) becomes*

$$\ddot{X}(t) + 0.1\dot{X}(t) + \epsilon X^2(t)\dot{X}(t) + X(t) = \xi(t). \tag{35}$$

As in the previous example, we are going to obtain approximations to the five first moments, $\mathbb{E}\{\widehat{X}^i(t)\}$, $i = 1, \ldots, 5$, the correlation function and the variance, $\mathbb{V}\{\widehat{X}(t)\}$, of the approximate solution $\widehat{X}(t)$ of Equation (35). To implement the corresponding formulas derived throughout Section 3.2 saving computational time in Mathematica, we have taken into account the following properties of Dirac delta function,

$$\int_{-\infty}^{\infty} h(t)\delta(t-s)\,\mathrm{d}t = h(s), \qquad \int_{-\infty}^{\infty} h(t)\delta'(t-s)\,\mathrm{d}t = -h'(s).$$

As mentioned in Example 1, the moments of odd order are null and using the positiveness of even order moments we can obtain some bounds for the perturbation parameter ϵ. First, using expression (17), the second-order moment is determined by

$$\mathbb{E}\{\widehat{X}^2(t)\} = \frac{1}{40} - \frac{\epsilon}{160}, \tag{36}$$

so we obtain the bound $\epsilon < 4$. Since $\mathbb{E}\{\widehat{X}(t)\} = 0$, expression (36) is also the variance of the first-order approximation. Second, using expression (23)–(25),

$$\mathbb{E}\{\widehat{X}^4(t)\} = \frac{3}{1600} - \frac{759}{644800}\epsilon. \tag{37}$$

This provides a stronger bound, $\epsilon < 1.59289$.

Now, applying (27), we obtain the following approximation of the correlation function,

$$\Gamma_{\widehat{X}\widehat{X}}(\tau) = \begin{cases} \dfrac{e^{-\tau/20}\left(-399(-399+5\epsilon\tau)\cos\left(\frac{\sqrt{399}\tau}{20}\right)+\sqrt{399}(399+100\epsilon)\sin\left(\frac{\sqrt{399}\tau}{20}\right)\right)}{6,368,040} & \text{if } \tau \geq 0, \\[2mm] \dfrac{e^{\tau/20}\left(-399(-399+5\epsilon\tau)\cos\left(\frac{\sqrt{399}\tau}{20}\right)+\sqrt{399}(-399+100\epsilon)\sin\left(\frac{\sqrt{399}\tau}{20}\right)\right)}{6,368,040} & \text{if } \tau < 0. \end{cases} \quad (38)$$

In Figure 3, we show the plot of the correlation function, $\Gamma_{\widehat{X}\widehat{X}}(\tau)$, given in the expression (38) for different values of ϵ satisfying the weaker and the stronger bounds previously determined. We can observe that for smaller values of ϵ the obtained approximation of the correlation function better preserves the symmetry as expected.

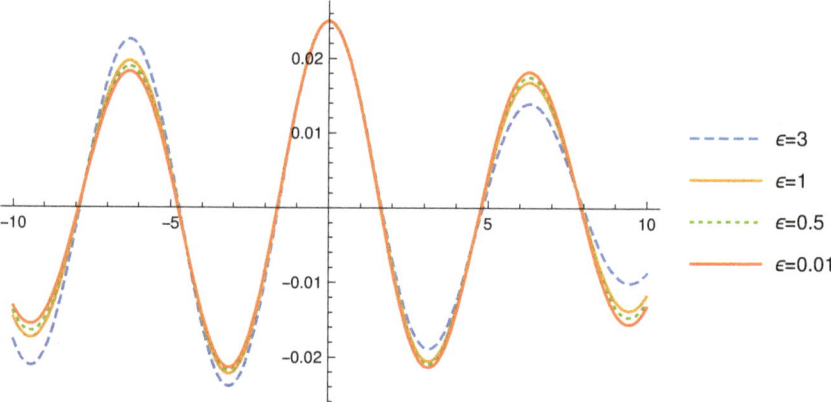

Figure 3. Correlation function $\Gamma_{\widehat{X}\widehat{X}}(\tau)$ of $X(t)$ for different values of ϵ. Example 2.

Applying the results presented in Section 4, we obtain the approximation of the PDF, $f_{\widehat{X}(t)}(x)$, for $\epsilon = 0.5$, which satisfies the stronger bound 1.59289. We first compute the approximation based on the three first moments

$$f_{\widehat{X}(t)}(x) = e^{-1+1.992+1.438\cdot 10^{-8}x-22.857x^2-2.197\cdot 10^{-7}x^3},$$

and, second, the approximation based on the five first moments

$$f_{\widehat{X}(t)}(x) = e^{-1+1.940-5.226\cdot 10^{-11}x-17.837x^2+1.580\cdot 10^{-9}x^3-42.679x^4-7.904\cdot 10^{-9}x^5}.$$

In Figure 4, we compare both graphical representations. We can observe, again, the similarity between them, thus showing full agreement in our numerical computations.

Finally, to check that our approximations are accurate, we compare the mean and standard deviation of $\widehat{X}(t)$ obtained via the perturbation method against the ones computed by Euler–Maruyama numerical scheme [35]. The results are shown in Table 2. We can observe that both approximations are similar.

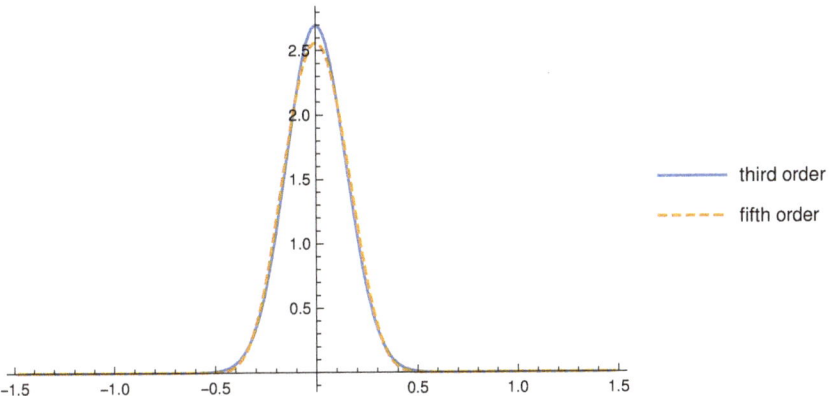

Figure 4. Approximation of PDF, $f_{\hat{X}(t)}(x)$, using until the third and the fifth order moment for $\epsilon = 0.5$ via the PME. Example 2.

Table 2. Comparison between perturbation method and Euler–Maruyama simulations using $\epsilon = 0.5$. Example 2.

	Perturbation Method	Euler-Maruyama (1000 Simulations)	Euler-Maruyama (10,000 Simulations)
Mean	0	0.00146868	0.00128423
Standard deviation	0.147902	0.157353	0.156475

6. Conclusions

We have studied, from a probabilistic standpoint, a family of oscillators subject to small perturbations on the nonlinear term that depends both upon the position and the velocity (cross-nonlinearity) and whose forcing source is driven by a mean square differentiable stationary zero-mean Gaussian process. Despite the hypothesis of differentiability for the stochastic excitation, we have checked, via a numerical example, the method also provides good results when this hypothesis is not fulfilled, but involved computations are performed using the concept of general differentiability in the sense of distributions. We must point out that the majority of contributions dealing with this type of stochastic oscillators focus on the computation of the mean, the variance and correlation function. Our main contribution is the computation of reliable approximations of the probability density function of the stationary solution, by combining the stochastic perturbation method and the principle of maximum entropy. In this manner, we provide a fuller probabilistic description of the solution since from the density one can determine any one-dimensional moment as well as further probabilistic information of the steady-state. The proposed approach can be very useful to open new avenues in the analysis to other kind of nonlinear oscillators subject to small fluctuations and whose forcing term is a stochastic process that satisfies certain hypotheses. In our future research, we will work to continue contributing in this direction.

Author Contributions: Funding acquisition, J.-C.C.; Methodology, J.-C.C., E.L.-N., J.-V.R. and M.-D.R.; Supervision, J.-V.R. and M.-D.R.; Writing—original draft, J.-C.C. and E.L.-N.; Writing—review & editing, E.L.-N., J.-V.R. and M.-D.R. All authors have read and agreed to the published version of the manuscript.

Funding: This work has been supported by the Spanish Ministerio de Economía, Industria y Competitividad (MINECO), the Agencia Estatal de Investigación (AEI) and Fondo Europeo de Desarrollo Regional (FEDER UE) grant MTM2017–89664–P. Elena López-Navarro has been supported by the European Union through the Operational Program of the [European Regional Development Fund (ERDF)/European Social Fund (ESF)] of the Valencian Community 2014-2020 (GJIDI/2018/A/010).

Institutional Review Board Statement: Not applicable.

Informed Consent Statement: Not applicable.

Data Availability Statement: Not applicable.

Conflicts of Interest: The authors declare that they do not have any conflict of interest.

References

1. Caraballo, T.; Colucci, R.; López-de-la Cruz, J.; Rapaport, A. A way to model stochastic perturbations in population dynamics models with bounded realizations. *Commun. Nonlinear Sci. Numer. Simul.* **2019**, *77*, 239–257. [CrossRef]
2. de la Cruz, H. Stabilized explicit methods for the approximation of stochastic systems driven by small additive noises. *Chaos Solitons Frac.* **2020**, *140*, 110195. [CrossRef]
3. Li, X.; Song, S. Research on synchronization of chaotic delayed neural networks with stochastic perturbation using impulsive control method. *Commun. Nonlinear Sci. Numer. Simul.* **2014**, *19*, 3892–3900. [CrossRef]
4. Shaikhet, L. Stability of the zero and positive equilibria of two connected neoclassical growth models under stochastic perturbations. *Commun. Nonlinear Sci. Numer. Simul.* **2019**, *68*, 86–93. [CrossRef]
5. Andronov, A.A.; Vitt, A.A.; Khaikin, S.E. *Theory of Oscillators*; Electrical Engineering, Dover Books: New York, NY, USA, 2011.
6. Cveticanin, L. *Strongly Nonlinear Oscillators: Analytical Solutions*; Springer: New York, NY, USA, 2014.
7. Yildirim, A.; Askari, H.; Saadatnia, Z.; KalamiYazdic, M.; Khan, Y. Analysis of nonlinear oscillations of a punctual charge in the electric field of a charged ring via a Hamiltonian approach and the energy balance method. *Comput. Math. Appl.* **2011**, *62*, 486–490. [CrossRef]
8. Khan, Y.; Mirzabeigy, A. Improved accuracy of He's energy balance method for analysis of conservative nonlinear oscillator. *Neural Comput. Appl.* **2014**, *25*, 889–895. [CrossRef]
9. Khan, Y.; Vázquez-Leal, H.; Hernández-Martínez, L. Removal of noise oscillation term Appearing in the nonlinear equation solution. *J. Appl. Math.* **2012**, 387365. [CrossRef]
10. Calatayud, J.; Cortés, J.C.; Jornet, M. The damped pendulum random differential equation: A comprehensive stochastic analysis via the computation of the probability density function. *Phys. A Stat. Mech. Appl.* **2018**, *512*, 261–279. [CrossRef]
11. Hinch, E. *Perturbation Methods*; Cambridge Texts in Applied Mathematics: New York, NY, USA, 1991.
12. Nayfeh, A.H. *Perturbation Methods*; Wiley Classics Library, Wiley VCH: New York, NY, USA, 2000.
13. Bellman, R. *Perturbation Techniques in Mathematics, Engineering and Physics*; Dover Books on Physics, Dover Publications: New York, NY, USA, 2003.
14. Simmonds, J.; Mann, J. *A First Look at Perturbation Theory*; Dover Books on Physics, Dover Publications: New York, NY, USA, 2013.
15. Gitterman, M. *The Noisy Oscillator: Random Mass, Frequency, Damping*; World Scientific: Singapore, 2013.
16. Crandall, S.H. *Random Vibration*; John Wiley & Sons: New York, NY, USA, 1958.
17. Crandall, S.H. Perturbation Techniques for Random Vibration of Nonlinear Systems. *J. Acoust. Soc. Am.* **1963**, *35*, 1700–1705. [CrossRef]
18. Ibrahim, R. *Parametric Random Vibration*; Dover Books on Engineering, Dover Publications: New York, NY, USA, 2008.
19. Newland, D. *An Introduction to Random Vibrations, Spectral and Wavelet Analysis*; Longman Group UK Ltd.: New York, NY, USA, 1994.
20. El-Tawil, M.; Al-Johany, A. Approximate solution of a mixed nonlinear stochastic oscillator. *Comput. Math. Appl.* **2009**, *58*, 2236–2259. [CrossRef]
21. El-Beltagy, M.; Al-Johany, A. Numerical approximation of higher-order solutions of the quadratic nonlinear stochastic oscillatory equation using WHEP technique. *J. Appl. Math.* **2013**, 685137. [CrossRef]
22. El-Beltagy, M.; Al-Johany, A. Toward a solution of a class of non-linear stochastic perturbed PDEs using automated WHEP algorithm. *Appl. Math. Model.* **2013**, *37*, 7174–7192. [CrossRef]
23. Gitterman, M. Stochastic oscillator with random mass: New type of Brownian motion. *Phys. A Stat. Mech. Appl.* **2014**, *395*, 11–21. [CrossRef]
24. Fua, C.; Xu, Y.; Yang, Y.; Lu, K.; Gu, F.; Ball, A. Response analysis of an accelerating unbalanced rotating system with both random and interval variables. *J. Sound Vib.* **2020**, *466*, 115047. [CrossRef]
25. Zhu, H.T.; Er, G.K.; Iu, V.P.; Kou, K.P. Probabilistic solution of nonlinear oscillators excited by combined Gaussian and Poisson white noises. *J. Sound Vib.* **2011**, *330*, 2900–2909. [CrossRef]
26. Chen, L.; Qian, J.; Zhu, H.; qiao Sun, J. The closed-form stationary probability distribution of the stochastically excited vibro-impact oscillators. *J. Sound Vib.* **2019**, *439*, 260–270. [CrossRef]
27. Ren, Z.; Xu, W.; Zhang, S. Reliability analysis of nonlinear vibro-impact systems with both randomly fluctuating restoring and damping terms. *Commun. Nonlinear Sci. Numer. Simul.* **2020**, *82*, 105087. [CrossRef]
28. Soong, T. *Random Differential Equations in Science and Engineering*; Academic Press: New York, NY, USA, 1973; Volume 103.
29. Lindgren, G. *Stationary Stochastic Processes. Theory and Applications*; Chapman and Hall/CRC: New York, NY, USA, 2012.
30. Lindgren, G.; Rootzen, H.; Sandsten, M. *Stationary Stochastic Processes for Scientists and Engineers*; Chapman and Hall/CRC: New York, NY, USA, 2013.
31. McLachlan, N. *Laplace Transforms and Their Applications to Differential Equations*; Dover Publication Inc.: New York, NY, USA, 2014; Volume 103.

32. Steidel, R.F. *An Introduction to Mechanical Vibrations*; Wiley: New York, NY, USA, 1989.
33. Michalowicz, J.V.; Nichols, J.M.; Bucholtz, F. *Handbook of Differential Entropy*; CRC Press: Abingdon, UK, 2018.
34. Strand, J. Random ordinary differential equations. *J. Differ. Equ.* **1970**, *7*, 538–553. [CrossRef]
35. Kloeden, P.; Platen, E. *Numerical Solution of Stochastic Differential Equations*; Volume 23 Stochastic Modelling and Applied Probability; Springer: Berlin/Heidelberg, Germany, 1992.

Article

Reliable Efficient Difference Methods for Random Heterogeneous Diffusion Reaction Models with a Finite Degree of Randomness

María Consuelo Casabán [†], Rafael Company [*,†] and Lucas Jódar [†]

Instituto Universitario de Matemática Multidisciplinar, Building 8G, Access C, 2nd Floor, Universitat Politècnica de València, Camino de Vera s/n, 46022 Valencia, Spain; macabar@imm.upv.es (M.C.C.); ljodar@imm.upv.es (L.J.)
* Correspondence: rcompany@imm.upv.es
† These authors contributed equally to this work.

Abstract: This paper deals with the search for reliable efficient finite difference methods for the numerical solution of random heterogeneous diffusion reaction models with a finite degree of randomness. Efficiency appeals to the computational challenge in the random framework that requires not only the approximating stochastic process solution but also its expectation and variance. After studying positivity and conditional random mean square stability, the computation of the expectation and variance of the approximating stochastic process is not performed directly but through using a set of sampling finite difference schemes coming out by taking realizations of the random scheme and using Monte Carlo technique. Thus, the storage accumulation of symbolic expressions collapsing the approach is avoided keeping reliability. Results are simulated and a procedure for the numerical computation is given.

Keywords: random mean square parabolic model; finite degree of randomness; monte carlo method; random finite difference scheme

MSC: 35R60; 60H15; 65M06; 65M12

1. Introduction

Dealing with deterministic partial differential equations (PDE), finite difference methods (FD) are probably the most used because they are easy to apply and fairly efficient, [1,2]. Trying to capture real world problems, the models introduced uncertainty in several ways, basically assuming that both data, parameters, initial and or boundary conditions are stochastic processes instead of deterministic functions, [3,4]. The uncertainty appears not only because of error measurements, but also considering heterogeneity of the media, material impurities, or even the lack of access to measurements [5,6]. Independently of the type of modelling the uncertainty, the consideration of partial differential equations models (PDEM) has particular challenges. In fact, it is necessary to compute not only the stochastic process solution or approximating stochastic process, but also their statistical moments, mainly the expectation and the variance. Integral transforms methods are efficient techniques to solve PDEM based on integration resources in fitting domains [7,8]. Another powerful technique suitable for models with complex geometries is the finite element method [9]. Iterative methods, for instance FD have particular troubles derived from the storage accumulation of intermediate levels when the computer manages symbolically the involved stochastic processes, [10–12]. This drawback of the iterative methods for solving PDEM occurs in both approaches, the one based on Itô calculus [13] the so-called stochastic differential approach (SDEA), as well as the one based on the mean square calculus [14] also called random differential equations approach (RDEA). To face this computational challenge we take a set of realizations of the model, then we solve each sampled problem using FD method. Finally, we use Monte Carlo technique, [15,16] to average the results of the deterministic sampled problems to compute the expectation and the variance of the approximating solution stochastic process. In our model the involved stochastic processes (s.p.'s) are defined in a complete probability space $(\Omega, \mathcal{F}, \mathbb{P})$ and have n degrees of randomness [14] (p. 37), i.e., they only depend on a finite number n of random variables (r.v.'s):

$$h(s) = F(s, A_1, A_2, \ldots, A_n), \tag{1}$$

where

$$\left.\begin{array}{l} A_i, \ i = 1, \ldots, n, \quad \text{are mutually independent r.v.'s;} \\[4pt] F \text{ is a differentiable real function of the variable } s \\ \text{(being } s \text{ the spatial variable } x \text{ or the temporal one } t). \end{array}\right\} \tag{2}$$

In addition, under this hypothesis, the s.p. $h(s)$ has sample differentiable trajectories (realizations), i.e., for a fixed event $\omega \in \Omega$, the real function $h(s, \omega) = F(s, A_1(\omega), A_2(\omega), \ldots, A_n(\omega))$ is a differentiable function of the real variable s. For the sake of clarity in the presentation and to save notational complexity, we will assume that involved s.p.'s in the coefficients and initial or boundary conditions, have one degree of randomness, i.e., they have the form

$$h(s) = F(s, A),$$

with A a r.v. and F a differentiable real function of the variable s. Then the s.p. $h(s)$ has sample differentiable trajectories, i.e., for a fixed event $\omega \in \Omega$, the real function $h(s, \omega) = F(s, A(\omega))$ is a differentiable function of the real variable s.

This paper deals with random parabolic partial differential models of the form

$$\frac{\partial u(x,t)}{\partial t} = \frac{\partial}{\partial x}\left[p(x)\frac{\partial u(x,t)}{\partial x}\right] - q(x)\,u(x,t), \quad 0 < x < 1, \quad t > 0, \tag{3}$$

$$u(0,t) = g_1(t), \quad t > 0, \tag{4}$$

$$u(1,t) = g_2(t), \quad t > 0, \tag{5}$$

$$u(x,0) = f(x), \quad 0 \leq x \leq 1, \tag{6}$$

where the diffusion coefficient $p(x)$, the reaction coefficient $q(x)$, the boundary conditions $g_i(t)$, $i = 1, 2$, and the initial condition $f(x)$ are s.p.'s with one degree of randomness in the sense as defined above. In addition we assume that $p(x)$, $q(x)$, $f(x)$ and $g_i(t)$, $i = 1, 2$ are mean square continuous s.p.'s in variables x and t, respectively, $p(x)$ is also a mean square differentiable s.p. and the sample realizations of the random inputs $p(x)$, $q(x)$, $g_i(t)$, $i = 1, 2$ and $f(x)$ satisfy the following conditions denoting $p'(x)$ as the mean square derivative of $p(x)$:

$$0 < a_1 \leq p(x, \omega) \leq a_2 < +\infty, \quad x \in [0,1], \text{ for almost every (a.e.) } \omega \in \Omega, \tag{7}$$

$$\frac{|p'(x,\omega)|}{p(x,\omega)} \leq b < +\infty, \quad x \in [0,1], \text{ for a.e. } \omega \in \Omega, \tag{8}$$

$$q_{\min} \leq q(x, \omega) \leq q_{\max}, \quad x \in [0,1], \text{ for a.e. } \omega \in \Omega, \tag{9}$$

$$g_i(t, \omega) \geq 0, \ i = 1, 2, \quad t > 0, \text{ for a.e. } \omega \in \Omega, \tag{10}$$

$$0 \le f(x,\omega) \le f_{\max}, \quad x \in [0,1], \text{ for a.e. } \omega \in \Omega, \tag{11}$$

This model is frequent in chemical engineering sciences and in heat and mass transfer theory for reaction-diffusion problems with parameters depending on the spatial variables as it occurs in heterogeneous and anisotropic solids, [17] (p. 455), [18] (p. 388), [19,20].

The paper is organized as follows. Section 2 deals with some preliminaries, definitions and notations about the mean square calculus as well as the construction of the random mean square finite difference scheme (RMSFDS) resulting from the discretization of model (3)–(6). Section 3 is addressed to the study of properties of the RMSFDS such as positivity, stability and consistency. Throughout a sample approach, sufficient conditions for stability and positivity of the random numerical solution s.p. in terms of the data and discretization step-sizes are found. Consistency of the RMSFDS with Equation (3) is also treated throughout a sample approach and the consistency of the corresponding realized deterministic scheme for each fixed event $\omega \in \Omega$. In Section 4 we construct an algorithm to perform the efficient computation of the expectation and the variance of the numerical solution s.p. using Monte Carlo method and the solution of the sampled scheme. Numerical simulations for a problem where the exact solution is known are performed showing the efficiency of the proposed numerical method as well as the computations of the expectation and the variance of the numerical approximated s.p. A conclusion Section 5 ends the paper.

2. Preliminaries and Construction of the Random Finite Difference Scheme

For the sake of clarity in the presentation, in this section we recall some definitions and concepts related to the L_p-calculus, [14]. In a probability space $(\Omega, \mathcal{F}, \mathcal{P})$, we denote $L_p(\Omega)$ the space of all real valued r.v.'s $U : \Omega \to \mathbb{R}$ of order p, endowed with the norm

$$\|U\|_p = (\mathbb{E}[|U|^p])^{1/p} = \left(\int_\Omega |U(\omega)|^p f_U(\omega) \, d\omega \right)^{1/p} < +\infty, \tag{12}$$

where $\mathbb{E}[\cdot]$ denotes the expectation operator, f_U the density function of the r.v. U and ω an event of sample space Ω.

Given $T \subset \mathbb{R}$, a family of t-indexed r.v.'s, say $\{V(t) : t \in T\}$, is called a stochastic process of order p (p-s.p.) if for each $t \in T$ fixed, the r.v. $V(t) \in L_p(\Omega)$. We say that a p-s.p. $\{V(t) : t \in T\}$ is p-th mean continuous at $t \in T$, if

$$\|V(t+h) - V(t)\|_p \to 0 \text{ as } h \to 0, \ t, t+h \in T.$$

Furthermore, if there exists a p-s.p. $V'(t)$, such that

$$\left\| \frac{V(t+h) - V(t)}{h} - V'(t) \right\|_p \to 0 \text{ as } h \to 0, \ t, t+h \in T,$$

then we say that the s.p. $\{V(t) : t \in T\}$ is p-th mean differentiable at $t \in T$ and $V'(t)$ is the p-derivative of $V(t)$. In the particular case that $p = 2$, $L_2(\Omega)$, definitions above leads to the corresponding concept of mean square (m.s.) continuity and m.s. differentiability, respectively.

In this section, we construct an explicit random finite difference scheme for solving problem (3)–(6). Firstly, let us write Equation (3) into the following form

$$\frac{\partial u(x,t)}{\partial t} = p(x) \frac{\partial^2 u(x,t)}{\partial x^2} + p'(x) \frac{\partial u(x,t)}{\partial x} - q(x) u(x,t), \tag{13}$$

where $p(x) \in L_p(\Omega)$ is p-th mean continuous and differentiable, $p'(x)$ is the p-derivative of $p(t)$ and $q(x) \in L_p(\Omega)$ is p-th mean continuous.

Let us consider the uniform partition of the spatial interval $[0,1]$, of the form $x_i = ih$, $0 \le i \le M$, with $Mh = 1$. For a fixed time horizon, T, we consider $N+1$ time levels

$t^n = nk$, $0 \leq n \leq N$ with $Nk = T$. The numerical approximation of the solution s.p. of the random problem (3)–(6) is denoted by u_i^n, i.e.,

$$u_i^n \approx u(x_i, t^n), \qquad 0 \leq i \leq M,\ 0 \leq n \leq N. \tag{14}$$

By using a forward first-order approximation of the time partial derivative and centred second-order approximations for the spatial partial derivatives in Equation (13) one gets the following random numerical scheme for the spatial internal mesh points

$$\frac{u_i^{n+1} - u_i^n}{k} = p_i \frac{u_{i-1}^n - 2u_i^n + u_{i+1}^n}{h^2} + p_i' \frac{u_{i+1}^n - u_{i-1}^n}{2h} - q_i u_i^n, \quad 1 \leq i \leq M-1,\ 0 \leq n \leq N-1, \tag{15}$$

where $p_i = p(x_i)$, $p_i' = p'(x_i)$ and $q_i = q(x_i)$. The resulting random discretized problem (3)–(6) can be rewritten in the following form

$$\left.\begin{array}{c} u_i^{n+1} = \frac{k}{h^2}\left(p_i - \frac{h}{2}p_i'\right)u_{i-1}^n + \left(1 - kq_i - \frac{2k}{h^2}p_i\right)u_i^n + \frac{k}{h^2}\left(p_i + \frac{h}{2}p_i'\right)u_{i+1}^n, \\ 1 \leq i \leq M-1,\ 1 \leq n \leq N-1, \\[4pt] u_0^n = g_1^n,\ u_M^n = g_2^n,\ 1 \leq n \leq N, \\[4pt] u_i^0 = f_i,\ 0 \leq i \leq M, \end{array}\right\} \tag{16}$$

where $g_1^n = g_1(t^n)$, $g_2^n = g_2(t^n)$, and $f_i = f(x_i)$. Please note that all the inputs of the random problem (16) are s.p.'s depending on one finite degree of randomness and lie in $L_p(\Omega)$.

3. Properties of the Random Numerical Scheme: Positivity, Stability and Consistency

We are going to prove the positivity of the random numerical solution $\{u_i^n,\ 0 \leq i \leq M,\ 0 \leq n \leq N\}$ of the random scheme (16) and its $\|\cdot\|_p$-stability in the sense of fixed station respect to the time. We extend this type of stability to the random field.

Definition 1. *A random numerical scheme is said to be $\|\cdot\|_p$-stable in the fixed station sense in the domain $[0,1] \times [0,T]$, if for every partition with $k = \Delta t$, $h = \Delta x$ such that $Nk = T$ and $Mh = 1$,*

$$\|u_i^n\|_p \leq C, \qquad 0 \leq i \leq M,\ 0 \leq n \leq N, \tag{17}$$

where C is independent of the step-sizes h, k and the time level n.

First, we are going to find sufficient conditions on the spatial step-size h and the temporal step-size k, so that the numerical solution $\{u_i^n(\omega)\}$ constructed by sampling random scheme (16) for a fixed $\omega \in \Omega$

$$\left.\begin{array}{c} u_i^{n+1}(\omega) = \\ \frac{k}{h^2}\left(p_i(\omega) - \frac{h}{2}p_i'(\omega)\right)u_{i-1}^n + \left(1 - kq_i(\omega) - \frac{2k}{h^2}p_i(\omega)\right)u_i^n + \frac{k}{h^2}\left(p_i(\omega) + \frac{h}{2}p_i'(\omega)\right)u_{i+1}^n, \\ 1 \leq i \leq M-1,\ 1 \leq n \leq N-1, \\[4pt] u_0^n(\omega) = g_1^n(\omega),\ u_M^n(\omega) = g_2^n(\omega),\ 1 \leq n \leq N, \\[4pt] u_i^0(\omega) = f_i(\omega),\ 0 \leq i \leq M, \end{array}\right\} \tag{18}$$

guarantee its positivity, i.e., $u_i^n(\omega) \geq 0$ for $0 \leq i \leq M$ and for each time level n, $0 \leq n \leq N$. We denote

$$\begin{aligned}
A_i(h,k)(\omega) &= \frac{k}{h^2}\left(p_i(\omega) - \frac{h}{2}p_i'(\omega)\right), \\
B_i(h,k)(\omega) &= 1 - k q_i(\omega) - \frac{2k}{h^2}p_i(\omega), \\
C_i(h,k)(\omega) &= p_i(\omega) + \frac{h}{2}p_i'(\omega),
\end{aligned} \quad (19)$$

then the sampling scheme (18) can be rewritten as follows

$$u_i^{n+1}(\omega) = A_i(h,k)(\omega)\, u_{i-1}^n(\omega) + B_i(h,k)(\omega)\, u_i^n(\omega) + C_i(h,k)(\omega)\, u_{i+1}^n(\omega).$$

To guarantee the positivity of the numerical approximation $\{u_i^n(\omega)\}$ it is sufficient to impose the positivity of coefficients defined in (19). Please note that the simultaneously positivity of coefficients $A_i(h,k)(\omega)$ and $C_i(h,k)(\omega)$ means that

$$-p_i(\omega) \leq \frac{h}{2}p_i'(\omega) \leq p_i(\omega),$$

that is

$$h \leq \frac{2 p_i(\omega)}{|p_i'(\omega)|}. \quad (20)$$

Taking into account the bound condition (8) it follows that coefficients $A_i(h,k)(\omega)$ and $C_i(h,k)(\omega)$, $1 \leq i \leq M-1$, are non-negative for a.e. $\omega \in \Omega$ under condition

$$h \leq \frac{2}{b}. \quad (21)$$

Please note that for the particular case where $p_i(\omega)$ is constant the positivity of coefficients $A_i(h,k)(\omega)$ and $C_i(h,k)(\omega)$ defined in (19), is established for $h > 0$. To guarantee the positivity of coefficient $B_i(h,k)(\omega)$ from (19) and bounds (7)–(9) note that

$$B_i(h,k)(\omega) = 1 - k q_i(\omega) - \frac{2k}{h^2}p_i(\omega) \geq 1 - k q_{\max} - \frac{2k}{h^2}a_2. \quad (22)$$

Thus, the positivity of $B_i(h,k)(\omega)$, $1 \leq i \leq M-1$, for a.e. $\omega \in \Omega$, is verified under the conditions

$$k \leq \frac{h^2}{2a_2}, \quad (\text{If } q_{\max} < 0), \quad (23)$$

$$k \leq \frac{h^2}{2a_2 + h^2 q_{\max}}, \quad (\text{If } q_{\max} \geq 0). \quad (24)$$

Then taking into account the sufficient conditions (21), (23) and (24) over the discretization step-sizes h and k, the positivity of all the coefficients (19) of sampling scheme (18) for a.e. $\omega \in \Omega$ is guaranteed and consequently the positivity of the numerical solution $\{u_i^n(\omega)\}$, $0 \leq i \leq M$, for each time level n, $0 \leq n \leq N$, ($T = kN$) is established.

Let us prove now that random numerical scheme (16) is $\|\cdot\|_p$-stable in the sense of Definition 1. In this study we need to distinguish two cases for the sampling parameter $q_i(\omega)$ for a fixed $\omega \in \Omega$.

Case 1. $\boxed{q_i(\omega) \geq 0}$, $0 \leq i \leq M$.

From (18) imposing conditions (21) and (24) one gets

$$
\begin{aligned}
u_i^{n+1}(\omega) &\leq (A_i(h,k)(\omega) + B_i(h,k)(\omega) + C_i(h,k)(\omega))\, u_{\mathrm{MAX}_i}^n(\omega) \\
&\leq (1 - k q_i(\omega))\, u_{\mathrm{MAX}_i}^n(\omega) \\
&\leq u_{\mathrm{MAX}_i}^n(\omega), \qquad 1 \leq i \leq M-1,\ 0 \leq n \leq N-1,
\end{aligned} \qquad (25)
$$

where

$$
u_{\mathrm{MAX}_i}^n(\omega) = \max_{0 \leq i \leq M} \{u_i^n(\omega)\}. \qquad (26)
$$

Using (11), the boundary conditions of (18) and (25) we have by recurrence

$$
\begin{aligned}
u_i^{n+1}(\omega) &\leq \max\{g_1^{n+1}(\omega),\, g_2^{n+1}(\omega),\, u_{\mathrm{MAX}_i}^n(\omega)\} \\
&\leq \max\{g_1^{n+1}(\omega),\, g_2^{n+1}(\omega),\, g_1^n(\omega),\, g_2^n(\omega),\, u_{\mathrm{MAX}_i}^{n-1}(\omega)\} \\
&\leq \cdots \\
&\leq \max_{1 \leq s \leq n+1}\left\{g_1^s(\omega),\, g_2^s(\omega),\, u_{\mathrm{MAX}_i}^0(\omega)\right\} \\
&\leq \max_{0 \leq t \leq (n+1)k}\left\{g_1(t,\omega),\, g_2(t,\omega),\, \max_{x \in [0,1]}\{f(x,\omega)\}\right\}.
\end{aligned} \qquad (27)
$$

We denote

$$
G(T) = \max_{0 \leq t \leq T}\{g_{1,\max}(T),\, g_{2,\max}(T),\, f_{\max}\}, \qquad (28)
$$

where

$$
g_{i,\max}(T) = \max_{0 \leq t \leq T}\{g_i(t,\omega),\ \text{for a.e. } \omega \in \Omega\},\ i = 1,2. \qquad (29)
$$

Thus, from (27) and (28) we obtain the following upper bound for the numerical solution of sampling scheme (18)

$$
0 \leq u_i^n(\omega) \leq G(T), \quad \text{for a.e. } \omega \in \Omega, \qquad (30)
$$

for each level n, $0 \leq n \leq N = Tk$, and for each spatial point x_i, $0 \leq i \leq M$.

Case 2. $\boxed{q_{\min} \leq \min_{0 \leq i \leq M}\{q_i(\omega)\} < 0}$

From (18) imposing conditions (21) and (23) and using (25) and (26) we obtain

$$
\begin{aligned}
u_i^{n+1}(\omega) &\leq (1 - k q_i(\omega))\, u_{\mathrm{MAX}_i}^n(\omega) \\
&\leq (1 + k|q_{\min}|)\, u_{\mathrm{MAX}_i}^n(\omega), \qquad 1 \leq i \leq M-1,\ 0 \leq n \leq N-1.
\end{aligned} \qquad (31)
$$

Then using the boundary conditions of (18) and applying recurrently the bound exhibits in (31) one gets

$$
\begin{aligned}
u_i^{n+1}(\omega) &\leq \max\{g_1^{n+1}(\omega),\, g_2^{n+1}(\omega),\, (1+k|q_{\min}|)u_{\mathrm{MAX}_i}^n(\omega)\} \\
&\leq \max\{g_1^{n+1}(\omega),\, g_2^{n+1}(\omega),\, (1+k|q_{\min}|)\max\{g_1^n(\omega),\, g_2^n(\omega),\, (1+k|q_{\min}|)u_{\mathrm{MAX}_i}^{n-1}(\omega)\}\} \\
&\leq (1+k|q_{\min}|)^2 \max\{g_1^{n+1}(\omega),\, g_2^{n+1}(\omega),\, g_1^n(\omega),\, g_2^n(\omega),\, u_{\mathrm{MAX}_i}^{n-1}(\omega)\} \\
&\leq \cdots \\
&\leq (1+k|q_{\min}|)^{n+1} \max_{1 \leq s \leq n+1}\left\{g_1^s(\omega),\, g_2^s(\omega),\, u_{\mathrm{MAX}_i}^0(\omega)\right\}.
\end{aligned} \qquad (32)
$$

Taking into account the following inequality

$$
(1+k|q_{\min}|)^s \leq (1+k|q_{\min}|)^N < e^{T|q_{\min}|}, \quad 0 \leq s \leq N,
$$

and the notation introduced in (28), we obtain this upper bound for the numerical solution of sample scheme (18)

$$0 \leq u_i^n(\omega) \leq e^{T|q_{\min}|} G(T), \quad \text{for a.e. } \omega \in \Omega, \tag{33}$$

for each level n, $0 \leq n \leq N = Tk$, and for each spatial point x_i, $0 \leq i \leq M$.

Please note that both bounds (30) and (33) are independent of n, h and k.

Finally, under discretization step-size conditions (21), (23) and (24) and from the upper bounds (30) and (33) it follows that

$$\|u_i^n\|_p = (\mathbb{E}[|u_i^n|^p])^{1/p} = \left(\int_\Omega |u_i^n(\omega)|^p f_{u_i^n}(\omega)\, d\omega\right)^{1/p} \leq \alpha(T)\, G(T) \underbrace{\left(\int_\Omega f_{u_i^n}(\omega)\, d\omega\right)^{1/p}}_{1}, \tag{34}$$

where $G(T)$ is defined in (28) and (29) and

$$\alpha(T) = \begin{cases} 1 & \text{if } q_{\min} \geq 0, \\ e^{T|q_{\min}|} & \text{if } q_{\min} < 0. \end{cases} \tag{35}$$

Consequently, random numerical scheme (16) is $\|\cdot\|_p$-stable in the sense of Definition 1. Summarizing, the following result was established.

Theorem 1. *With the previous notation under conditions (21), (23) and (24) on the discretized step-sizes $h = \Delta x$ and $k = \Delta t$, the random numerical solution s.p. $\{u_i^n\}$ of the RMSFDS (16) for the random partial differential model (3)–(11) is positive for $0 \leq i \leq M$ at each time-level $0 \leq n \leq N$ with $T = kN$. Furthermore the RMSFDS (16) is $\|\cdot\|_p$-stable in the fixed station sense taking the value*

$$C = \alpha(T)\, G(T),$$

where constants $G(T)$ and $\alpha(T)$ are defined in (28) and (35), respectively.

To prove the consistency of the random finite difference scheme (16) with the random partial differential Equation (13) let us introduce the following definition inspired in the well-known concept of consistency for deterministic PDEs, see [2].

Definition 2. *Let us consider a RMSFDS $F(u_i^n) = 0$ for a random partial differential equation (RPDE) $\mathcal{L}(u) = 0$ and let the local truncation error $T_i^n(U(\omega))$ for a fixed event $\omega \in \Omega$ be defined by*

$$T_i^n(U(\omega)) = F(U_i^n(\omega)) - \mathcal{L}(U_i^n(\omega)),$$

where $U_i^n(\omega)$ denotes the theoretical solution of $\mathcal{L}(u)(\omega) = 0$ evaluated at (x_i, t^n). We call $T_i^n(U)$ by

$$\|T_i^n(U)\|_p = (\mathbb{E}[|T_i^n(U)|^p])^{1/p} = \left(\int_\Omega |T_i^n(U(\omega))|^p f_{T_i^n(U)}(\omega)\, d\omega\right)^{1/p}.$$

With previous notation, the RMSFDS $F(u_i^n) = 0$ is said to be $\|\cdot\|_p$-consistent with the RPDE $\mathcal{L}(u) = 0$ if

$$\|T_i^n(U)\|_p \to 0 \text{ as } h = \triangle x \to 0,\; k = \triangle t \to 0.$$

Next result shows the consistency in the p-norm of RFDS (16) with RPDE (13).

Theorem 2. *The RFDS (16) is $\|\cdot\|_p$- consistent with the RPDE (13).*

Proof. Please note that for each fixed $\omega \in \Omega$ the local truncation error using (13) and (15) is given by

$$T_i^n(U(\omega)) = \frac{U_i^{n+1}(\omega) - U_i^n(\omega)}{k} - \frac{\partial U(\omega)}{\partial t}(x_i, t^n) - p_i \frac{U_{i-1}^n(\omega) - 2U_i^n(\omega) + U_{i+1}^n(\omega)}{h^2} +$$
$$p_i \frac{\partial^2 U(\omega)}{\partial x^2}(x_i, t^n) - p_i' \frac{U_{i+1}^n(\omega) - U_{i-1}^n(\omega)}{2h} + p_i' \frac{\partial U(\omega)}{\partial x}(x_i, t^n).$$

Assuming that $U(x,t)(\omega)$ is four times continuously differentiable with respect to x and two times continuously differentiable with respect to t and using Taylor expansions of $U(x,t)(\omega)$ at (x_i, t^n) one gets

$$T_i^n(U(\omega)) = \frac{k}{2} \frac{\partial^2 U(\omega)}{\partial t^2}(x_i, \delta) - p_i \frac{h^2}{12} \frac{\partial^4 U(\omega)}{\partial x^4}(\eta_1, t^n) - p_i' \frac{h^2}{6} \frac{\partial^3 U(\omega)}{\partial x^3}(\eta_2, t^n), \quad (36)$$

where $t^n < \delta < t^{n+1}$, $x_{i-1} < \eta_j < x_{i+1}$, $j = 1, 2$.

Let us denote

$$E_1(i,n)(\omega) = \max\left\{\left|\frac{\partial^2 U(\omega)}{\partial t^2}(x_i, t)\right|, \ t^n < t < t^{n+1}\right\}, \quad (37)$$

$$E_2(i,n)(\omega) = \max\left\{\left|\frac{\partial^4 U(\omega)}{\partial x^4}(x, t^n)\right|, \ x_{i-1} < x < x_{i+1}\right\}, \quad (38)$$

$$E_3(i,n)(\omega) = \max\left\{\left|\frac{\partial^3 U(\omega)}{\partial x^3}(x, t^n)\right|, \ x_{i-1} < x < x_{i+1}\right\}. \quad (39)$$

As we are in the scenario of finite degree of randomness and the involved variables have a truncated range, there exist $D_j(i,n)$, $j = 1, 2, 3$, positive constants such that

$$E_j(i,n)(\omega) \leq D_j(i,n), \ 1 \leq j \leq 3, \text{ a.e. } \omega \in \Omega. \quad (40)$$

From Definition 2, condition (7) and (36)–(40) it follows that

$$\|T_i^n(U)\|_p \leq \left(\int_\Omega \left[D_1(i,n)\frac{k}{2} + \frac{h^2}{12}(p_i D_2(i,n) + 2|p_i'| D_3(i,n))\right]^p f_{T_i^n(U)}(\omega) \, d\omega\right)^{1/p}$$
$$= \frac{k}{2} D_1(i,n) + \frac{h^2}{12}(p_i D_2(i,n) + 2|p_i'| D_3(i,n)) = O(k) + O(h^2). \quad (41)$$

□

4. Algorithm

From a computational point of view, as it was commented on in the Introduction Section, the handling of the random scheme (16) in a direct way makes unavailable the computation of approximations beyond a few first temporal levels. This is because, throughout the iterative temporal levels, $n = 1, \cdots, N$, it is necessary to store the symbolic expressions of all the previous levels of the iteration process collecting big and complex random expressions with which the expectation and the standard deviation must be computed. Furthermore, although the random expressions can be stored it does not guarantee that the two first statistical moments could be computed in a numerical way. For this reason, we propose using the random numerical scheme (16) together with the Monte Carlo technique avoiding the described computational drawbacks. The procedure is as follows: to take a number K of realizations of the random data involved in the random PDE (3)–(6) according to their probability distributions; to compute the numerical solution, $u_i^n(\omega_j), j = 1, \cdots, K$, of the sampling deterministic difference schemes (18); to obtain the mean and the standard deviation of these K numerical solutions evaluated in the mesh points $i = 1, \cdots, M-1$, at the last time-level N, denoted respectively by

$$\mathbb{E}_{\text{MC}}^{K}[u_i^N] = \mu\left(u_i^N(\omega_1), u_i^N(\omega_2), \cdots, u_i^N(\omega_K)\right). \tag{42}$$

$$\sqrt{\text{Var}_{\text{MC}}^{K}[u_i^N]} = \sigma\left(u_i^N(\omega_1), u_i^N(\omega_2), \cdots, u_i^N(\omega_K)\right). \tag{43}$$

Algorithm 1 summarizes the steps to compute the stable approximations of the expectation and the standard deviation of the solution s.p., u_i^n, generated by means of the sampling difference scheme (18) and the MC method.

Algorithm 1 Procedure to compute the approximations to the expectation and the standard deviation of the numerical solution u_i^N of the problem (3)–(6).

1: Consider random inputs $p(x), q(x), g_i(t), i = 1, 2$, and $f(x)$ as s.p.'s taking the form described in conditions (1) and (2).
2: Check that $p(x) \in L_p(\Omega)$ is m.s. continuous and m.s. differentiable for $0 < x < 1$. Verify condition (7) and compute the bounds a_1 and a_2.
3: Compute the m.s. derivative of $p(x)$, $p'(x)$, verify condition (8) and compute the bound b.
4: Check that coefficient $q(x) \in L_p(\Omega)$ being m.s. continuous s.p.'s for $0 < x < 1$ and verifying condition (9).
5: Check that boundary conditions $g_i(t) \in L_p(\Omega)$, $i = 1, 2$, being m.s. continuous s.p.'s for $t > 0$ and verifying condition (10).
6: Check that initial condition $f(x) \in L_p(\Omega)$ being m.s. continuous s.p.'s for $0 \le x \le 1$ and verifying condition (11).
7: Select a spatial stepsize $h = \Delta x$ verifying condition (21).
8: Consider a partition of the spatial domain $[0,1]$ of the form $x_i = ih$, $i = 0, \ldots, M$, where the integer $M = \dfrac{1}{h}$ is the number of discrete points in $[0,1]$.
9: Select a temporal step-size $k = \Delta t$ verifying condition (23) or (24).
10: Choose a time horizon T.
11: Consider a partition of the temporal interval $[0, T]$ of the form $t^n = nk$, $n = 0, \ldots, N$, where the integer $N = \frac{T}{k}$ is the number of levels necessary to achieve the time T;
12: Take and carry out a number K of MC realizations, ω_i, $1 \le i \le K$, over the r.v.'s involved in the random data of the problem (3)–(6).
13: **for** each realization ω_ℓ, $1 \le \ell \le K$ **do**
14:   **for** $i = 0$ to M **do**
15:     Evaluations of $p(x_i; \omega_\ell)$, $p'(x_i; \omega_\ell)$, $q(x_i; \omega_\ell)$, $f(x_i, \omega_\ell)$.
16:   **end for**
17: **end for**
18: **for** each realization ω_ℓ, $1 \le \ell \le K$ **do**
19:   **for** $n = 0$ to N **do**
20:     Evaluations of $g_1(t^n; \omega_\ell)$, $g_2(t^n; \omega_\ell)$.
21:   **end for**
22: **end for**
23: **for** each realization ω_ℓ, $1 \le \ell \le K$ **do**
24:   **for** $n = 0$ to N **do**
25:     Compute $u_i^n(\omega_\ell)$ using the sampling deterministic difference scheme (18).
26:   **end for**
27: **end for**
28: **for** $i = 0$ to M **do**
29:   Compute the mean, μ, of the K-deterministic solutions obtained in the time level N using (42).
30:   Compute the standard deviation, σ, of the K-deterministic solutions obtained in the time level N using (43).
31: **end for**

4.1. Numerical Example

To illustrate the efficiency of our proposed method in this subsection we present a test example where the exact solution s.p. is available. We consider the problem (3)–(6) with the random coefficients

$$p(x) = a\,e^{-x}, \quad q(x) = -c, \tag{44}$$

and the following boundary and initial conditions

$$g_1(t) = e^{ct}\left(\frac{1}{2} + at\right), \quad g_2(t) = e^{ct}\left(\frac{e^2}{2} + aet\right), \quad f(x) = \frac{e^{2x}}{2}, \tag{45}$$

that is,

$$\frac{\partial u(x,t)}{\partial t} = a\,e^{-x}\frac{\partial^2 u(x,t)}{\partial x^2} - a\,e^{-x}\frac{\partial u(x,t)}{\partial x} + c\,u(x,t), \quad 0 < x < 1, \quad t > 0, \tag{46}$$

$$u(0,t) = e^{ct}\left(\frac{1}{2} + at\right), \quad t > 0, \tag{47}$$

$$u(1,t) = e^{ct}\left(\frac{e^2}{2} + aet\right), \quad t > 0, \tag{48}$$

$$u(x,0) = \frac{e^{2x}}{2}, \quad 0 \le x \le 1, \tag{49}$$

where the r.v. a follows a Gaussian distribution of mean $\mu = 0.5$ and standard deviation $\sigma = 0.1$ truncated on the interval $[0.4, 0.6]$, that is $a \sim N_{[0.4,0.6]}(0.5; 0.1)$, and the r.v. $c > 0$ has a beta distribution of parameters $(2;4)$ truncated on the interval $[0.45; 0.55]$, that is $c \sim \text{Beta}_{[0.45,0.55]}(2;4)$. Hereinafter, we will assume that a and c are independent r.v.'s. Please note that $p(x)$ in (44) is a s.p. with one degree of randomness verifying condition (2) and $g_i(t)$, $i = 1, 2$, in (45) are s.p.'s with two degree of randomness (because they involve both r.v.'s a and c) verifying condition (2). Furthermore all random input data $p(x)$, $q(x)$, $g_1(t)$, $g_2(t)$ and $f(x)$ lie in $L_2(\Omega)$ and they are m.s. continuous and $p(x)$ is m.s. differentiable too. In addition, conditions (7)–(11) are satisfied with

$$a_1 = 0.4\,e^{-1}, \quad a_2 = 0.6\,e^0, \quad -0.55 \le q(x,\omega) \le -0.45, \ \omega \in \Omega, \quad 0 \le f(x,\omega) \le 3.69453, \ \omega \in \Omega.$$

From [18] (Section 3.8.5.) the exact solution of problem (46)–(49) when both parameters a and c are deterministic, is given by

$$u(x,t) = e^{ct}\left(a\,e^x t + \frac{e^{2x}}{2}\right). \tag{50}$$

In our context, both a and c are r.v.'s, and expression (50) must be interpreted as a s.p. Then, using the independence between r.v.'s a and c, the expectation and the standard deviation of s.p. (50) are given by

$$\mathbb{E}[u(x,t)] = \mathbb{E}[e^{ct}]\left(\mathbb{E}[a]\,e^x t + \frac{e^{2x}}{2}\right), \tag{51}$$

$$\sqrt{\text{Var}[u(x,t)]} = \sqrt{\mathbb{E}[(u(x,t))^2] - (\mathbb{E}[u(x,t)])^2}, \tag{52}$$

being

$$\mathbb{E}[(u(x,t))^2] = \mathbb{E}\left[e^{2ct}\right]\left(\mathbb{E}[a^2]\,e^{2x}t^2 + \mathbb{E}[a]\,e^{3x}t + \frac{e^{4x}}{4}\right). \tag{53}$$

Figure 1 shows the evolution of the expectation (51) and the standard deviation (52) and (53) of the exact solution s.p. (50) when both parameters a and c are considered as the r.v.'s described above.

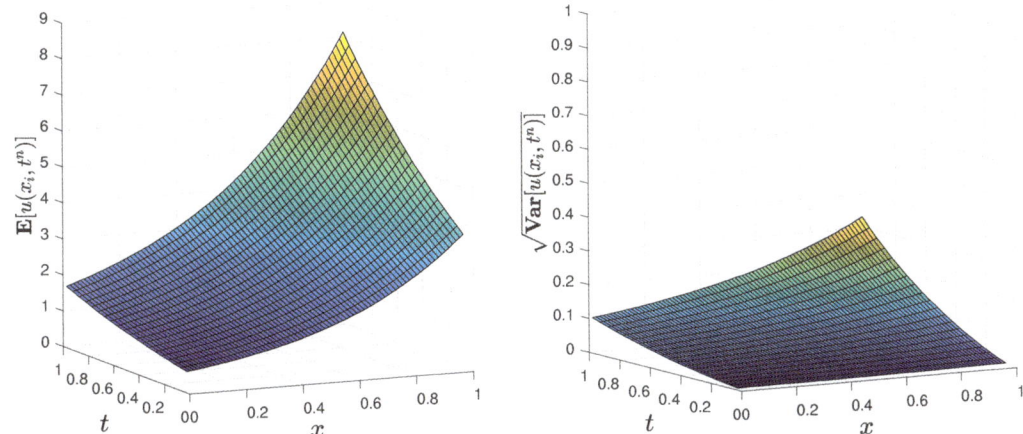

Figure 1. (**Left**): Surface of the expectation of the exact solution (50), $\mathbb{E}[u(x_i, t^n)]$, computed according to (51). (**Right**): Surface of the standard deviation of the exact solution (50), $\sqrt{\mathrm{Var}[u(x_i, t^n)]}$, computed according to (52) and (53). For both statistical moments the parameters considered in (50) are $a \sim N_{[0.4, 0.6]}(0.5; 0.1)$, $c \sim \mathrm{Beta}_{[0.45, 0.55]}(2; 4)$ and the plotted domain corresponds to $(x_i = ih, t^n = nk) \in [0 + h = 0.0125, 1 - h = 0.9875] \times [0.05, 1]$ with the step-sizes $h = \Delta x = 0.0125$, $1 \leq i \leq 79$, and $k = \Delta t = 0.05$, $1 \leq n \leq 20$.

Numerical convergence of the expectation and the standard deviation of the approximate solution s.p. generated by means the sampling difference scheme (18) using the Monte Carlo (MC) technique shown in Algorithm 1, is illustrated in the following way. In the first study, with a fixed time T, we have chosen both the spatial and temporal step-sizes h and k, respectively, according to the stability conditions (21) and (23) and we have varied the number of realizations, K, of the r.v.'s a and c involved in the random problem (46)–(49). Then, at the temporal level N where the time T is achieved, we have computed the expectation (mean), $\mathbb{E}_{\mathrm{MC}}^K[u_i^N]$, and the standard deviation, $\sqrt{\mathrm{Var}_{\mathrm{MC}}^K[u_i^N]}$, of the K-deterministic solutions, u_i^N, obtained to solve the K-deterministic difference schemes (18). Table 1 collects the RMSEs (Root Mean Square Errors) computed at the time instant $T = Nk = 1$ with the temporal step-size $k = 0.0001$ ($N = 10{,}000$) for $M - 1 = 79$ internal spatial points $x_i = ih$, $1 \leq i \leq 79$ with $h = \Delta x = 0.0125$ in the domain $[0.0125, 1]$, using the following expressions

$$\mathrm{RMSE}\left[\mathbb{E}_{\mathrm{MC}}^K[u_i^N]\right] = \sqrt{\frac{1}{M-1} \sum_{i=1}^{M-1} \left(\mathbb{E}[u(x_i, t^N)] - \mathbb{E}_{\mathrm{MC}}^K[u_i^N]\right)^2}, \quad (54)$$

$$\mathrm{RMSE}\left[\sqrt{\mathrm{Var}_{\mathrm{MC}}^K[u_i^N]}\right] = \sqrt{\frac{1}{M-1} \sum_{i=1}^{M-1} \left(\sqrt{\mathrm{Var}[u(x_i, t^N)]} - \sqrt{\mathrm{Var}_{\mathrm{MC}}^K[u_i^N]}\right)^2}, \quad (55)$$

where $\mathbb{E}[u(x_i, t^N)]$ and $\sqrt{\mathrm{Var}[u(x_i, t^N)]}$ are given by (51)–(53), respectively.

The good behaviour of both approximations the expectation and the standard deviation as the K simulations increases was observed. That is, the accuracy of the approximations to both statistical moments increases when the number of MC simulations is growing. In this sense, Figure 2 reflects the improvement of the approximations considering the study of the relative errors computed by the expressions

$$\mathrm{RelErr}\left[\mathbb{E}_{\mathrm{MC}}^K[u_i^N]\right] = \left|\frac{\mathbb{E}[u(x_i, t^N)] - \mathbb{E}_{\mathrm{MC}}^K[u_i^N]}{\mathbb{E}[u(x_i, t^N)]}\right|, \quad (56)$$

$$\text{RelErr}\left(\sqrt{\text{Var}_{\text{MC}}^{K}[u_i^N]}\right) = \left|\frac{\sqrt{\text{Var}[u(x_i,t^N)]} - \sqrt{\text{Var}_{\text{MC}}^{K}[u_i^N]}}{\sqrt{\text{Var}[u(x_i,t^N)]}}\right|. \quad (57)$$

Table 1. Root mean square errors (RMSEs) at $T = Nk = 1$ with $k = 0.0001$ (N0 = 1000) for the numerical expectation and the numerical standard deviation computed after solving the K-deterministic difference scheme (18) for several Monte Carlo (MC) realizations $K \in \{50, 200, 800, 3200, 12{,}800\}$. The spatial discretization have been considered on the domain $[0+h = 0.0125, 1-h = 0.9875]$ with $x_i = ih$, $1 \leq i \leq 79$, $h = 0.0125$.

K	RMSE $[\mathbb{E}_{\text{MC}}^{K}[u_i^N]]$	RMSE $\left[\sqrt{\text{Var}_{\text{MC}}^{K}[u_i^N]}\right]$
50	1.45604×10^{-2}	1.32856×10^{-2}
200	1.11710×10^{-2}	1.84435×10^{-3}
800	1.08512×10^{-2}	1.06139×10^{-3}
3200	4.20138×10^{-3}	6.01374×10^{-3}
12,800	2.07183×10^{-4}	1.69504×10^{-3}

Table 2 shows the second complementary study, where we have fixed the number of MC simulations K, $K = 1600$, and we have refined the step-sizes h and k attending to the stability conditions (21) and (23). It is observed the decrease of the RMSEs of the expectation (54) and an apparent stabilization in the RMSEs behaviour of the standard deviation (55). Computations have been carried out by Mathematica© software version 12.0.0.0, [21] for Windows 10Pro (64-bit) AMD Ryzen Threadripper 2990WX, 3.00 GHz 32 kernels. The CPU times (in seconds) spent in the Wolfram Language kernel to compute, in both experiments, the expectation (mean) and the standard deviation are included in Tables 3 and 4. As a result, a good strategy to study the convergence of approximations consists of choosing step-sizes h and k verifying the stability conditions and take a big enough number of realizations K such that the error does not vary significantly when one increases the number of realizations. For problems with no available solution the error is changed by the deviation between two successive numerical solutions.

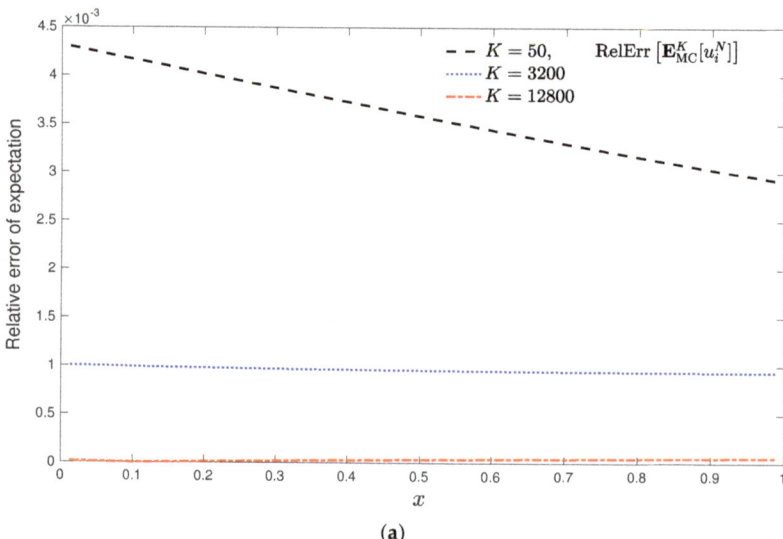

(a)

Figure 2. Cont.

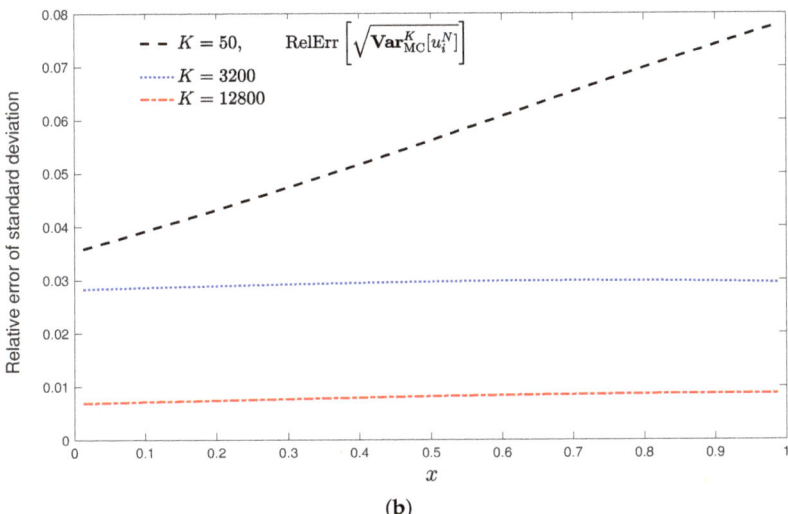

(b)

Figure 2. Plot (**a**): Relative errors of the approximations to the expectation (mean), $\mathbb{E}_{MC}^K[u_i^N]$, (56). Plot (**b**): Relative errors of the approximations to the standard deviation, $\sqrt{\text{Var}_{MC}^K[u_i^N]}$, (57). For both graphics the fixed time horizon is $T = 1 = Nk$ with the temporal step-size $k = 0.0001$ ($N = 10,000$), the spatial domain $x_i \in [0 + h, 1 - h]$ with $x_i = ih$, $h = 0.0125$, but varying the number of MC simulations $K \in \{50, 3200, 12,800\}$.

Table 2. RMSEs at $T = 1$ and $K = 1600$ (MC simulations) for the expectation (54) and the standard deviation (55). The considered step-sizes h and k verify stability conditions (21) and (23). $T = Nk = 1$, $N \in \{125, 500, 2000, 8000\}$, the spatial domain is $[0 + h, 1 - h]$ considering $M - 1$ internal points $x_i = ih$, $1 \le i \le M - 1$ with $M = 1/h$.

h	k	RMSE $[\mathbb{E}_{MC}^K[u_i^N]]$	RMSE $\left[\sqrt{\text{Var}_{MC}^K[u_i^N]}\right]$
0.1	0.008	5.29465×10^{-2}	2.36145×10^{-3}
0.05	0.002	3.19431×10^{-3}	2.49070×10^{-3}
0.025	0.0005	2.72452×10^{-3}	2.52301×10^{-3}
0.0125	0.000125	2.61957×10^{-3}	2.53168×10^{-3}

Table 3. CPU time (in seconds) spent to compute the approximations to the expectation (mean), $\mathbb{E}_{MC}^K$, and the standard deviation, $\sqrt{\text{Var}_{MC}^K}$ in Table 1, for a fixed time horizon $T = 1$ and the step-sizes $h = 0.0125$ and $k = 0.0001$ while the number of MC simulations, K, varies.

K	CPU,s $\left[\mathbb{E}_{MC}^K / \sqrt{\text{Var}_{MC}^K}\right]$
50	630.516
200	982.375
800	2052.330
3200	6209.480
12,800	22,600.100

The use of MC method has allowed the obtainment of approximations to the expectation and the standard deviation of the solution s.p. u_i^N of the random difference scheme (16) at time horizon $T = Nk$ for N not necessarily small. However, if we use the random numerical scheme (16) directly in this example with the step-sizes $h = 0.05$ ($M = 20$) and $k = 0.002$ verifying the stability conditions (21) and (23), troubles appear in the early time-level $n = 3$, that corresponds to time $t^n = 0.006$. In this case the symbolic expressions for the random numerical solution u_i^n and $(u_i^n)^2$, for $n = 3$ are available and their correspond expectations too. However, the Mathematica© software can not compute the numerical expectation of $(u_i^n)^2$ for $2 \leq i \leq 18$, $n = 3$, in consequence it is not possible to compute the approximation of the standard deviation for these internal points at $t^n = 0.006$ and hence at no other later time.

Table 4. CPU time (in seconds) spent to compute the approximations to the expectation (mean), $\mathbb{E}_{MC}^K$, and the standard deviation, $\sqrt{Var_{MC}^K}$ in Table 2, for a fixed time horizon $T = 1$ and $K = 1600$ MC simulations but varying the temporal step-size k and the spatial step-size h in the domain $[0 + h, 1 - h]$.

h	k	CPU,s $\left[\mathbb{E}_{MC}^K / \sqrt{Var_{MC}^K}\right]$
0.1	0.008	11.4688
0.05	0.002	56.2344
0.025	0.0005	341.6410
0.0125	0.000125	2438.70000

5. Conclusions

The main target of this paper is to solve the challenge of storage accumulation and further computational breakdown dealing with FD methods for solving random PDEM. Our approach is based on a combination of Monte Carlo method and the solution of sampled deterministic methods using explicit FD schemes. Explicitness is necessary to compute the statistical moments of the approximate solution what disregards the implicit FD methods. We here use the explicit classic difference method, but the Crank-Nicolson semi-implicit approach could be tried, making an ad hoc analysis. Numerical analysis provides sufficient conditions for positivity, stability and consistency for the proposed RMSFDS. Numerical experiments illustrate the reliability of the approach.

Author Contributions: M.C.C., R.C. and L.J. contributed equally to this work. All authors have read and agreed to the published version of the manuscript.

Funding: This work was supported by the Spanish Ministerio de Economía, Industria y Competitividad (MINECO), the Agencia Estatal de Investigación (AEI) and Fondo Europeo de Desarrollo Regional (FEDER UE) grant MTM2017-89664-P.

Data Availability Statement: Not applicable.

Conflicts of Interest: The authors declare no conflict of interest.

References

1. Hundsdorfer, W.; Verwer, J.G. *Numerical Solution of Time-Dependent Advection-Diffusion-Reaction Equations*; Springer: New York, NY, USA, 2003.
2. Smith, G.D. *Numerical Solution of Partial Differential Equations: Finite Difference Methods*, 3rd ed.; Clarendon Press: Oxford, UK, 1985.
3. Bharucha-Reid, A.T. On the Theory of Random Equations. In *Stochastic Processes in Mathematical Physics and Engineering. Proceedings of Symposia in Applied Mathematics*; American Mathematical Society: Providence, PI, USA, 1964; Volume XVI, pp. 40–69.
4. Bharucha-Reid, A.T. *Probabilistic Methods in Applied Mathematics*; Academic Press, Inc.: New York, NY, USA, 1973.
5. Torquato, S. *Random Heterogeneous Materials: Microstructure and Macroscopic Properties*; Springer: New York, NY, USA, 2002.
6. Vanmarcke, E. *Random Fields: Analysis and Synthesis*; World Scientific Publishing Co. Inc.: London, UK, 2010.
7. Farlow, S.J. *Partial Differential Equations for Scientists and Engineers*; Dover: New York, NY, USA, 1993.
8. Casabán, M.-C.; Company, R.; Egorova, V.; Jódar, L. Integral transform solution of random coupled parabolic partial differential models. *Math. Meth. Appl. Sci.* **2020**, *43*, 8223–8236. [CrossRef]
9. Gunzburger, M.D.; Webster, C.G.; Zhang, G. Stochastic finite element methods for partial differential equations with random input data. *Acta Numer.* **2014**, *23*, 521–650. [CrossRef]
10. Casabán, M.-C.; Company, R.; Jódar, L. Numerical solutions of random mean square Fisher-KPP models with advection. *Math. Meth. Appl. Sci.* **2020**, *43*, 8015–8031. [CrossRef]

11. Cortés, J.-C.; Romero, J.-V.; Roselló, M.-D.; Sohaly, M.A. Solving random boundary heat model using the finite difference method under mean square convergence. *Comp. Math. Meth.* **2019**, *1*, e1026. [CrossRef]
12. Sohaly, M.A. Random difference scheme for diffusion advection model. *Adv. Differ. Equ.* **2019**, *2019*, 54. [CrossRef]
13. Øksendal, B. *Stochastic Differential Equations. An Introduction with Applications*, 5th ed.; Springer: Berlin/Heidelberg, Germany, 2003.
14. Soong, T.T. *Random Differential Equations in Science and Engineering*; Academic Press: New York, NY, USA, 1973.
15. Kroese, D.P.; Taimre, T.; Botev, Z.I. *Handbook of Monte Carlo Methods*; Wiley Series in Probability and Statistics; John Wiley & Sons: New York, NY, USA, 2011.
16. Asmussen, S.; Glynn, P.W. *Stochastic Simulation: Algorithms and Analysis*; Rozovskii, B., Grimmett, G., Eds.; Springer Science & Business Media, LLC: New York, NY, USA, 2007.
17. Özişik, M.N. *Boundary Value Problems of Heat Conduction*; Dover Publications, Inc.: New York, NY, USA, 1968.
18. Polyanin, A.D.; Nazaikinskii, V.E. *Handbook of Linear Partial Differential Equations for Engineers and Scientists*; Taylor & Francis Group: Boca Raton, FL, USA, 2016.
19. Jódar, L.; Caudillo, L.A. A low computational cost numerical method for solving mixed diffusion problems. *Appl. Math. Comput.* **2005**, *170*, 673–685. [CrossRef]
20. Coatléven, J. A virtual volume method for heterogeneous and anisotropic diffusion-reaction problems on general meshes. *Esaim Math. Model. Numer. Anal.* **2017**, *51*, 797–824. [CrossRef]
21. Wolfram Research, Inc. *Mathematica*; Version 12.0; Mathematica Wolfram Research, Inc.: Champaign, IL, USA. 2020. Available online: https://www.wolfram.com/mathematica (accessed on 7 December 2020).

Article

The Relativistic Harmonic Oscillator in a Uniform Gravitational Field

Michael M. Tung

Instituto Universitario de Matemática Multidisciplinar, Universitat Politècnica de Valencia, Camino de Vera, s/n, 46022 Valencia, Spain; mtung@mat.upv.es

Abstract: We present the relativistic generalization of the classical harmonic oscillator suspended within a uniform gravitational field measured by an observer in a laboratory in which the suspension point of the spring is fixed. The starting point of this analysis is a variational approach based on the Euler–Lagrange formalism. Due to the conceptual differences of mass in the framework of special relativity compared with the classical model, the correct treatment of the relativistic gravitational potential requires special attention. It is proved that the corresponding relativistic equation of motion has unique periodic solutions. Some approximate analytical results including the next-to-leading-order term in the non-relativistic limit are also examined. The discussion is rounded up with a numerical simulation of the full relativistic results in the case of a strong gravity field. Finally, the dynamics of the model is further explored by investigating phase space and its quantitative relativistic features.

Keywords: relativistic harmonic oscillator; kinematics of a particle; special relativity; nonlinear problems in mechanics; equations of motion in gravitational theory

MSC: 70B05; 83A05; 70K42; 70K99; 83C10

Citation: Tung, M.M. The Relativistic Harmonic Oscillator in a Uniform Gravitational Field. *Mathematics* **2021**, *9*, 294. https://doi.org/10.3390/math9040294

Academic Editors: Rami Ahmad El-Nabulsi; Andrea Scapellato and Lucas Jódar

Received: 9 January 2021
Accepted: 1 February 2021
Published: 3 February 2021

Publisher's Note: MDPI stays neutral with regard to jurisdictional clai-ms in published maps and institutio-nal affiliations.

Copyright: © 2021 by the author. Licensee MDPI, Basel, Switzerland. This article is an open access article distributed under the terms and conditions of the Creative Commons Attribution (CC BY) license (https://creativecommons.org/licenses/by/4.0/).

1. Introduction

The harmonic oscillator or mechanical spring, implementing Hooke's Law, is one of the standard textbook examples for introducing the student to Newtonian mechanics [1]. Treating classical motion under a Hookean potential is simplest, in spite of additional difficulties when, e.g., velocity-dependent forces (as friction) are added. Remarkably, the relativistic counterpart of an oscillator or a pendulum—which approximates to a harmonic oscillator for small amplitudes—stationed within some supplementary force field has so far been dealt with only scarcely.

A detailed discussion of relativistic effects on a simple pendulum without any additional forces has been carried out by Erkal in 2000 [2]. In 2008, Torres shows that the relativistic pendulum with friction possesses periodic solutions which are absent in the classical case [3]. In a more recent publication of this area of research, in 2017, de la Fuente and Torres focuses on relativistic extensions for the motion of the harmonic oscillator from the view of the oscillating body, but without including any gravitational effects [4]. In all these publications an appropriate laboratory frame is chosen where characteristic, preferred points are fixed, i.e., the suspension points of the pendulum and the spring representing the oscillator, respectively.

The present work fills an outstanding gap in the existing research literature by examining the relativistic effects of a harmonic oscillator in a uniform gravitational field, adopting and extending the approach by Goldstein and Bender [5]. In the classical model, the maximum velocity of the mass in motion can be arbitrarily large depending on the displacement with respect to the equilibrium point of the spring. Relativistic mechanics will adjust this behavior by only allowing a maximum speed less than the speed of light, as accelerating a mass to higher velocities will in like manner increase its inertial mass (equivalent to an increase in kinetic energy). Although the oscillating mass point is not considered to generate its own gravitational field and the uniform gravitational field shall be caused by an external object, any variable inertial mass will also be subject to gravity. This will inescapably lead to further nontrivial complications when studying relativistic effects of gravity for the harmonic oscillator—even from the perspective of special relativity without considering general relativity. Nonetheless, the present model may provide a legitimate and satisfactory approximation for an oscillator close to a black hole, as on a small scale the Schwarzschild spacetime at a sufficiently large distance from the event horizon (where gravitational tidal effects can be ignored) approximates well a strong uniform gravitational field.

The core of this paper is organized as follows. In Section 2, after a short introduction to the variational principle [6,7] for the classical case of the harmonic oscillator, the Euler–Lagrange formalism in the framework of special relativity [8,9] is used to set up the equations of motion for a harmonic oscillator which will be subject to an external uniform gravitational field and measured by an observer in a laboratory where the suspension point of the spring is fixed. Generalizing from the classical to the relativistic regime is not as trivial as it appears at first sight due to the particular, distinct nature of relativistic mass—mass which attains a dynamical quality—and the fact that variable kinetic energy itself is equivalent to additional mass which is susceptible to the external gravitational field. Special care has to be taken to take these entirely relativistic effects into account.

Therefore, Section 3 concentrates on the full derivation of the correct relativistic potential for the uniform gravitational field surrounding the harmonic oscillator. As approximation in the case of weak gravitational fields, we consider the Taylor expansion of the potential in the non-relativistic limit and some of its particular properties. Furthermore, we examine the full relativistic results for the potential with strong gravity and, in particular, identify its physically allowed regions. Although we are able to derive the relativistic gravitational potential for the case at hand, and in closed analytical form, the final results for the equation of motion become intangible for analytical evaluation.

In Section 4, we perform the numerical integration of the equation of motion to simulate the dynamics of this model and explore some significant characteristics of the system. In general, we prove that the equation of motion for the relativistic harmonic oscillator also has unique periodic solutions. In the strong gravity case, we compare relativistic with classical estimates for the oscillating amplitudes. For further analysis, we present the corresponding phase-space trajectories and discuss its most prominent characteristics.

2. Variational Principle and Equation of Motion

Robert Hooke (1635–1703) first pointed out that the mathematical description for small oscillations of a body with mass $m_0 > 0$ attached to an elastic spring with position $x = x(t)$ takes the form: $m_0 \ddot{x} = -kx$. The positive constant $k > 0$ depends on the elastic properties of the spring in question. As a natural length scale serves the length of the spring at its maximum elongation, denoted as $\ell > 0$. This mechanical system is termed "harmonic oscillator". Such systems are of utmost relevance in physics and engineering, as any mass particle subject to a force in stable equilibrium will effectively operate as a harmonic oscillator for small fluctuations—small fluctuations being displacements with only a fraction of length ℓ. Additional importance emerges in the dynamics of a continuous classical field as it may be formulated as the dynamics of an infinite number of harmonic oscillators. Furthermore, the quantum harmonic oscillator describes some of the most important model systems in quantum mechanics.

Already for the elementary classical case of a spring extended in a uniform Newtonian gravitational field, the most efficient and powerful approach is the framework of

Lagrangian mechanics [6,7], which is based on a variational principle. Accordingly, the deterministic equations of motion will result from the following principle of least action

$$\delta \int dt\, L = \delta \int dt\left[\tfrac{1}{2}m_0\dot{x}^2 - \tfrac{1}{2}kx^2 - m_0 g x\right] = 0 \qquad (1)$$

by varying over all possible paths $x = x(t)$ and keeping the end points fixed. From Equation (1), it becomes clear that we exclude the oscillating mass point (which has negligible mass) as possible source of a gravitational field. For practical purposes and as is customary, the position x of the mass point and the external uniform gravitational field with strength $g > 0$ are measured in the laboratory frame, where the suspension of the spring at one end is fixed.

The integrand in Equation (1) is the Lagrangian function, L, which includes kinetic energy $T = \tfrac{1}{2}m_0\dot{x}^2$, the spring potential $V_s(x) = \tfrac{1}{2}kx^2$, and the gravitational potential $V_g(x) = m_0 g x$. Note that we consider spring elongations with respect to position $x = 0$ which is also the suspension point of the spring. In this laboratory frame, with the suspension point at rest, the oscillating device is stationed within a uniform gravitational field (determined by the gravitational constant $g > 0$) in such a way that both are aligned. This dynamical system is one-dimensional, and the corresponding Euler–Lagrange equation amounts to solving a simple linear second-order differential equation deriving from

$$\left(\frac{d}{dt}\frac{\partial}{\partial \dot{x}} - \frac{\partial}{\partial x}\right) L = 0. \qquad (2)$$

By substituting the Lagrangian L from Equation (1) into Equation (2), it is straightforward to reproduce the well-known general solution in closed analytical form:

$$x(t) = C_1 \cos\left(\sqrt{\frac{k}{m_0}}\, t\right) + C_2 \sin\left(\sqrt{\frac{k}{m_0}}\, t\right) - \frac{m_0 g}{k}, \qquad (3)$$

where C_1 and C_2 are the two integration constants depending on the initial values for the differential equation.

However, the classical result, Equation (3), does not contemplate strong gravitational fields and when velocities $\dot{x}$ get closer to the speed of light $c > 0$. Furthermore, in special relativity the mass is a dynamical quantity, dependent on the relative velocity $\dot{x}$ of the observer, such that $m = \gamma m_0$, where the usual relativistic factor is $\gamma(\dot{x}) = 1/\sqrt{1 - \dot{x}^2/c^2}$. Consequently, Equation (3) will utterly fail in giving a faithful description of the physical effects in the relativistic domain.

In order to generalize to a correct description in the relativistic domain, the best starting point is to modify the classical principle of least action, Equation (1). As expected, the rest mass m_0 in Equation (1) will have to be divided by the factor γ to correctly incorporate both rest mass and kinetic energy. Moreover, the spring potential V_s is unaltered, and the relativistic gravitational potential V_g, however, is hitherto undetermined. Therefore, we postulate the relativistic Lagrangian

$$L(x,\dot{x}) = -\frac{m_0 c^2}{\gamma(\dot{x})} - \tfrac{1}{2}kx^2 - V_g(x), \qquad (4)$$

which readily yields the relativistic action principle

$$\delta \int dt\left[-m_0 c^2 \sqrt{1 - \frac{\dot{x}^2}{c^2}} - \tfrac{1}{2}kx^2 - V_g(x)\right] = 0. \qquad (5)$$

Note that $V_g(x)$, as stressed before, gives the relativistic gravitational potential as measured in the laboratory frame with fixed strength $g > 0$ and as a function of varying spring elongation x.

Complications for identifying V_g in Equation (5) arise, because mass in special relativity will change with its variable kinetic energy and thus simultaneously cause a change of the mass which is subject to the external gravitational field. This effect can be taken into account by using the relativistic ansatz

$$V_g(x) = \int dx\, mg = m_0 g \int \frac{dx}{\sqrt{1 - \frac{\dot{x}^2}{c^2}}}. \tag{6}$$

For a rigorous general definition of this potential see Goldstein & Bender [5].

As V_g, and thus L, do not explicitly depend on time, according to Noether's theorem total energy must be conserved. For this purpose, we carry out the Legendre transformation [6,7] of Equation (4) yielding the Lagrangian energy function

$$E(x, \dot{x}) = \frac{m_0 c^2}{\sqrt{1 - \frac{\dot{x}^2}{c^2}}} + \tfrac{1}{2} k x^2 + V_g(x) =: E, \tag{7}$$

which is just the constant total energy E of the system. Without loss of generality, but for convenience, we assume for the remainder of the derivation that at position $x = 0$ the mass particle be at rest, i.e., $E = m_0 c^2$, or equivalently, we will assume that the following initial conditions hold:

$$x(0) = 0 \quad \text{and} \quad \dot{x}(0) = 0. \tag{8}$$

Furthermore, this immediately implies via Equation (7) that

$$V_g(0) = 0. \tag{9}$$

Observe that condition Equation (9) is chosen to agree with the definition of the classical potential and is nothing more than just fixing the arbitrary, and unphysical, integration constant in Equation (6).

Applying the Euler–Lagrange formalism to Equation (5) produces the following relativistic equation of motion,

$$\frac{d}{dt}\left(\frac{m_0 \dot{x}}{\sqrt{1 - \frac{\dot{x}^2}{c^2}}} \right) + kx + \frac{d}{dx} V_g(x) = 0, \tag{10}$$

where $V_g(x)$ is still unknown and needs to be determined. The next section focuses on uncovering the explicit relativistic form of gravitational potential $V_g(x)$.

3. Relativistic Potential for Uniform Gravitational Field

As already stressed, the problem of dealing with the relativistic model of the harmonic oscillator in a uniform gravitational field is appreciably more complicated than the classical case. The core problem originates from the fundamentally different concept of mass in the classical or the relativistic description of the physical phenomena. The ansatz for the gravitational potential given in Equation (6) naturally considers relativistic corrections of a mass in relative motion with respect to the laboratory.

Moreover, from energy conservation, viz. Equation (7), with $E = m_0 c^2$, and from solving for the relativistic factor γ, it directly follows that

$$\gamma = \frac{1}{\sqrt{1 - \frac{\dot{x}^2}{c^2}}} = \frac{m_0 c^2 - \tfrac{1}{2} k x^2 - V_g(x)}{m_0 c^2} \geq 1, \tag{11}$$

which constrains the physically admissible range of potential V_g.

Now, substituting the relativistic factor of Equation (11) into Equation (6) gives the integral equation

$$V_g(x) = \frac{g}{c^2} \int dx \left[m_0 c^2 - \tfrac{1}{2} k x^2 - V_g(x) \right], \tag{12}$$

or correspondingly, the differential equation

$$\frac{d}{dx} V_g(x) = \frac{g}{c^2} \left[m_0 c^2 - \tfrac{1}{2} k x^2 - V_g(x) \right] > 0, \quad V_g(0) = 0, \tag{13}$$

where we have considered integration condition Equation (9). Furthermore, we indicated the correct sign of gradient dV_g/dx, since the gravitational force, $F_g = -dV_g/dx$, is oriented downwards the x-axis. It is important to notice that Equation (13) contains the constant $k > 0$, considering that obviously the elastic property of the spring affects the speed of the oscillating mass and enters V_g via the relativistic factor γ. This is in full agreement with previous published results using a similar approach, e.g., for the relativistic pendulum with a gravitational potential energy also depending on length of the pendulum, viz. (see in [2], Equation (6)).

Equation (13) is an ordinary differential equation of type $f'(x) = a + bx^2 + cf(x)$ which can easily be solved. The solution is

$$V_g(x) = \left[m_0 c^2 - k \left(\frac{c^2}{g} \right)^2 \right] \left(1 - e^{-\frac{g}{c^2} x} \right) + \tfrac{1}{2} k \left(\frac{2c^2}{g} - x \right) x \tag{14}$$

with derivative

$$\frac{d}{dx} V_g(x) = \left[m_0 g - k \frac{c^2}{g} \right] e^{-\frac{g}{c^2} x} + k \left(\frac{c^2}{g} - x \right). \tag{15}$$

Some checks of Equation (15) are in order: Note that in the absence of any spring ($k = 0$), the result $F_g = -V'_g = -m_0 g \, e^{-\frac{g}{c^2} x}$ is recovered, which is in full agreement with the calculations by Goldstein and Bender [5]. As a consequence, the classical result, $F_g = -m_0 g$ is obtained for $x = 0$, before the mass particle is set in motion, viz. Equation (8). Moreover, in the weak gravity domain ($g\ell/c^2 \ll 1$), to lowest order the gravitational force obviously has to be independent of spring constant k, and it is $F_g \approx -m_0 g + \mathcal{O}(g\ell/c^2)$, with ℓ being the natural length scale of the spring, viz. Section 2. However, as gravity becomes stronger, the harmonic force will entangle with gravity in the potential V_g due to the subtle relativistic effects already mentioned. To see this, we expand Equation (14) in a power series and obtain the expansion of V_g up to first order in the dimensionless scale parameter $g\ell/c^2$:

$$V_g(x) = m_0 g \, x - \frac{k x^3}{6 \ell} \left(\frac{g \ell}{c^2} \right) + \mathcal{O}\left(\frac{g \ell}{c^2} \right)^2. \tag{16}$$

Observe that the quantity g/c^2 is a Lorentz scalar, representing an invariant for all inertial frames. Obviously, the speed of light c is a Lorentz scalar, and g being an acceleration is measured the same in all inertial frames with relative motion to the rest/laboratory frame. Thus, the ratio g/c^2 is also invariant under Lorentz transformations.

In Equation (16), the term of order $\mathcal{O}(g\ell/c^2)$ already contains spring constant k. Therefore, in first approximation for weak gravity, the odd powers indicate a symmetric result, more precisely rotational symmetry with respect to rotations of 180° about the origin, see Figure 1. In the full relativistic regime, including all orders of the expansion, symmetry is broken. Observe also that in this expansion of the gravitational potential, the additional next-to-leading-order term represents the main correction of special relativity. It bears some similarity with the post-Newtonian approximation in general relativity. However, here in our approach—within the framework of special relativity—the underlying spacetime is of course Minkowskian, and thus flat.

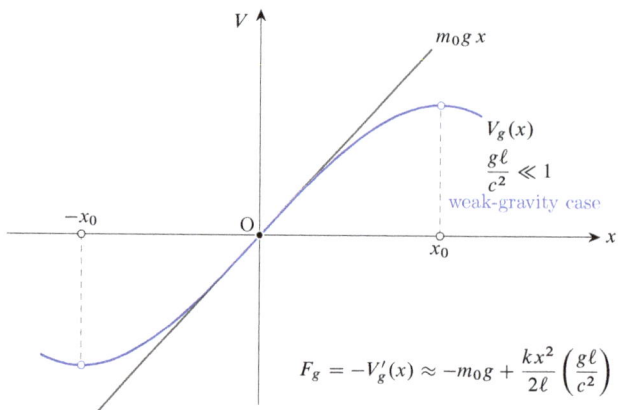

Figure 1. In the weak gravity case, when $g\ell/c^2 \ll 1$, the approximation for the relativistic gravitational potential V_g in the non-relativistic limit is most suitable (blue curve), see Equation (16). This approximate result is symmetric with respect to rotations of 180° about the origin—a symmetry property which is lost in the full relativistic case. The approximate model is only physically valid well within the interval $[-x_0, x_0]$, where $x_0 = \sqrt{\frac{2m_0}{k}}\,c$, so that sign-flips of the force field are avoided. The non-relativistic, classical case is indicated by the straight gray line.

Figure 2 displays the full relativistic result for V_g in a strong gravitational field. Here, to achieve $g/c^2 = 1\,\mathrm{m}^{-1}$ (measured in physical units m^{-1}), all parameters are set to unity, except for the spring constant which we chose to be $k = 2\,\mathrm{kg/s^2}$ (measured in physical units $\mathrm{kg/s^2}$). We also represent the gradient, $V_g' = dV_g/dx$, for any position $x \in [-1, 1]$. Note that the gravitational force therefore will flip sign at x_0, satisfying $V_g'(x_0) = 0$ for Equation (13), and is given by

$$x_0 = \frac{c^2}{g}\left[W_0\left(\frac{\frac{m_0 g^2}{kc^2} - 1}{e}\right) + 1\right], \tag{17}$$

where W_0 is the principal branch of Lambert's W function (see in [10], §4.13). By including higher orders up to $\mathcal{O}(g\ell/c^2)^3$ in Equation (16), a reasonably good approximation for Equation (17) is obtained: $x_0 \approx \frac{m_0 g}{k}\left(\sqrt{\frac{2c^2 k}{m_0 g^2} + 1} - 1\right)$.

For the data in Figure 2, it is $x_0 \approx 0.77\,\mathrm{m}$, and thus all estimates for $x > x_0$ are unphysical. However, with initial conditions Equation (8) the mass particle will move only on the negative axis, $x \leq 0$, and thus will safely be in the physical region.

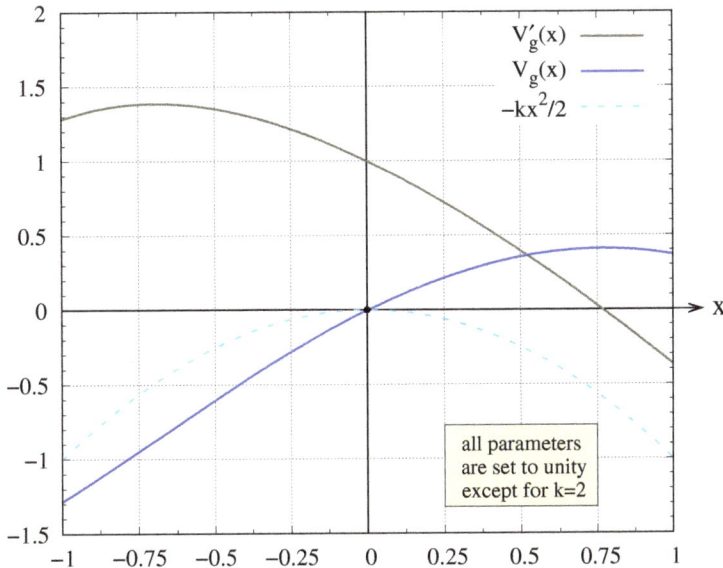

Figure 2. The relativistic gravitational potential, $V_g(x)$, for an oscillating body posted in a strong uniform gravitational field. For the graphical representation we use $m_0 = 1\,\text{kg}$, $g = 1\,\text{kg m/s}^2$, and $k = 2\,\text{kg/s}^2$. Further, the speed of light is normalized to $c = 1\,\text{m/s}$. The dashed line marks below the physically allowed region for V_g, viz. Equation (20). The gradient, $V'_g(x)$, is also shown.

4. Model Dynamics and Numerical Simulation

With the exact analytical result for the relativistic gravitational potential V_g, given by Equation (14), we are now in the position to complete the description of the dynamics of the model at hand. Substituting its derivative V'_g, given by Equation (15), into the equation of motion, Equation (10), readily yields the Euler–Lagrange equation—a nonlinear second-order differential equation—which governs the physical system

$$\gamma^3 m_0 \ddot{x} + \left[m_0 g - k\frac{c^2}{g} \right] e^{-\frac{g}{c^2}x} + k\frac{c^2}{g} = 0, \tag{18}$$

where γ and $e^{-\frac{g}{c^2}x}$ can be eliminated via Equations (11) and (14), respectively. After some lengthy but straightforward simplification, we arrive at the following equivalent and for numerical implementation more convenient form

$$\alpha(x)\ddot{x} + \beta(x)g = 0, \tag{19a}$$

$$\begin{cases} \alpha(x) = \left(1 - \dfrac{\frac{1}{2}kx^2 + V_g(x)}{m_0 c^2}\right)^3, \\[1em] \beta(x) = \dfrac{\frac{1}{2}k\left(\dfrac{2c^2}{g} - x\right)x + m_0 c^2 - V_g(x)}{m_0 c^2}, \end{cases} \tag{19b}$$

where V_g is given by Equation (14). Observe that α, β are dimensionless factors, and in particular it is $\alpha(0) = \beta(0) = 1$, such that $\ddot{x} = -g$ at the initial position $x = 0$, as is expected. Furthermore, we obtain $\ddot{x} = -g$ at all positions, when $k = 0$ and $V_g \equiv 0$. This represents the non-relativistic case in the absence of harmonic forces, that is, classical free fall. Similarly, for $k > 0$ and $V_g \equiv 0$, it is easily checked that the result reduces to that of a classical spring in a gravitational field: $m_0 \ddot{x} = -kx - m_0 g$ with general solutions Equation (3). Anyhow, Equations (19a) and (19b) embraces the full relativistic case provided that constraint Equation (11) is satisfied. The constraint Equation (11) may simply be rewritten as

$$V_g(x) \leq -\tfrac{1}{2}kx^2. \tag{20}$$

Therefore, the physically relevant gravitational potential V_g always has to lie below this concave down parabola. Figure 2 shows that this will approximately hold for the range $-x_0 \leq x \leq 0$, where no sign-flip for the gravitational force occurs. Recall that with the aforementioned parameters (all set to unity except for $k = 2 \text{ kg/s}^2$), we found $x_0 \approx 0.77$ m, viz. Equation (17).

The particular structure of the relativistic equation of motion, Equations (19a) and (19b), also implies existence and uniqueness of periodic solutions. During the past two decades, considerable progress has been made in the study of second-order differential equations and the periodic properties of their solutions [11]. For a closer analysis, we recast Equation (19a) into the form

$$\ddot{x}(t) = -\frac{\beta(x)}{\alpha(x)}g =: f(x), \tag{21}$$

where in this case the continuous function f (assuming $\alpha(x) \neq 0$) does not explicitly depend on the variables t or $\dot{x}$. Moreover, recall that $f(x(0)) = -g$. To guarantee existence and uniqueness, we have to check that f' is bounded by two continuous functions for all $t \in [0, T]$, with period $T > 0$ [11]. For this purpose, we also use the following general results for the limits of Equation (19b) and its derivatives:

$$\lim_{x \to \infty} \alpha = -\infty, \qquad \lim_{x \to \infty} \beta = \frac{k}{m_0}\left(\frac{c}{g}\right)^2,$$

$$\lim_{x \to \infty} \alpha' = -\infty, \qquad \lim_{x \to \infty} \beta' = 0, \tag{22}$$

$$\text{and} \qquad \lim_{x \to \infty} \frac{\alpha'}{\alpha} = 0.$$

Now, we can conclude that

$$\lim_{x \to \infty} f' = \lim_{x \to \infty} \frac{\alpha'\beta - \alpha\beta'}{\alpha^2} g = 0. \tag{23}$$

As $f'(x)$ and $x(t)$ are continuous, f' necessarily has to be bounded by a lower and an upper continuous function for $t \geq 0$, and we affirm that the equation of motion, Equations (19a) and (19b), has unique periodic solutions.

The series expansion in the non-relativistic limit of Equations (19a) and (19b) up to first order in the dimensionless parameter $g\ell/c^2$ gives

$$m_0 \ddot{x} + m_0 g + kx + \left(\frac{2m_0 g}{\ell}x + \frac{4k}{\ell}x^2 + \frac{3k^2}{2m_0 g\ell}x^3\right)\frac{g\ell}{c^2} = 0, \tag{24}$$

and analytical integration is possible but will already yield a rather convoluted list of elliptic integrals. Therefore, for the full relativistic result—including all higher orders—numerical integration is the most appropriate, if not the only possible, open pathway.

For the actual numerical integration of Equations (19a) and (19b), we decided to implement the code in the Julia programming language due to its efficiency and high performance [12]. Figure 3 displays the computed amplitudes for the classical and relativistic case over the time interval $t \in [0, 50]$ in SI units of seconds, again with all parameters set equal to unity, except for the spring constant $k = 2 \text{ kg/s}^2$. Using the Julia library

DifferentialEquations.jl [13], we chose the algorithm based on first-order interpolation with at least stepsize $\Delta t = 0.05\,\text{s}$ and a relative tolerance of 10^{-8} in the adaptive timestepping. Note that a long-time analysis has shown that no significant numerical errors materialize.

As shown in Figure 3, the difference between the simple classical result, resulting from Equation (3) with conditions Equation (8),

$$x(t) = \frac{m_0 g}{k}\left[\cos\left(\sqrt{\frac{k}{m_0}}\,t\right) - 1\right] \quad (25)$$

and the relativistic numerical estimate is quite pronounced—not only in amplitude but also in phase, with a positive phase shift. The physical results for the relativistic oscillator always possess longer amplitude and period than the classical analogue by cause of the relativistic mass increase.

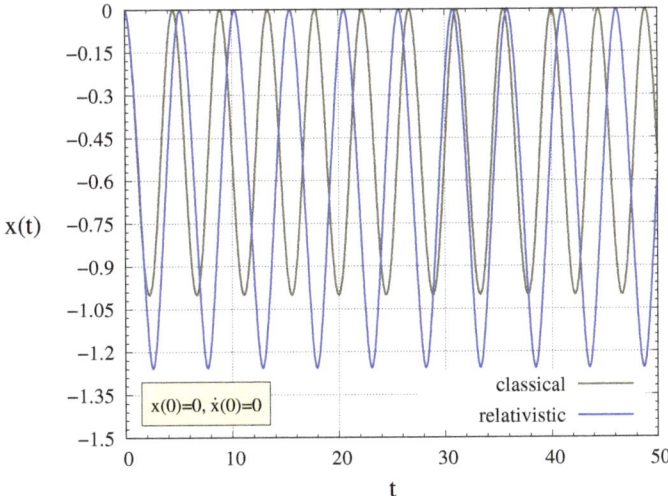

Figure 3. Amplitudes for the relativistic harmonic oscillator compared to the classical model, both suspended in a uniform gravitational field, either using the simple classical solution, viz. Equation (3), or the relativistic estimates resulting by numerical integration of Equations (19a) and (19b). The gravitational field is strong with $g/c^2 \sim 1\,\text{m}^{-1}$, choosing again all parameters unity, except for $k = 2\,\text{kg/s}^2$, see Figure 2. The two initial conditions are $x(0) = \dot{x}(0) = 0$, see Equation (8).

For an extended time interval $t \in [0, 100]$ in units of seconds, Figure 4 shows the difference between relativistic and classical predictions, $\Delta x(t) = x_{rel}(t) - x_{clas}(t)$, as already individually shown in Figure 3. As found before from the two curves in Figure 3, the relativistic corrections—now more easily recognized—propagate with a positive phase while significantly modulating the amplitude of the classical model. These contributions are substantial and cannot be neglected. Remarkably, these relativistic corrections take the shape of pronounced wave packets.

Figure 5 depicts the phase space for the relativistic harmonic oscillator with potential V_g in comparison with the classical model, using the same parameters as before, viz. Figure 2. This phase portrait provides a global overview about the dynamics of the oscillating system and shows the quantitative difference between the two models.

The relativistic phase-space trajectories emerge from the equation of motion, Equation (18), with the preserved quantity

$$\frac{m_0 c^2}{\sqrt{1 - \frac{\dot{x}^2}{c^2}}} + \tfrac{1}{2} k x^2 + V_g(x) = E_0 + m_0 c^2, \tag{26}$$

which represents energy conservation with total energy $E_0 + m_0 c^2$ in Equation (7). In the non-relativistic limit, we may use $\gamma \approx 1 + \tfrac{1}{2}\frac{\dot{x}^2}{c^2}$ and $V_g(x) \approx m_0 g x$ in Equation (26) to obtain

$$\tfrac{1}{2} m_0 \dot{x}^2 + \tfrac{1}{2} k x^2 + m_0 g x = E_0, \tag{27}$$

which is just the well-known result for the harmonic oscillator with uniform Newtonian gravity.

The classical solutions are then reproduced in the phase plane $(x, \dot{x})$ as the level curves of Equation (27) by varying $E_0 \geq 0$. Similarly, we obtain the relativistic solutions in the phase plane by drawing the level curves of Equation (26). With the chosen initial conditions in Equation (8), we have $E_0 = 0$ for both cases (which is equivalent to $E = m_0 c^2$). Figure 5 shows the two corresponding curves. Observe that Equations (26) and (27) always produce phase-space trajectories (relativistically and classically) which are closed. As a consequence, both types of solutions have to be periodic. The phase portraits of both cases represent center stable dynamics. Another characteristic effect is that the phase-space path for the relativistic case is larger than for the classical path—an effect which also has been observed for the relativistic pendulum [2] and a harmonic oscillator satisfying a relativistic isochronicity principle [4].

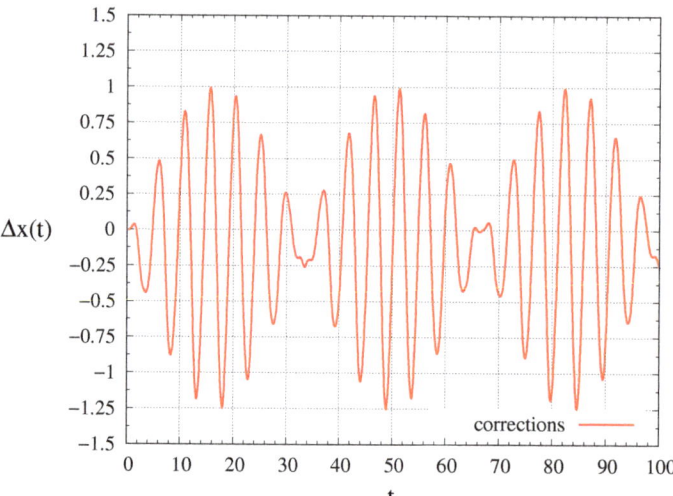

Figure 4. Amplitude changes, $\Delta x(t) = x_{rel}(t) - x_{clas}(t)$, characterizing the relativistic corrections for the classical harmonic oscillator in a strong uniform gravitational field, corresponding to the physical configuration of Figure 3, but for the larger time interval $t \in [0, 100]$ in units of seconds. These corrections are significant and modulate in both amplitude and phase.

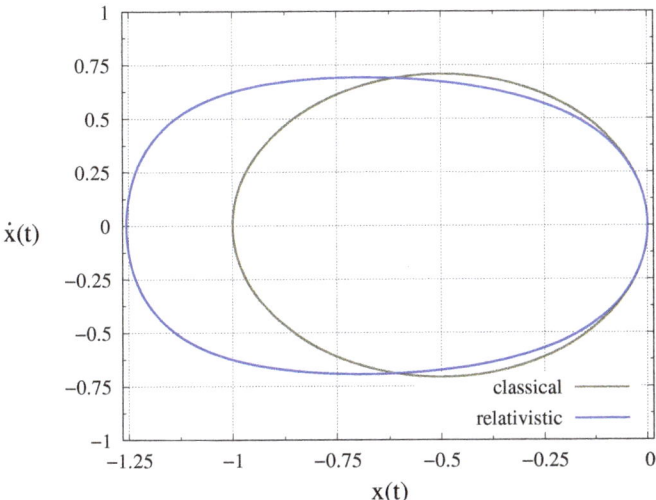

Figure 5. The phase-space diagrams for the harmonic oscillator in a uniform gravitational field corresponding to the classical and the relativistic case. We are using the same parameters as in Figure 2. The characteristics for both cases are similar: the closed trajectories represent periodic solutions, and their phase portraits represent center stable dynamics.

5. Conclusions

We have derived and studied the relativistic generalization of an oscillating body in motion under a Hookean potential and in parallel alignment with a uniform gravitational field.

If gravity is strong, these relativistic corrections differ substantially from the classical predictions and are relevant. These amplitude corrections can be pictured as fluctuating wave packets traveling on top of the classically predicted oscillations.

Quantitatively, the relativistic oscillator always acquires longer amplitudes and periods than the classical analogue. This is the dynamical effect due to the increase in mass of a moving object as dictated by special relativity. In spite of this fundamental difference between the relativistic and non-relativistic framework, we have proven that the corresponding relativistic equation of motion still maintains unique periodic solutions, similar to the well-known classical case.

For practical purposes, we have also presented approximations in the non-relativistic limit (keeping the next-to-leading-order term) for the relativistic gravitational potential and for the equation of motion of the harmonic oscillator—an approximate equation which could still be solved analytically in terms of elliptic integrals. However, all estimates in the fully relativistic regime have to be solved by numerical integration. Toward this end, we implemented the integration code with high precision in the Julia programming language by using efficient first-order interpolation. This simulation also allowed to represent the classical and relativistic phase space and to explore the dynamics of both models with their common and quantitatively different features.

Funding: This work has been supported by the Spanish *Ministerio de Economía y Competitividad*, the European Regional Development Fund (ERDF) under grant TIN2017-89314-P, and the *Programa de Apoyo a la Investigación y Desarrollo 2018* (PAID-06-18) of the Universitat Politècnica de València under grant SP20180016.

Institutional Review Board Statement: Not applicable.

Informed Consent Statement: Not applicable.

Data Availability Statement: Not applicable.

Conflicts of Interest: The author declares no conflicts of interest.

References

1. Goldstein, H.; Poole, C.; Safko, J. *Classical Mechanics*; Addison Wesley: Boston, MA, USA, 2001.
2. Erkal, C. The simple pendulum: A relativistic revisit. *Eur. J. Phys.* **2000**, *21*, 377–384. [CrossRef]
3. Torres, P.J. Periodic oscillations of the relativistic pendulum with friction. *Phys. Lett. A* **2008**, *372*, 6386–6387. [CrossRef]
4. de la Fuente, D.; Torres, P.J. A new relativistic extension of the harmonic oscillator satisfying an isochronicity principle. *Qual. Theory Dyn. Syst.* **2017**, *34*, 579–589. [CrossRef]
5. Goldstein, H.F.; Bender, C.M. Relativistic brachistochrone. *J. Math. Phys.* **1986**, *27*, 507–511. [CrossRef]
6. Lanczos, C. *The Variational Principles of Mechanics*; Dover Publications: New York, NY, USA, 1970.
7. Mann, P. *Lagrangian and Hamiltonian Dynamics*; Oxford University Press: Oxford, UK, 2018.
8. Christodoulides, C. *The Special Theory of Relativity: Foundations, Theory, Verification, Applications*; Springer: New York, NY, USA, 2016.
9. Rahaman, F. *The Special Theory of Relativity: A Mathematical Approach*; Springer: Berlin/Heidelberg, Germany, 2014.
10. Olver, F.J.W.; Lozier, D.W.; Boisvert, R.F.; Clark, C.W. *NIST Digital Library of Mathematical Functions*; National Institute of Standards and Technology: Gaithersburg, MD, USA, 2020. Release 1.0.27 of 2020-06-15; Section 4.13: Lambert W-Function. Available online: https://dlmf.nist.gov/4.13 (accessed on 9 January 2021).
11. Wei, Y. Existence and Uniqueness of Periodic Solutions for Second Order Differential Equations. *J. Funct. Spaces* **2014**, *2014*, 246258. [CrossRef]
12. Bezanson, J.; Edelman, A.; Karpinski, S.; Shah, V.B. Julia: A fresh approach to numerical computing. *SIAM Rev.* **2017**, *59*, 65–98. [CrossRef]
13. Rackauckas, C.; Nie, Q. DifferentialEquations.jl—A performant and feature-rich ecosystem for solving differential equations in Julia. *J. Open Res. Softw.* **2017**, *5*, 15. [CrossRef]

Article

Stochastic Modeling of Plant Virus Propagation with Biological Control

Benito Chen-Charpentier

Department of Mathematics, University of Texas at Arlington, Arlington, TX 76019, USA; bmchen@uta.edu

Abstract: Plants are vital for man and many species. They are sources of food, medicine, fiber for clothes and materials for shelter. They are a fundamental part of a healthy environment. However, plants are subject to virus diseases. In plants most of the virus propagation is done by a vector. The traditional way of controlling the insects is to use insecticides that have a negative effect on the environment. A more environmentally friendly way to control the insects is to use predators that will prey on the vector, such as birds or bats. In this paper we modify a plant-virus propagation model with delays. The model is written using delay differential equations. However, it can also be expressed in terms of biochemical reactions, which is more realistic for small populations. Since there are always variations in the populations, errors in the measured values and uncertainties, we use two methods to introduce randomness: stochastic differential equations and the Gillespie algorithm. We present numerical simulations. The Gillespie method produces good results for plant-virus population models.

Keywords: virus propagation; stochastic modeling; Gillespie algorithm

Citation: Chen-Charpentier, B. Stochastic Modeling of Plant Virus Propagation with Biological Control. *Mathematics* **2021**, *9*, 456. https://doi.org/10.3390/math9050456

Academic Editor: Rafael Company

Received: 18 December 2020
Accepted: 16 February 2021
Published: 24 February 2021

Publisher's Note: MDPI stays neutral with regard to jurisdictional claims in published maps and institutional affiliations.

Copyright: © 2021 by the author. Licensee MDPI, Basel, Switzerland. This article is an open access article distributed under the terms and conditions of the Creative Commons Attribution (CC BY) license (https://creativecommons.org/licenses/by/4.0/).

1. Introduction

Viruses cause a great number of diseases in plants. In [1] the authors gave a review of the top ten viruses. In plants the most common way that viruses are propagated is by means of a vector, usually insects. A carrier insect will bite an infected plant, thereby infecting it [2]. Mathematical models of virus-caused diseases in plants can be used to better understand the processes involved [3–7]. Viruses take time to propagate and replicate. These effects can be taken into account by considering latent populations [8,9]. In this paper we present three different mathematical models of virus propagation in plants with a predator used as a biological control of the insects that transmit the virus [10]. The description of the model is given in terms of differential equations and also biochemical style reactions. Models based on differential equations are built on the assumption that the number of individuals is very large, which is not always the case. The model that we use incorporating delays in the spread of the virus in both plants and vectors is based on [11]. An extension to optimal control through the use of predators and insecticide is in [12]. The authors in [13] introduced a delay to the model in [8] to account for the incubation period of the plants. In [14] there is a different model with delays. Populations have variabilities and there are errors in measuring and estimating the parameters involved. The variations and uncertainties in the populations and environmental conditions can be modeled by introducing randomness or stochasticity into the models. One common way is to use stochastic differential equations [15,16]. A second way is to consider that some of the coefficients in the model are random variables, as in the method of polynomial chaos [17,18]. Another method is to consider discrete populations and work with Markov chains [19–21], including working with the master equation [22]. Models based on continuous time Markov chains usually use the stochastic simulation algorithm of Gillespie [23,24]. Very complex models of plant viral assembly have been presented in [25,26]. Mathematical models of populations involve a series of simplifying hypotheses [27]. One of them is that the individuals in each population group have the same properties. Another usual one is that the populations are homogeneously distributed. A third one is that the number of individuals is large. Based on these simplifications, the populations can be assumed to be continuous and the model can be described in terms of differential equations or maybe delay differential equations. An alternative method of developing mathematical population models is to consider the interactions between individuals to be a reaction. For example, an individual of the class susceptible S dying can be described by the following reaction: $S \xrightarrow{k}$. Here the blank space denotes an empty population. Using differential equations, it would be $\frac{dS}{dt} = -kS$. Additionally, the interaction between a susceptible S and an infective I leading to a new infective is $S + I \xrightarrow{\beta/N} I + I$, while the corresponding differential equation is $\frac{dS}{dt} = -\beta SI/N$. k is the death rate, β/N is the infection rate divided by the total population N in the differential equations and both are the reaction rates in the corresponding reactions. This is the formulation preferred by biology and chemistry researchers. By dividing the time period of interest into subintervals and considering that all the different reactions happen at the same instant in each subinterval, the reaction system can be converted into a system of discrete equations giving the change of each population in the time subinterval. By letting the length of the time subintervals go to zero, a system of differential equations is obtained. However, if the number of individuals is not very large, as is the case in our plant-virus propagation problem or in cell processes in systems biology, the limiting process introduces errors, and the assumption that all reactions occur simultaneously is not a good one. An alternative is to consider the reactions as continuous time Markov chains [28]. This assumption leads to the stochastic simulation algorithm, or Gillespie algorithm [23,24], and its variations.

We apply this method to a simple susceptibles (S), infectives (I) and recovereds (R) (SIR) epidemic model and to a predator–prey model. The objective of presenting these two simple models is to introduce the basic ideas of writing ordinary differential equations as biochemical reactions. Then we apply them to our main interest, which is a model of virus propagation consisting of six populations, susceptible plants, infective plants, recovered plants, susceptible insects, infective insects and predators. For this virus propagation model, we consider three cases: all the interactions are mass actions—some of them are saturated and some are saturated with delays. The saturated interactions are modeled using Holling type 2 functionals [29]. We also write the models as a system of reactions. Simulations were performed using the stochastic simulation algorithm. For comparison purposes, the corresponding system of differential equations is solve numerically, and finally white noise is added to the differential equation systems to obtain stochastic differential equations and simulate them numerically. The rest of the paper is organized as follows. In Section 2, the development of the models is presented, as is the addition of stochasticity to the models to take into account the variability in the populations and other uncertainties always present in the processes modeled. Next, the virus propagation models and the numerical methods are described. Section 3 shows sample numerical simulations. In Section 4, the results are discussed. Finally there is a conclusions section.

2. Materials and Methods

2.1. Mathematical Modeling

Mathematical models for populations are usually given in terms of differential equations or difference equations. The differential equations may be ordinary, delay, partial or even fractional. However, the models may also be given by describing the interactions between the different population groups and the rate at which they happen. That is, they may be described in terms of reactions with the same structure as biochemical reactions. Therefore, we can write differential equations arising from mathematical biology as reactions. For example $\frac{dS}{dt} = -kS$ can be written as $S \xrightarrow{kS}$, where the ban space means that the population disappears. As a second example, $\frac{dS}{dt} = -\beta SI$ is equivalent to $S + I \xrightarrow{\beta SI} I + I$. Sometimes the quantity written on top of the arrow is the number of times the reaction occurs in a population (propensity), and sometimes only the reaction rate constant is included. We use the propensities.

As an illustration on the basic ideas of converting differential equations into biochemical reactions, consider the well known SIR model giving the interaction of susceptibles (S), infectives (I) and recovereds (R) [27]:

$$\begin{aligned}
\frac{dS}{dt} &= -\frac{\beta SI}{N} \\
\frac{dI}{dt} &= \frac{\beta SI}{N} - \gamma I \\
\frac{dR}{dt} &= \gamma I.
\end{aligned} \quad (1)$$

In writing the model as a system of reactions, we obtain a reaction for each distinct term on the right hand-side of the differential equations. In this example we have two such terms, $\frac{\beta SI}{N}$ and γI. The first term has the reaction of S and I. The second gives the change of I to another population. If the differential equation that has the derivative of a population includes the given term on the right-hand side with a positive sign, then the product of the reaction is that population. Hence, by writing the model as reactions between the populations we have

$$\begin{aligned}
I + S &\xrightarrow{SI\beta/N} 2I \\
I &\xrightarrow{\gamma I} R.
\end{aligned} \quad (2)$$

The first reaction is the conversion of one S into one I, and the second one is the conversion of one I into one R.

The system of reactions is not unique. This non-uniqueness has been established by [30]. In this case a more complex system is

$$\begin{aligned}
I + S &\xrightarrow{SI\beta/N} I \\
I + S &\xrightarrow{SI\beta/N} S + 2I \\
I &\xrightarrow{\gamma I} R.
\end{aligned} \quad (3)$$

In this second set of reactions, S loses one member first and in the next reaction I gains one. The reaction happens in two steps. We use the simpler formulation with S converting directly into I. However, below we will show that both formulations give the same system of differential equations.

To convert the model given in terms of reactions to a differential equation model, consider a time interval $[t, t + \delta t]$ and assume that all the reactions occur at the same instant in this interval. For the reaction system (3) we have

$$\begin{aligned}
I(t + \delta t) &= I(t) + \delta t (S(t)I(t))\beta/N \\
S(t + \delta t) + 2I(t + \delta t) &= S(t) + 2I(t) + \delta t (S(t)I(t))\beta/N \\
R(t + \delta t) &= R(t) + I(t)\gamma.
\end{aligned}$$

Assuming that the three populations are large enough, so that all the reactions happen even for very small δt, we can take the limit as $t \to 0$ to get

$$\frac{dI}{dt} = (IS)\beta/N$$
$$\frac{d(S+2I)}{dt} = (IS)\beta/N$$
$$\frac{dR}{dt} = \gamma I.$$

By subtracting the first equation from the second, we obtain the original system of differential Equation (1). Similarly, using the reactions in (2), we obtain (1) directly. In [31], the authors used biochemical reactions to model SIR processes. Reference [32] also included implementations of the Gillespie algorithm for the SIR model.

As a second example, consider a simple predator–prey model as given by Lotka and Volterra [27]:

$$\begin{aligned}\frac{dx}{dt} &= (b - py)x \\ \frac{dy}{dt} &= (rx - d)y,\end{aligned} \tag{4}$$

where x is the prey and y is the predator. The corresponding system of reactions is

$$\begin{aligned} x &\xrightarrow{bx} 2x \\ x+y &\xrightarrow{pxy} y \\ x+y &\xrightarrow{rxy} x+2y \\ y &\xrightarrow{dy} \end{aligned} \tag{5}$$

The first reaction is the birth of new prey, the second reaction is the elimination of prey by predators, the third one is the conversion of dead prey into new predators and the last one is the death of predators. The conversion of prey into predators needs to be in two reactions, since one killed prey does not convert into one new predator.

Even though it is straightforward to convert a population model based on differential equations to one described by reactions, there exist software packages that will automate the process. Biocham [33] http://lifeware.inria.fr/biocham4/ (accessed on 12 December 2020) will convert systems of ordinary differential equations with general interactions between populations written in xppauto format [34] to biochemical reactions. Moccasin [35] uses biocham and adds an interface for the MATLAB format of differential equations. Both programs write the equations in SBML [36], a widely used machine-readable description of biochemical reactions.

2.2. Stochastic Modeling

In order to produce more realistic results, mathematical models need to take into account the existence of errors in the observed or measured population data; variability in the populations; and uncertainties such as missing data and lack of knowledge.

These uncertainties can be modeled using random differential equations wherein it is considered that the parameters are random variables [37]. Another method is to use discrete or continuous time Markov chain models [38,39]. A different method consists of introducing the uncertainties as white noise and obtaining stochastic differential equations [16,40]. In this paper we only consider the second and third methods.

2.2.1. Continuous Time Markov Chain Models

As mentioned above, population models can be described in terms of biochemical-type reactions. A Markov chain is a stochastic process in which the probability of an event happening depends on a sequence of possible events, in which the probability of each event depends only on the previous event [28,39]. In a discrete time Markov chain, the changes of state happen at fixed points in time. In a continuous time Markov chain, the changes of state can happen at any time. Each reaction is a random event that has a given probability of happening. This probability is a function of the reaction rate and of the numbers of individuals of the populations involved. Since the reactions can occur at any time, the reactions are continuous time Markov chains. The master chemical equation for a reaction is the time evolution equation for the probability distribution over the state space of a Markov process. This is derived by substituting the transitional probability of the Markov process into the Chapman–Kolmogorov equation [41,42]. For a system of reactions, the master chemical equation is a system of ordinary differential equations that is hard to solve either analytically or numerically [43,44]. An alternative is the stochastic simulation algorithm (SSA) proposed by Gillespie [23] which produces numerical realizations. The next reaction and the time until it occurs are determined by Monte Carlo simulations involving the propensities of the reactions, and the process is repeated. The processes are Poisson processes with exponentially distributed transition times. Improvements on Gillespie's direct method are given in [24,45]. The SSA works with populations which should not be very small. Numerical implementations are, for example, in [46–48]

2.2.2. Stochastic Differential Equations

Randomness can be added to an ordinary differential equation $dx = f(x,t)dt, x(t_0) = x_0$ by including a white noise process [16,40]. A stochastic differential equation is given by

$$dX(t,\omega) = f(X(t,\omega),t) + g(X(t,\omega),t)dW(t,\omega),$$

where ω is an element of the sample space and $X = X(t,\omega)$ is a stochastic process. The initial condition $X(0,\omega) = X_0$ is taken to be known with probability one. A Brownian motion or Wiener process is formed by a sequence of random variables parameterized by time that are independent and identically distributed (iid). A stochastic process $W(t), t \in [0,\infty]$ is a Wiener process (or a standard Brownian motion) if it satisfies: (i) It is defined for $t \geq 0$ with $W_0 = 0$; (ii) if $0 \leq s < t < \infty$, then $W(t) - W(s)$ is normally distributed with mean 0 and variance $t - s$, that is, $W(t) - W(s) \sim N(0, t-s)$; (iii) if $0 \leq r < s < t < \infty$, the increments $W(t) - W(s)$ and $W(s) - W(r)$ are independent.

The noise term is called additive if it is independent of the population. This noise is also called environmental noise. If the noise term is proportional to the population, it is called multiplicative. These two are the most common types of white noise used, but they can have more complicated forms [49–52]. For uniform populations the environmental noise may be dominant. However, populations usually have variations. Demographic stochasticity is usually defined as the variation in the time evolution of a small population due to the randomness of individual birth, death, infection and other rates. However, the term can also be used for populations of any size. It can be related to stochasticity with multiplicative white noise by considering that the rates have a deterministic part plus a stochastic one. Additionally, environmental fluctuations may be related to overpopulation in ways such as shortage of food, increased aggression toward each other, etc. Therefore, a reasonable way to modify the deterministic equation is to consider introducing randomness that is proportional to the size of the population. However, this is just an assumption that needs to be verified, even though it has been commonly used, for example, in [53,54]. The best choice of white noise regarding its magnitude depends on the particular problem and is still an open research question.

2.3. A Plant-Virus Model with Biological Control

In this paper, we consider the application of the SSA and stochastic differential approaches to the model in [10]. We consider six different population groups: susceptible plants $S(t)$, infected plants $I(t)$, recovered plants $R(t)$, susceptible vectors $X(t)$, infected vectors $Y(t)$ and predators $P(t)$. Each variable represents the number of individuals in the respective population group at time t. Susceptible plants are healthy but could get the disease if infected with the virus. The infected plants have the virus but can only infect a susceptible plant through a vector. Additionally, the death rate of infected plants is higher than that of susceptible plants, since infected plants can also die from the viral infection. We also assume that farm workers replace any dead plant immediately with a new susceptible plant. Therefore, we can assume that the total plant population remains constant. We will denote this constant by K. Using this assumption, we can simplify the model in the deterministic case, since $K = S(t) + I(t) + R(t)$ can be used to eliminate the recovered population from the system of equations, and thus we can work with only five populations. This cannot be done in the stochastic case. The virus is not present in susceptible insects but they can be infected with it if they bite an infected plant. By biting a susceptible plant, infected insects can transmit viruses to it. Another assumption is that there is no vertical transmission of the virus in either plants or vectors. Moreover, we assume that the the vector does not get sick from the virus and thus it does not defend against the virus and will remain infected for the rest of its life. Therefore, there are no recovered vectors. The predators feed on the vectors and use the resulting energy to increase the number of predators. We also assume predators do not get infected by the virus even if they eat an infected vector. There is also intra-species competition between predators for the insects. A consequence of the infected vectors not being sick is that the predators feed on the infected insects and susceptible insects at the same rate. The interactions between vector and plant and predator and vector are assumed to have a limitation modeled by a predator–prey Holling type 2 functional [29]. For a large number of vectors, the number of infected plants due to the infected vectors tends to saturate as the number of infected vectors increases, which is the behavior of the Holling type 2 functional. In other words, for small number of vectors, doubling the population doubles the number of plants that are infected by the vectors, but for large number of vectors, doubling this number does not double the number of infected plants since there are not enough plants to be infected. Even though there are many other functionals that can be used to model this saturation effect, the Holling type 2 is a simple one.

After an infected vector bites a susceptible plant, it takes time for the plant to be infected, since the virus has to enter the plant cells, replicate, burst the cell and spread in the plant. It also takes time for the virus to spread inside a susceptible insect after it bites an infected plant. Hence, we introduce two discrete delays: τ_1, which is the time it takes a plant to become infected after an infected bite, and τ_2, the time it takes a vector to become infected after biting an infected plant. τ_1 is much larger than τ_2 since the infection process is more complex for plants. The assumptions used in the model are: the number of plants is constant, so the recovered plant population can be eliminated from the system of equations; plants die and are infected by infected vectors and converted into infected plants after a delay; infected plants can recover and also die; susceptible vectors are recruited at a constant rate, can die, are infected after biting an infected plant after a delay and can be eaten by a predator; infected insects can die and be eaten by a predator; predators are recruited at a constant rate, grow due to they eating vectors and can die; the interactions between populations saturate according to a Holling type 2 functional.

The model with the two discrete delays is

$$\frac{dS}{dt} = \mu(K - S) + dI - \frac{\beta Y(t - \tau_1)}{1 + \alpha Y(t - \tau_1)} S(t - \tau_1)$$

$$\frac{dI}{dt} = \frac{\beta Y(t - \tau_1)}{1 + \alpha Y(t - \tau_1)} S - (d + \mu + \gamma) I$$

$$\frac{dX}{dt} = \Lambda - \frac{\beta_1 I(t - \tau_2)}{1 + \alpha_1 I(t - \tau_2)} X(t - \tau_2) - \frac{c_1 X}{1 + \alpha_3 X} P - mX \qquad (6)$$

$$\frac{dY}{dt} = \frac{\beta_1 I(t - \tau_2)}{1 + \alpha_1 I(t - \tau_2)} X(t - \tau_2) - \frac{c_2 Y}{1 + \alpha_3 Y} P - mY$$

$$\frac{dP}{dt} = \Lambda_p + \frac{\alpha_4 c_1 X}{1 + \alpha_3 X} P + \frac{\alpha_4 c_2 Y}{1 + \alpha_3 Y} P - \delta P$$

The meanings of the parameters and the values used in the simulations to obtain the results presented in Section 3 are in Table 1, which is based on the data in [11]. P-unit is the number of individuals in the population group.

Table 1. Values for the parameters of the virus model.

Parameter Name	Description	Value
K	Total plant host population	63 P-unit
β	Infection rate of plants due to vectors	0.01/day/P-unit
β_1	Infection rate of vectors due to plants	0.01/day/P-unit
α	Saturation constant of plants due to vectors	0.2/P-unit
α_1	Saturation constant of vectors due to plants	0.1/P-unit
μ	Natural death rate of plants	0.01/day
m	Natural death rate of vectors	0.2974/day
γ	Recovery rate of plants	0.01/day
Λ	Replenishing rate of vectors	10 P-unit/day
d	Death rate of infected plants due to the disease	0.2/day
c_1	Contact rate between predators and healthy insects	0.05/day/P-unit
c_2	Contact rate between predators and infected insects	0.05/day/P-unit
δ	Natural death rate of predators	0.05/day
α_3	Saturation of predators due to insects	0.1/P-unit
Λ_p	Recruiting rate of predators	0.4 P-unit/day
α_4	Conversion rate of predators due to insects	0.1
τ_1	Delay for plants	24 days
τ_2	Delay for vectors	1 day

The equivalent reactions-based system is

$$S \xrightarrow{\mu K}$$
$$S \xrightarrow{\mu S}$$
$$Y + S \xrightarrow{k_1} Y + I, \quad \text{where} \quad k_1 = \frac{\beta Y S}{1 + \alpha Y}$$
$$I \xrightarrow{\mu I}$$
$$I \xrightarrow{dI} S$$
$$\xrightarrow{\Lambda} X$$
$$X + P \xrightarrow{k_4} P, \quad \text{where} \quad k_4 = \frac{\alpha_4 c_1 X}{1 + \alpha_3 X} P$$
$$X \xrightarrow{mX}$$
$$I + X \xrightarrow{k_3} I + Y, \quad \text{where} \quad k_3 = \frac{\beta_1 I X}{1 + \alpha_1 I} \quad (7)$$
$$Y + P \xrightarrow{k_5} P, \quad \text{where} \quad k_5 = \frac{\alpha_4 c_2 Y}{1 + \alpha_3 Y} P$$
$$Y \xrightarrow{mY}$$
$$\xrightarrow{\Lambda_p} P$$
$$X + P \xrightarrow{k_4} 2P + X$$
$$Y + P \xrightarrow{k_5} 2P + Y$$
$$P \xrightarrow{\delta}$$
$$I \xrightarrow{\gamma I} R$$
$$R \xrightarrow{\mu R} .$$

Note that the delays do not appear explicitly in the reactions.

The stochastic differential equations for the virus model are given system (8).

$$\begin{aligned}
dS &= (\mu(K - S) + dI - \frac{\beta Y(t - \tau_1)}{1 + \alpha Y(t - \tau_1)} S(t - \tau_1)) dt + \sigma_1 S dW_1 \\
dI &= (\frac{\beta Y(t - \tau_1)}{1 + \alpha Y(t - \tau_1)} S - (d + \mu + \gamma) I) dt + \sigma_2 dW_2 \\
dR &= (\gamma I - \mu R) dt + \sigma_3 R dW_3 \\
dX &= (\Lambda - \frac{\beta_1 I(t - \tau_2)}{1 + \alpha_1 I(t - \tau_2)} X(t - \tau_2) - \frac{c_1 X}{1 + \alpha_3 X} P - mX) dt + \sigma_4 X dW_4 \\
dY &= (\frac{\beta_1 I(t - \tau_2)}{1 + \alpha_1 I(t - \tau_2)} X(t - \tau_2) - \frac{c_2 Y}{1 + \alpha_3 Y} P - mY) dt + \sigma_5 Y dW_5 \\
dP &= (\Lambda_p + \frac{\alpha_4 c_1 X}{1 + \alpha_3 X} P + \frac{\alpha_4 c_2 Y}{1 + \alpha_3 Y} P - \delta P) dt + \sigma_6 P dW_6
\end{aligned} \quad (8)$$

2.4. Numerical Methods

Numerical methods for ordinary differential equations can be modified for delay differential equations by integrating piece-wise over time intervals chosen such that they are multiples of the delays [55,56]. A very good numerical solver based on Runge–Kutta methods is given in [56]. In our numerical simulations we used the following values for our parameters: $K = 63$, $\beta = 0.01$, $\beta_1 = 0.01$, $\alpha = 0.2$, $\alpha_1 = 0.1$, $\mu = 0.01$, $m = 0.2974$, $\gamma = 0.01$, $\Lambda = 10$,

$d = 0.2$, $c_1 = 0.05$, $c_2 = 0.05$, $\delta = 0.05$, $\Lambda_p = 0.4$, $\alpha_3 = 0.1$ and $\alpha_4 = 0.1$. Additionally, we chose the following initial conditions: $S(0) = 59.8478$, $I(0) = 1.57612$, $X(0) = 14.6247478$, $Y(0) = 19.5$ and $P(0) = 2$. We considered the values of the delays to be $\tau_1 = 24$ and $\tau_2 = 1$. The history for the delay equations is usually taken to be constant and equal to the initial values. However, if we consider the model in terms of reactions, the delayed reactions do not happen during the history, so we took the history to be equal to zero. The values of most the parameters were taken from [7] and those referring to the delays and the predator were from [10]. These are not implied to apply to specific plants, vectors and viruses. We were not able to find real values for many of the parameters. Even for the well studied maize streak virus, the model in [57], which uses the parameters presented in [58–60], some of the values are assumed.

For stochastic differential equations using Ito calculus, two common methods are the Euler–Murayama and Milstein methods [15]. Both can be easily modified to include delays in a similar way as for ordinary differential equations [61]. The SSA can be modified to include delays in the reactions by adding the corresponding delay to the time when the reaction happens [62,63].

For the deterministic ordinary differential equations we used an Euler method implemented in GNU octave [64]. For the stochastic differential equations the Milstein method was also implemented in octave. Both methods are first order. For the reaction-based method, we used the software stochPy [65], an open source program implemented in Python.

3. Results

The first simulation is of the SIR model given by differential Equation (1) and reactions (3). The values of the parameters were: $N = 63$, $\beta = 0.2$ and $\gamma = 0.1$. The initial values were $S(0) = 60$, $I(0) = 2$ and $R(0) = 1$. Figure 1 shows the deterministic simulation on the left and the SSA simulation on the right. For the SSA simulation, only the average values of the populations are plotted. For the values of the parameters used, the populations tended to the disease-free equilibrium for a large amount of time.

The next simulation was of the predator–prey model described by Equation (4) and reactions (5). The parameter values used were $p = 1, r = 1, b = 1$ and $d = 1$. The initial conditions were $X(0) = 2$ and $Y(0) = 2$. Figure 2 shows on the left the deterministic simulation and on the right the simulation using SSA. For the SSA simulation only the average values are plotted. Note that the deterministic solution is periodic but SSA is not quite periodic.

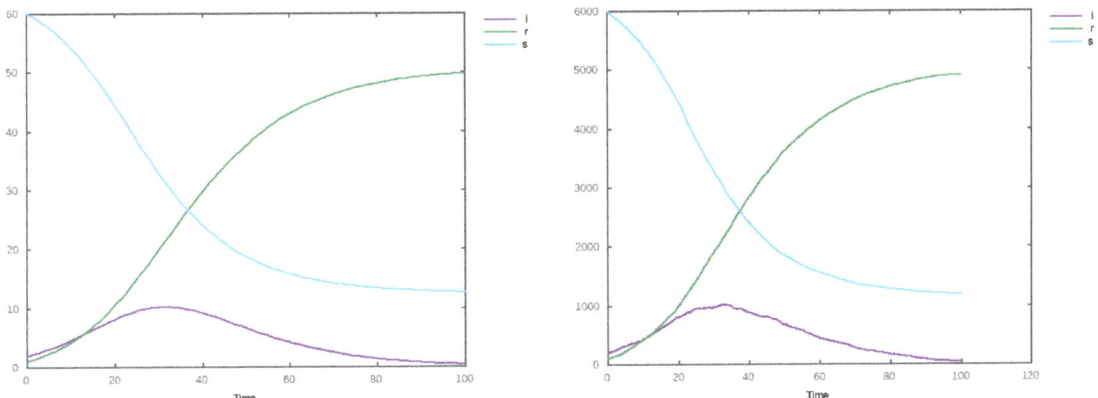

Figure 1. Simulation of susceptibles (S), infectives (I) and recovereds (R) (SIR) model. (**Left**) deterministic. (**Right**) Stochastic simulation algorithm (SSA); only the average values are shown.

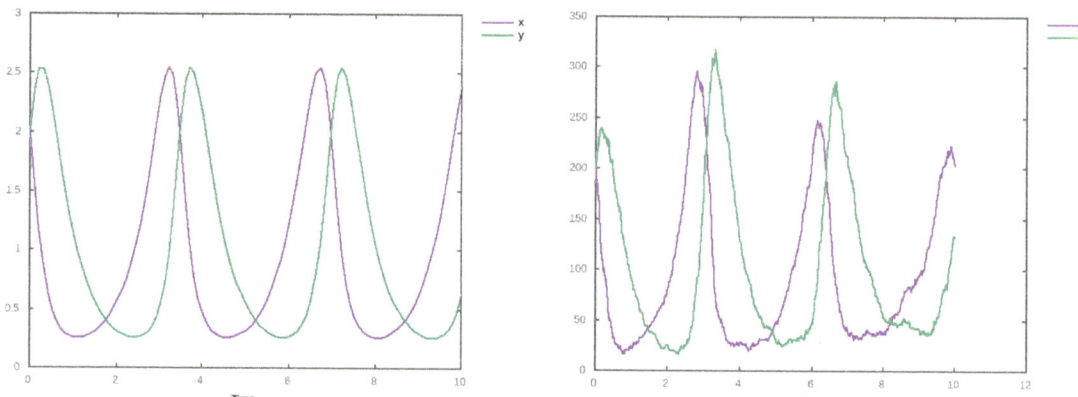

Figure 2. Simulation of the predator–prey model. (**Left**) deterministic. (**Right**) SSA; only the average values are shown.

The next simulations were for the plant-virus model given by Equation (6) and reactions (7). The stochastic differential equations used were (8). The values of the parameters used in the simulations are in Table 1.

We did three simulations of the plant-virus model. The first one was with mass action kinetics and no delays. That is, $\alpha_1, \alpha_2, \alpha_3, \alpha_4, \tau_1$ and τ_2 all being equal to zero. All the other parameters were as given in Table 1, with the exception of σ_1 to σ_6. An open question is how to determine the values of the σs. We did one simulation of the deterministic model, one of the reaction model using the SSA and three of the stochastic differential equation model (8) using three different values of the σ, 0.01. 0.025 and 0.05. Figure 3 shows the results of the deterministic simulation on the left and of the SSA simulation on the right. The stochastic simulation shows the plots for the mean and the mean +/− standard deviation for 1000 realizations. Figure 4 shows the stochastic differential equation simulations results for the mean and the mean +/− one standard deviation for 1000 simulations with all σs equal to 0.01 on the left and with the σs equal to 0.025 on the right. Figure 5 shows on the left the mean and the mean +/− one standard deviation plots for the σs equal to 0.05. From these figures we see that the σs equal to 0.025 gives a stochastic effect that is important but does not overcome the deterministic part. The plots for the mean values of the stochastic simulations were taken for 1000 realizations since the results for 500 and 1000 realizations agree to at least three significant figures.

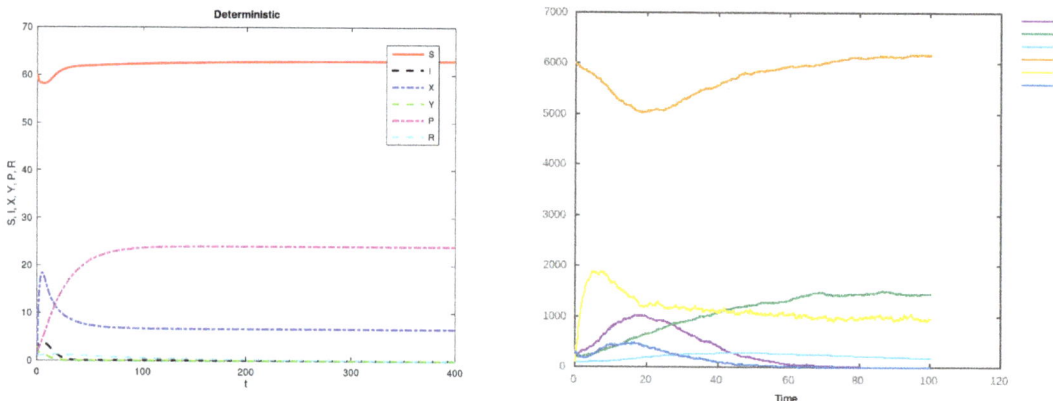

Figure 3. Plots for the virus plant model with mass action interactions and no delays. (**Left**) deterministic. (**Right**) SSA with only the average values included.

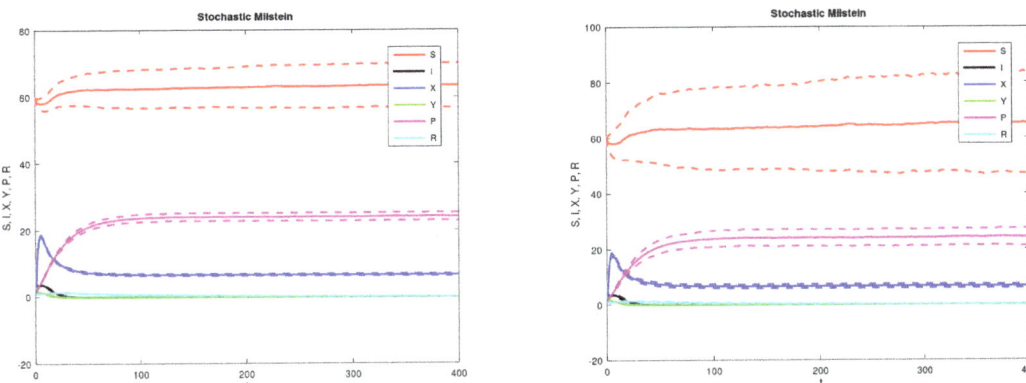

Figure 4. Plots for the plant-virus model with mass action interactions and no delays using the stochastic differential equations model. The solid lines are the mean values, and the dashed lines are the mean value plus or minus one standard deviation. (**Left**) σs equal to 0.01. (**Right**) σs equal to 0.025.

The next simulation for the plant-virus model is with Holling type 2 saturation kinetics and no delays. So $\alpha_1, \ldots, \alpha_4$ are nonzero with their values and the values of the other parameters given in Table 1 and $\tau_1 = \tau_2 = 0$. Figure 6 has the plots of the deterministic simulation. Figure 7 has the plots of simulations using the SSA (left) and stochastic differential equation s (right). For the SSA, the vertical lines represent the intervals [mean − 1 standard deviation, mean + 1 standard deviation]. For the stochastic differential equations plot the mean is given by the solid lines and the dashed lines give the mean +/− one standard deviation for 1000 realizations for the stochastic simulations. On the left for the SSA simulation and on the right for the stochastic differential equation run with σs equal to 0.025.

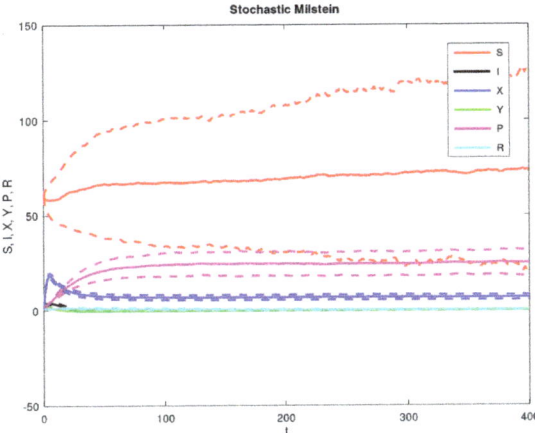

Figure 5. Plots for the plant-virus model for mass action interactions and no delays using the stochastic differential model. The solid lines are the mean values, and the dashed lines are the mean values +/− one standard deviation. The σs are equal to 0.05.

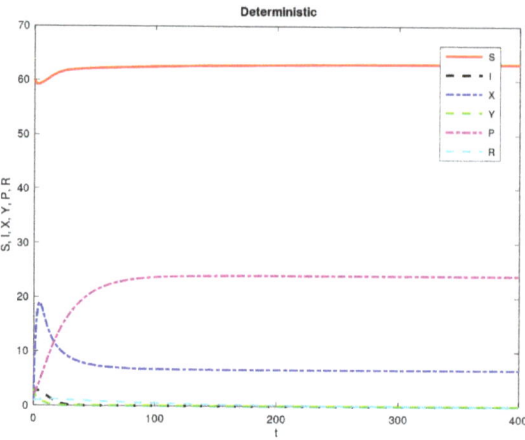

Figure 6. Plots for the virus plant model with saturations and no delays for the deterministic model.

The final simulation was for the plant-virus model with both saturation and delays. All the parametrer values are given in Table 1 and the delays were $\tau_1 = 24$ and $\tau_2 = 1$. Figure 8 has the plot of the deterministic simulation. Figure 9 has the plots of the mean and of the mean +/− one standard deviation for 1000 realizations for the stochastic simulations. On the left are the results for the SSA simulation, and the vertical lines represent the intervals (mean − 1 standard deviation, mean + 1 standard deviation). On the right are the plots for the stochastic differential equation run with σ equal to 0.025. The solid lines represent the mean values and the dashed lines the mean +/− one standard deviation.

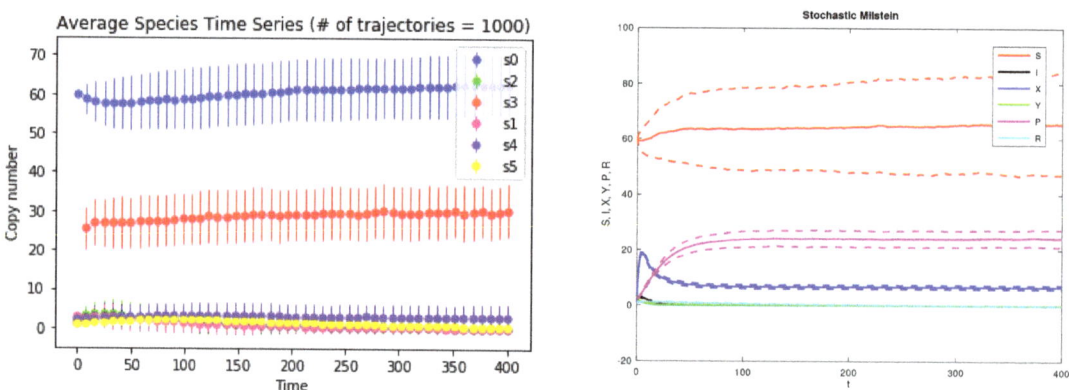

Figure 7. Plots for virus plant models with saturation and no delays. (**Left**) SSA; the vertical lines denote the intervals (mean − 1 standard deviation, mean + 1 standard deviation). (**Right**) Stochastic differential equations simulation—mean values (solid lines) and mean values +/− 1 standard deviation (dashed lines).

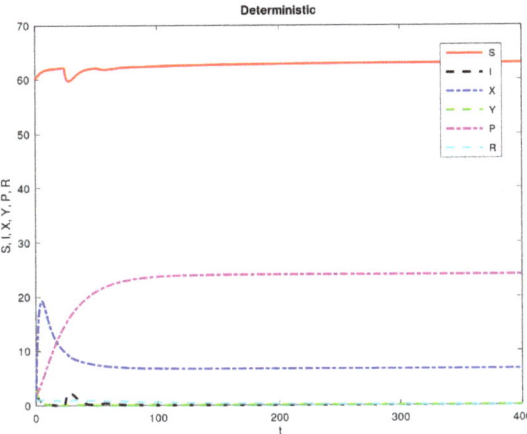

Figure 8. Plots for the virus plant model with saturations and delays for the deterministic model.

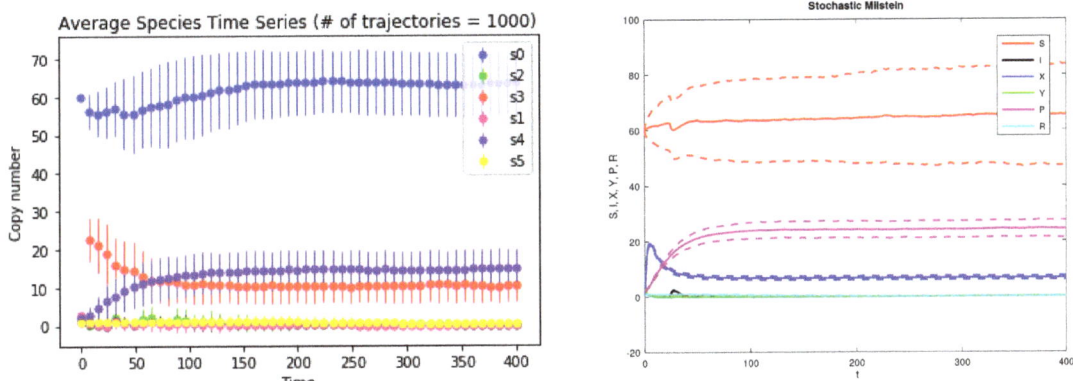

Figure 9. Plots for mean values and the mean values +/− 1 standard deviation for the virus plant model with saturation and delays. (**Left**) SSA with the vertical lines giving the intervals (mean − 1 standard deviation, mean + 1 standard deviation). (**Right**) Stochastic differential equations with the solid lines giving the mean values and the dashed lines the means +/− 1 standard deviation.

4. Discussions

The numerical simulations in all cases gave similar results for the differential equation, the stochastic differential and the reaction models. The differential equation simulations give one trajectory and do not take into account the variations in the populations and environment, or the errors in measurements of parameters. However, they are the easiest simulations to implement and the fastest. The stochastic differential equations include variations of the populations when multiplicative noise is added, as we did. By doing multiple simulations, they give the mean values and the standard deviations of the solutions, which are more realistic. However, there are other types of white noise and we need to estimate the size of the noise. The reaction model simulations using the Gillespie algorithm also give the means and standard deviations of the solutions when run for multiple simulations. It has a more solid theoretical base for small populations, and this is our case. Through both the stochastic differential equation method and the Gillespie algorithm there are some outcomes with different behavior than what comes from the deterministic method. For some realizations the disease may not disappear for very long time periods, and in future work it may be worth estimating the probability.

5. Conclusions

Mathematical population models based on differential equations give realistic results even when the populations are not very large. However, models based on biochemical reactions are more realistic. Models given in terms of reactions are usually easier to understand for biologists, ecologists and other specialists without backgrounds in mathematics. The stochastic simulation algorithm has solid theoretical backing for its use for small populations, and there is a wide variety of existing software implementing it. As with all stochastic simulations, there is the question of how many realizations to use. In our case we did one hundred and then one thousand realizations, and the results have at least three significant digits of agreement. The results for the stochastic differential equations depend on the type and magnitude of the white noise term, and it is usually not obvious which one should be used.

Even though from the first epidemic models it has been known that the models can be written in terms of biochemical reactions [66], their use and the use of Gillespie's algorithm should be more frequent, since in epidemic models the populations are many times not very large and the continuity hypothesis is not justified. Second, this method is easy to implement and does not require one—as the stochastic differential method does—to decide on the type and magnitude of the white noise. There are many papers dealing with the use of Gillespie's algorithm for epidemic models and others on using biochemical reactions and other methods. For example, in [67] the authors used a biochemical reactions formulation and cellular automata. In [68], the authors mentioned that epidemic models can be described as chemical reactions but then used a stochastic model based on binomial drawing for certain terms. In [31] the equivalent reactions for a SIR model are given. In some other papers the Gillespie algorithm was applied to variations of the SIR method [69–71]. Reference [72] presents continuous time Markov chain and stochastic differential equation models, including an SIR model and a four-population model for malaria. For this last model, the author approximated the branching process by working with the forward and backward Kolmogorov equations. In [70] the model used was an SEIR (Susceptible, Exposed, Infectious, Recovered) with added numbers of patients hospitalized and killed regarding Ebola. To the best of my knowledge, the publications about the applications of the Gillespie algorithm to epidemic models do not include models with saturation interaction terms, particularly with Holling type 2 functionals, or epidemic models with discrete delays; they also do not involve plant diseases caused bv viruses and spread by vectors. The comparison of the results for the model using stochastic equation simulations with different sizes for the white noise is also new. Future work will include applying the application to more complex models and to models with time varying rates.

Funding: This research received no external funding.

Institutional Review Board Statement: Not applicable.

Informed Consent Statement: Not applicable.

Data Availability Statement: Not applicable.

Conflicts of Interest: The author declares no conflict of interest.

References

1. Scholthof, K.B.G.; Adkins, S.; Czosnek, H.; Palukaitis, P.; Jacquot, E.; Hohn, T.; Hohn, B.; Saunders, K.; Candresse, T.; Ahlquist, P.; et al. Top 10 plant viruses in molecular plant pathology. *Mol. Plant Pathol.* **2011**, *12*, 938–954. [CrossRef] [PubMed]
2. Fereres, A. Insect vectors as drivers of plant virus emergence. *Curr. Opin. Virol.* **2015**, *10*, 42–46. [CrossRef]
3. Jeger, M.; Van Den Bosch, F.; Madden, L.; Holt, J. A model for analysing plant-virus transmission characteristics and epidemic development. *Math. Med. Biol. A J. IMA* **1998**, *15*, 1–18. [CrossRef]

4. Jeger, M.; Holt, J.; Van Den Bosch, F.; Madden, L. Epidemiology of insect-transmitted plant viruses: modelling disease dynamics and control interventions. *Physiol. Entomol.* **2004**, *29*, 291–304. [CrossRef]
5. Van Maanen, A.; Xu, X.M. Modelling plant disease epidemics. *Eur. J. Plant Pathol.* **2003**, *109*, 669–682. [CrossRef]
6. Anguelov, R.; Lubuma, J.; Dumont, Y. Mathematical analysis of vector-borne diseases on plants. In Proceedings of the 2012 IEEE 4th International Symposium on Plant Growth Modeling, Simulation, Visualization and Applications, Shanghai, China, 31 October–3 November 2012; pp. 22–29.
7. Shi, R.; Zhao, H.; Tang, S. Global dynamic analysis of a vector-borne plant disease model. *Adv. Differ. Equ.* **2014**, *2014*, 59. [CrossRef]
8. Meng, X.; Li, Z. The dynamics of plant disease models with continuous and impulsive cultural control strategies. *J. Theor. Biol.* **2010**, *266*, 29–40. [CrossRef]
9. Al Basir, F.; Adhurya, S.; Banerjee, M.; Venturino, E.; Ray, S. Modelling the effect of incubation and latent periods on the dynamics of vector-borne plant viral diseases. *Bull. Math. Biol.* **2020**, *82*, 1–22. [CrossRef]
10. Jackson, M.; Chen-Charpentier, B.M. Modeling plant virus propagation with delays. *J. Comput. Appl. Math.* **2017**, *309*, 611–621. [CrossRef]
11. Jackson, M.; Chen-Charpentier, B.M. A model of biological control of plant virus propagation with delays. *J. Comput. Appl. Math.* **2018**, *330*, 855–865. [CrossRef]
12. Chen-Charpentier, B.M.; Jackson, M. Direct and indirect optimal control applied to plant virus propagation with seasonality and delays. *J. Comput. Appl. Math.* **2020**, *380*, 112983. [CrossRef]
13. Zhang, T.; Meng, X.; Song, Y.; Li, Z. Dynamical analysis of delayed plant disease models with continuous or impulsive cultural control strategies. *Abstr. Appl. Anal.* **2012**, *2012*. [CrossRef]
14. Al Basir, F.; Takeuchi, Y.; Ray, S. Dynamics of a delayed plant disease model with Beddington-DeAngelis disease transmission. *Math. Biosci. Eng.* **2021**, *18*, 583–599. [CrossRef] [PubMed]
15. Kloeden, P.E.; Platen, E. *Numerical Solution of Stochastic Differential Equations*; Springer Science & Business Media: Berlin/Heidelberg, Germany, 2013; Volume 23.
16. Evans, L.C. *An Introduction to Stochastic Differential Equations*; American Mathematical Soc.: Providence, RI, USA, 2012; Volume 82.
17. Ghanem, R.; Spanos, P.D. Polynomial Chaos in Stochastic Finite Elements. *J. Appl. Mech.* **1990**, *57*, 197–202. [CrossRef]
18. Xiu, D.; Karniadakis, G.E. The Wiener–Askey polynomial chaos for stochastic differential equations. *SIAM J. Sci. Comput.* **2002**, *24*, 619–644. [CrossRef]
19. Gibson, G.J. Markov chain Monte Carlo methods for fitting spatiotemporal stochastic models in plant epidemiology. *J. R. Stat. Soc. Ser. C Appl. Stat.* **1997**, *46*, 215–233. [CrossRef]
20. Keeling, M.J.; Ross, J.V. On methods for studying stochastic disease dynamics. *J. R. Soc. Interface* **2008**, *5*, 171–181. [CrossRef]
21. Qi, H.; Meng, X.; Chang, Z. Markov semigroup approach to the analysis of a nonlinear stochastic plant disease model. *Electron. J. Differ. Equ.* **2019**, *2019*, 1–19.
22. Stollenwerk, N.; Briggs, K.M. Master equation solution of a plant disease model. *Phys. Lett. A* **2000**, *274*, 84–91. [CrossRef]
23. Gillespie, D.T. Exact stochastic simulation of coupled chemical reactions. *J. Phys. Chem.* **1977**, *81*, 2340–2361. [CrossRef]
24. Cao, Y.; Gillespie, D.T.; Petzold, L.R. Efficient step size selection for the tau-leaping simulation method. *J. Chem. Phys.* **2006**, *124*, 044109. [CrossRef]
25. Hemberg, M.; Yaliraki, S.N.; Barahona, M. Stochastic kinetics of viral capsid assembly based on detailed protein structures. *Biophys. J.* **2006**, *90*, 3029–3042. [CrossRef]
26. Perlmutter, J.D.; Hagan, M.F. Mechanisms of virus assembly. *Annu. Rev. Phys. Chem.* **2015**, *66*, 217–239. [CrossRef] [PubMed]
27. Allen, L. *An Introduction to Mathematical Biology*; Pearson-Prentice Hall: Upper Saddle River, NJ, USA, 2007.
28. Gagniuc, P.A. *Markov Chains: From Theory to Implementation and Experimentation*; John Wiley & Sons: Hoboken, NJ, USA, 2017.
29. Holling, C. The Components of Predation as Revealed by a Study of Small Mammal Predation of the European Pine Sawfly. *Can. Entomol.* **1959**, *91*, 293–320. [CrossRef]
30. Fages, F.; Gay, S.; Soliman, S. Inferring reaction systems from ordinary differential equations. *Theor. Comput. Sci.* **2015**, *599*, 64–78. [CrossRef]
31. Simon, C.M. The SIR dynamic model of infectious disease transmission and its analogy with chemical kinetics. *PeerJ Phys. Chem.* **2020**, *2*, e14. [CrossRef]
32. Keeling, M.J.; Rohani, P. *Modeling Infectious Diseases in Humans and Animals*; Princeton University Press: Princeton, NJ, USA, 2011.
33. Fages, F.; Soliman, S. On robustness computation and optimization in BIOCHAM-4. In Proceedings of the International Conference on Computational Methods in Systems Biology, Brno, Czech Republic, 12–14 September 2018; Springer: Berlin/Heidelberg, Germany, 2018; pp. 292–299.
34. Ermentrout, B. *Simulating, Analyzing, and Animating Dynamical Systems: A Guide to XPPAUT for Researchers and Students*; Siam: Philadelphia, PA, USA, 2002; Volume 14.
35. Gómez, H.F.; Hucka, M.; Keating, S.M.; Nudelman, G.; Iber, D.; Sealfon, S.C. Moccasin: converting matlab ode models to sbml. *Bioinformatics* **2016**, *32*, 1905–1906. [CrossRef] [PubMed]
36. Hucka, M.; Finney, A.; Sauro, H.M.; Bolouri, H.; Doyle, J.C.; Kitano, H.; Arkin, A.P.; Bornstein, B.J.; Bray, D.; Cornish-Bowden, A.; et al. The systems biology markup language (SBML): A medium for representation and exchange of biochemical network models. *Bioinformatics* **2003**, *19*, 524–531. [CrossRef]

37. Soong, T.T. *Random Differential Equations in Science and Engineering*; Academic Press: New York, NY, USA, 1973.
38. Pinsky, M.; Karlin, S. *An Introduction to Stochastic Modeling*; Academic Press: Burlington, MA, USA, 2010.
39. Allen, L.J. *An Introduction to Stochastic Processes with Applications to Biology*; CRC Press: Hoboken, NJ, USA, 2010.
40. Oksendal, B. *Stochastic Differential Equations: An Introduction with Applications*; Springer Science & Business Media: Berlin/Heidelberg, Germany, 2013.
41. McQuarrie, D.A. Stochastic approach to chemical kinetics. *J. Appl. Probab.* **1967**, *4*, 413–478. [CrossRef]
42. Van Kampen, N.G. *Stochastic Processes in Physics and Chemistry*; Elsevier: Amsterdam, The Netherlands, 1992; Volume 1.
43. Gillespie, D.T. Stochastic Simulation of Chemical Kinetics. *Annu. Rev. Phys. Chem.* **2007**, *58*, 35–55. [CrossRef] [PubMed]
44. Weber, M.F.; Frey, E. Master equations and the theory of stochastic path integrals. *Rep. Prog. Phys.* **2017**, *80*, 046601. [CrossRef] [PubMed]
45. Gibson, M.A.; Bruck, J. Efficient exact stochastic simulation of chemical systems with many species and many channels. *J. Phys. Chem. A* **2000**, *104*, 1876–1889. [CrossRef]
46. Van Gend, C.; Kummer, U. STODE-automatic stochastic simulation of systems described by differential equations. In Proceedings of the 2nd International Conference on Systems Biology, Pasadena, CA, USA, 1–5 November 2001; Volume 326, p. 333.
47. COmplex PAthway SImulator (COPASI). Available online: http://copasi.org/ (accessed on 12 December 2020).
48. Higham, D.J. Modeling and Simulating Chemical Reactions. *SIAM Rev.* **2008**, *50*, 347–368. [CrossRef]
49. Allen, E. *Modeling with Itô Stochastic Differential Equations*; Springer: Berlin/Heidelberg, Germany, 2007; Volume 22.
50. Gray, A.; Greenhalgh, D.; Hu, L.; Mao, X.; Pan, J. A stochastic differential equation SIS epidemic model. *SIAM J. Appl. Math.* **2011**, *71*, 876–902. [CrossRef]
51. Zhao, Y.; Jiang, D.; Mao, X.; Gray, A. The threshold of a stochastic SIRS epidemic model in a population with varying size. *Discret. Contin. Dyn. Syst. Ser. B* **2015**, *20*, 1277–1295. [CrossRef]
52. Zhu, L.; Hu, H. A stochastic SIR epidemic model with density dependent birth rate. *Adv. Differ. Equ.* **2015**, *2015*, 330. [CrossRef]
53. Rao, F. Dynamics analysis of a stochastic SIR epidemic model. *Abstr. Appl. Anal.* **2014**, *2014*, 356013. [CrossRef]
54. Chang, Z.; Meng, X.; Zhang, T. A new way of investigating the asymptotic behaviour of a stochastic SIS system with multiplicative noise. *Appl. Math. Lett.* **2019**, *87*, 80–86. [CrossRef]
55. Bellen, A.; Zennaro, M. *Numerical Methods for Delay Differential Equations*; Oxford University Press: Oxford, UK, 2013.
56. Shampine, L.F.; Thompson, S.; Kierzenka, J. Solving Delay Differential Equations with dde23. 2000. Available online: http://www.runet.edu/~thompson/webddes/tutorial.pdf (accessed on 12 December 2020).
57. Alemneh, H.T.; Makinde, O.D.; Theuri, D.M. Optimal Control Model and Cost Effectiveness Analysis of Maize Streak Virus Pathogen Interaction with Pest Invasion in Maize Plant. *Egypt. J. Basic Appl. Sci.* **2020**, *7*, 180–193. [CrossRef]
58. Bosque-Pérez, N.A. Eight decades of maize streak virus research. *Virus Res.* **2000**, *71*, 107–121. [CrossRef]
59. Magenya, O.; Mueke, J.; Omwega, C. Significance and transmission of maize streak virus disease in Africa and options for management: A review. *Afr. J. Biotechnol.* **2008**, *7*, 4897–4910.
60. Alemneh, H.T.; Makinde, O.D.; Mwangi Theuri, D. Ecoepidemiological Model and Analysis of MSV Disease Transmission Dynamics in Maize Plant. *Int. J. Math. Math. Sci.* **2019**, *2019*. [CrossRef]
61. Mao, X.; Sabanis, S. Numerical solutions of stochastic differential delay equations under local Lipschitz condition. *J. Comput. Appl. Math.* **2003**, *151*, 215–227. [CrossRef]
62. Barrio, M.; Burrage, K.; Leier, A.; Tian, T. Oscillatory Regulation of Hes1: Discrete Stochastic Delay Modelling and Simulation. *PLoS Comput. Biol.* **2006**, *2*, e117. [CrossRef] [PubMed]
63. Barbuti, R.; Caravagna, G.; Milazzo, P.; Maggiolo-Schettini, A. On the Interpretation of Delays in Delay Stochastic Simulation of Biological Systems. *Electron. Proc. Theor. Comput. Sci.* **2009**, *6*, 17–29. [CrossRef]
64. GNU. GNU Octave. Library Catalog. Available online: www.gnu.org (accessed on 12 December 2020).
65. Maarleveld, T.R.; Olivier, B.G.; Bruggeman, F.J. StochPy: A Comprehensive, User-Friendly Tool for Simulating Stochastic Biological Processes. *PLoS ONE* **2013**, *8*, e79345. [CrossRef] [PubMed]
66. Kermack, W.O.; McKendrick, A.G. A contribution to the mathematical theory of epidemics. *Proc. R. Soc. Lond. Ser. A Contain. Pap. Math. Phys. Character* **1927**, *115*, 700–721.
67. Mondal, S.; Mukherjee, S.; Bagchi, B. Mathematical modeling and cellular automata simulation of infectious disease dynamics: Applications to the understanding of herd immunity. *J. Chem. Phys.* **2020**, *153*, 114119. [CrossRef]
68. Cummings, D.A.; Lessler, J. Infectious disease dynamics. In *Infectious Disease Epidemiology: Theory and Practice*; Nelson, K.E., Masters Williams, C., Eds.; Jones & Bartlett Publishers: Burlington, MA, USA, 2014; pp. 131–166.
69. Funk, S.; Salathé, M.; Jansen, V.A. Modelling the influence of human behaviour on the spread of infectious diseases: A review. *J. R. Soc. Interface* **2010**, *7*, 1247–1256. [CrossRef]
70. Rivers, C.M.; Lofgren, E.T.; Marathe, M.; Eubank, S.; Lewis, B.L. Modeling the impact of interventions on an epidemic of Ebola in Sierra Leone and Liberia. *PLoS Curr.* **2014**, *6*. [CrossRef]
71. Begon, M.; Bennett, M.; Bowers, R.G.; French, N.P.; Hazel, S.; Turner, J. A clarification of transmission terms in host-microparasite models: numbers, densities and areas. *Epidemiol. Infect.* **2002**, *129*, 147–153. [CrossRef] [PubMed]
72. Allen, L.J. A primer on stochastic epidemic models: Formulation, numerical simulation, and analysis. *Infect. Dis. Model.* **2017**, *2*, 128–142. [CrossRef] [PubMed]

Article

Conservative Finite Volume Schemes for Multidimensional Fragmentation Problems

Jitraj Saha [1] and Andreas Bück [2],*

[1] Department of Mathematics, National Institute of Technology Tiruchirappalli, Tiruchirappalli 620 015, Tamil Nadu, India; jitraj@nitt.edu

[2] Institute of Particle Technology (LFG), Friedrich-Alexander University Erlangen-Nürnberg, D-91058 Erlangen, Germany

* Correspondence: andreas.bueck@fau.de

Abstract: In this article, a new numerical scheme for the solution of the multidimensional fragmentation problem is presented. It is the first that uses the conservative form of the multidimensional problem. The idea to apply the finite volume scheme for solving one-dimensional linear fragmentation problems is extended over a generalized multidimensional setup. The derivation is given in detail for two-dimensional and three-dimensional problems; an outline for the extension to higher dimensions is also presented. Additionally, the existing one-dimensional finite volume scheme for solving conservative one-dimensional multi-fragmentation equation is extended to solve multidimensional problems. The accuracy and efficiency of both proposed schemes is analyzed for several test problems.

Keywords: conservative formulation; multidimensional fragmentation equation; weight functions; finite volume scheme

1. Introduction

Fragmentation, breakage or attrition, describe processes in which a single object is separated into at least two new objects. The reasons for breakage can be manifold but are often linked to some kind of stress exerted on the object, for instance thermal stress from heating and rapid cooling (or vice versa and cyclically)—a natural process specifically observed in deserts, leading to disintegration of rocks; or mechanical stress, applied for millenia in the process of grain milling. Nowadays, fragmentation plays a key role in several industrial sectors like mineral processing (e.g., comminution of ores [1–4]), reaction engineering (e.g., break-up of bubbles in reacting bubble columns for separation processes [5–8] or steel-casting [9]) or pharmaceutical industries (e.g., milling of active pharmaceutical ingredients to increase their solubility and uptake capacity in the human or animal body [10–12]).

Many objects, e.g., particles, bubbles or even rain drops, consist of different components resulting in their anisotropic structure. The probability of fragmentation upon stress therefore depends on the distribution of the components within the objects, i.e., each component adds an independent dimension to the fragmentation problem. In an attempt to describe these complex processes and make them accessible for model- and knowledge-based process design, optimization and control, multidimensional fragmentation equations have been proposed and used in different fields of application, see, for instance, the works [13–17].

Theoretical aspects on the existence of scaling solutions and their behavior at the onset of "shattering" transition have been discussed for instance in the works of [18–21]. Fragmentation models are particularly challenging as they consist of partial-integro differential equations as will be shown in the following. Analytical results are scarce and often of very limited practical relevance, strongly motivating the development of numerical methods for approximation of the solution to (multidimensional) fragmentation problems.

As a prototype, consider the conservative formulation of the multiple fragmentation equation given by [22,23]: The initial value problem for $t \geq 0$ is formulated as

$$\frac{\partial g(t,x)}{\partial t} = \frac{\partial \mathcal{H}(t,x)}{\partial x}, \quad \text{where} \quad x \in \mathbb{R}^+ := (0, \infty), \quad (1)$$

with the initial condition

$$g(x,0) = g_0(x) \quad (\geq 0), \quad \text{for} \quad x \in \mathbb{R}^+. \quad (2)$$

The *flux function* $\mathcal{H}(t,x)$ is defined by

$$\mathcal{H}(t,x) := \int_x^\infty \int_0^x \frac{u}{v} b(u,v) S(v) g(t,v) \, du \, dv, \quad x \in \mathbb{R}^+. \quad (3)$$

In Equation (1), the internal variables x and t denote the particle property and the time component, respectively. On the left hand side, the function $g(t,x)$ is defined by $g(t,x) := x f(t,x)$, where $f(t,x)$ denotes the distribution of particle volume x in a system at time t. The rate of selection of an x-volume cluster to undergo breakage is denoted by $S(x)$, and the distribution of daughter particles y due to the breakage of large particle x is denoted by $b(y,x)$. The breakage function $b(y,x)$ satisfies the following relations:

$$\int_0^\infty b(y,x) \, dy = \nu(x), \quad \text{and} \quad \int_0^x y b(y,x) \, dy = x. \quad (4)$$

The first relation defines that $\nu(x)$ number of fragments are produced during the breakup of a large x-cluster, and the second relation defines that the total volume of the daughter y-clusters is exactly the same as the volume of the mother x-cluster. Note that the formulation (1) is well-known in the literature as the volume conservative form. Integration of Equation (1) over the volume variable x from 0 to ∞, with the help of relation (4), yields

$$\frac{d}{dt} \int_0^\infty g(t,x) \, dx = 0. \quad (5)$$

It should be noted that Equation (1) is a first-order hyperbolic, initial value partial differential equation. In this regard, the representation (1) gathered importance because the divergent nature allows the model to obey the volume conservation laws. The coefficient S belongs to $L_{loc}^\infty([0,\infty))$ and $g_0, b \in L^1((0,\infty)) \cap L^1((0,\infty), x dx)$. Here and below, the notation $L^1(\mathbb{R}^+, x dx)$ stands for the space of the Lebesgue measurable real-valued functions on $\mathbb{R}^+$ which are integrable with respect to the measure $x dx$.

In most of the previous studies it is assumed that a single parameter, which is usually volume, mass or size of the particle, is sufficient to describe the particle property (readers can refer to [24] for further details). However, a single parameter is not always sufficient to describe various physical systems. For example, fragment mass distribution obtained by crushing gypsum or glass depends on the initial geometry of the particles. On the other hand, the degradation of polyelectrolyte may depend upon both their mass and excitation (or kinetic) energy. Therefore, the fragmentation dynamics need to be represented by including additional variables to the mathematical model. These variables are equivalently classified as the degrees of freedom of the dynamical system and hence, the multidimensional formulation of the fragmentation equations becomes necessary to represent such cases. The purpose of this article is to take in account more than one particle property and present an efficient numerical model which estimates them with high accuracy. In particular, we present the mathematical representations of two-dimensional and three-dimensional volume conservative linear fragmentation equations. Further extension of the mathematical formulation can be done in a similar manner.

For the population balance models, the moment functions of the particle property distribution play a major role as some of them describe a significant physical property of the system. Therefore, before we proceed further, let us first gather some important information about the moment functions in a generalized multidimensional setup.

1.1. Moment Functions

Let $\mathbf{x} := \{x_1, x_2, \ldots, x_n\}$, with x_i-s representing different particle properties like, mass, entropy, moisture content, shape factor, etc. and thus the function $f(t, \mathbf{x})$ denotes the distribution of particle property $\mathbf{x}$ at some instance t. The formal definition of the moment functions for a general n-dimensional population balance problem is written as follows:

$$\mathcal{M}_{p_1, p_2, \ldots, p_n}(t) := \int_0^\infty \left(\prod_{r=1}^n x_r^{p_r} \right) f(t, \mathbf{x}) d\mathbf{x}, \tag{6}$$

where the integrations are defined as

$$\int_0^\infty (\cdot) d\mathbf{x} := \underbrace{\int_0^\infty dx_1 \cdots \int_0^\infty (\cdot) dx_n}_{n-\text{times}}.$$

In Equation (6), $p_1, p_2, \ldots, p_n$ are nonnegative integers. As mentioned earlier, the moment functions play an important role to define various physical properties of the system. Like the zeroth moment, $\mathcal{M}_{0,0,\ldots,0}(t)$ defines the total number of particles present in the system. The first-order moment $\mathcal{M}_{0,\ldots,1,\ldots,0}(t)$ (1 is the rth position) denotes the total content of the x_rth component in the system, which can be equivalently represented as the total volume of x_rth property. Hence, for a multidimensional system, the volume conservation of the system can be defined as the total conservation of all the first-order moments taken together. Thus, defining $\phi(\mathbf{x}) := \sum_{r=1}^n x_r$, the volume conservation law for the n-dimensional system is expressed as

$$\frac{d}{dt} \int_0^\infty \phi(\mathbf{x}) f(t, \mathbf{x}) d\mathbf{x} = 0. \tag{7}$$

Furthermore, the n-th order cross moment is defined by $\mathcal{M}_{1,\ldots,1,\ldots,1}(t)$ and it represents the particle geometry or hypervolume. Therefore, to preserve the initial geometry of the particles, we need to preserve the cross moments; hence, the hypervolume preservation law is written as

$$\frac{d}{dt} \int_0^\infty \psi(\mathbf{x}) f(t, \mathbf{x}) d\mathbf{x} = 0, \tag{8}$$

where $\psi(\mathbf{x}) := \prod_{r=1}^n x_r$.

Similarly, other higher order moments can be defined using the formulation (6), and depending upon the problem they may correlate to some physical properties of the system. For example, in a pipeline flow for the transport of natural gas from seabed, the breakage of hydrate particles often takes place. In this event, if the first moment $\mathcal{M}_{1,\ldots,0}(t)$ is proportional to the mean radius of the hydrate particle, then the corresponding second order moment $\mathcal{M}_{2,\ldots,0}(t)$ and third order moment $\mathcal{M}_{3,\ldots,0}(t)$ are proportional to the total area and the volume concentration of the hydrate particles, respectively. In general, only the zeroth, first-order and the cross-moments bear the same meaning for any population balance models. However, it should not be misunderstood that higher order moments should always correspond to certain physical characteristics.

In the literature, a limited number of articles are dedicated to the numerical study of multidimensional fragmentation events, and therefore several aspects of study still remain unexplored. The articles of [25–29] discuss the development of different numerical schemes to approximate the fragmentation problems. To note that, unlike the methods, e.g., cell average technique, fixed pivot techniques, method of moments, etc., the finite volume methods have gained popularity because the latter are robust to be applied on a multidimensional setup. Moreover, the underlying stencil of the finite volume scheme is simpler, and easy to compute (the readers can refer to the articles of [23,29] for further details on the computational advantage of finite volume schemes).

The article is organized in the following manner. In the next section, we present the mathematical representations of the continuous two- and three-dimensional equations. In this regard, the three-dimensional model is represented using vector notation, which will also provide an outline to extend the equations into further higher dimensions. In Section 3, step-by-step derivation of the numerical schemes are presented. An interesting outcome of this presentation includes the multidimensional extension of the finite volume scheme presented in [23]. Section 4 contains the numerical validation of the proposed models over some standard empirical test problems. Finally some conclusions and a summary of the work are presented.

2. Continuous Equations in Two- and Three-Dimensions

2.1. Conservative Formulations in Two-Dimensions

In Equation (1), the variable x represents a single particle property which can be considered as the particle volume. Therefore, the first moment always corresponds to the total volume of the particle in the system, and hence Equation (1) is simply coined as the volume-conservative model. However, the representation is not that simple in the case of a multidimensional fragmentation event. Depending on the definition of volume and hypervolume, the mathematical model changes, and thus we get two different mathematical equations representing the two conservative formulations in the multidimensional setup. For example, consider two independent particle properties kinetic energy and moisture content that are defined by the variables x and y, respectively and we set $\mathbf{x} := (x, y)$. Then $f(t, \mathbf{x})$ is the two-dimensional particle properties distribution function at time t. Now referring to the Equations (7) and (8), the solutions corresponding to the volume-conservative and hypervolume conservative formulations are defined as $n(t, \mathbf{x}) := \phi(\mathbf{x}) f(t, \mathbf{x})$ and $m(t, \mathbf{x}) := \psi(\mathbf{x}) f(t, \mathbf{x})$, respectively.

In accordance with the above definition, the two-dimensional or bivariate volume-conservative fragmentation equation is written as,

$$\frac{\partial n(t,\mathbf{x})}{\partial t} = \frac{\partial \mathcal{F}(t,\mathbf{x})}{\partial x} + \frac{\partial \mathcal{G}(t,\mathbf{x})}{\partial y} - \frac{\partial^2 \mathcal{H}(t,\mathbf{x})}{\partial x \partial y}, \qquad (9)$$

with the initial data

$$n(0, \mathbf{x}) = n_0(\mathbf{x}) \geq 0, \quad \text{for all} \quad \mathbf{x} > 0. \qquad (10)$$

Here, the functions $\mathcal{F}, \mathcal{G}$ and $\mathcal{H}$ denote the flux flow at the cell boundaries. In this regard, we first define the following notations to be used for defining the fluxes. Let $\mathbf{u} := (u, v)$, $\boldsymbol{\epsilon} := (\epsilon, \xi)$, then

$$\int_{\mathbf{x}}^{\infty} (\cdot) d\mathbf{u} := \int_{x}^{\infty} \int_{y}^{\infty} (\cdot) dv du, \quad \text{and} \quad \int_{0}^{\mathbf{x}} (\cdot) d\mathbf{u} := \int_{0}^{x} \int_{0}^{y} (\cdot) dv du.$$

With the help of the above notations, the flux functions are defined as follows:

$$\mathcal{F}(t,\mathbf{x}) := \int_{\mathbf{x}}^{\infty} \int_{0}^{x} \frac{\phi(\epsilon, y)}{\phi(\mathbf{u})} b(\epsilon, y | \mathbf{u}) S(\mathbf{u}) n(t, \mathbf{u}) d\epsilon du, \qquad (11)$$

$$\mathcal{G}(t,\mathbf{x}) := \int_{\mathbf{x}}^{\infty} \int_{0}^{y} \frac{\phi(x, \xi)}{\phi(\mathbf{u})} b(x, \xi | \mathbf{u}) S(\mathbf{u}) n(t, \mathbf{u}) d\xi du, \qquad (12)$$

and
$$\mathcal{H}(t,\mathbf{x}) := \int_{\mathbf{x}}^{\infty} \int_{0}^{\mathbf{x}} \frac{\phi(\epsilon)}{\phi(\mathbf{u})} b(\epsilon|\mathbf{u}) S(\mathbf{u}) n(t,\mathbf{u}) d\epsilon d\mathbf{u}. \tag{13}$$

In the above expressions, $S(\mathbf{x})$ is the selection function which defines the rate at which particles of properties $\mathbf{x}$ to undergo further fragmentation, and the breakage function $b(\epsilon|\mathbf{u})$ corresponds to the distribution of daughter fragments ϵ formed due to the fragmentation of $\mathbf{u}$-cluster. In the multidimensional fragmentation setup, the breakage function b plays a key role to govern the system to obey either volume-conservation (7) or hypervolume conservation (8) laws. For the volume conservative formulation, it is assumed that the breakage function $b(\epsilon|\mathbf{u})$ should satisfy

$$\int_{0}^{\mathbf{u}} \phi(\epsilon) b(\epsilon|\mathbf{u}) d\epsilon = \phi(\mathbf{u}). \tag{14}$$

The relation (14) is significant as it controls the system to follow volume conservation property (7). Therefore, with the above assumption it can easily be calculated that

$$\frac{d}{dt} \int_{0}^{\infty} n(t,\mathbf{x}) d\mathbf{x} = \frac{d}{dt}[\mathcal{M}_{1,0}(t) + \mathcal{M}_{0,1}(t)] = 0, \tag{15}$$

that is, the volume conservation laws are perfectly obeyed.

Note that the flux function $\mathcal{H}$ (13) in the bivariate Equation (9) is the straightforward extension of the flux in univariate model (1). Additionally, the bivariate model (9) contains two flux functions $\mathcal{F}$ (11) and $\mathcal{G}$ (12) as compared to its one-dimensional counterpart Equation (1). Here, $\mathcal{F}$ defines the distribution of the daughter particles along the x-component, while keeping the y-component fixed. Similarly, the flux $\mathcal{G}$ is defined along y-component.

We now present the continuous hypervolume conservative formulation of the pure bivariate fragmentation model. It is expressed in a manner similar to the volume conservative model (9), and reads as

$$\frac{\partial m(t,\mathbf{x})}{\partial t} = \frac{\partial \bar{\mathcal{F}}(t,\mathbf{x})}{\partial x} + \frac{\partial \bar{\mathcal{G}}(t,\mathbf{x})}{\partial y} - \frac{\partial^2 \bar{\mathcal{H}}(t,\mathbf{x})}{\partial x \partial y}, \tag{16}$$

with the flux functions $\bar{\mathcal{F}}$, $\bar{\mathcal{G}}$ and $\bar{\mathcal{H}}$ redefined as follows

$$\bar{\mathcal{F}}(t,\mathbf{x}) := \int_{\mathbf{x}}^{\infty} \int_{0}^{x} \frac{\psi(\epsilon,y)}{\psi(\mathbf{u})} b(\epsilon,y|\mathbf{u}) S(\mathbf{u}) m(t,\mathbf{u}) d\epsilon d\mathbf{u}, \tag{17}$$

$$\bar{\mathcal{G}}(t,\mathbf{x}) := \int_{\mathbf{x}}^{\infty} \int_{0}^{y} \frac{\psi(x,\xi)}{\psi(\mathbf{u})} b(x,\xi|\mathbf{u}) S(\mathbf{u}) m(t,\mathbf{u}) d\xi d\mathbf{u}, \tag{18}$$

and

$$\bar{\mathcal{H}}(t,\mathbf{x}) := \int_{\mathbf{x}}^{\infty} \int_{0}^{\mathbf{x}} \frac{\psi(\epsilon)}{\psi(\mathbf{u})} b(\epsilon|\mathbf{u}) S(\mathbf{u}) m(t,\mathbf{u}) d\epsilon d\mathbf{u}. \tag{19}$$

In this case, the breakage function satisfies condition

$$\int_{0}^{\mathbf{u}} \psi(\epsilon) b(\epsilon|\mathbf{u}) d\epsilon = \psi(\mathbf{u}), \tag{20}$$

and hence, integrating Equation (16), one can easily obtain that the hypervolume conservation law is properly obeyed, that is

$$\frac{d}{dt} \int_0^\infty m(t,x) dx = \frac{d\mathcal{M}_{1,1}(t)}{dt} = 0. \tag{21}$$

2.2. Conservative Formulations in Three-Dimensions

In a similar manner as discussed above, we now present the three-dimensional representations of the fragmentation equations which obey the (i) volume conservation laws, and (ii) hypervolume conservation laws. In this part, we present the mathematical model using vector notation to pave the way for higher dimensional extension.

Let the particle property distribution be written as $f(t,\mathbf{x})$ where the vector $\mathbf{x} := \{x_1, x_2, x_3\}$ represents different particle properties, and $n(t,\mathbf{x}) := \phi(\mathbf{x}) f(t,\mathbf{x})$. Using the extended form of all the above-mentioned notations, the three-dimensional volume conservative formulation is written as follows

$$\begin{aligned}\frac{\partial n(t,\mathbf{x})}{\partial t} &= \frac{\partial \mathcal{F}^{(1)}(t,\mathbf{x})}{\partial x_1} + \frac{\partial \mathcal{F}^{(2)}(t,\mathbf{x})}{\partial x_2} + \frac{\partial \mathcal{F}^{(3)}(t,\mathbf{x})}{\partial x_3} \\ &- \frac{\partial^2 \mathcal{G}^{(1)}(t,\mathbf{x})}{\partial x_2 \partial x_3} - \frac{\partial^2 \mathcal{G}^{(2)}(t,\mathbf{x})}{\partial x_1 \partial x_3} - \frac{\partial^2 \mathcal{G}^{(3)}(t,\mathbf{x})}{\partial x_1 \partial x_2} + \frac{\partial^3 \mathcal{H}(t,\mathbf{x})}{\partial x_1 \partial x_2 \partial x_3},\end{aligned} \tag{22}$$

with the flux flows being functions of both t, $\mathbf{x}$ and are defined as

$$\mathcal{F}^{(1)}(t,\mathbf{x}) = \int_{\mathbf{x}}^\infty \int_0^{x_1} \frac{(u_1 + x_2 + x_3)}{\phi(\mathbf{y})} b(u_1, x_2, x_3|\mathbf{y}) S(\mathbf{y}) n(t,\mathbf{y}) du_1 d\mathbf{y}, \tag{23}$$

$$\mathcal{F}^{(2)}(t,\mathbf{x}) = \int_{\mathbf{x}}^\infty \int_0^{x_2} \frac{(x_1 + u_2 + x_3)}{\phi(\mathbf{y})} b(x_1, u_2, x_3|\mathbf{y}) S(\mathbf{y}) n(t,\mathbf{y}) du_2 d\mathbf{y}, \tag{24}$$

$$\mathcal{F}^{(3)}(t,\mathbf{x}) = \int_{\mathbf{x}}^\infty \int_0^{x_3} \frac{(x_1 + x_2 + u_3)}{\phi(\mathbf{y})} b(x_1, x_{2,3}|\mathbf{y}) S(\mathbf{y}) n(t,\mathbf{y}) du_3 d\mathbf{y}, \tag{25}$$

$$\mathcal{G}^{(1)}(t,\mathbf{x}) = \int_{\mathbf{x}}^\infty \int_0^{x_2} \int_0^{x_3} \frac{(x_1 + u_2 + u_3)}{\phi(\mathbf{y})} b(x_1, u_2, u_3|\mathbf{y}) S(\mathbf{y}) n(t,\mathbf{y}) du_2 du_3 d\mathbf{y}, \tag{26}$$

$$\mathcal{G}^{(2)}(t,\mathbf{x}) = \int_{\mathbf{x}}^\infty \int_0^{x_1} \int_0^{x_3} \frac{(u_1 + x_2 + u_3)}{\phi(\mathbf{y})} b(u_1, x_2, u_3|\mathbf{y}) S(\mathbf{y}) n(t,\mathbf{y}) du_3 du_1 d\mathbf{y}, \tag{27}$$

$$\mathcal{G}^{(3)}(t,\mathbf{x}) = \int_{\mathbf{x}}^\infty \int_0^{x_1} \int_0^{x_2} \frac{(u_1 + u_2 + x_3)}{\phi(\mathbf{y})} b(u_1, u_2, x_3|\mathbf{y}) S(\mathbf{y}) n(t,\mathbf{y}) du_2 du_1 d\mathbf{y}, \tag{28}$$

$$\mathcal{H}(t,\mathbf{x}) = \int_{\mathbf{x}}^\infty \int_0^{\mathbf{x}} \frac{\phi(\mathbf{u})}{\phi(\mathbf{y})} b(\mathbf{u}|\mathbf{y}) S(\mathbf{y}) n(t,\mathbf{y}) d\mathbf{u} d\mathbf{y}. \tag{29}$$

In a similar manner, we can represent the three-dimensional hypervolume conservation equations. Consider that $m(t,\mathbf{x}) := \psi(\mathbf{x}) f(t,\mathbf{x})$ is the solution function, and the breakage function follows the relation (20). Then simply by replacing the function ϕ by ψ and simultaneously taking care of all the corresponding changes in the Equation (22), one can easily represent the three-dimensional hypervolume conservative model. Furthermore, one can extend the conservative formulations for problems with n-number of particle property components.

3. Numerical Formulations

3.1. Two-Dimensional Model

In this section, we present the discretized form of Equation (9). For this purpose, the truncated rectangular domain considered is $V :=]0, X_1] \times]0, X_2]$. Let I_1 and I_2 be two positive integers, and V is further discretized in $(I_1 \times I_2)$ number of rectangular subcells $V_\mathbf{i} :=]x_{i_1-1/2}, x_{i_1+1/2}] \times]x_{i_2-1/2}, x_{i_2+1/2}]$, where $\mathbf{i} := (i_1, i_2)$, $\mathbf{I} := (I_1, I_2)$ such that $\mathbf{1} \leq \mathbf{i} \leq \mathbf{I}$ along with $x_{1/2} = y_{1/2} = 0$, and $x_{I_1+1/2} = X_1$, $x_{I_2+1/2} = X_2$. Let $\Delta_{i_1} := x_{i_1+1/2} - x_{i_1-1/2}$, $\Delta_{i_2} := x_{i_2+1/2} - x_{i_2-1/2}$ and $\Delta_\mathbf{i} := \Delta_{i_1}\Delta_{i_2}$. Further, let $\mathbf{x_i} := (x_{i_1}, x_{i_2})$ be the pivot or representative of the cell $V_\mathbf{i}$, and the components of $\mathbf{x_i}$ are defined by

$$x_{i_1} := \frac{x_{i_1+1/2} - x_{i_1-1/2}}{2}, \qquad x_{i_2} := \frac{x_{i_2+1/2} - x_{i_2-1/2}}{2}.$$

Under the above considerations, the flux flow at the right boundaries of the cell are given by $\mathcal{F}(x_{i_1+1/2}, x_{i_2}, t)$, $\mathcal{G}(x_{i_1}, x_{i_2+1/2}, t)$ and $\mathcal{H}(x_{i_1+1/2}, x_{i_2+1/2}, t)$, and similarly the flux flow at the other boundaries are defined.

Let $n_\mathbf{i}$ be the average value of the solution $n(t, \mathbf{x})$ over the cell $V_\mathbf{i}$, and is defined by

$$n_\mathbf{i} = \frac{1}{\Delta_\mathbf{i}} \int_{V_\mathbf{i}} n(t, \mathbf{x}) d\mathbf{x}. \tag{30}$$

Consider that $\hat{n}_\mathbf{i}(t)$ denotes the numerical approximation of $n_\mathbf{i}$. For notational convenience, we drop the argument of the t from $\hat{n}_\mathbf{i}(t)$ in further discussions and simply denote it as $\hat{n}_\mathbf{i}$.

Let us now evaluate the numerical approximation of the flux $\mathcal{F}(x_{i_1+1/2}, x_{i_2}, t)$.

$$\begin{aligned}
\mathcal{F}(x_{i_1+1/2}, x_{i_2}, t) &= \int_{x_{i_1+1/2}}^{X_1} \int_{x_{i_2}}^{X_2} \left[\int_0^{x_{i_1+1/2}} (\epsilon + x_{i_2}) b(\epsilon, x_{i_2}|\mathbf{u}) d\epsilon \right] \frac{S(\mathbf{u})}{\phi(\mathbf{u})} n(t, \mathbf{u}) d\mathbf{u} \\
&= \sum_{k_1=i_1+1}^{I_1} \int_{x_{k_1-1/2}}^{x_{k_1+1/2}} \sum_{k_2=i_2}^{I_2} \int_{\beta(i_2, k_2)}^{x_{k_2+1/2}} \left[\sum_{l_1=1}^{i_1} \int_{x_{l_1-1/2}}^{x_{l_1+1/2}} (\epsilon + x_{i_2}) b(\epsilon, x_{i_2}|\mathbf{u}) d\epsilon \right] \\
&\quad \times \frac{S(\mathbf{u})}{\phi(\mathbf{u})} n(t, \mathbf{u}) d\mathbf{u},
\end{aligned}$$

Here,

$$\beta(i_2, k_2) := \begin{cases} x_{k_2}, & \text{when } i_2 = k_2, \\ x_{k_2-1/2}, & \text{otherwise.} \end{cases}$$

Applying quadrature formulae to the integrals, the numerical flux is given by

$$\mathcal{F}_{i_1+1/2, i_2} := \sum_{k_1=i_1+1}^{I_1} \sum_{k_2=i_2}^{I_2} \hat{n}_\mathbf{k} \mathcal{A}_\mathbf{k}^\beta \sum_{l_1=1}^{i_1} \mathcal{B}_{l_1, i_2|\mathbf{k}} \Delta_{l_1} \Delta_\mathbf{k}, \tag{31}$$

where $\mathbf{k} := (k_1, k_2)$, $\mathbf{l} := (l_1, l_2)$,

$$\mathcal{B}_{l_1, i_2|\mathbf{k}} := (x_{l_1} + y_{i_2}) b(x_{l_1}, y_{i_2}|x_\mathbf{k}), \quad \text{and} \quad \mathcal{A}_\mathbf{k}^\beta := \int_{x_{k_1-1/2}}^{x_{k_1+1/2}} \int_{\beta(i_2, k_2)}^{y_{k_2+1/2}} \frac{S(\mathbf{u})}{\phi(\mathbf{u})} d\mathbf{u}.$$

In a similar manner, under the following notations

$$\mathcal{B}_{i_1, l_2|\mathbf{k}} := (x_{i_1} + x_{l_2}) b(x_{i_1}, x_{l_2}|x_\mathbf{k}), \quad \mathcal{B}_{\mathbf{l}|\mathbf{k}} := (x_{l_1} + x_{l_2}) b(x_{l_1}, y_{l_2}|x_\mathbf{k}),$$

$$\mathcal{A}_\mathbf{k}^\alpha := \int_{\alpha(i_1, k_1)}^{x_{k_1+1/2}} \int_{x_{k_2-1/2}}^{x_{k_2+1/2}} \frac{S(\mathbf{u})}{\phi(\mathbf{u})} d\mathbf{u}, \quad \text{and} \quad \mathcal{A}_\mathbf{k} := \int_{x_{k_1-1/2}}^{x_{k_1+1/2}} \int_{y_{k_2-1/2}}^{y_{k_2+1/2}} \frac{S(\mathbf{u})}{\phi(\mathbf{u})} d\mathbf{u},$$

with

$$\alpha(i_1, k_1) := \begin{cases} x_{k_1}, & \text{when } i_1 = k_1, \\ x_{k_1-1/2}, & \text{otherwise.} \end{cases}$$

the other numerical fluxes at the cell interfaces are written as

$$\mathcal{G}_{i_1,i_2+1/2} := \sum_{k_1=i_1}^{I_1} \sum_{k_2=i_2+1}^{I_2} \hat{n}_{\mathbf{k}} \mathcal{A}_{\mathbf{k}}^{\alpha} \sum_{l_2=1}^{i_2} \mathcal{B}_{i_1,l_2|\mathbf{k}} \Delta_{l_2} \Delta_{\mathbf{k}}, \tag{32}$$

and

$$\mathcal{H}_{i+1/2} := \sum_{\mathbf{k}=i+1}^{I} \hat{n}_{\mathbf{k}} \mathcal{A}_{\mathbf{k}} \sum_{l=1}^{i} \mathcal{B}_{l|\mathbf{k}} \Delta_l \Delta_{\mathbf{k}}. \tag{33}$$

Therefore, the semi-discrete finite volume representation of Equation (9) is written as

$$\frac{d\hat{n}_{\mathbf{i}}}{dt} \Delta_{\mathbf{i}} = \left[\mathcal{F}_{i_1+1/2,i_2} - \mathcal{F}_{i_1-1/2,i_2} \right] \Delta_{i_2} + \left[\mathcal{G}_{i_1,i_2+1/2} - \mathcal{G}_{i_1,i_2-1/2} \right] \Delta_{i_1} \\ - \left[\mathcal{H}_{i_1+1/2,i_2+1/2} - \mathcal{H}_{i_1+1/2,i_2-1/2} - \mathcal{H}_{i_1-1/2,i_2+1/2} + \mathcal{H}_{i_1-1/2,i_2-1/2} \right]. \tag{34}$$

The above scheme (34) obeys the discrete volume conservation law (the detailed calculations are given in Appendix A). However, in the subsequent section, we shall numerically validate that scheme (34) fails to predict the evolution of total number of fragments with good accuracy. In this context, the flux functions are redefined by introducing a weight function which enables the model to obey volume conservation laws, as well as predict the zeroth moment with high accuracy. The newly proposed semi-discrete formulation is written as follows:

$$\frac{d\hat{n}_{\mathbf{i}}}{dt} \Delta_{\mathbf{i}} = \left[\hat{\mathcal{F}}_{i_1+1/2,i_2} - \hat{\mathcal{F}}_{i_1-1/2,i_2} \right] \Delta_{i_2} + \left[\hat{\mathcal{G}}_{i_1,i_2+1/2} - \hat{\mathcal{G}}_{i_1,i_2-1/2} \right] \Delta_{i_1} \\ - \left[\hat{\mathcal{H}}_{i_1+1/2,i_2+1/2} - \hat{\mathcal{H}}_{i_1+1/2,i_2-1/2} - \hat{\mathcal{H}}_{i_1-1/2,i_2+1/2} + \hat{\mathcal{H}}_{i_1-1/2,i_2-1/2} \right], \tag{35}$$

where the modified fluxes at the cell boundaries are defined as follows:

$$\hat{\mathcal{F}}_{i_1+1/2,i_2} := \sum_{k_1=i_1+1}^{I_1} \sum_{k_2=i_2}^{I_2} \hat{n}_{\mathbf{k}} \delta_{\mathbf{k}} \mathcal{A}_{\mathbf{k}}^{\beta} \sum_{l_1=1}^{i_1} \mathcal{B}_{l_1,i_2|\mathbf{k}} \Delta_{l_1} \Delta_{\mathbf{k}}, \tag{36}$$

$$\hat{\mathcal{G}}_{i_1,i_2+1/2} := \sum_{k_1=i_1}^{I_1} \sum_{k_2=i_2+1}^{I_2} \hat{n}_{\mathbf{k}} \delta_{\mathbf{k}} \mathcal{A}_{\mathbf{k}}^{\alpha} \sum_{l_2=1}^{i_2} \mathcal{B}_{i_1,l_2|\mathbf{k}} \Delta_{l_2} \Delta_{\mathbf{k}}, \tag{37}$$

$$\hat{\mathcal{H}}_{i+1/2} := \sum_{\mathbf{k}=i+1}^{I} \hat{n}_{\mathbf{k}} \delta_{\mathbf{k}} \mathcal{A}_{\mathbf{k}} \sum_{l=1}^{i} \mathcal{B}_{l|\mathbf{k}} \Delta_l \Delta_{\mathbf{k}}, \tag{38}$$

and δ is the weight factor, defined by

$$\delta_{\mathbf{k}} := \frac{S_{\mathbf{k}}[\nu(\mathbf{x}_{\mathbf{k}}) - 1]}{\mathcal{A}_{\mathbf{k}} \sum_{i=1}^{k} (\phi(\mathbf{x}_{\mathbf{k}}) - \phi(\mathbf{x}_i)) \mathcal{B}_{\mathbf{i}|\mathbf{k}}}, \tag{39}$$

along with $\delta_{1,1} = 1$. In the above definition of the weight, the terms $S_{\mathbf{k}}$ and $\nu(\mathbf{x}_{\mathbf{k}})$ denote the discrete selection function and the number of fragments, respectively.

In the following section, we will numerically validate that the two-dimensional scheme (35) is consistent with the zeroth moment and it also obeys the volume conservation law. The proof of this claim follows similar to that of the model (34) and can easily be followed from the outline given in Appendix A.

In a similar manner, a new scheme preserving the cluster hypervolume and estimating the continuous model (16) along with the zeroth moment can be defined as follows:

$$\frac{d\hat{m}_{\mathbf{i}}}{dt}\Delta_{\mathbf{i}} = \left[\hat{\mathcal{F}}_{i_1+1/2,i_2} - \hat{\mathcal{F}}_{i_1-1/2,i_2}\right]\Delta_{i_2} + \left[\hat{\mathcal{G}}_{i_1,i_2+1/2} - \hat{\mathcal{G}}_{i_1,i_2-1/2}\right]\Delta_{i_1} \\ - \left[\hat{\mathcal{H}}_{i_1+1/2,i_2+1/2} - \hat{\mathcal{H}}_{i_1+1/2,i_2-1/2} - \hat{\mathcal{H}}_{i_1-1/2,i_2+1/2} + \hat{\mathcal{H}}_{i_1-1/2,i_2-1/2}\right], \tag{40}$$

where the discrete flux functions at the cell boundaries are defined by

$$\hat{\mathcal{F}}_{i_1+1/2,i_2} := \sum_{k_1=i_1+1}^{I_1} \sum_{k_2=i_2}^{I_2} \hat{m}_{\mathbf{k}}\omega_{\mathbf{k}}\mathcal{A}_{\mathbf{k}}^{\beta} \sum_{l_1=1}^{i_1} \tilde{\mathcal{B}}_{l_1,i_2|\mathbf{k}}\Delta_{l_1}\Delta_{\mathbf{k}}, \tag{41}$$

$$\hat{\mathcal{G}}_{i_1,i_2+1/2} := \sum_{k_1=i_1}^{I_1} \sum_{k_2=i_2+1}^{I_2} \hat{m}_{\mathbf{k}}\omega_{\mathbf{k}}\mathcal{A}_{\mathbf{k}}^{\alpha} \sum_{l_2=1}^{i_2} \tilde{\mathcal{B}}_{i_1,l_2|\mathbf{k}}\Delta_{l_2}\Delta_{\mathbf{k}}, \tag{42}$$

$$\hat{\mathcal{H}}_{\mathbf{i}+1/2} := \sum_{\mathbf{k}=\mathbf{i}+1}^{I} \hat{m}_{\mathbf{k}}\omega_{\mathbf{k}}\mathcal{A}_{\mathbf{k}} \sum_{l=1}^{\mathbf{i}} \tilde{\mathcal{B}}_{l|\mathbf{k}}\Delta_{l}\Delta_{\mathbf{k}}, \tag{43}$$

with the discrete breakage function as

$$\tilde{\mathcal{B}}_{\mathbf{i}|\mathbf{k}} := \psi(\mathbf{x_i})b(\mathbf{x_i}|\mathbf{x_k}),$$

and ω is the weight factor, defined by

$$\omega_{\mathbf{k}} := \frac{S_{\mathbf{k}}[\nu(\mathbf{x_k})-1]}{\mathcal{A}_{\mathbf{k}}\sum_{\mathbf{i}=1}^{\mathbf{k}}(\psi(\mathbf{x_k})-\psi(\mathbf{x_i}))\tilde{\mathcal{B}}_{\mathbf{i}|\mathbf{k}}}, \tag{44}$$

Remark 1. *In the proposed two-dimensional model (10), there are three numerical fluxes operating at the cell boundaries. An interesting feature of the new two-dimensional model is that a single weight function is sufficient for redefining the modified scheme to become consistent with the zeroth- and the first-order moments.*

Remark 2. *It is to be noted that the finite volume scheme (34) is obtained by direct application of the midpoint quadrature rules to the continuous Equation (9). Thus, it represents the numerical model of [23] with two degrees of freedom.*

3.2. Three-Dimensional Model

In this part, we present the three-dimensional finite volume scheme approximating the multi-fragmentation model. Here, the scheme is expressed using vector notation, which will give an outline for further extension of the proposed scheme in higher dimensions.

Similar to the two-dimensional model, the computational domain considered is $V := \prod_{r=1}^{3}]0, X_r]$ which is further divided into a finite number of sub-cells

$$V_{\mathbf{i}} := \prod_{r=1}^{3}[x_{i_r-1/2}, x_{i_r+1/2}]$$

with $i_r = 1, 2, \ldots, I_r$. Let $\hat{n}_{\mathbf{i}}$ be the numerical approximation of $n(t, \mathbf{x})$ over the cell $V_{\mathbf{i}}$. Further, let $\Delta_{\mathbf{i}}$ denote the volume of the cell $V_{\mathbf{i}}$, and the the cell representative is defined by $\mathbf{x_i} := \{x_{i_1}, x_{i_2}, x_{i_3}\}$. Consider that the breakage function obeys the conservative Formula (14), then the three-dimensional extension of the newly proposed volume-conservative formulation (33) is written as

$$\frac{d\hat{n}_\mathbf{i}}{dt}\Delta_\mathbf{i} = \left[\mathcal{F}^{(1)}_{i_1+1/2,i_2,i_3} - \mathcal{F}^{(1)}_{i_1-1/2,i_2,i_3}\right]\Delta_{i_2,i_3} + \left[\mathcal{F}^{(2)}_{i_1,i_2+1/2,i_3} - \mathcal{F}^{(2)}_{i_1,i_2-1/2,i_3}\right]\Delta_{i_1,i_3}$$
$$+ \left[\mathcal{F}^{(3)}_{i_1,i_2,i_3+1/2} - \mathcal{F}^{3}_{i_1,i_2,i_3-1/2}\right]\Delta_{i_1,i_2}$$
$$- \left[\mathcal{G}^{(1)}_{i_1,i_2+1/2,i_3+1/2} - \mathcal{G}^{(1)}_{i_1,i_2-1/2,i_3+1/2} - \mathcal{G}^{(1)}_{i_1,i_2+1/2,i_3-1/2} + \mathcal{G}^{(1)}_{i_1,i_2-1/2,i_3-1/2}\right]\Delta_{i_1}$$
$$- \left[\mathcal{G}^{(2)}_{i_1+1/2,i_2,i_3+1/2} - \mathcal{G}^{(2)}_{i_1-1/2,i_2,i_3+1/2} - \mathcal{G}^{(2)}_{i_1+1/2,i_2,i_3-1/2} + \mathcal{G}^{(2)}_{i_1-1/2,i_2,i_3-1/2}\right]\Delta_{i_2}$$
$$- \left[\mathcal{G}^{(3)}_{i_1+1/2,i_2+1/2,i_3} - \mathcal{G}^{(3)}_{i_1-1/2,i_2+1/2,i_3} - \mathcal{G}^{(3)}_{i_1+1/2,i_2-1/2,i_3} + \mathcal{G}^{(3)}_{i_1-1/2,i_2-1/2,i_3}\right]\Delta_{i_3}$$
$$+ \left[\hat{\mathcal{H}}_{i_1+1/2,i_2+1/2,i_3+1/2} - \hat{\mathcal{H}}_{i_1+1/2,i_2+1/2,i_3-1/2} - \hat{\mathcal{H}}_{i_1+1/2,i_2-1/2,i_3+1/2}\right.$$
$$+ \hat{\mathcal{H}}_{i_1+1/2,i_2-1/2,i_3-1/2} - \hat{\mathcal{H}}_{i_1-1/2,i_2+1/2,i_3+1/2} + \hat{\mathcal{H}}_{i_1-1/2,i_2-1/2,i_3+1/2}$$
$$\left. + \hat{\mathcal{H}}_{i_1-1/2,i_2+1/2,i_3-1/2} - \hat{\mathcal{H}}_{i_1-1/2,i_2-1/2,i_3-1/2}\right]. \tag{45}$$

Considering $\mathbf{k} := (k_1, k_2, k_3)$ and $\mathbf{l} := (l_1, l_2, l_3)$, the redefined flux functions are written as follows:

$$\mathcal{F}^{(1)}_{i_1+1/2,i_2,i_3} := \sum_{k_1=i_1+1}^{I_1} \sum_{(k_2,k_3)=(i_2,i_3)}^{(I_2,I_3)} \hat{n}_\mathbf{k}\omega_\mathbf{k}\mathcal{A}^{\beta,\gamma}_\mathbf{k}\Delta_\mathbf{k} \sum_{l_1=1}^{i_1} \mathcal{B}_{l_1,i_2,i_3|\mathbf{k}}\Delta_{l_1}, \tag{46}$$

$$\mathcal{F}^{(2)}_{i_1,i_2+1/2,i_3} := \sum_{k_2=i_2+1}^{I_2} \sum_{(k_1,k_3)=(i_1,i_3)}^{(I_1,I_3)} \hat{n}_\mathbf{k}\omega_\mathbf{k}\mathcal{A}^{\alpha,\gamma}_\mathbf{k}\Delta_\mathbf{k} \sum_{l_2=1}^{i_2} \mathcal{B}_{i_1,l_2,i_3|\mathbf{k}}\Delta_{l_2}, \tag{47}$$

$$\mathcal{F}^{(3)}_{i_1,i_2,i_3+1/2} := \sum_{(k_1,k_2)=(i_1,i_2)}^{(I_1,I_2)} \sum_{k_3=i_3+1}^{I_3} \hat{n}_\mathbf{k}\omega_\mathbf{k}\mathcal{A}^{\alpha,\beta}_\mathbf{k}\Delta_\mathbf{k} \sum_{l_3=1}^{i_3} \mathcal{B}_{i_1,i_2,l_3|\mathbf{k}}\Delta_{l_3}, \tag{48}$$

$$\mathcal{G}^{(1)}_{i_1,i_2+1/2,i_3+1/2} := \sum_{k_1=i_1}^{I_1} \sum_{(k_2,k_3)=(i_2+1,i_3+1)}^{(I_2,I_3)} \hat{n}_\mathbf{k}\omega_\mathbf{k}\mathcal{A}^{\alpha}_\mathbf{k}\Delta_\mathbf{k} \sum_{(l_2,l_3)=1}^{i_2,i_3} \mathcal{B}_{i_1,l_2,l_3|\mathbf{k}}\Delta_{l_2,l_3}, \tag{49}$$

$$\mathcal{G}^{(2)}_{i_1+1/2,i_2,i_3+1/2} := \sum_{(k_1,k_3)=(i_1+1,i_3+1)}^{(I_1,I_3)} \sum_{k_2=i_2}^{I_2} \hat{n}_\mathbf{k}\omega_\mathbf{k}\mathcal{A}^{\beta}_\mathbf{k}\Delta_\mathbf{k} \sum_{(l_1,l_2)=1}^{i_1,i_3} \mathcal{B}_{l_1,i_2,l_3|\mathbf{k}}\Delta_{l_1,l_2}, \tag{50}$$

$$\mathcal{G}^{(3)}_{i_1+1/2,i_2+1/2,i_3} := \sum_{k_1=i_1+1}^{I_1} \sum_{(k_2,k_3)=(i_2+1,i_3+1)}^{(I_2,I_3)} \hat{n}_\mathbf{k}\omega_\mathbf{k}\mathcal{A}^{\gamma}_\mathbf{k}\Delta_\mathbf{k} \sum_{(l_1,l_2)=1}^{i_1,i_2} \mathcal{B}_{l_1,l_2,i_3|\mathbf{k}}\Delta_{l_2,l_3}, \tag{51}$$

$$\hat{\mathcal{H}}_{\mathbf{i}+1/2} := \sum_{\mathbf{k}=\mathbf{i}}^{I} \hat{n}_\mathbf{k}\omega_\mathbf{k}\mathcal{A}_\mathbf{k}\Delta_\mathbf{k} \sum_{\mathbf{l}=1}^{\mathbf{i}} \mathcal{B}_{\mathbf{l}|\mathbf{k}}\Delta_\mathbf{l}. \tag{52}$$

Here, $\delta_\mathbf{k}$ is the weight factor, defined by

$$\delta_\mathbf{k} := \frac{S_\mathbf{k}[\nu(\mathbf{x}_\mathbf{k})-1]}{\mathcal{A}_\mathbf{k}\sum_{\mathbf{i}=1}^{\mathbf{k}}[\phi(\mathbf{x}_\mathbf{k})-\phi(\mathbf{x}_\mathbf{i})]\mathcal{B}_{\mathbf{i}|\mathbf{k}}}, \tag{53}$$

and the factors $\mathcal{A}^{\alpha,\beta}_\mathbf{k}$, $\mathcal{A}^{\beta,\gamma}_\mathbf{k}$ and $\mathcal{A}^{\alpha,\gamma}_\mathbf{k}$ are defined as follows:

$$\mathcal{A}^{\alpha,\beta}_\mathbf{k} := \int_{\alpha(i_1,l_1)}^{x_{l_1+1/2}} \int_{\beta(i_2,l_2)}^{x_{l_2+1/2}} \int_{x_{l_3-1/2}}^{x_{l_3+1/2}} \frac{S(\mathbf{u})}{\phi(\mathbf{u})}d\mathbf{u},$$

$$\mathcal{A}_{\mathbf{k}}^{\beta,\gamma} := \int_{x_{l_1-1/2}}^{x_{l_1}+1/2} \int_{\beta(i_2,l_2)}^{x_{l_2}+1/2} \int_{\gamma(i_3,l_3)}^{x_{l_3}+1/2} \frac{S(\mathbf{u})}{\phi(\mathbf{u})} d\mathbf{u},$$

$$\mathcal{A}_{\mathbf{k}}^{\alpha,\gamma} := \int_{\alpha(i_1,l_1)}^{x_{l_1}+1/2} \int_{x_{l_2-1/2}}^{x_{l_2}+1/2} \int_{\gamma(i_3,l_3)}^{x_{l_3}+1/2} \frac{S(\mathbf{u})}{\phi(\mathbf{u})} d\mathbf{u},$$

along with $\mathcal{A}_{\mathbf{k}}^{\alpha}, \mathcal{A}_{\mathbf{k}}^{\beta}, \mathcal{A}_{\mathbf{k}}^{\gamma}, \mathcal{A}_{\mathbf{k}}$ and $\alpha(i_1, l_1), \beta(i_2, l_2), \gamma(i_3, l_3)$ being defined in a similar manner as done before.

Following the same trail, one can easily define the three-dimensional hypervolume preserving numerical model. In this case, the numerical solution will be given by $\hat{m}_\mathbf{i}$, which is the approximation of $m(t, \mathbf{x})$ over the cell $V_\mathbf{i}$ and the breakage function will obey the hypervolume conservation law (20). Thus, the weight function will be defined as

$$\omega_{\mathbf{k}} := \frac{S_{\mathbf{k}}[\nu(\mathbf{x_k}) - 1]}{\mathcal{A}_{\mathbf{k}} \sum_{i=1}^{k} [\psi(\mathbf{x_k}) - \psi(\mathbf{x_i})] \check{\mathcal{B}}_{\mathbf{i}|\mathbf{k}}}. \tag{54}$$

4. Results

In this section, we validate the efficiency of the newly proposed finite volume models with the standard finite volume scheme over several test problems. Since the new models are defined with the help of a weight factor, we call it weighted finite volume scheme (WFVS). On the other hand, Remark 2 indicates that the standard forms of the schemes which are directly derived from the continuous equations were initially proposed by [23] for fragmentation models with one degree of freedom. Therefore, for future reference, we call the standard models the existing finite volume schemes (EFVS). However, we need to mention that the two-dimensional extension of EFVS is not available in the literature to date, and this article not only proposes an improved model, but also extends the existing finite volume schemes for multidimensional fragmentation events.

For one-dimensional fragmentation problems, Ref. [30,31] has obtained the exact solutions for a certain class of fragmentation kinetics. However, exact solutions in closed forms are very rare in the multidimensional setup. In order to validate the accuracy of the proposed schemes, we choose four test problems with two degrees of freedom, and two test problems with three degrees of freedom. For all the test problems, exact solutions in closed form are not always available in the literature. However, the zeroth and the first-order moments can be computed exactly, which is sufficient to validate the accuracy of the new schemes. Therefore, in the following section we discuss the efficiency of the WFV scheme to predict the different physically important moment functions over the EFV scheme. Our study builds up on both qualitative and quantitative assessments. The qualitative accuracy is represented through graphical representation of the different entities whereas, the qualitative analysis is performed by computing relative errors of the moment functions over different grid points.

The computational domain $V = [0, 1] \times [0, 1]$ is considered for all the two-dimensional test problems. The domain V is discretized into 15×15 nonuniform subintervals bearing the geometric relation $x_{i+1/2} = r x_{i-1/2}$ where $r := 3.9811$ is the geometric ratio. Additionally, all the test problems are supported by a mono-dispersed initial condition

$$f(x_1, x_2, 0) = \delta(x_1 - 1)\delta(x_2 - 1).$$

Similar extensions of the above data are considered for the three dimensional models. The semi-discrete schemes (7) and (8) are solved using MATLAB-2019B software in a standard PC with i5-7500 CPU processor @ 3.41 GHz and 8 GB RAM.

4.1. Examples in Two-Dimensions

4.1.1. Volume Conservation Problems

In the first instance, we consider test problems with constant particle selection rate, that is $S(x,y) = 1$ and two different daughter distribution functions

$$b_1(x,y|x',y') = \frac{2}{x'y'} \quad \text{and} \quad b_2(x,y|x',y') = 2\delta\left(x - \frac{x'}{2}\right)\delta\left(y - \frac{y'}{2}\right).$$

The first breakage function b_1 is a size-independent function of its arguments and physically represents random scission of particles. On the other hand, the second breakage function b_2 represents size-dependent distribution of daughter fragments, choosing the daughter-particles exactly half the size of the parent particle. The exact solution in closed form for the above set of fragmentation kernels are not available in the literature. However, we can calculate the zeroth $\mathcal{M}_{0,0}(t)$, first $\mathcal{M}_{1,0}(t) + \mathcal{M}_{0,1}(t)$ and the cross moments $\mathcal{M}_{1,1}(t)$ exactly for both the problems (calculations of the exact moments are given in Appendix B). In this regard, the exact moments are given in the Table 1.

Table 1. Exact moment functions for $S=1$ and breakage functions b_1, b_2.

Selection Function	Breakage Function	Exact Moments
$S(x,y) = 1$	$b_1(x,y\|x',y') = \frac{2}{x'y'}$	$\mathcal{M}_{k,l}(t) = \mathcal{M}_{k,l}(0) \exp\left[\left(\frac{2}{(k+1)(l+1)} - 1\right)t\right]$
$S(x,y) = 1$	$b_2(x,y\|x',y') = 2\delta\left(x - \frac{x'}{2}\right)\delta\left(y - \frac{y'}{2}\right)$	$\mathcal{M}_{k,l}(t) = \mathcal{M}_{k,l}(0) \exp\left[\left(2^{1-k-l} - 1\right)t\right]$

Figures 1 and 2 represent the numerical moments obtained from EFVS and WFVS against their exact values with breakage functions b_1 and b_2, respectively. More precisely, Figures 1a and 2a present the comparison of zeroth and first-order moment functions, and Figures 1b and 2b presents the first-order cross moment function $\mathcal{M}_{1,1}(t)$. In order to obtain a clear visibility of different markers, we plot Figures 1a and 2a on a semilogarithmic scale with respect to the y-axis. In both the cases, it is observed that WFVS estimates the zeroth-order, first-order and the cross moments with high accuracy, whereas the EFVS conserves the total volume but fails to produce a good estimate of the other moments.

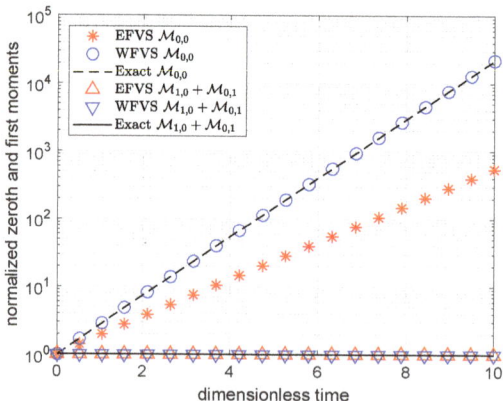

(**a**) Zeroth- and first-order moments.

Figure 1. *Cont.*

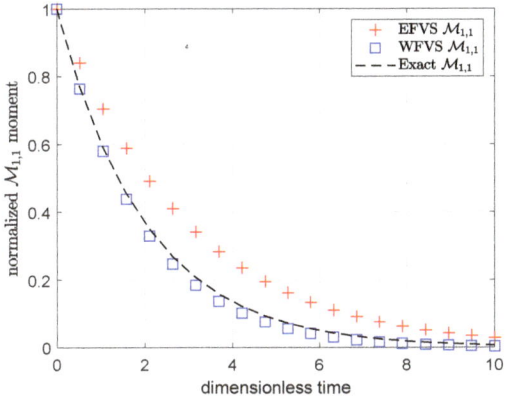

(b) First-order cross moments.

Figure 1. Comparison of different moments with selection function $S = 1$, and breakage function b_1.

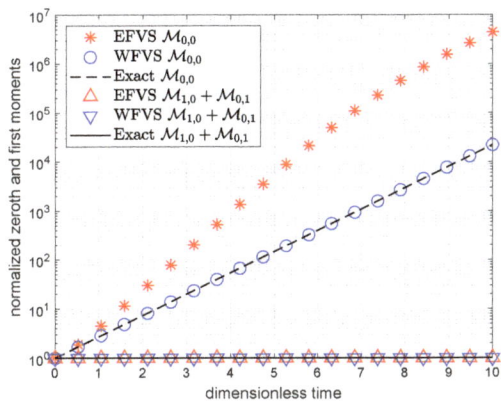

(a) Zeroth- and first-order moments.

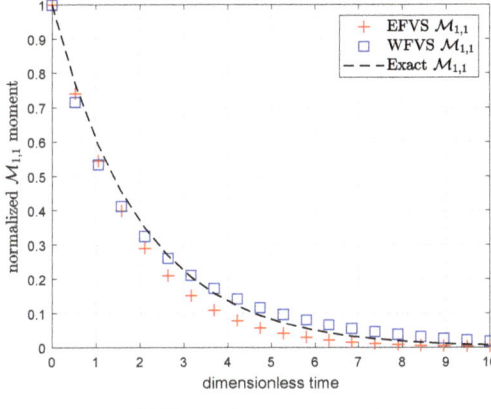

(b) First-order cross moments

Figure 2. Comparison of different moments with selection function $S = 1$, and breakage function b_2.

In the Table 2, the relative error of the moment functions for $S = 1$ and b_1 are calculated at $t = 10$ over three different grid sizes. The geometric ratios to generate 10×10, 15×15 and 20×20 grids are 7.9433, 3.9811 and 2.8184, respectively. The discrete L^1-error norm

$$\text{error} := \sum_{i=1}^{I} \left| \frac{\mathcal{M}^{exact} - \mathcal{M}^{num}}{\mathcal{M}^{exact}} \right|$$

is used to calculate the errors. Similarly, the relative error acquired while computing the moments for $S = 1$ and b_2 over different grid points are represented in Table 3.

Table 2. Relative error for the weighted moments at different grid points for the test case with $S = 1$ and breakage functions b_1 at $t = 10$.

Grids	WFVS				EFVS			
	$\mathcal{M}_{0,0}(t)$	$\mathcal{M}_{1,0}(t) + \mathcal{M}_{0,1}(t)$	$\mathcal{M}_{1,1}(t)$	CPU Time	$\mathcal{M}_{0,0}(t)$	$\mathcal{M}_{1,0}(t) + \mathcal{M}_{0,1}(t)$	$\mathcal{M}_{1,1}(t)$	CPU Time
10×10	2.2714×10^{-9}	1.2589×10^{-1}	6.6331×10^{-3}	3 s	9.9946×10^{-1}	1.2589×10^{-1}	3.6744×10^{-2}	2 s
15×15	1.3878×10^{-9}	2.5123×10^{-1}	6.2981×10^{-3}	11 s	9.9492×10^{-1}	2.5119×10^{-1}	9.0704×10^{-3}	7 s
20×20	1.1027×10^{-9}	3.5483×10^{-1}	5.7431×10^{-3}	22 s	9.7599×10^{-1}	3.5481×10^{-1}	2.8711×10^{-3}	11 s

Table 3. Relative error for the weighted moments at different grid points for the test case with $S = 1$ and breakage functions b_2 at $t = 10$.

Grids	WFVS				EFVS			
	$\mathcal{M}_{0,0}(t)$	$\mathcal{M}_{1,0}(t) + \mathcal{M}_{0,1}(t)$	$\mathcal{M}_{1,1}(t)$	CPU Time	$\mathcal{M}_{0,0}(t)$	$\mathcal{M}_{1,0}(t) + \mathcal{M}_{0,1}(t)$	$\mathcal{M}_{1,1}(t)$	CPU Time
10×10	4.8517×10^{-8}	3.5481×10^{-1}	6.7058×10^{-3}	41 s	2.5560×10^{-1}	3.5481×10^{-1}	6.0140×10^{-3}	2 s
15×15	4.8517×10^{-8}	3.8986×10^{-1}	6.7163×10^{-3}	60 s	2.1835×10^{-1}	3.8986×10^{-1}	5.6563×10^{-3}	31 s
20×20	4.8517×10^{-8}	4.3652×10^{-1}	6.7229×10^{-3}	82 s	1.2736×10^{-1}	4.3652×10^{-1}	4.8956×10^{-3}	40 s

In the second instance, we choose two problems by setting the size-dependent selection function $S(x,y) = x + y$, along with the previously chosen particle daughter distribution functions b_1 and b_2. In this case, also the exact solutions are not available in the literature, however only the zeroth- and first-order moment functions can be calculated exactly. In both the cases, the moment functions are given as

$$\mathcal{M}_{1,0}(t) + \mathcal{M}_{0,1}(t) = 1, \quad \text{and} \quad \mathcal{M}_{0,0}(t) = 1 + 2t.$$

The following Figure 3a,b represent the efficiency of the WFV scheme over the EFV scheme to estimate the zeroth- and the first-order moments. It is observed that both the schemes obey the volume conservation laws with high accuracy, but the WFV scheme is highly accurate to predict the evolution of total number of particles. Tables 4 and 5 represent the relative errors over different grid points at time $t = 10$.

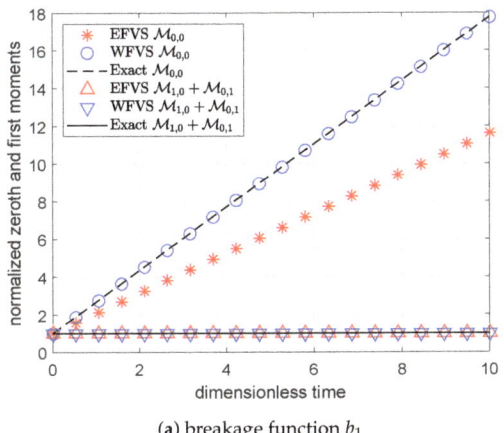

(a) breakage function b_1

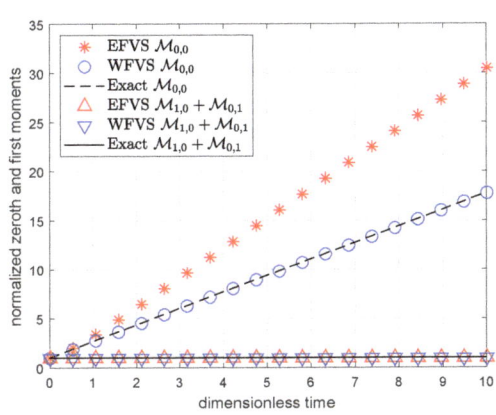

(b) breakage function b_2

Figure 3. Comparison of zeroth- and first-order moments with selection function $S = x + y$, and breakage functions b_1, b_2.

Table 4. Relative error for the weighted moments at different grid points for the test case with $S = x + y$ and breakage function b_1 at $t = 10$.

Grids	WFVS			EFVS		
	$\mathcal{M}_{0,0}(t)$	$\mathcal{M}_{1,0}(t) + \mathcal{M}_{0,1}(t)$	CPU Time	$\mathcal{M}_{0,0}(t)$	$\mathcal{M}_{1,0}(t) + \mathcal{M}_{0,1}(t)$	CPU Time
10×10	9.9944×10^{-1}	1.2589×10^{-1}	2 s	9.9983×10^{-1}	1.2589×10^{-1}	1 s
15×15	9.9939×10^{-1}	2.5119×10^{-1}	5 s	9.9969×10^{-1}	2.5119×10^{-1}	2 s
20×20	9.9934×10^{-1}	3.5481×10^{-1}	11 s	9.9957×10^{-1}	3.5481×10^{-1}	5 s

Table 5. Relative error for the weighted moments at different grid points for the test case with $S = x + y$ and breakage function b_2 at $t = 10$.

Grids	WFVS			EFVS		
	$\mathcal{M}_{0,0}(t)$	$\mathcal{M}_{1,0}(t) + \mathcal{M}_{0,1}(t)$	CPU Time	$\mathcal{M}_{0,0}(t)$	$\mathcal{M}_{1,0}(t) + \mathcal{M}_{0,1}(t)$	CPU Time
10×10	1.8625×10^{-1}	3.5481×10^{-1}	21 s	7.6213×10^{-1}	3.5481×10^{-1}	10 s
15×15	1.8658×10^{-1}	3.8986×10^{-1}	28 s	5.2710×10^{-1}	3.8986×10^{-1}	17 s
20×20	1.8698×10^{-1}	4.3652×10^{-1}	41 s	2.7194×10^{-1}	4.3652×10^{-1}	21 s

4.1.2. Hypervolume Conservation Problems

In this part, the considered breakage functions b should satisfy the hypervolume conservation rule (8). In this regard, we choose the following breakage functions

$$b_3(x,y|x',y') = \frac{4}{x'y'}, \quad \text{and} \quad b_4(x,y|x',y') = \frac{x'\delta(x-x') + y'\delta(y-y')}{x'y'}. \tag{55}$$

The breakage function b_3 is independent of the daughter-particle size, whereas b_4 represents the particle breakage along the longer side of the rectangular structure. Similar to the examples of volume conservation models, we choose two types of selection functions: (i) size-independent kernels $S(x,y) = 1$, and (ii) size-dependent kernels $S(x,y) = xy$.

Like before, the exact solutions are not available in the literature, however we can calculate three moment functions exactly, and they are given in Table 6.

Table 6. Exact moment functions for $S = 1$ and breakage functions b_3, b_4.

Selection Function	Breakage Function	Exact Moments
$S(x,y) = 1$	$b_3(x,y\|x',y') = \dfrac{4}{x'y'}$	$\mathcal{M}_{k,l}(t) = \mathcal{M}_{k,l}(0)\exp\left[\left(\dfrac{4}{(k+1)(l+1)} - 1\right)t\right]$
$S(x,y) = 1$	$b_4(x,y\|x',y') = \dfrac{x'\delta(x-x') + y'\delta(y-y')}{x'y'}$	$\mathcal{M}_{1,1}(t) = 1, \mathcal{M}_{0,0}(t) = \exp(t),$ $\mathcal{M}_{1,0}(t) + \mathcal{M}_{0,1}(t) = \exp(t/2)$

In Figure 4a, we plot the zeroth- and the cross-moment functions and observe that the new WFV scheme predicts the corresponding moments with high accuracy. On the other hand, we plot the first-order moment in Figure 4b. In this case also, the weighted scheme exhibits high accuracy to estimate the moment compared to the standard model. In Figure 4b, we take the the axes in loglog scale for a distinct visibility of the plots. In this problem, Figure 4c represents the comparison of hypervolume distribution functions with the numerical values obtained from the two schemes. We follow the flat pictorial representation to plot the hypervolume distribution as presented in [32]. For the other problems, only the exact moment functions can be calculated for comparison with the numerical models.

The relative errors over different grid points are presented at Table 7.

In the second instance, we consider the size-dependent selection function $S(x,y) = 1$ and b_4 as the daughter distribution function. The exact solution is not available for this problem, but we can evaluate the zeroth and cross moments exactly. From the Figure 5 and Table 8, we can see that the newly proposed WFV scheme predicts the moments with high accuracy.

Table 7. Relative error for the weighted moments at different grid points for the test case with $S = 1$ and breakage functions b_3 at $t = 10$.

Grids	WFVS				EFVS			
	$\mathcal{M}_{0,0}(t)$	$\mathcal{M}_{1,0}(t) + \mathcal{M}_{0,1}(t)$	$\mathcal{M}_{1,1}(t)$	CPU Time	$\mathcal{M}_{0,0}(t)$	$\mathcal{M}_{1,0}(t) + \mathcal{M}_{0,1}(t)$	$\mathcal{M}_{1,1}(t)$	CPU Time
10×10	1.9052×10^{-8}	9.5538	6.8292×10^{-1}	6 s	1.0000	2.2013×10^{1}	6.8309×10^{-1}	2 s
15×15	1.7344×10^{-8}	6.5432	6.0859×10^{-1}	19 s	1.0000	2.1890×10^{1}	6.0863×10^{-1}	8 s
20×20	1.6682×10^{-8}	4.9080	5.4110×10^{-1}	32 s	1.0000	2.1353×10^{1}	5.4112×10^{-1}	14 s

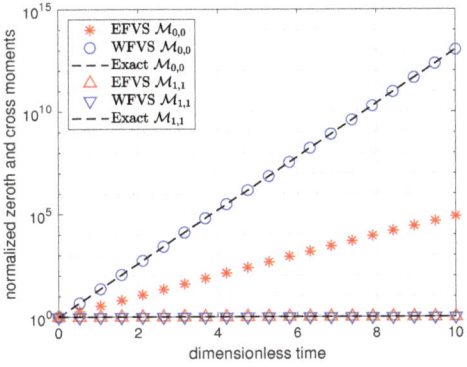

(**a**) Zeroth- and first-order cross moment.

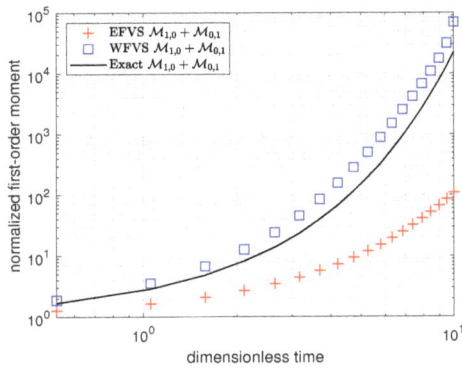

(**b**) First-order moments

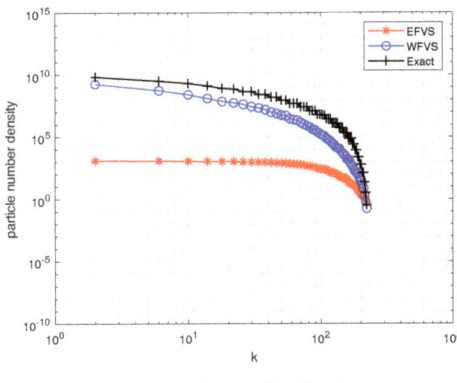

(**c**) Particle size distribution.

Figure 4. Comparison of different moments with selection function $S = 1$, and breakage function b_3.

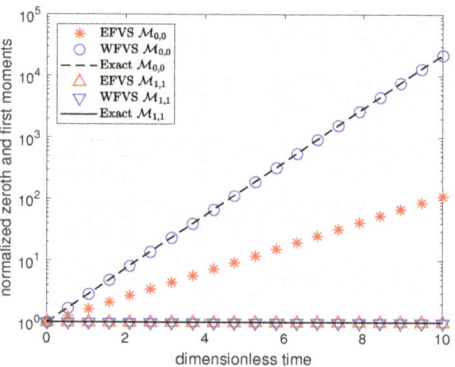

(a) Zeroth- and first-order cross moment

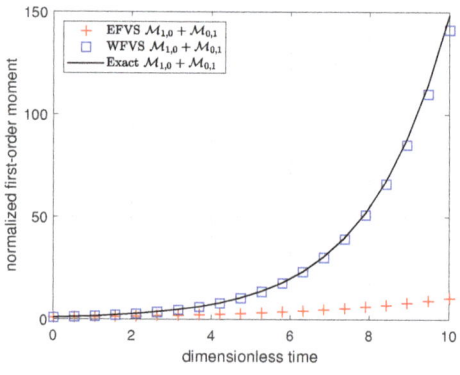

(b) First-order moments

Figure 5. Comparison of different moments with selection function $S = 1$, and breakage function b_4.

Table 8. Relative error for the weighted moments at different grid points for the test case with $S = 1$ and breakage functions b_4 at $t = 10$.

Grids	WFVS				EFVS			
	$\mathcal{M}_{0,0}(t)$	$\mathcal{M}_{1,0}(t) + \mathcal{M}_{0,1}(t)$	$\mathcal{M}_{1,1}(t)$	CPU Time	$\mathcal{M}_{0,0}(t)$	$\mathcal{M}_{1,0}(t) + \mathcal{M}_{0,1}(t)$	$\mathcal{M}_{1,1}(t)$	CPU Time
10×10	4.5211×10^{-10}	2.2694	6.8309×10^{-1}	4 s	9.9946×10^{-1}	1.4453×10^{2}	6.8309×10^{-1}	2 s
15×15	2.2036×10^{-10}	2.8531×10^{-1}	6.0863×10^{-1}	10 s	9.9489×10^{-1}	1.3513×10^{2}	6.0863×10^{-1}	5 s
20×20	1.6944×10^{-10}	4.7015×10^{-2}	5.4112×10^{-1}	22 s	9.7532×10^{-1}	1.1682×10^{2}	5.4112×10^{-1}	10 s

Next, we consider two problems with size-dependent selection function $S(x,y) = xy$ and the breakage functions b_3 and b_4. Only the zeroth- and first-order moment functions can be calculated in closed form, and are given in the Table 9.

Table 9. Exact moment functions for $S = xy$ and breakage functions b_3, b_4.

Selection Function	Breakage Function	Exact Moments
$S(x,y) = xy$	$b_3(x,y|x',y') = \dfrac{4}{x'y'}$	$\mathcal{M}_{1,1}(t) = 1, \mathcal{M}_{0,0}(t) = 1 + 3t$
$S(x,y) = xy$	$b_4(x,y|x',y') = \dfrac{x'\delta(x-x') + y'\delta(y-y')}{x'y'}$	$\mathcal{M}_{1,1}(t) = 1, \mathcal{M}_{0,0}(t) = 1 + t$

Numerical evaluation of the moments using the WFV and EFV schemes are presented in Figure 6 and Tables 10 and 11. The improved accuracy of the new scheme is observed.

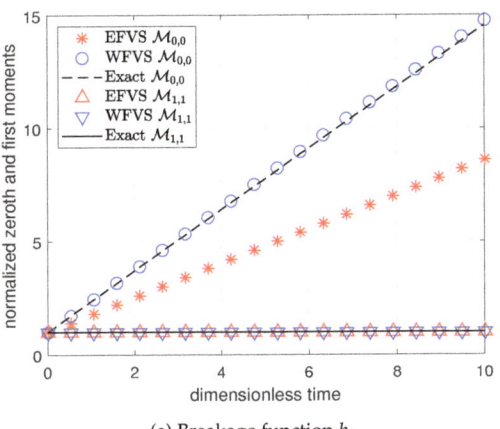

(a) Breakage function b_3.

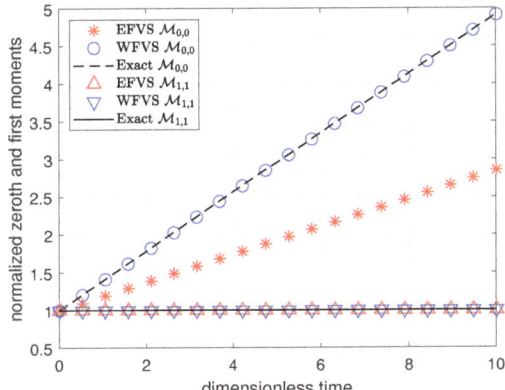

(b) breakage function b_4

Figure 6. Comparison of different moments with selection function $S = xy$, and breakage functions b_3, b_4.

Table 10. Relative error for the weighted moments at different grid points for the test case with $S = xy$ and breakage function b_3 at $t = 10$.

Grids	WFVS			EFVS		
	$\mathcal{M}_{0,0}(t)$	$\mathcal{M}_{1,1}(t)$	CPU Time	$\mathcal{M}_{0,0}(t)$	$\mathcal{M}_{1,1}(t)$	CPU Time
10×10	5.4678×10^{-1}	6.8309×10^{-1}	1 s	8.8071×10^{-1}	6.8309×10^{-1}	1 s
15×15	5.5130×10^{-1}	6.0863×10^{-1}	1 s	8.0374×10^{-1}	6.0863×10^{-1}	1 s
20×20	5.5418×10^{-1}	5.4112×10^{-1}	3 s	7.4013×10^{-1}	5.4112×10^{-1}	2 s

Table 11. Relative error for the weighted moments at different grid points for the test case with $S = xy$ and breakage function b_4 at $t = 10$.

Grids	WFVS			EFVS		
	$\mathcal{M}_{0,0}(t)$	$\mathcal{M}_{1,1}(t)$	CPU Time	$\mathcal{M}_{0,0}(t)$	$\mathcal{M}_{1,1}(t)$	CPU Time
10 × 10	3.3663×10^{-1}	6.8309×10^{-1}	1 s	7.6014×10^{-1}	6.8309×10^{-1}	1 s
15 × 15	1.3416×10^{-1}	6.0863×10^{-1}	1 s	7.9649×10^{-1}	6.0863×10^{-1}	1 s
20 × 20	5.4011×10^{-1}	5.4112×10^{-1}	2 s	7.4013×10^{-1}	5.4112×10^{-1}	2 s

4.2. Three-Dimensional Examples

Volume conservative problems: In this section, we consider two test problems with size-dependent selection function $S(\mathbf{y}) = \phi(\mathbf{y})$. The three-dimensional extension of the above-mentioned breakage functions b_1 and b_2, that is,

$$b_5(\mathbf{x}|\mathbf{y}) = \frac{2}{\psi(\mathbf{y})}, \quad \text{and} \quad b_6(\mathbf{x}|\mathbf{y}) = 2\delta\left(x_1 - \frac{y_1}{2}\right)\delta\left(x_2 - \frac{y_2}{2}\right)\delta\left(x_3 - \frac{y_3}{2}\right)$$

are considered here. Like before, we can only calculate the zeroth and first moments exactly and they are given in the following Table 12.

Table 12. Exact moment functions for $S = \phi(\mathbf{y})$ and breakage functions b_5, b_6.

Selection Function	Breakage Function	Exact Moments
$S(\mathbf{y}) = \phi(\mathbf{y})$	$b_5(\mathbf{x}|\mathbf{y}) = \frac{2}{\psi(\mathbf{y})}$	$\mathcal{M}_{k_1,k_2,k_3}(t) = \mathcal{M}_{k_1,k_2,k_3}(0) \exp\left[\left(\frac{8}{k_1 k_2 k_3} - 1\right)t\right]$
$S(\mathbf{y}) = \phi(\mathbf{y})$	$b_6(\mathbf{x}|\mathbf{y}) = 2\delta(x_1 - \frac{y_1}{2})\delta(x_2 - \frac{y_2}{2})\delta(x_3 - \frac{y_3}{2})$	$\mathcal{M}_{1,0,0}(t) + \mathcal{M}_{1,0,0}(t) + \mathcal{M}_{1,0,0}(t) = 1$, $\mathcal{M}_{0,0}(t) = 1 + 3t$

From Figure 7 and Tables 13 and 14, we can observe that the WFV scheme estimates the moment functions more accurately as compared to the EFV scheme.

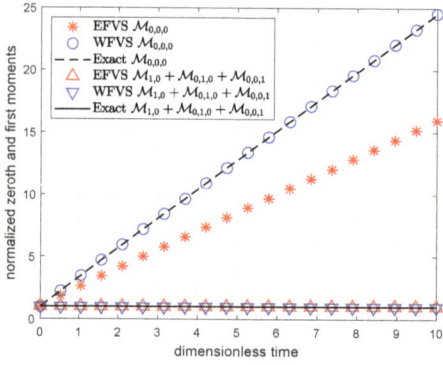

(**a**) Breakage function b_5

Figure 7. *Cont.*

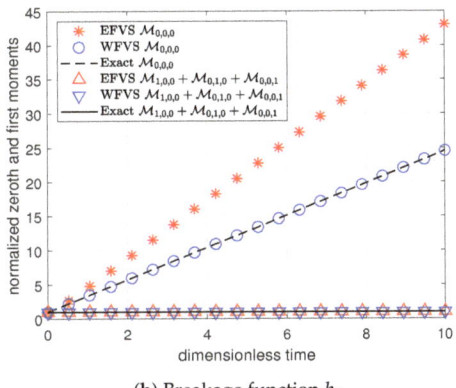

(b) Breakage function b_6

Figure 7. Comparison of zeroth- and first-order moments with selection function $S = \phi(\mathbf{x})$, and different breakage functions b_5, b_6.

Table 13. Relative error for the weighted moments at different grid points for the test case with $S = \phi(\mathbf{x})$ and breakage function b_5 at $t = 10$.

Grids	WFVS			EFVS		
	$\mathcal{M}_{0,0,0}(t)$	$\mathcal{M}_{1,0,0}(t) + \mathcal{M}_{0,1,0}(t) + \mathcal{M}_{0,0,1}(t)$	CPU Time	$\mathcal{M}_{0,0,0}(t)$	$\mathcal{M}_{1,0,0}(t) + \mathcal{M}_{0,1,0}(t) + \mathcal{M}_{0,0,1}(t)$	CPU Time
$10 \times 10 \times 10$	9.9302×10^{-15}	6.8884×10^{-1}	52 s	7.1010×10^{-1}	6.8884×10^{-1}	34 s
$15 \times 15 \times 15$	1.1192×10^{-15}	8.0464×10^{-1}	198 s	5.7788×10^{-1}	8.0464×10^{-1}	78 s
$20 \times 20 \times 20$	2.8756×10^{-16}	8.7678×10^{-1}	505 s	5.0076×10^{-1}	8.7678×10^{-1}	292 s

Table 14. Relative error for the weighted moments at different grid points for the test case with $S = \phi(\mathbf{x})$ and breakage function b_6 at $t = 10$.

Grids	WFVS			EFVS		
	$\mathcal{M}_{0,0,0}(t)$	$\mathcal{M}_{1,0,0}(t) + \mathcal{M}_{0,1,0}(t) + \mathcal{M}_{0,0,1}(t)$	CPU Time	$\mathcal{M}_{0,0,0}(t)$	$\mathcal{M}_{1,0,0}(t) + \mathcal{M}_{0,1,0}(t) + \mathcal{M}_{0,0,1}(t)$	CPU Time
$10 \times 10 \times 10$	3.7623×10^{-1}	1.8500×10^{-1}	73 s	2.9725×10^{-1}	5.0569×10^{-1}	36 s
$15 \times 15 \times 15$	1.4588×10^{-1}	3.5055×10^{-2}	294 s	2.0743×10^{-1}	3.8846×10^{-1}	139 s
$20 \times 20 \times 20$	9.6814×10^{-2}	3.0046×10^{-2}	660 s	1.1094×10^{-1}	1.9000×10^{-1}	335 s

Hypervolume conservation: In this instance, we consider the size-independent daughter distribution function $b_7(\mathbf{x}|\mathbf{y}) = \dfrac{8}{\psi(\mathbf{y})}$ along with the constant selection $S = 1$ and size-dependent selection $S(\mathbf{y}) = \psi(\mathbf{y})$. The exact moments are calculated in the Table 15.

Table 15. Exact moment functions for $S(\mathbf{y}) = 1$, $S = \psi(\mathbf{y})$ and breakage function b_7.

Selection Function	Breakage Function	Exact Moments
$S(\mathbf{y}) = 1$	$b_7(\mathbf{x}|\mathbf{y}) = \dfrac{8}{\psi(\mathbf{y})}$	$\mathcal{M}_{0,0,0}(t) = \exp(7t)$, $\mathcal{M}_{1,1,1}(t) = 1$, $\mathcal{M}_{1,0,0}(t) + \mathcal{M}_{0,1,0}(t) + \mathcal{M}_{0,0,1}(t) = \exp(3t)$
$S(\mathbf{y}) = \psi(\mathbf{y})$	$b_7(\mathbf{x}|\mathbf{y}) = \dfrac{8}{\psi(\mathbf{y})}$	$\mathcal{M}_{0,0,0}(t) = 1 + 7t$, $\mathcal{M}_{1,1,1}(t) = 1$

Figure 8 and Table 16 exhibit the improved accuracy obtained from the WFV scheme over the standard scheme to predict the above mentioned three moments.

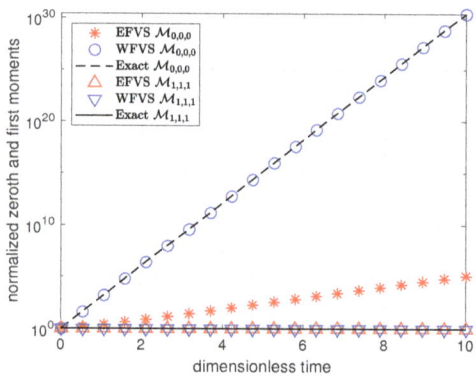

(a) Zeroth and cross moments.

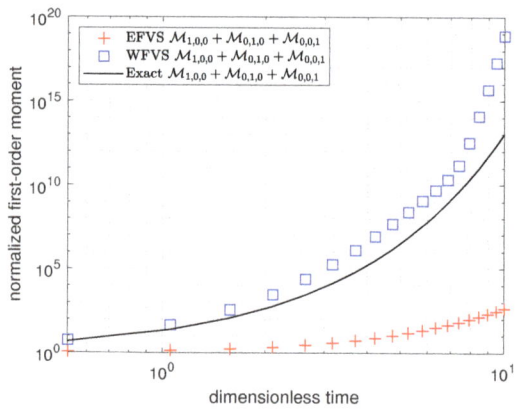

(b) First-order moments.

Figure 8. Comparison of zeroth- and first-order moments with selection function $S = 1$, and breakage functions b_7.

Table 16. Relative error for the weighted moments at different grid points for the test case with $S = 1$ and breakage functions b_7 at $t = 10$.

Grids	WFVS				EFVS			
	$\mathcal{M}_{0,0,0}(t)$	$\mathcal{M}_{1,0,0}(t) + \mathcal{M}_{0,1,0}(t) + \mathcal{M}_{0,0,1}(t)$	$\mathcal{M}_{1,1,1}(t)$	CPU Time	$\mathcal{M}_{0,0,0}(t)$	$\mathcal{M}_{1,0,0}(t) + \mathcal{M}_{0,1,0}(t) + \mathcal{M}_{0,0,1}(t)$	$\mathcal{M}_{1,1,1}(t)$	CPU Time
$10 \times 10 \times 10$	5.8045×10^{-8}	2.8046	1.5759×10^{-1}	164 s	1.0000	1.0013×10^{1}	8.2160×10^{-1}	44 s
$15 \times 15 \times 15$	5.7040×10^{-8}	1.7385×10^{-1}	3.7535×10^{-2}	605 s	1.0000	1.0000	7.8233×10^{-1}	272 s
$20 \times 20 \times 20$	5.6751×10^{-8}	1.4056×10^{-1}	1.9838×10^{-2}	1500 s	1.0000	9.9999×10^{-1}	7.5516×10^{-1}	650 s

On the other hand, Figure 9 and Table 17 represent the comparison of the above moments as predicted by the WFV and EFV schemes.

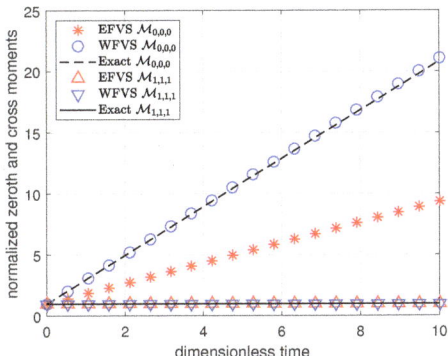

Figure 9. Comparison of zeroth and cross moments with selection function $S = \psi(\mathbf{x})$, and breakage function b_7.

Table 17. Relative error for the weighted moments at different grid points for the test case with $S = \psi(\mathbf{x})$ and breakage function b_7 at $t = 10$.

Grids	WFVS			EFVS		
	$\mathcal{M}_{0,0,0}(t)$	$\mathcal{M}_{1,1,1}(t)$	CPU Time	$\mathcal{M}_{0,0,0}(t)$	$\mathcal{M}_{1,1,1}(t)$	CPU Time
$10 \times 10 \times 10$	3.9509×10^{-16}	8.2160×10^{-1}	5 s	8.0112×10^{-1}	8.2160×10^{-1}	3 s
$15 \times 15 \times 15$	6.5640×10^{-16}	7.8233×10^{-1}	24 s	7.1237×10^{-1}	7.8233×10^{-1}	11 s
$20 \times 20 \times 20$	9.7932×10^{-16}	7.5516×10^{-1}	70 s	6.4883×10^{-1}	7.5516×10^{-1}	30 s

5. Conclusions

In this article, we have proposed finite volume schemes for solving multidimensional fragmentation problems. In addition to the one-dimensional scheme of [23], it is also extended in the multidimensional setup. It is observed that a careful reconstruction of the standard multidimensional scheme leads to the development of very accurate schemes. The newly proposed schemes obey the conservation laws and also predict several physical moment functions with high accuracy. Several empirical test problems in two- and three-dimensions are collected from the literature to validate the efficiency of the new schemes.

Author Contributions: Conceptualization, J.S. and A.B.; methodology, J.S.; software, J.S.; validation, J.S. and A.B.; formal analysis, J.S.; investigation, J.S.; resources, J.S and A.B.; data curation, J.S.; writing—original draft preparation, J.S.; writing—review and editing, J.S. and A.B.; project administration, J.S. and A.B.; funding acquisition, J.S. and A.B. All authors have read and agreed to the published version of the manuscript.

Funding: J.S. thanks the NITT seed grant (NITT/R&C/SEED GRANT/19-20/P-13/MATHS/JS/E1) for their funding support during this work. A.B. gratefully acknowledges funding by the Deutsche Forschungsgemeinschaft (DFG, German Research Foundation)—Project-ID 416229255—SFB 1411. The authors acknowledge support by Deutsche Forschungsgemeinschaft and Friedrich-Alexander-Universität Erlangen-Nürnberg (FAU) within the funding program Open Access Publishing.

Institutional Review Board Statement: Not applicable.

Informed Consent Statement: Not applicable.

Acknowledgments: The authors are thankful to the anonymous reviewers for providing valuable suggestions which have helped to improve the quality of the manuscript.

Conflicts of Interest: The authors declare no conflict of interest.

Appendix A. Proof of Conservation Laws in Two-Dimensions

We first calculate the difference $\mathcal{F}_{i+1/2} - \mathcal{F}_{i-1/2}$.

$$\mathcal{F}_{i+1/2,j} - \mathcal{F}_{i-1/2,j} = \sum_{l_1=i+1}^{I} \sum_{m_1=j}^{J} n_{l_1,m_1} \mathcal{A}^{\beta}_{l_1,m_1} \sum_{l_2=1}^{i} \mathcal{B}_{l_2,j|l_1,m_1} \Delta_{l_2} \Delta_{l_1,m_1}$$

$$- \sum_{l_1=i}^{I} \sum_{m_1=j}^{J} n_{l_1,m_1} \mathcal{A}^{\beta}_{l_1,m_1} \sum_{l_2=1}^{i-1} \mathcal{B}_{l_2,j|l_1,m_1} \Delta_{l_2} \Delta_{l_1,m_1}$$

$$= \sum_{l_1=i}^{I} \sum_{m_1=j}^{J} n_{l_1,m_1} \mathcal{A}^{\beta}_{l_1,m_1} \mathcal{B}_{i,j|l_1,m_1} \Delta_i \Delta_{l_1,m_1}$$

$$- \sum_{m_1=j}^{J} n_{i,m_1} \mathcal{A}^{\beta}_{i,m_1} \sum_{l_2=1}^{i} \mathcal{B}_{l_2,j|i,m_1} \Delta_{l_2} \Delta_{i,m_1}. \tag{A1}$$

In a similar manner, calculating and simplifying the difference $\mathcal{G}_{i+1/2} - \mathcal{G}_{i-1/2}$.

$$\mathcal{G}_{i,j+1/2} - \mathcal{G}_{i,j-1/2} = \sum_{l_1=i}^{I} \sum_{m_1=j+1}^{J} n_{l_1,m_1} \mathcal{A}^{\alpha}_{l_1,m_1} \sum_{m_2=1}^{j} \mathcal{B}_{i,m_2|l_1,m_1} \Delta_{m_2} \Delta_{l_1,m_1}$$

$$- \sum_{l_1=i}^{I} \sum_{m_1=j}^{J} n_{l_1,m_1} \mathcal{A}^{\alpha}_{l_1,m_1} \sum_{m_2=1}^{j-1} \mathcal{B}_{i,m_2|l_1,m_1} \Delta_{m_2} \Delta_{l_1,m_1}$$

$$= \sum_{l_1=i}^{I} \sum_{m_1=j}^{J} n_{l_1,m_1} \mathcal{A}^{\alpha}_{l_1,m_1} \mathcal{B}_{i,j|l_1,m_1} \Delta_j \Delta_{l_1,m_1}$$

$$- \sum_{l_1=i}^{I} n_{l_1,j} \mathcal{A}^{\alpha}_{l_1,j} \sum_{m_2=1}^{j} \mathcal{B}_{i,m_2|l_1,j} \Delta_{m_2} \Delta_{l_1,j}. \tag{A2}$$

Next, we calculate the following flux,

$$\mathcal{H}_{i+1/2,j+1/2} - \mathcal{H}_{i+1/2,j-1/2} - \mathcal{H}_{i-1/2,j+1/2} - \mathcal{H}_{i-1/2,j-1/2}$$

$$= \sum_{l_1=i+1}^{I} \sum_{m_1=j+1}^{J} n_{l_1,m_1} \mathcal{A}_{l_1,m_1} \sum_{l_2=1}^{i} \sum_{m_2=1}^{j} \mathcal{B}_{l_2,m_2|l_1,m_1} \Delta_{l_2,m_2} \Delta_{l_1,m_1}$$

$$- \sum_{l_1=i+1}^{I} \sum_{m_1=j}^{J} n_{l_1,m_1} \mathcal{A}_{l_1,m_1} \sum_{l_2=1}^{i} \sum_{m_2=1}^{j-1} \mathcal{B}_{l_2,m_2|l_1,m_1} \Delta_{l_2,m_2} \Delta_{l_1,m_1}$$

$$- \sum_{l_1=i}^{I} \sum_{m_1=j+1}^{J} n_{l_1,m_1} \mathcal{A}_{l_1,m_1} \sum_{l_2=1}^{i-1} \sum_{m_2=1}^{j} \mathcal{B}_{l_2,m_2|l_1,m_1} \Delta_{l_2,m_2} \Delta_{l_1,m_1}$$

$$+ \sum_{l_1=i}^{I} \sum_{m_1=j}^{J} n_{l_1,m_1} \mathcal{A}_{l_1,m_1} \sum_{l_2=1}^{i-1} \sum_{m_2=1}^{j-1} \mathcal{B}_{l_2,m_2|l_1,m_1} \Delta_{l_2,m_2} \Delta_{l_1,m_1}.$$

$$= - \sum_{l_1=i+1}^{I} n_{l_1,j} \mathcal{A}_{l_1,j} \sum_{l_2=1}^{i} \sum_{m_2=1}^{j} \mathcal{B}_{l_2,m_2|l_1,j} \Delta_{l_2,m_2} \Delta_{l_1,j}$$

$$+ \sum_{l_1=i+1}^{I} \sum_{m_1=j}^{J} n_{l_1,m_1} \mathcal{A}_{l_1,m_1} \sum_{l_2=1}^{i} \mathcal{B}_{l_2,j|l_1,m_1} \Delta_{l_2,j} \Delta_{l_1,m_1}$$

$$+ \sum_{l_1=i}^{I} n_{l_1,j} \mathcal{A}_{l_1,j} \sum_{l_2=1}^{i-1} \sum_{m_2=1}^{j} \mathcal{B}_{l_2,m_2|l_1,j} \Delta_{l_2,m_2} \Delta_{l_1,j}$$

$$- \sum_{l_1=i}^{I} \sum_{m_1=j}^{J} n_{l_1,m_1} \mathcal{A}_{l_1,m_1} \sum_{l_2=1}^{i-1} \mathcal{B}_{l_2,j|l_1,m_1} \Delta_{l_2,j} \Delta_{l_1,m_1}. \tag{A3}$$

Further rearrangement and simplification of the terms gives

$$
\begin{aligned}
&\mathcal{H}_{i+1/2,j+1/2} - \mathcal{H}_{i+1/2,j-1/2} - \mathcal{H}_{i-1/2,j+1/2} - \mathcal{H}_{i-1/2,j-1/2} \\
&= \sum_{l_1=i}^{I}\sum_{m_1=j}^{J} n_{l_1,m_1} \mathcal{A}_{l_1,m_1} \mathcal{B}_{i,j|l_1,m_1} \Delta_{i,j} \Delta_{l_1,m_1} - \sum_{l_1=i}^{I} n_{l_1,j} \mathcal{A}_{l_1,j} \sum_{m_2=1}^{j} \mathcal{B}_{i,m_2|l_1,j} \Delta_{i,m_2} \Delta_{l_1,j} \\
&\quad - \sum_{m_1=j}^{J} n_{i,m_1} \mathcal{A}_{i,m_1} \sum_{l_2=1}^{i} \mathcal{B}_{l_2,j|i,m_1} \Delta_{l_2,j} \Delta_{i,m_1} + n_{i,j}\mathcal{A}_{i,j} \sum_{l_2=1}^{i}\sum_{m_2=1}^{j} \mathcal{B}_{l_2,m_2|i,j} \Delta_{l_2,m_2} \Delta_{i,j}. \quad (A4)
\end{aligned}
$$

Therefore, substituting relations (A1), (A2) and (A4) in the discrete formulation (34) and simplifying, we get

$$
\begin{aligned}
\frac{dn_{i,j}\Delta_{i,j}}{dt} &= \sum_{l_1=i}^{I}\sum_{m_1=j}^{J} n_{l_1,m_1} \mathcal{A}^{\beta}_{l_1,m_1} \mathcal{B}_{i,j|l_1,m_1}\Delta_{i,j}\Delta_{l_1,m_1} - \sum_{m_1=j}^{J} n_{i,m_1}\mathcal{A}^{\beta}_{i,m_1} \sum_{l_2=1}^{i} \mathcal{B}_{l_2,j|l_1,m_1}\Delta_{l_2,j}\Delta_{l_1,m_1} \\
&+ \sum_{l_1=i}^{I}\sum_{m_1=j}^{J} n_{l_1,m_1}\mathcal{A}^{\alpha}_{l_1,m_1}\mathcal{B}_{i,j|l_1,m_1}\Delta_{i,j}\Delta_{l_1,m_1} - \sum_{l_1=i}^{I} n_{l_1,j}\mathcal{A}^{\alpha}_{l_1,j}\sum_{m_2=1}^{j}\mathcal{B}_{i,m_2|l_1,m_1}\Delta_{i,m_2}\Delta_{l_1,m_1} \\
&- \sum_{l_1=i}^{I}\sum_{m_1=j}^{J} n_{l_1,m_1}\mathcal{A}_{l_1,m_1}\mathcal{B}_{i,j|l_1,m_1}\Delta_{i,j}\Delta_{l_1,m_1} + \sum_{l_1=i}^{I} n_{l_1,j}\mathcal{A}_{l_1,j}\sum_{m_2=1}^{j}\mathcal{B}_{i,m_2|l_1,j}\Delta_{i,m_2}\Delta_{l_1,j} \\
&+ \sum_{m_1=j}^{J} n_{i,m_1}\mathcal{A}_{i,m_1}\sum_{l_2=1}^{i}\mathcal{B}_{l_2,j|i,m_1}\Delta_{l_2,j}\Delta_{i,m_1} - n_{i,j}\mathcal{A}_{i,j}\sum_{l_2=1}^{i}\sum_{m_2=1}^{j}\mathcal{B}_{l_2,m_2|i,j}\Delta_{l_2,m_2}\Delta_{i,j} \\
&= \sum_{l_1=i+1}^{I}\sum_{m_1=j+1}^{J} n_{l_1,m_1}\mathcal{A}_{l_1,m_1}\mathcal{B}_{i,j|l_1,m_1}\Delta_{i,j}\Delta_{l_1,m_1} - n_{i,j}\mathcal{A}_{i,j}\sum_{l_2=1}^{i-1}\sum_{m_2=1}^{j-1}\mathcal{B}_{l_2,m_2|i,j}\Delta_{l_2,m_2}\Delta_{i,j}. \quad (A5)
\end{aligned}
$$

Temporal evolution of total volume: Taking sum over i and j of Equation (A5), we get

$$
\begin{aligned}
\frac{d}{dt}[\mathcal{M}_{1,0} + \mathcal{M}_{0,1}] &= \sum_{i=1}^{I}\sum_{j=1}^{J}\sum_{l_1=i+1}^{I}\sum_{m_1=j+1}^{J} n_{l_1,m_1}\mathcal{A}_{l_1,m_1}\mathcal{B}_{i,j|l_1,m_1}\Delta_{i,j}\Delta_{l_1,m_1} \\
&\quad - \sum_{i=1}^{I}\sum_{j=1}^{J} n_{i,j}\mathcal{A}_{i,j}\sum_{l_2=1}^{i-1}\sum_{m_2=1}^{j-1}\mathcal{B}_{l_2,m_2|i,j}\Delta_{l_2,m_2}\Delta_{i,j}.
\end{aligned}
$$

Changing the order of summation in the first term, we get

$$
\begin{aligned}
\frac{d}{dt}[\mathcal{M}_{1,0} + \mathcal{M}_{0,1}] &= \sum_{i=1}^{I}\sum_{l_1=i+1}^{I}\sum_{m_1=1}^{J}\sum_{i=1}^{l_1-1}\sum_{j=1}^{m_1-1} n_{l_1,m_1}\mathcal{A}_{l_1,m_1}\mathcal{B}_{i,j|l_1,m_1}\Delta_{i,j}\Delta_{l_1,m_1} \\
&\quad - \sum_{i=1}^{I}\sum_{j=1}^{J} n_{i,j}\mathcal{A}_{i,j}\sum_{l_2=1}^{i-1}\sum_{m_2=1}^{j-1}\mathcal{B}_{l_2,m_2|i,j}\Delta_{l_2,m_2}\Delta_{i,j} \\
&= 0.
\end{aligned}
$$

Hence, the volume conservation law is obeyed.

Temporal evolution of zeroth moment: Dividing both sides of Equation (A5) by $(x_i + y_j)$, and taking sum over i, j, we get

$$
\begin{aligned}
\frac{d\mathcal{M}_{0,0}}{dt} &= \sum_{i=1}^{I}\sum_{j=1}^{J}\frac{1}{(x_i+y_j)}\sum_{l_1=i+1}^{I}\sum_{m_1=j+1}^{J} n_{l_1,m_1}\mathcal{A}_{l_1,m_1}\mathcal{B}_{i,j|l_1,m_1}\Delta_{i,j}\Delta_{l_1,m_1} \\
&\quad - \sum_{i=1}^{I}\sum_{j=1}^{J}\frac{n_{i,j}\mathcal{A}_{i,j}}{(x_i+y_j)}\sum_{l_2=1}^{i-1}\sum_{m_2=1}^{j-1}\mathcal{B}_{l_2,m_2|i,j}\Delta_{l_2,m_2}\Delta_{i,j}.
\end{aligned}
$$

Changing order of summation in the first term, we get

$$\frac{d\mathcal{M}_{0,0}}{dt} = \sum_{l_1=1}^{I} \sum_{m_1=1}^{J} n_{l_1,m_1} \mathcal{A}_{l_1,m_1} \Delta_{l_1,m_1} \sum_{i=1}^{l_1-1} \sum_{j=1}^{m_1-1} \left[\frac{1}{x_i + y_j} - \frac{1}{x_{l_1} + y_{m_1}} \right] \mathcal{B}_{i,j|l_1,m_1} \Delta_{i,j}$$

$$= \sum_{l_1=1}^{I} \sum_{m_1=1}^{J} \frac{n_{l_1,m_1} \mathcal{A}_{l_1,m_1}}{\phi(x_{l_1}, y_{m_1})} \Delta_{l_1,m_1} \sum_{i=1}^{l_1-1} \sum_{j=1}^{m_1-1} \left[\frac{\phi(x_{l_1}, y_{m_1})}{\phi(x_i, y_j)} - 1 \right] \mathcal{B}_{i,j|l_1,m_1} \Delta_{i,j}.$$

Appendix B. Exact Moments

The temporal evolution of zeroth moment $\mathcal{M}_{0,0}(t)$ is given by

$$\frac{d\mathcal{M}_{0,0}(t)}{dt} = \int_0^\infty \int_0^\infty \frac{1}{x+y} \left[\frac{\partial \mathcal{F}(t,x,y)}{\partial x} + \frac{\partial \mathcal{G}(t,x,y)}{\partial y} - \frac{\partial^2 \mathcal{H}(t,x,y)}{\partial x \partial y} \right] dy dx. \quad (A6)$$

For $S(x,y) = 1$ and $b(x,y|x',y') = \dfrac{2}{x'y'}$, we have

$$\frac{\partial \mathcal{F}(t,x,y)}{\partial x} = -\int_0^x \int_y^\infty \frac{\epsilon+y}{x+v} \frac{2}{xv} n(t,x,v) dv d\epsilon + \int_x^\infty \int_y^\infty \frac{x+y}{u+v} \frac{2}{uv} n(t,u,v) dv du,$$

$$\frac{\partial \mathcal{G}(t,x,y)}{\partial y} = -\int_x^\infty \int_0^y \frac{x+\xi}{u+y} \frac{2}{uy} n(t,u,y) d\xi du + \int_x^\infty \int_y^\infty \frac{x+y}{u+v} \frac{2}{uv} n(t,u,v) dv du,$$

and

$$\frac{\partial^2 \mathcal{H}(t,x,y)}{\partial x \partial y} = \int_0^x \int_0^y \frac{\epsilon+\xi}{x+y} \frac{2}{xy} n(t,x,y) dx id\epsilon - \int_x^\infty \int_0^y \frac{x+\xi}{u+y} \frac{2}{uy} n(t,u,y) d\xi du$$
$$- \int_0^x \int_y^\infty \frac{\epsilon+y}{x+v} \frac{2}{xv} n(t,x,v) dv d\epsilon + \int_x^\infty \int_y^\infty \frac{x+y}{u+v} \frac{2}{uv} n(t,u,v) dv du.$$

Now substituting in Equation (A6) and simplifying, we get

$$\frac{d\mathcal{M}_{0,0}(t)}{dt} = \int_0^\infty \int_0^\infty \frac{1}{x+y} \left[\int_x^\infty \int_y^\infty \frac{2(x+y)}{uv(u+v)} n(t,u,v) dv du \right.$$
$$\left. - \int_0^x \int_0^y \frac{2(\epsilon+\xi)}{xy(x+y)} n(t,x,y) d\xi d\epsilon \right] dy dx$$
$$= \int_0^\infty \int_0^\infty \int_0^u \int_0^v \frac{2}{uv} dx dy f(t,u,v) dv du - \int_0^\infty \int_0^\infty f(t,x,y) dy dx$$
$$= \mathcal{M}_{0,0}(t).$$

Solving these equations, we get

$$\mathcal{M}_{0,0}(t) = \mathcal{M}_{0,0}(0) \exp(t). \quad (A7)$$

Similarly, one can calculate the exact moment functions corresponding to the different selection and breakage functions.

References

1. Hauk, T.; Bonaccurso, E.; Roisman, I.; Tropea, C. Ice crystal impact onto a dry solid wall. Particle fragmentation. *Proc. R. Soc. A Math. Phys. Eng. Sci.* **2015**, *471*, 20150399. [CrossRef]
2. Kang, D.H.; Ko, C.K.; Lee, D.H. Attrition characteristics of iron ore by an air jet in gas-solid fluidized beds. *Powder Technol.* **2017**, *316*, 69–78. [CrossRef]
3. Cavalcanti, P.P.; de Carvalho, R.M.; Anderson, S.; da Silveira, M.W.; Tavares, L.M. Surface breakage of fired iron ore pellets by impact. *Powder Technol.* **2019**, *342*, 735–743. [CrossRef]
4. Fulchini, F.; Ghadiri, M.; Borissova, A.; Amblard, B.; Bertholin, S.; Cloupet, A.; Yazdanpanah, M. Development of a methodology for predicting particle attrition in a cyclone by CFD-DEM. *Powder Technol.* **2019**, *357*, 21–32. [CrossRef]

5. Jain, D.; Kuipers, J.; Deen, N.G. Numerical study of coalescence and breakup in a bubble column using a hybrid volume of fluid and discrete bubble model approach. *Chem. Eng. Sci.* **2014**, *119*, 134–146. [CrossRef]
6. Li, L.; Kang, Y.T. Effects of bubble coalescence and breakup on CO2 absorption performance in nanoabsorbents. *J. CO_2 Util.* **2020**, *39*, 101170. [CrossRef]
7. Foroushan, H.K.; Jakobsen, H.A. On the Dynamics of Fluid Particle Breakage Induced by Hydrodynamic Instabilities: A Review of Modelling Approaches. *Chem. Eng. Sci.* **2020**, *2020*, 115575. [CrossRef]
8. Herø, E.H.; Forgia, N.L.; Solsvik, J.; Jakobsen, H.A. Determination of Breakage Parameters in Turbulent Fluid-Fluid Breakage. *Chem. Eng. Technol.* **2019**, *42*, 903–909. [CrossRef]
9. Yang, W.; Luo, Z.; Gu, Y.; Liu, Z.; Zou, Z. Simulation of bubbles behavior in steel continuous casting mold using an Euler-Lagrange framework with modified bubble coalescence and breakup models. *Powder Technol.* **2020**, *361*, 769–781. [CrossRef]
10. Nakach, M.; Authelin, J.R.; Chamayou, A.; Dodds, J. Comparison of various milling technologies for grinding pharmaceutical powders. *Int. J. Miner. Process.* **2004**, *74*, S173–S181. [CrossRef]
11. Flach, F.; Konnerth, C.; Peppersack, C.; Schmidt, J.; Damm, C.; Breitung-Faes, S.; Peukert, W.; Kwade, A. Impact of formulation and operating parameters on particle size and grinding media wear in wet media milling of organic compounds—A case study for pyrene. *Adv. Powder Technol.* **2016**, *27*, 2507–2519. [CrossRef]
12. Braig, V.; Konnerth, C.; Peukert, W.; Lee, G. Enhanced dissolution of naproxen from pure-drug, crystalline nanoparticles: A case study formulated into spray-dried granules and compressed tablets. *Int. J. Pharm.* **2019**, *554*, 54–60. [CrossRef]
13. Sauvageot, H.; Koffi, M. Multimodal raindrop size distributions. *J. Atmos. Sci.* **2000**, *57*, 2480–2492. [CrossRef]
14. Fuerstenau, D.; De, A.; Kapur, P. Linear and nonlinear particle breakage processes in comminution systems. *Int. J. Miner. Process.* **2004**, *74*, S317–S327. [CrossRef]
15. Oddershede, L.; Dimon, P.; Bohr, J. Self-organized criticality in fragmenting. *Phys. Rev. Lett.* **1993**, *71*, 3107. [CrossRef]
16. Ishii, T.; Matsushita, M. Fragmentation of long thin glass rods. *J. Phys. Soc. Jpn.* **1992**, *61*, 3474–3477. [CrossRef]
17. Lee, K.F.; Dosta, M.; McGuire, A.D.; Mosbach, S.; Wagner, W.; Heinrich, S.; Kraft, M. Development of a multi-compartment population balance model for high-shear wet granulation with discrete element method. *Comput. Chem. Eng.* **2017**, *99*, 171–184. [CrossRef]
18. Maslov, D. Absence of self-averaging in shattering fragmentation processes. *Phys. Rev. Lett.* **1993**, *71*, 1268. [CrossRef] [PubMed]
19. Rodgers, G.; Hassan, M. Fragmentation of particles with more than one degree of freedom. *Phys. Rev. E* **1994**, *50*, 3458. [CrossRef]
20. Singh, P.; Hassan, M. Kinetics of multidimensional fragmentation. *Phys. Rev. E* **1996**, *53*, 3134. [CrossRef] [PubMed]
21. Hernández, G. Discrete model for fragmentation with random stopping. *Phys. A Stat. Mech. Its Appl.* **2001**, *300*, 13–24. [CrossRef]
22. Saha, J.; Bück, A. Improved accuracy and convergence analysis of finite volume methods for particle fragmentation models. *Math. Methods Appl. Sci.* **2021**, *44*, 1913–1930. [CrossRef]
23. Kumar, R.; Kumar, J. Numerical simulation and convergence analysis of a finite volume scheme for solving general breakage population balance equations. *Appl. Math. Comput.* **2013**, *219*, 5140–5151. [CrossRef]
24. Boyer, D.; Tarjus, G.; Viot, P. Exact solution and multifractal analysis of a multivariable fragmentation model. *J. Phys. I* **1997**, *7*, 13–38. [CrossRef]
25. Nandanwar, M.N.; Kumar, S. A new discretization of space for the solution of multi-dimensional population balance equations: Simultaneous breakup and aggregation of particles. *Chem. Eng. Sci.* **2008**, *63*, 3988–3997. [CrossRef]
26. Kumar, J.; Saha, J.; Tsotsas, E. Development and convergence analysis of a finite volume scheme for solving breakage equation. *SIAM J. Numer. Anal.* **2015**, *53*, 1672–1689. [CrossRef]
27. Buffo, A.; Alopaeus, V. Solution of bivariate population balance equations with high-order moment-conserving method of classes. *Comput. Chem. Eng.* **2016**, *87*, 111–124. [CrossRef]
28. Saha, J.; Kumar, J.; Heinrich, S. On the approximate solutions of fragmentation equations. *Proc. R. Soc. A Math. Phys. Eng. Sci.* **2018**, *474*, 20170541. [CrossRef]
29. Saha, J.; Das, N.; Kumar, J.; Bück, A. Numerical solutions for multidimensional fragmentation problems using finite volume methods. *Kinet. Relat. Model.* **2019**, *12*, 79–103. [CrossRef]
30. Ziff, R.M.; McGrady, E. The kinetics of cluster fragmentation and depolymerisation. *J. Phys. A Math. Gen.* **1985**, *18*, 3027. [CrossRef]
31. Ziff, R.M. New solutions to the fragmentation equation. *J. Phys. A Math. Gen.* **1991**, *24*, 2821. [CrossRef]
32. Chakraborty, J.; Kumar, S. A new framework for solution of multidimensional population balance equations. *Chem. Eng. Sci.* **2007**, *62*, 4112–4125. [CrossRef]

Article

Non-Stationary Contaminant Plumes in the Advective-Diffusive Regime

Iván Alhama [1], Gonzalo García-Ros [1,*] and Matteo Icardi [2]

[1] Mining and Civil Engineering Department, Technical University of Cartagena, 30201 Murcia, Spain; ivan.alhama@upct.es
[2] School of Mathematical Sciences, University of Nottingham, Nottingham NG7 2RD, UK; matteo.icardi@nottingham.ac.uk
* Correspondence: gonzalo.garcia@upct.es; Tel.: +34-968-32-5743

Abstract: Porous media with low/moderate regional velocities can exhibit a complex dynamic of contamination plumes, in which advection and molecular diffusion are comparable. In this work, we present a two-dimensional scenario with a constant concentration source and impermeable upper and lower boundaries. In order to characterise the plume patterns, a detailed discriminated dimensionless technique is used to obtain the dimensionless groups that govern the problem: an aspect ratio of the domain including characteristic lengths, and two others relating time and the horizontal length of the spread of contamination. The monomials are related to each other to enable their dependences to be translated into a set of new universal abacuses. Extensive numerical simulations were carried out to check the monomials and to plot these type curves. The abacuses provide a tool to directly manage the contamination process, covering a wide spectrum of possible real cases. Among other applications of interest, they predict the maximum horizontal and transversal plume extensions and the time-spatial dependences of iso-concentration patterns according to the physical parameters of the problem.

Keywords: contamination plume; advection-diffusion; universal curves

Citation: Alhama, I.; García-Ros, G.; Icardi, M. Non-Stationary Contaminant Plumes in the Advective-Diffusive Regime. *Mathematics* **2021**, *9*, 725. https://doi.org/10.3390/math9070725

Academic Editors: Marco Pedroni and Lucas Jódar

Received: 4 February 2021
Accepted: 25 March 2021
Published: 27 March 2021

Publisher's Note: MDPI stays neutral with regard to jurisdictional claims in published maps and institutional affiliations.

Copyright: © 2021 by the authors. Licensee MDPI, Basel, Switzerland. This article is an open access article distributed under the terms and conditions of the Creative Commons Attribution (CC BY) license (https://creativecommons.org/licenses/by/4.0/).

1. Introduction

There are many non-stationary scenarios in large extension water-saturated porous media in which the existence of both the regional velocity and molecular diffusion of a solute in the fluid are combined. The primary interest in eventual processes of pollutant spreading is to determine the spatial and temporal evolution of the contaminant plumes in order to plan actions of control. The environmental effects involve aquifer contamination, a problem that has been widely treated in the scientific literature since the middle of the last century. Specific aspects that range from the mathematical-physical theoretical description of the process [1] to real cases that generally cause socio-economic impacts [2] have been addressed.

The objective of this work, as in other works where similar methodologies have been applied [3–8], is to obtain dimensionless groups that govern the expansive dynamics of the plumes caused by the simultaneous effects of advection and diffusion. These groups will collect the preponderance of one phenomenon over another, and they are the monomials according to which the temporal evolution of the horizontal and transversal plume extensions can be described and represented graphically by means of a universal abacus.

From a practical point of view, these type curves or universal abacuses allows hydrogeologists and engineers to confront contamination problems in groundwater systems through direct analyses to gain rapid predictions using curve matching and interpolation techniques, making computer simulations or field testing unnecessary. Although the accuracy of the results depends, to a great extent, on the experimental parameters, which could be difficult to obtain when time or economic variables are present, the universal curves can establish a range of spatial-temporal contamination extensions based on the reliability of the field data introduced as input, which must be set as a starting hypothesis. These type curves can also be used to obtain a preliminary idea of the spread of the contamination when planning in–situ trace tests, which require setting the location of control points [9,10] as well as delimiting the time-dependent perimeters when establishing different types of land use [11,12].

To limit the scope of the work, large 2D rectangular scenarios in which the extension of the contamination is far from reaching the boundaries of the domain have been devised. The regional flow has been implemented from left to right and the horizontal faces of the domain are impervious to both flow and solute, simulating a narrowly-confined layer of 1m in depth parallel to the horizontal surface. The methodology described in this work to obtain the proper monomials consisted of, firstly, applying the discriminated dimensionless technique to the governing equations to obtain the dimensionless groups (a standard objective of dimensional analysis [13–18]), secondly, defining the interdependences among monomials or unknowns using the Pi theorem [19] and, finally, providing extensive numerical simulations to verify the reliability of the groups and to depict the universal abacuses.

It is worth mentioning that in the first step of the protocol, the use of a standard non-discriminated technique to obtain dimensionless groups in the mathematical model that govern this problem is a topic of heated debate among researchers [20,21]. The classic techniques used to obtain the dimensionless groups, once the governing equation has been written in its dimensionless form, would not, in fact, lead to a proper characterisation of the problem. These techniques do not generally introduce the hidden magnitudes (which are the unknowns of interest) or the time factor into the process, so that such unknowns cannot be derived from the groups by applying the Pi theorem. Instead, the proposed dimensionless protocol (discriminated and normalised nondimensionalisation of the governing equation) leads directly to the least number of independent groups and to the function of dependence between the unknowns and the physical and geometric parameters of the problem. By means of mathematical approximations and manipulation, the protocol associates to each addend of the equation a numerical coefficient of a dimensional character which balances with the rest, deducing the groups as independent ratios between these coefficients.

It should also be noted that the dimensionless group that characterises this type of coupled problem (advection and molecular diffusion) is the so-called Peclet number [22]. A dimensional study of the equation has been carried out in some research with different transport and flow conditions in dispersive scenarios, depicting type curves that are dependent on the Peclet number [23,24]. Despite the fact that hydrodynamic dispersion effects can be neglected when regional velocities are small enough, in the scenario presented here, this number cannot be defined a priori since the extensions of the domain (length or width) are not relevant to the study of the dynamics of contamination. As mentioned above, in this work we consider instead the time needed for the contamination plume to reach the boundaries. The dimensionless groups deduced, all of which contain unknowns, allow us to establish the functional relationships that we are interested in.

As for the second step in the methodology, once the Pi theorem has been applied and the dependences of the unknowns have been deduced, the direct quotient between the regional velocity (v_o) and the molecular diffusivity (D) emerges as a determining factor in the prevalence of the transport phenomenon over diffusion. It is a dimensional factor (v_o/D) whose unit is the inverse of a length (m^{-1}) with no apparent physical meaning. Indeed, the unknowns (horizontal and transversal extensions of the contamination plume or global isoline pattern) are also dependent on a time factor. The expression of this dependence allows us to separate the effect of the quotient (v_o/D) and the product $v_o\tau$ (τ is time) to substantially simplify the universal representation.

Finally, the numerical simulations are performed with SEAWAT [25], a widely recognised and reliable software package used for theoretical [26,27] and practical-technical purposes [28,29]. A large number of numerical simulations has allowed us to develop a wide set of universal abacuses that provide relevant information on the space-time dynamics of the contamination plume. The physical variables of the problem cover a range of solutions which are fully representative of all the cases that may occur in practice, including those asymptotic cases of negligible diffusion versus advection, and vice versa. The limit of negligible diffusivity allows us to represent new universal curves to characterise pure advective processes. The use of universal graphs is illustrated with examples.

The hypotheses assumed in the physical model are the following: (i) velocities small enough to neglect the effect of hydrodynamic dispersion, (ii) the Darcy flow (negligible inertial forces and laminar flow), (iii) isotropic and uniform hydraulic conductivity, (iv) viscosity independent of concentration, (v) two-dimensional geometry and the absence of gravitational effects, (vi) isothermal conditions, (vii) water-saturated porous media, (viii) the absence of sources and/or sinks for flow, (ix) fully miscible single-phase fluid with negligible compressibility for both the fluid and the porous matrix (constant porosity) and (x) non-reactive solute transport.

2. Nomenclature

a_o	numerical constant (dimensionless)
c	fluid concentration (kg/m^3)
c_o	constant fluid concentration at a certain point (kg/m^3)
D	molecular diffusivity coefficient (m^2/d)
g	gravity acceleration (m/s^2)
h	hydraulic head (energy per unit of specific weight)
H	height of the domain (m)
h_l, h_r	hydraulic head at the left and right boundaries, respectively
k	permeability (m^2)
K	hydraulic conductivity (m/d)
l	length (m)
L	horizontal length of the domain (m)
L	length of the domain (m)
H	vertical length of the domain (m)
p	pressure at the point (Pa)
q	specific discharge or Darcy velocity (m/d)
v	fluid velocity (m/d)
v_o	regional Darcy velocity (m/d)
x_o	location of the constant concentration point (m)
x, y	horizontal and transversal coordinates (m)
θ	porosity (dimensionless)
μ	dynamic viscosity (kg m^{-1}s^{-1})
π	dimensionless groups
ρ	fluid density (kg/m^3)
ρ_o	density of the fluid with zero concentration (Kg/m^3)
τ	time (days)
τ_c	time factor (days)
Ψ	identifies a general unknown function
$\sim$	denotes the same order of magnitude
$\equiv$	denotes equivalence for dimensionless groups

Subscripts

1,2,...	used to identify the dimensionless groups
x, y	related with coordinate direction
left	refers to the left of the focus

Superscripts

$\prime$	denotes the dimensionless variable
*	denotes characteristic values of a given quantity

3. Mathematical Model

Under the conditions mentioned at the end of Section 1, the mathematical model is composed of the mass conservation equations for the fluid and for the contaminant [30,31]. In mathematical terms, these read as follows:

$$-[\nabla \cdot (\rho \mathbf{q})] = \theta \frac{\partial (\rho)}{\partial \tau} \quad (1)$$

$$[\nabla \cdot D(\nabla c)] - \nabla \cdot (\mathbf{v}c) = \frac{\partial c}{\partial \tau} \quad (2)$$

The relationship between the specific discharge and the actual fluid velocity is given by $\mathbf{q} = \mathbf{v}\theta$, with θ being the porosity of the medium. Thanks to the coupling, according to Equation (2), the velocity of the fluid in the porous medium causes continuous redistribution of the concentration and, therefore, of the density of the solution, which affects its movement through Equation (1).

Equation (1) can be solved in terms of hydraulic potential or pressure by means of Darcy's law (i.e., a stationary momentum equation), expressed by Muskat [32] as follows:

$$\mathbf{q} = -\frac{k}{\mu}\nabla p \quad (3)$$

Since the global pressure distribution depends on the global density distribution, the q_x and q_y components of the specific discharge are affected by spatial variations in density. In terms of the hydraulic head, $h = \frac{P}{\rho g}$, and assuming that the fluid viscosity is not dependent on concentration, Darcy's equation is written as $\mathbf{q} = -K\frac{\partial h}{\partial x}$, with K being hydraulic conductivity, a physical parameter of the porous medium, which is decisive for velocity, included in the mathematical model.

We consider a two-dimensional water-saturated rectangular domain with a 2:1 aspect ratio, according to the schematics of Figure 1.

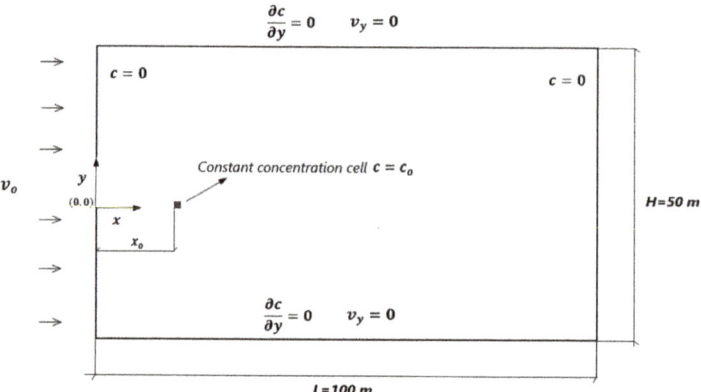

Figure 1. Porous domain and boundary conditions.

The parameters of the soil that will later be shown to influence the values of the monomials have values of between 2.5 and 20 m/d (equivalent to fine sand or silty sand) for the hydraulic conductivity K, an effective porosity of 0.15 and a molecular diffusivity coefficient D ranging between 0.0003 and 0.0012 m^2/d (representative of different salt species).

A zero-concentration regional flow, with a constant velocity v_0 and constant density ρ_0, enters from the boundary on the left due to a fixed hydraulic potential drop between the left and right-hand boundaries. The upper and lower horizontal boundaries are impermeable (zero normal velocity) and do not let the concentration go through by diffusion (zero normal concentration gradient). In the position (x = x_o, y = 0), which we will call focus, there is a cell of constant concentration c_o, representing the source of contaminant. The boundary conditions can be summarised as follows:

Hydraulic head:

$$h_{(x=0,y,\tau)} = h_l, \quad h_{(x=L,\,y,\tau)} = h_r, \quad \frac{\partial h}{\partial y}\bigg|_{(y=\pm\frac{H}{2},\tau)} = 0 \quad (4)$$

Concentration:

$$c_{(x=x_o,\,y=y_o,\tau=0)} = c_o, \quad c_{(x,y,\tau=0)} = 0 \quad \text{(at the rest of the domain)} \quad (5)$$

Under these conditions, we will study the non-stationary pattern of concentrations when H is large compared to the transversal dimension of the contamination plume. The isolines of the concentration pattern are deformed ellipsoids that widen in the horizontal and vertical directions of the domain due to the coupled processes of diffusion and advection. These ellipsoids surround the focus and finally keep its left end in a fixed position. As long as the contamination plume does not reach the right and upper (or lower) boundaries of the domain, the problem is not stationary (except for the small region to the left of the focus) so we expect the vertical and horizontal extent of the plume to depend on the parameters of the problem and on time. The interaction between the advection and diffusion phenomena is not easily predictable, even though each equation is linear in its unknown, as the position of the plume and isolines are complex non-linear functions of the velocities, pressure, and their gradients.

In the following section, we rewrite this mathematical model in its dimensionless form and, after some algebraic manipulations and the application of the Pi theorem, the dependences between the variables of interests and the physical parameters of the problem are obtained.

4. Dimensionless Groups

4.1. Deduction of the Dimensionless Groups

Before deriving the dimensionless groups for the coupled diffusion and advection problems or for purely advective problems, it is worth mentioning the easier problem of pure diffusion for which analytical solutions have been established [33]. Some comments should be made regarding the applied protocol.

The procedure followed to deduce the dimensionless groups that rule a given problem based on the Pi theorem consists of reducing the governing equations to their dimensionless normalised forms and obtaining such groups by comparing the coefficients that multiply the derivatives of each addend of the equation. These coefficients, which are physically or dimensionally homogeneous, are of the same order of magnitude in the normalised equation so their ratios are the dimensionless groups that are sought. In recent years, a formal procedure of nondimensionalising has been proposed and successfully applied to many complex problems in different areas to obtain new universal solutions. For example, it is worth citing the consolidation problem in soil mechanics, flow and (heat or solute) transport in porous media, and a variety of mechanical problems, all of which are coupled and nonlinear [7,8,34]. It should be noted that since the dimensionless numbers for most engineering or scientific problems are already found in the literature, the former procedure is generally avoided, and classical known numbers are directly used in the Pi theorem. This, however, generally leads to poorer predictive capabilities compared to the approach presented here.

The first step carried out to deduce the dimensionless governing equation is to choose a suitable list of reference quantities to define the dependent and independent dimensionless variables. These are chosen either from the input parameters of the problem or (as in this case) by introducing suitable unknowns whose order of magnitude will later be deduced as a consequence of the application of the Pi theorem. The only criterion with which these reference quantities should be chosen is that the range of values of the normalised variables is as bounded as possible to the interval [0, 1]. This criterion allows the derivatives of these variables to be averaged to a value of the order of the unit throughout the entire length of the scenario. In addition, the discrimination must be applied [35,36]. This means that vector variables such as position or velocity must be made dimensionless by means of different references according to their spatial direction. In this way, dimensionless groups commonly called "aspect ratios" or "form factors" do not necessarily emerge directly as independent groups.

As in other problems of similar complexity, the procedure requires good physical knowledge of the problem and some experience to find the unknowns introduced as references, their order of magnitude (by application of the Pi theorem) and their exact solution through the universal curves obtained by a large number of precise numerical simulations. This procedure has some advantages over the classical dimensional analysis. Firstly, the dimensionless groups emerging in the process, some containing unknowns and others without unknowns, are formally obtained and the relationship between them constitutes the direct application of the Pi theorem. Secondly, the groups have a unitary order of magnitude since they are obtained as balances between the addends of the governing equation. Groups of an order of magnitude higher (or lower) than unity, can be neglected in the governing equation, which simplifies the problem. Thirdly, this procedure incorporates the dimensionless physical parameters into the deduced monomials, thus reducing the global number of groups and making the characterisation more precise. The resulting groups are the proper parameters to represent universal solutions.

4.2. Coupled Advection and Diffusion Case

We assume isotropic molecular diffusivity, the porosity θ to be constant, and we neglect the transversal velocity (v_y) and its spatial derivatives (since they are of an order of magnitude much lower than the regional velocity and its changes). Therefore, it follows that $v_x \approx v_o$ and $\frac{\partial v_x}{\partial x}$ is negligible, and Equation (2) in rectangular 2D coordinates is reduced to:

$$\frac{\partial c}{\partial \tau} = -v_o \frac{\partial c}{\partial x} + D \frac{\partial^2 c}{\partial x^2} + D \frac{\partial^2 c}{\partial y^2} \qquad (6)$$

Dimensionless variables for x, y, c, v_x, and τ are defined (discriminately) as

$$x' = \frac{x}{l_x^*} \quad y' = \frac{y}{l_y^*} \quad c' = \frac{c}{c_o} \quad \tau' = \frac{\tau}{\tau^*} \qquad (7)$$

In these definitions, c_o and v_o are known parameters, while l_x^*, l_y^* and τ^* are unknown parameters related to each other. For a given time characteristic (τ^*), l_x^* and l_y^* define the extension of the solute plume from the constant concentration point. For example, they can be defined as the region in which the concentrations are above a certain percentage of c_o. With this choice, dimensionless variables may be considered normalised since they vary in the range [0, 1]. Substituting (7) in (6), the last equation becomes dimensionless:

$$\frac{c_o}{\tau^*}\left(\frac{\partial c'}{\partial \tau'}\right) = -\frac{c_o v_o}{l_x^*}\left(\frac{\partial c'}{\partial x'}\right) + \frac{c_o D}{l_x^{*2}}\left(\frac{\partial^2 c'}{\partial x'^2}\right) + \frac{c_o D}{l_y^{*2}}\left(\frac{\partial^2 c'}{\partial y'^2}\right) \qquad (8)$$

Assuming the derivatives between brackets to be of the order of one, four coefficients are found to describe the solution (or patterns) of the equation in the domain defined by l_x^*, l_y^* and time τ^*. These are:

$$\frac{c_o}{\tau^*}, \quad \frac{c_o v_o}{l_x^*}, \quad \frac{c_o D}{l_x^{*2}}, \quad \frac{c_o D}{l_y^{*2}} \qquad (9)$$

These coefficients have to be of the same order of magnitude since their terms in the equation balance each other. The independent ratios between these coefficients, chosen for suitability, are the dimensionless groups. We choose them as follows:

$$\pi_1 = \frac{\frac{c_o D}{l_y^{*2}}}{\frac{c_o D}{l_x^{*2}}} = \frac{l_x^{*2}}{l_y^{*2}} \equiv \frac{l_x^*}{l_y^*} \quad \pi_2 = \frac{v_o l_x^*}{D} \quad \pi_3 = \frac{v_o \tau^*}{l_x^*} \tag{10}$$

To characterise the solution in the better way, each of the unknowns l_x^*, l_y^* and τ^* must appear only in one group, unless this is not possible (which is the case of l_x^*). Therefore, group π_3 can be substituted for the product of groups π_2 and π_3. This finally leads to the alternative solution:

$$\pi_1 = \frac{l_y^*}{l_x^*} \quad \pi_2 = \frac{v_o l_x^*}{D} \quad \pi_3 = \frac{v_o^2 \tau^*}{D} \tag{11}$$

Group π_1 is an aspect ratio of the domain and provides information about the relation between l_x^* and l_y^*. The other groups allow us to relate time and the horizontal length of the transition or salt contaminated region measured from the focus. According to the Pi theorem, the solution $\pi_2 = \Psi(\pi_3)$ leads to:

$$l_x^* = \frac{D}{v_o} \Psi\left(\frac{v_o^2 \tau^*}{D}\right) \tag{12}$$

with Ψ an unknown function of its arguments. Writing the solution in the following way:

$$l_x^* = \left(\frac{D}{v_o}\right) \Psi\left\{\left(\frac{v_o}{D}\right) v_o \tau^*\right\} \tag{13}$$

we obtain interesting and useful information. As expected, each iso-concentrated line of the pattern (defined by the dimensional value c') depends on time (τ^*), regional velocity (v_o) and molecular diffusivity (D), and for the same values of $\frac{v_o}{D}$ and $v_o \tau^*$, the concentration isoline is the same. Equation (13) shows detailed information about this kind of dependence. Firstly, keeping the ratio $\frac{v_o}{D}$ constant, the patterns for each c' are the same for all the times so that $v_o \tau^*$ is also constant. This means that if we take two scenarios, the first with the pair of values (v_o, D) and the second with the pair $(a_o v_o, a_o D)$, the patterns of dimensionless isolines (c') at a given time τ^* for the first scenario is the same as that of the second one at time $\frac{\tau^*}{a_o}$. In this way, scenarios of the same value of the ratio $\frac{v_o}{D}$, for a given time τ^* contain the information from the patterns corresponding to all the regional velocity values. This allows us to depict a set of abacuses, one for each value of the ratio $\frac{v_o}{D}$, in which the extension of each concentration isoline, $l_{x(c')}^*$ may be represented as a function of time τ^*, choosing c' as the parameter of the abacus. To use this information, the time has to be suitably corrected according to the value of the regional velocity. In the following section related to the construction of the universal curves, this will be further clarified.

With regards to the characteristic transversal length, the group $\pi_1 = \frac{l_y^*}{l_x^*}$ allows us to write $l_y^* \sim l_x^*$, which is equivalent to the dependence

$$l_y^* = \left(\frac{D}{v_o}\right) \Psi\left\{\left(\frac{v_o}{D}\right) v_o \tau^*\right\} \tag{14}$$

with Ψ being a new function of its argument. The same discussion made above for l_x^* applies here.

Finally, we can expect that the part of the pattern which is located to the left of the focus (x_o, $y = 0$) will be a steady-state pattern after a relatively short time characteristic for which each isoline defined by c' is located at a characteristic distance from $(x_o, 0)$, $l_{x(left)}^*$. Indeed, deleting the addend $\frac{\partial c}{\partial t}$ from Equation (6) and assuming changes only in the x coordinate, the only emerging dimensionless group is $\pi = \frac{v_o l_{x(left)}^*}{D}$. Then, the solution is given as:

$$l_{x(left)}^* = \left(\frac{D}{v_o}\right) \tag{15}$$

The solutions provided by Equations (13)–(15), and the universal curves that make use of them, constitute an important management tool since they provide the level of global contamination of the domain and its gradation, at every instant.

4.3. Pure Advection

For this scenario, with the same assumptions as the coupled problem and using the same dimensionless variables defined in (7), Equation (2) is reduced to:

$$-\frac{c_o v_o}{l_x^*}\left(\frac{\partial c'}{\partial x'}\right) = \frac{c_o \partial c'}{\tau^* \partial t'} \tag{16}$$

which leads to two coefficients, $\frac{c_o v_o}{l_x^*}$ and $\frac{c_o}{\tau^*}$ and one single dimensionless group $\pi_1 = \frac{v_o \tau^*}{l_x^*}$. This results in the following dependence:

$$l_x^* = v_o \tau^* \tag{17}$$

As in the former analysis, each isoline defined by its dimensionless concentration c' has its particular solution. So, $l_{x(c')}^*$ is the horizontal extension at time τ^*, going from the furthest point of the isoline to the constant concentration position imposed by the inner boundary condition. This length is proportional to the regional velocity but changes from one isoline to another

$$\frac{l_x^*}{\tau^*} = v_o \, \Psi(c') \tag{18}$$

The function $\Psi(c')$ is universal and may be depicted by a single numerical simulation. The region of concentration is a slender arrow in which the advancing fronts of the lower concentration isolines are ahead of the higher concentration fronts. The distance between the fronts of any pair of isolines c_1' and c_2', increases with time and depends on their concentrations according to the expression:

$$l_{x\,(c_1')}^* - l_{x\,(c_2')}^* = v_o \tau^* \left\{\Psi(c_1') - \Psi(c_2')\right\} \tag{19}$$

This result can also be conveniently represented by a universal abacus.

4.4. Perspectives

It is interesting to discuss here what can be deduced from the previous results in finite domains and large time periods in which contamination reaches the boundaries of the scenario. Although this is not the main subject of this work, the treatment of finite scenarios adds new dimensionless groups and makes the solution more complex. For example, a finite scenario in the horizontal direction (but very large in the transversal direction) introduces the extension L in the dimensionalising process, giving rise to a new dimensionless group. In contrast, if the scenario is finite in the transversal direction (and very extensive in the horizontal one), the extension H of the domain must be considered, which also gives rise to the appearance of a new group. Finally, if the scenario is finite in both directions, the introduction of L and H would create two new dimensionless groups.

In this way, the introduction of any new condition like a sink or source (an injection or abstraction well), or reactive transport (degradation tracers), increases the number of monomials and, thus, the universal solutions are translated into a set of abacuses in a way that each one is set according to a specific value of a selected dimensionless group. A scenario similar to the one here presented but introducing an abstraction well located at a specific distance would be an interesting case for further research. From the practical point of view, this represents an often-used procedure in field testing to estimate the physical parameters of soils. The use of universal abacuses will contribute to obtaining these parameters by means of an inverse protocol.

5. Verification by Numerical Simulations

The numerical simulations have been carried out using the programme SEAWAT V.4 [37]. The following figures illustrate some of the simulated scenarios created to obtain the universal graphs. For example, Figure 2 shows the iso-concentration patterns in a large 2D scenario with a focus or constant concentration point located near the left-hand boundary, in which advection and diffusion effects are coupled. The data are: $v_o = 0.0006$ m/d, $D = 0.0006$ m^2/d and $c_o = 1000$ kg/m^3. The concentrations of the isolines are 10, 200... kg/m^3, which correspond to 10%, 20%... of c_o. The vertical extension of each pattern has been trimmed to reduce the size of the figure. As shown, the patterns progressively extend the contaminated region in vertical and horizontal directions. The deformation of the isolines (more pronounced for small concentration isolines) with respect to the circular shape that they would have with pure diffusion is due to the advective effect.

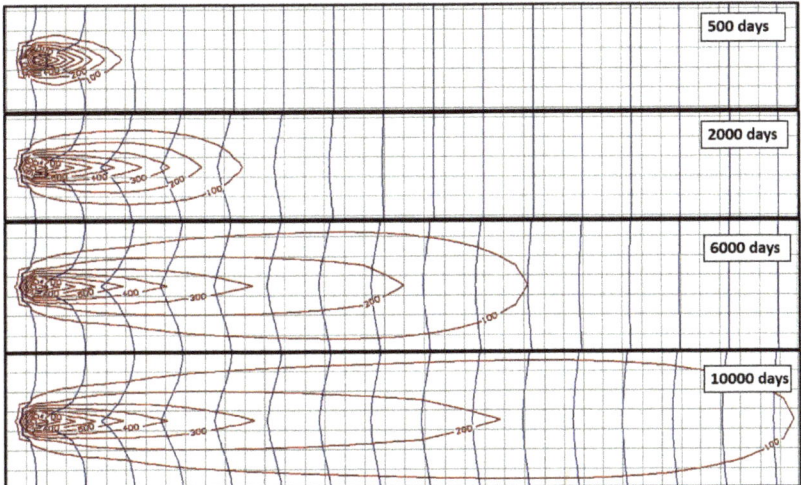

Figure 2. Iso-concentration patterns for different times in a large scenario with advection and diffusion. t = 500, 2000, 6000 and 10,000 days. The blue lines and grey network represent piezometric lines and cell extension (1 m^2), respectively.

In contrast, the patterns for a scenario with only the advection effect is shown in Figure 3. The data are: $v_o = 0.05$ m/d, $D = 0$ and $c_o = 1000$ kg/m^3. In this scenario, the focus has been converted into a vertical segment to better appreciate the progress of the concentration fronts (of equal length to that of the segment) of each isoline. It can be seen that the velocities of each front depend on the concentration of the isoline, with values greater than the regional velocity (v_o) for the small concentration isolines and lower for those with higher concentrations.

Figures 4 and 5 show the evolution of the isoline fronts in x direction (for $v_o = 0.05$ m/d) when $D = 0$ and advection dominates. The first one shows the temporal evolution of the width of the region of variable concentration; that is, the area affected by significant contamination, defined as the distance between the isolines of concentration 100 and 900 (10% and 90% of c_o, respectively). The solution for an identical scenario with a $2c_o$ concentration has been superimposed. The separation between isolines of the same relative concentration (10% and 90%) has the same evolution. The different lines depicted in Figure 5 establish the distance to the focus of any iso-concentration line (defined as a percentage in respect to the contamination source) as a function of time. As previously mentioned, the velocity of each front is constant but increases as the concentration diminishes; that is, the variation in the inclination of different lines indicates that low concentration isolines spread faster than high concentration values.

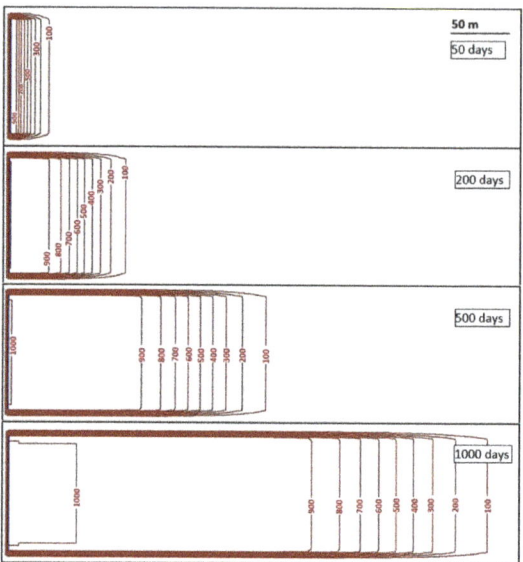

Figure 3. Iso-concentration patterns for times 50, 200, 500 and 1000 days in a large horizontal scenario due to advection (without diffusion).

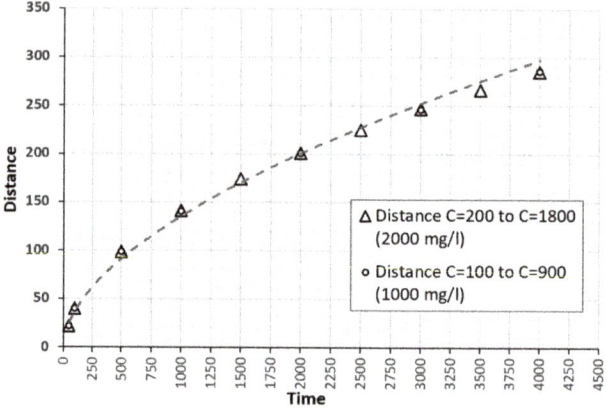

Figure 4. Distance between the fronts of concentrations $0.1c_o$ and $0.9c_o$ as a function of time. c_o = 1000 and 2000 kg/m^3, v_o = 0.05 m/d.

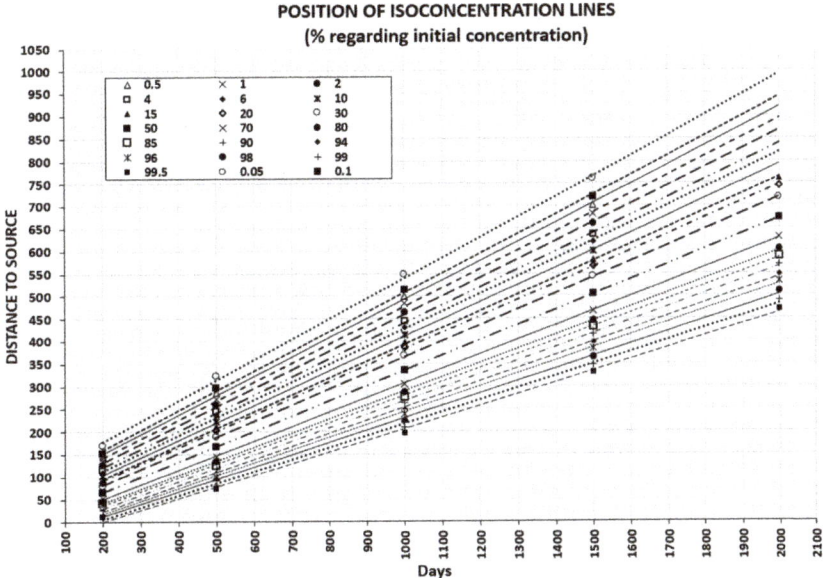

Figure 5. Location of the isoline fronts as a function of time. $v_o = 0.05$ m/d.

Continuing with the illustration of the coupled diffusion and advection effects, Figure 6a–c show the typical concentration profiles for different ratios of v_o/D and $c_o = 1000$ kg/m^3. These figures can be contemplated as a different and more complete configuration of Figures 2 and 3; the x-y axis distribution of concentration for specific times is now plotted on the vertical axis of concentration (z coordinate) in Figure 6a–c. Figure 6a is the case of no diffusion, with c = 1000 kg/m^3 and $v_o = 0.0006$ m/a, a velocity that corresponds to the front of the isoline of concentration $0.5c_o$. The profiles gradually decrease their negative slope, increasing the distance from the small concentration isolines to those of higher concentration. Below the legend, the spreading of contamination in the x-y coordinates corresponding to the curve t = 6000 days is represented. Positions of concentrations 900 and 100 are represented as normalised with the 0.9 and 0.1 values in coordinate z (vertical axis in the chart). Figure 6b is a comparison between the profiles after 80 days for a pure diffusion process (D = 0.0006 m^2/d), pure advection ($v_o = 0.0006$ m/d) and coupled advection-diffusion (D = 0.0006 m^2/d, $v_o = 0.0006$ m/d). The profile corresponding to the combination of both effects presents two marked inflection points, depending on the v_o/D ratio. This is a consequence of the coupling between both effects.

Finally, Figure 6c shows the profiles for a coupled problem with D = 0.0006 m^2/d and $v_o = 0.0003$ m/d for different times. For this ratio, $v_o/D = 2$ as well as for ratios that are larger than unity, the profiles present an interesting result. If we consider the same concentration value in all the profiles (for example 0.3, which corresponds to c = 300 kg), the point of the domain with this concentration moves to the right more and more slowly until it stops at a certain distance from the focus at a certain time. This distance is set by the curve corresponding to the last time studied. The lower the chosen concentration, the greater the distance and time at which the concentration is fixed, and vice versa. As will be seen in the next section, similar phenomena occur in the transversal direction. As in Figure 6a, an x-y coordinate image of iso-concentration lines for t=14,000 days has been included below the legend so it can be compared with the concentration values of the same line in the z coordinate (or vertical axis in the chart).

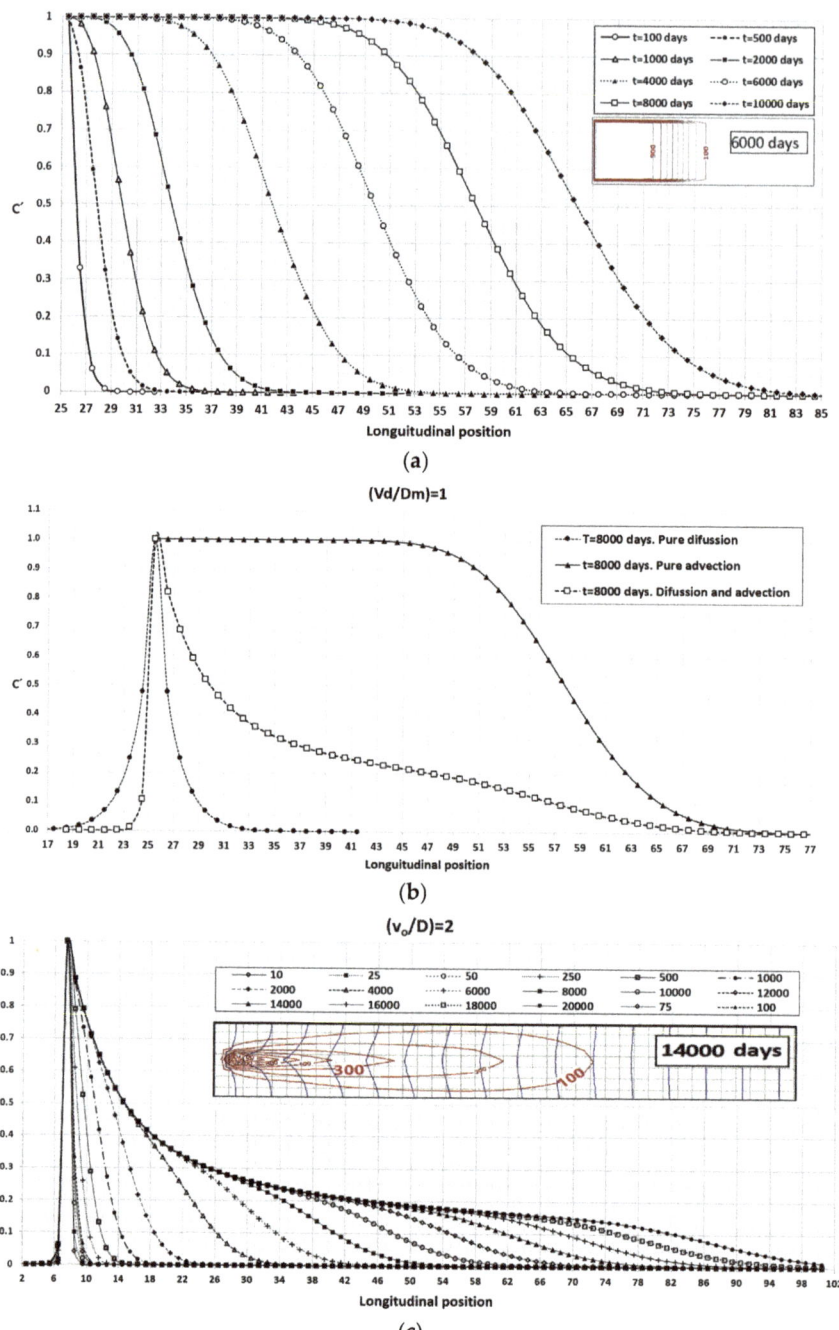

Figure 6. Concentration profiles. (**a**) Only advection, c = 1000 kg/m^3 and v$_o$ = 0.0006 m/d. (**b**) For τ = 80 days, pure diffusion (D = 0.0006 m^2/d), pure advection (v$_o$ = 0.0006 m/d) and advection-diffusion (D = 0.0006 m^2/d, v$_o$ = 0.0006 m/d). (**c**) Coupled problem, D = 0.0003 m^2/d and v$_o$ = 0.0006 m/d.

All these results, which have been qualitatively described, will be represented by universal curves in the following section, after a large number of numerical simulations have been made. All this will verify the mathematical dependences derived through the non-dimensionalisation process followed in Section 4.

6. Universal Solutions

6.1. Scenarios with Only Advective Flow

For these scenarios, the only universal curve comes from expression (18), which we rewrite in the form of:

$$\frac{l_x^*/\tau^*}{v_o} = \frac{v_{c'}}{v_o} = \Psi(c') \qquad (20)$$

This relationship represents the ratio between the velocity of the front of the dimensionless concentration c', and the regional velocity v_o. c' is the ratio between the actual concentration of the front (c) and the constant concentration of the source. Figure 7 shows the universal dependence $\frac{v_{c'}}{v_o}$, or $\Psi(c')$, on c'. The $c' = 0.5$ front travels at the regional Darcy velocity. For a wide range of concentrations, $c' \in [0.15–0.85]$, the velocities of the fronts deviate very little from the value of v_o (less than 10%). Only for very small or very large values of c', the velocities are somewhat higher (up to $1.3v_o$) or lower ($0.7v_o$) than v_o, respectively.

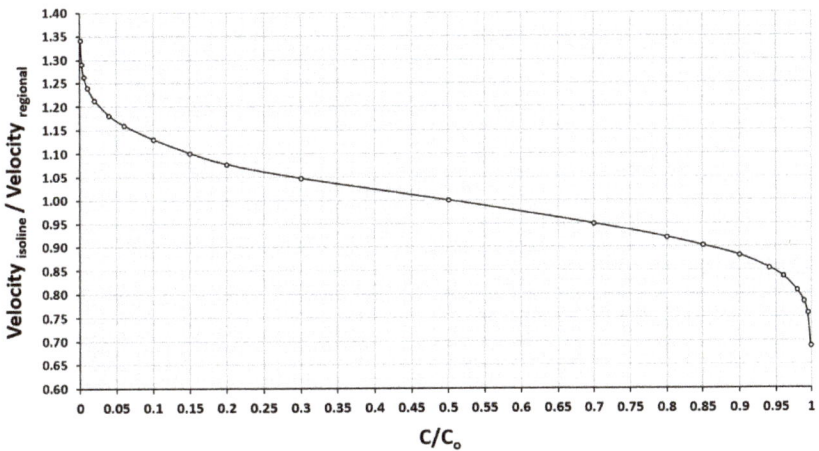

Figure 7. Universal dependence $\frac{v_{c'}}{v_o} = \Psi(c')$ (only advective effect).

6.2. Scenarios with Coupled Diffusion and Advection

The universal curves presented in this sub-section derive from the expressions (13)–(15). The first expresses the longitudinal extension measured from the source of contamination of each concentration isoline, the second, the maximum transversal extension, and the third, the horizontal extension of the polluted region to the left of the focus. Each isoline is characterised by its dimensionless concentration, taking that of the focus as a reference. The way in which these characteristic lengths depend on the physical parameters of the problem, v_o and D, allows us to organise the universal curves in the form of an abacus, with a specific value for the relationship v_o/D for each one. Every curve in Figure 8 represents the horizontal extension (vertical axis) of the isolines of dimensionless concentration c' ($l_{x(c')}^*$) against a time factor τ_c (horizontal axis), for $\frac{v_o}{D} = 1$. The simulations to determine the function of the dependence (13) have been carried out for values $v_o = 0.0006$ m/d and $D = 0.0006$ m^2/d. In this way, the time factor (τ_c) is related to real time (τ^*) by

$$\tau_c = \left(\frac{0.0006}{v_o}\right)\tau^* \qquad (21)$$

Thus, for the same ratio ($\frac{v_o}{D}=1$) and different values of v_o and D, the extension of any isoline ($l^*_{x(c')}$) associated with a real time τ^* is obtained by entering the abscissa axis with the value τ_c given by expression (21). For greater detail, the time factor has been separated into two intervals with ranges [100–2000 days] and [10–100 days], Figure 8a,b, respectively. For longer times, the contaminated regions stabilise for successively increasing values of c'. The lower the concentration is, the sooner the stabilisation occurs. Figure 8c,d show the extensions of some isolines for times of [2000–20,000 days] and [2000–8000 days], respectively.

The abacus corresponding to the monomials $\frac{v_o}{D} = 2, 5$ and 10 are shown in Figures 9–11, respectively, with details similar to the previous ones. Numerical solutions have been obtained by retaining the regional velocity ($v_o = 0.0006$ m/d) and changing the diffusivity to successive values $D = 3\cdot10^{-4}, 1.2\cdot10^{-4}$ and $3\cdot10^{-4}$ m^2/d. In this way, the time factor is related to real time through the same expression as the former abacus (21).

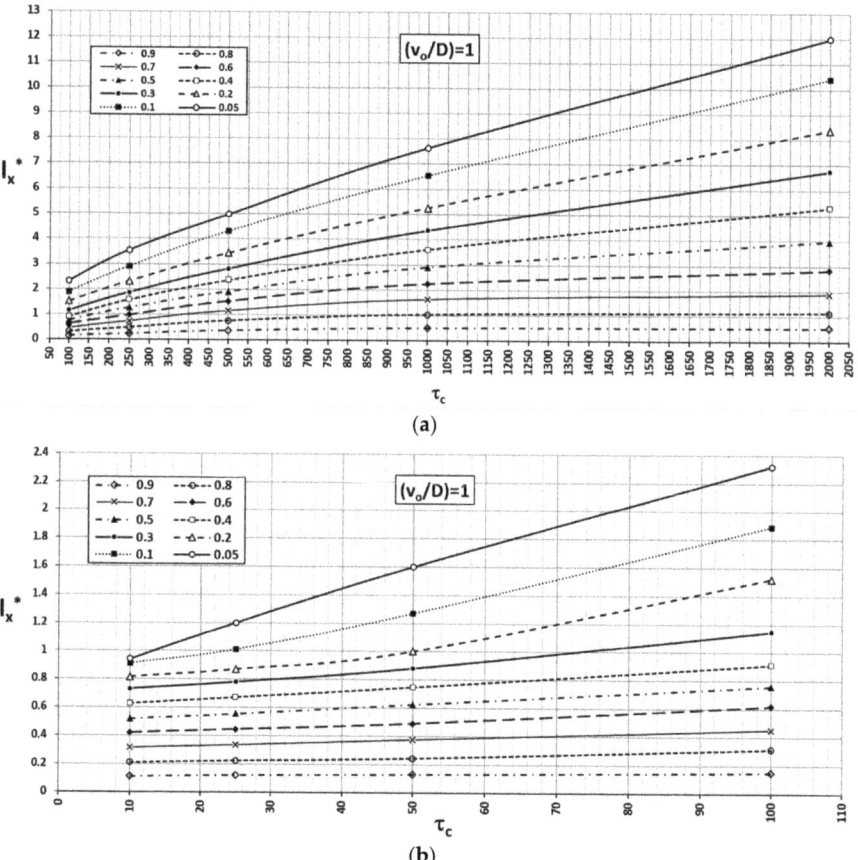

Figure 8. Cont.

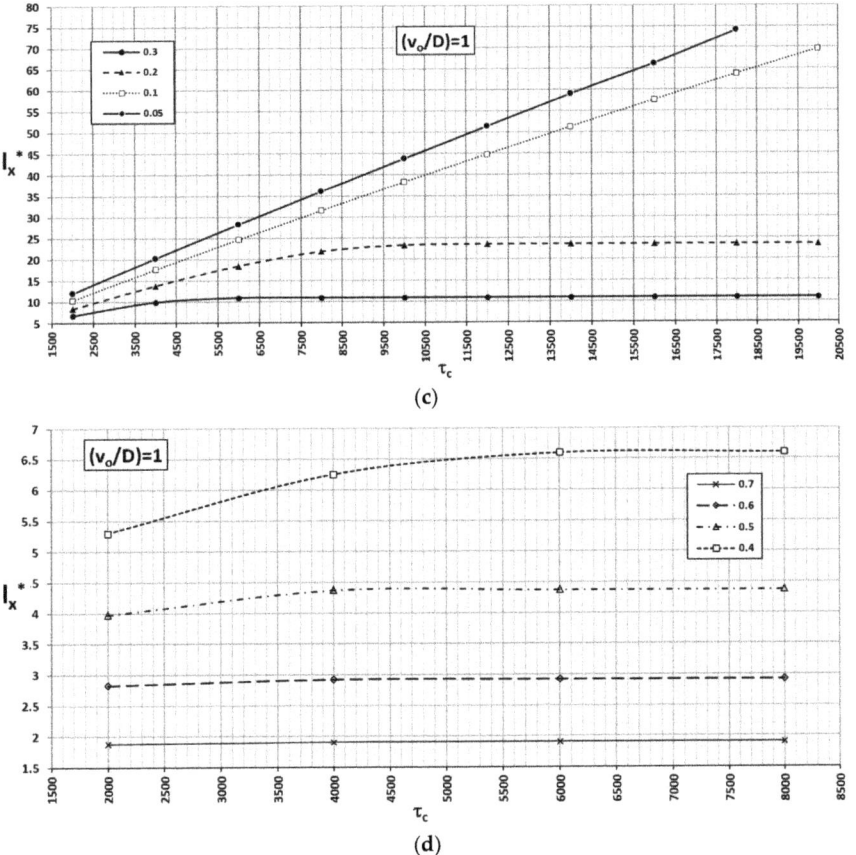

Figure 8. Universal curve $l^*_{x(c')}$ as a function of time factor τ_c. Parameter of the abacus $\frac{v_o}{D} = 1$, $v_o = 0.0006$ m/d. (**a**) $\tau_c \in$ [100–2000 days], (**b**) $\tau_c \in$ [10–100 days], (**c**) $\tau_c \in$ [2000–20,000 days], (**d**) $\tau_c \in$ [2000–8000 days].

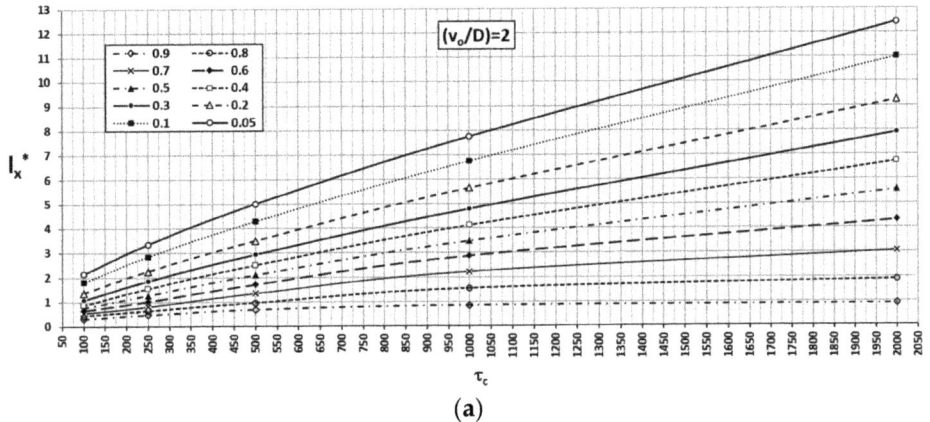

Figure 9. *Cont.*

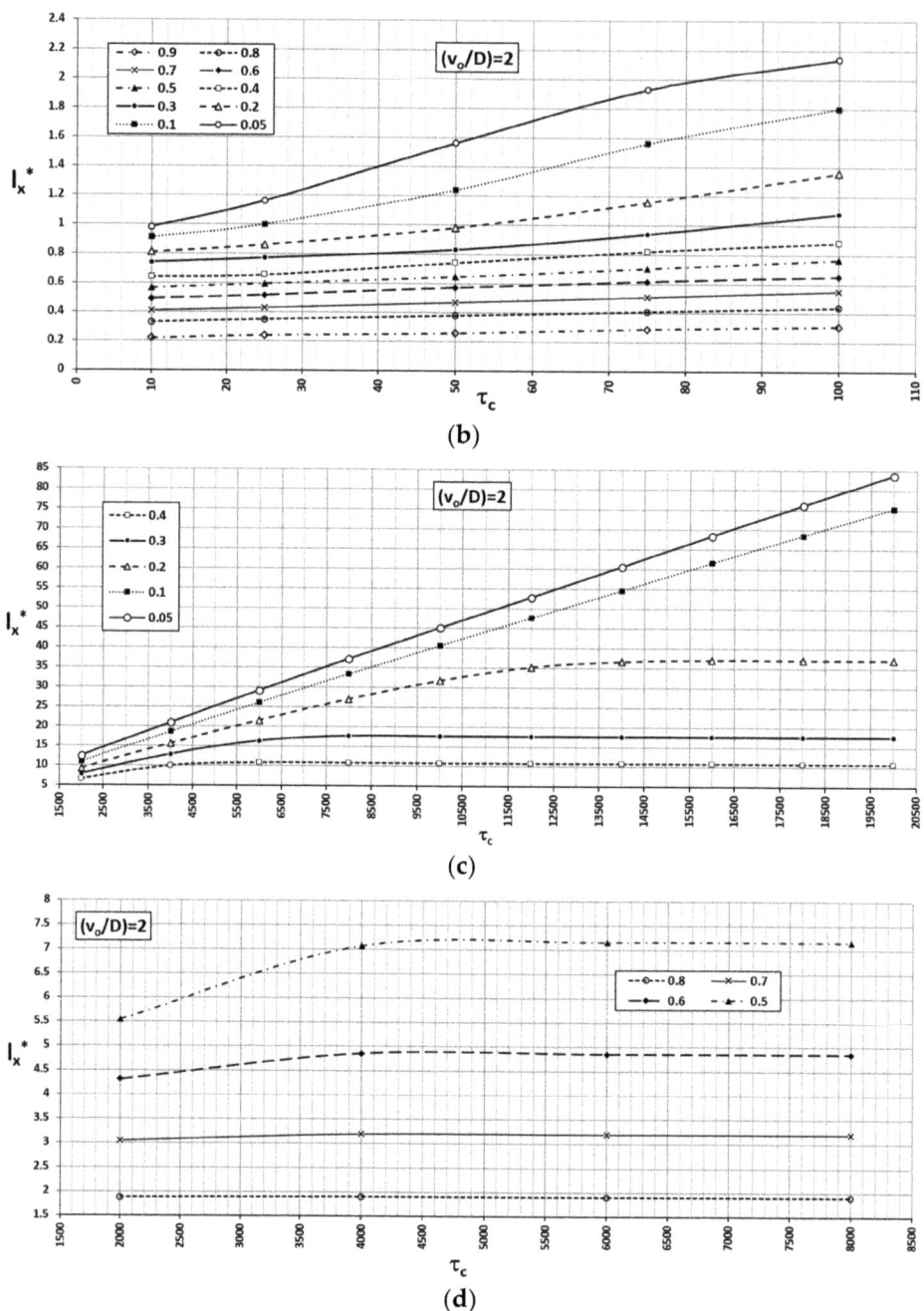

Figure 9. Universal curve $l^*_{x(c')}$ as a function of time factor τ_c. Parameter of the abacus $\frac{v_o}{D} = 2$. $v_o = 0.0006$ m/d. (**a**) $\tau_c \in [100\text{–}2000 \text{ days}]$, (**b**) $\tau_c \in [10\text{–}100 \text{ days}]$, (**c**) $\tau_c \in [2000\text{–}20{,}000 \text{ days}]$, (**d**) $\tau_c \in [2000\text{–}8000 \text{ days}]$.

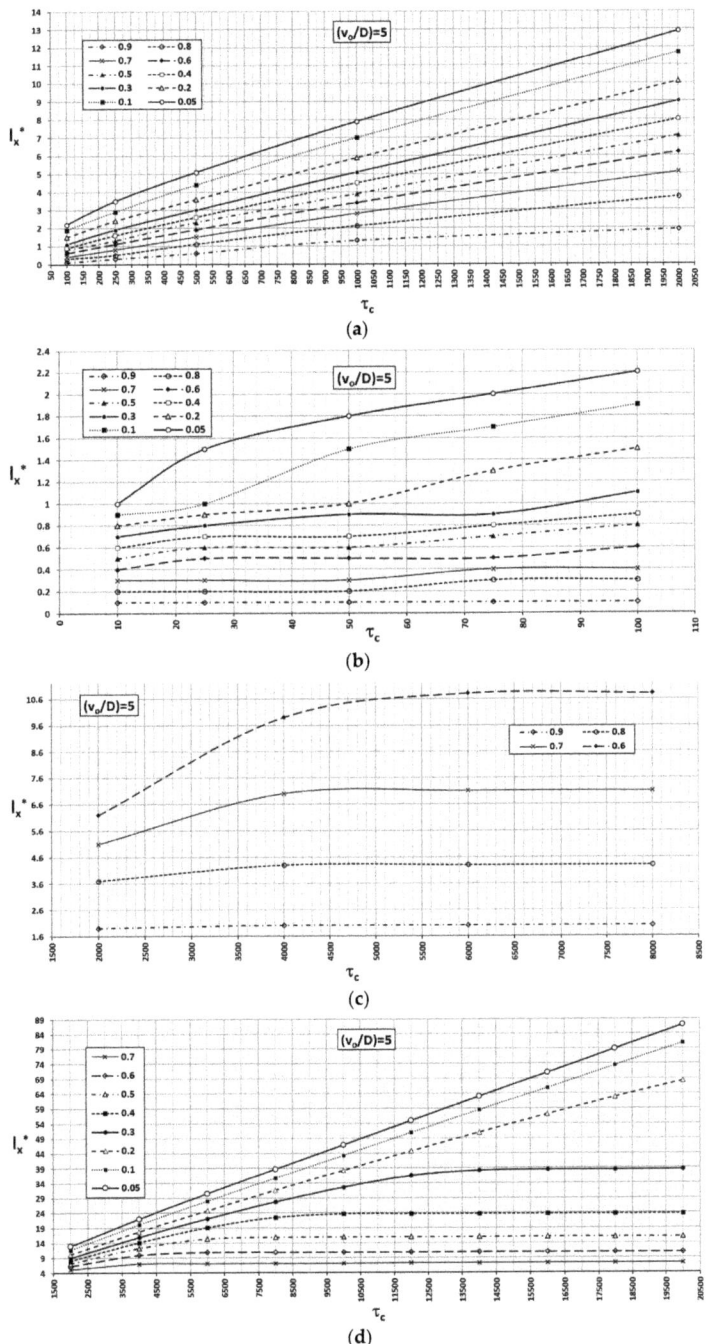

Figure 10. Universal curve $l^*_{x(c')}$ as a function of time factor τ_c. Parameter of the abacus $\frac{v_0}{D} = 5$. $v_0 = 0.0006$ m/d. (**a**) $\tau_c \in [100–2000$ days], (**b**) $\tau_c \in [10–100$ days], (**c**) $\tau_c \in [2000–8000$ days], (**d**) $\tau_c \in [2000–20,000$ days].

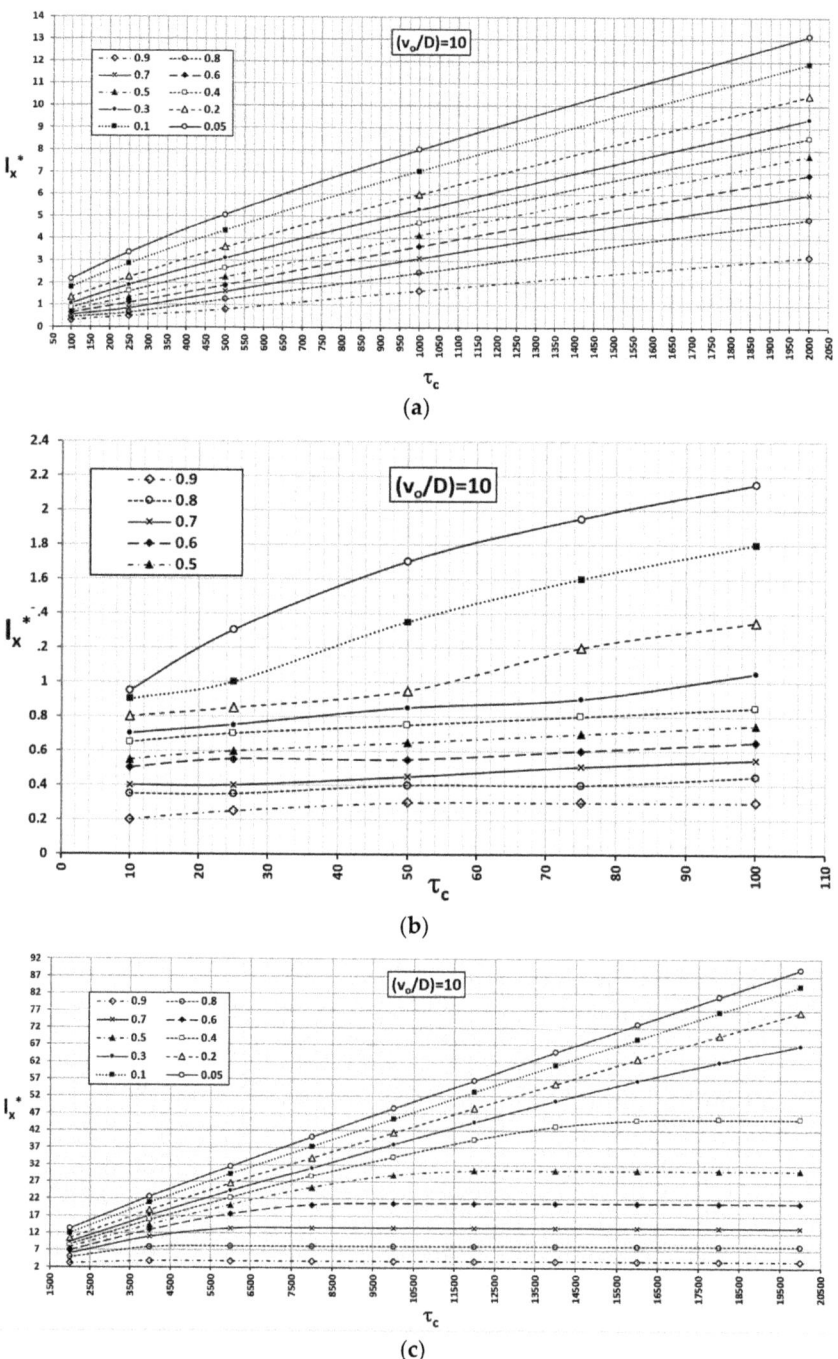

Figure 11. Universal curve $l^*_{x(c')}$ as a function of time factor τ_c. Parameter of the abacus $\frac{v_o}{D} = 10$. $v_o = 0.0006$ m/d. (a) $\tau_c \in$ [100–2000 days], (b) $\tau_c \in$ [10–100 days], (c) $\tau_c \in$ [2000–20,000 days].

Similarly, the abacus corresponding to values $\frac{v_o}{D} = 0.5$ and 0.1, made with $v_o = 0.0003$ and 0.00006 m/d, are shown in Figures 12 and 13, respectively. Accordingly, the time factors and real time are related by:

$$\tau_c = \left(\frac{0.0003}{v_o}\right)\tau^*, \quad \tau_c = \left(\frac{0.00006}{v_o}\right)\tau^* \quad (22)$$

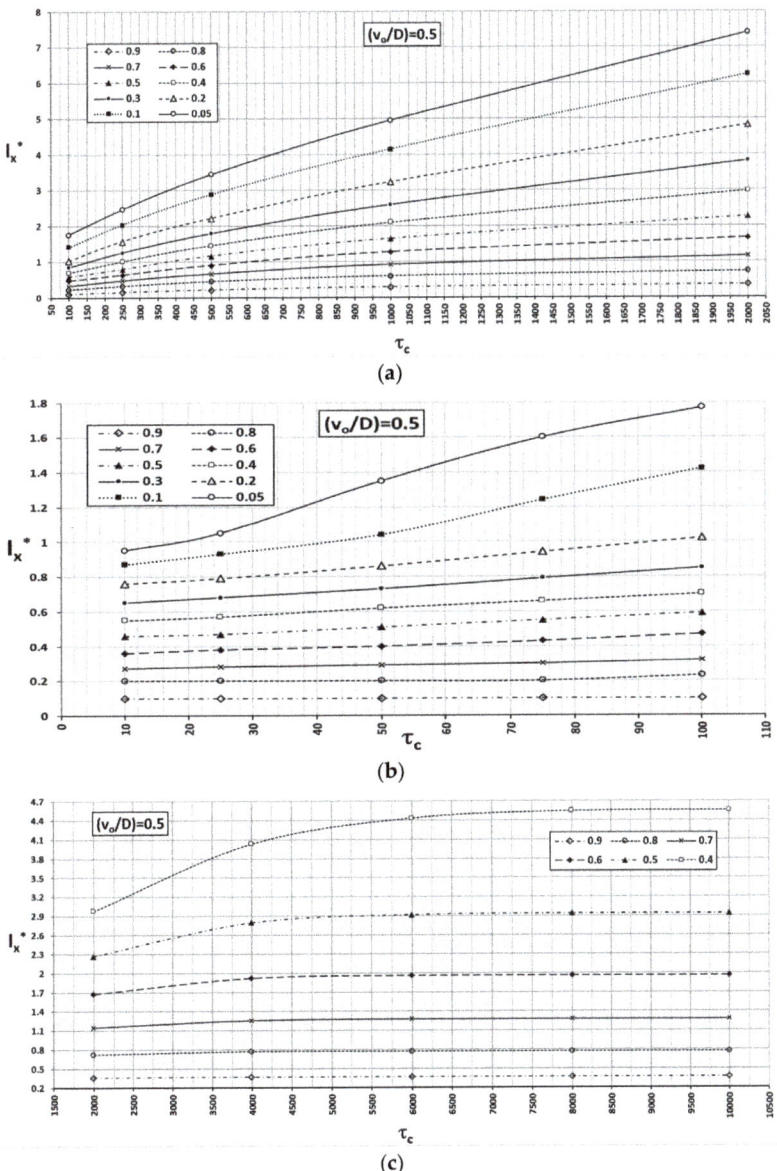

Figure 12. Cont.

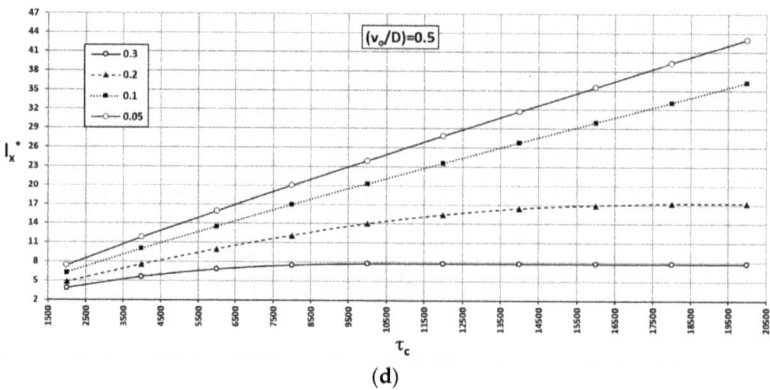

(d)

Figure 12. Universal curve $l^*_{x(c')}$ as a function of time factor τ_c. Parameter of the abacus $\frac{v_o}{D} = 0.5$. $v_o = 0.0003$ m/d. (**a**) $\tau_c \in$ [100–2000 days], (**b**) $\tau_c \in$ [10–100 days], (**c**) $\tau_c \in$ [2000–10,000 days], (**d**) $\tau_c \in$ [2000–20,000 days].

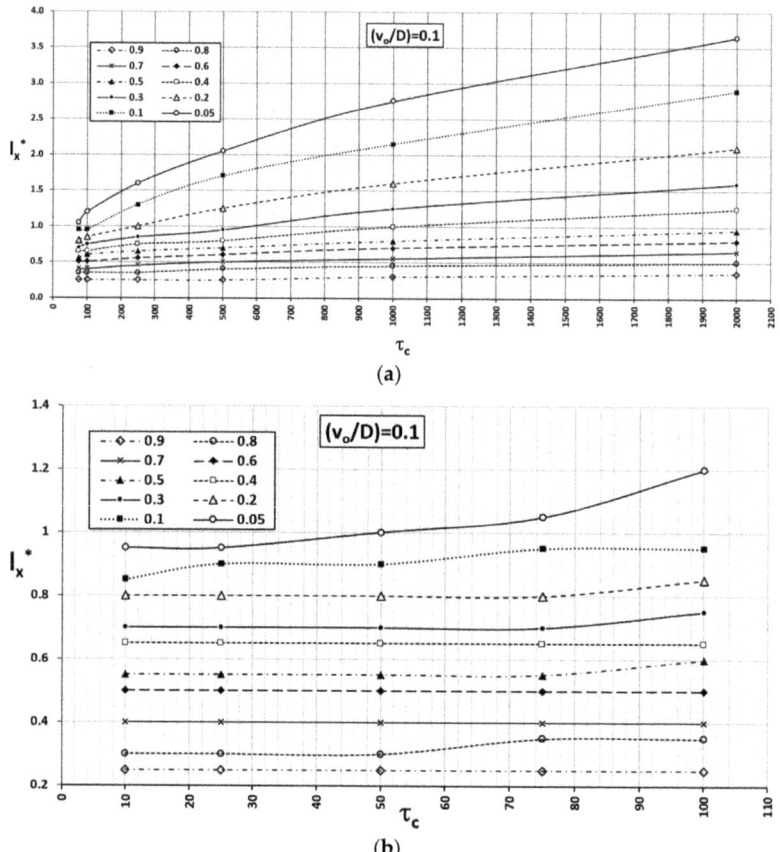

Figure 13. Cont.

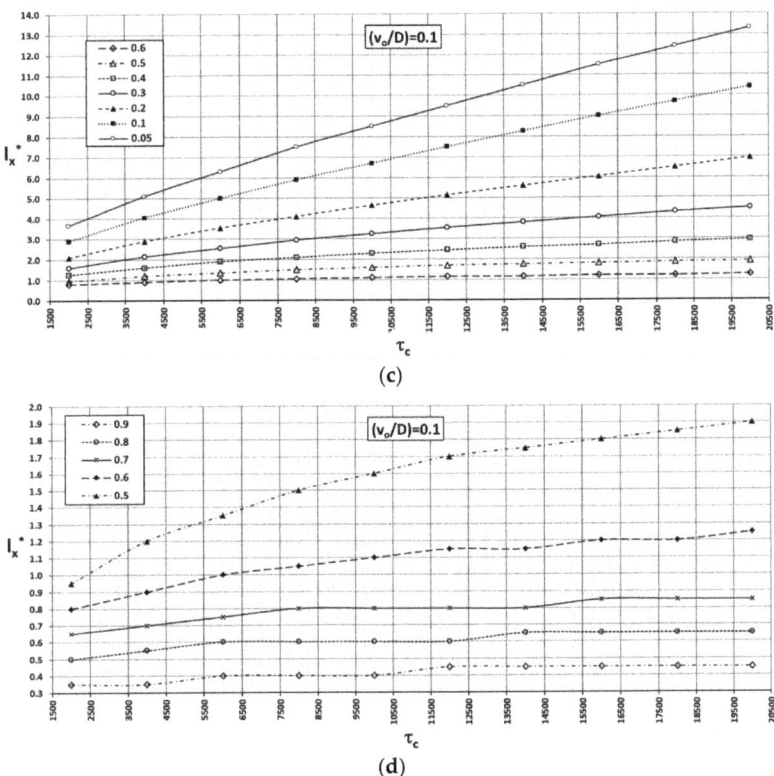

Figure 13. Universal curve $l^*_{x(c')}$ as a function of time factor τ_c. Parameter of the abacus $\frac{v_o}{D} = 0.1$. $v_o = 0.00006$ m/d. (**a**) $\tau_c \in$ [100–2000 days], (**b**) $\tau_c \in$ [10–100 days], (**c**) $\tau_c \in$ [2000–20,000 days], (**d**) $\tau_c \in$ [2000–20,000 days].

In relation to the largest transversal extension ($l^*_{y(c'=0.1)}$) given by dependence (17), the curves depicted in Figure 14 show this parameter (vertical axis) for isoline $c' = 0.1$, which represents 10% of the concentration of the focus, as a function of time factor τ_c (horizontal axis) for $\frac{v_o}{D} = 0.1, 0.5, 1, 2, 5$ and 10. As in the former figures, the time factor scale and real time are related by expressions $\tau_c = \left(\frac{0.0006}{v_o}\right)\tau^*$ for the curves $\frac{v_o}{D} = 1, 2, 5$ and 10, $\tau_c = \left(\frac{0.0003}{v_o}\right)\tau^*$ for the curve $\frac{v_o}{D} = 0.5$ and $\tau_c = \left(\frac{0.00006}{v_o}\right)\tau^*$ for the curve $\frac{v_o}{D} = 0.1$. The lack of monotony in the slope of the curves is a consequence of the coupling between advection and diffusion, which occurs at different times according to the relative influence of one effect or the other. The patterns tend to stabilise when the advective and diffusive processes balance each other, while the preponderance of the diffusive process clearly establishes an unsteady pattern. There is a relationship between this maximum transversal extension ($l^*_{y(c'=0.1)}$) and the horizontal location ($l^{**}_{x(c'=0.1)}$) measured from the focus (f) at which such extension occurs. Figure 15 shows this dependence, in addition to that of the dimensionless concentration $c' = 0.1$ and the same ratios of $\frac{v_o}{D}$.

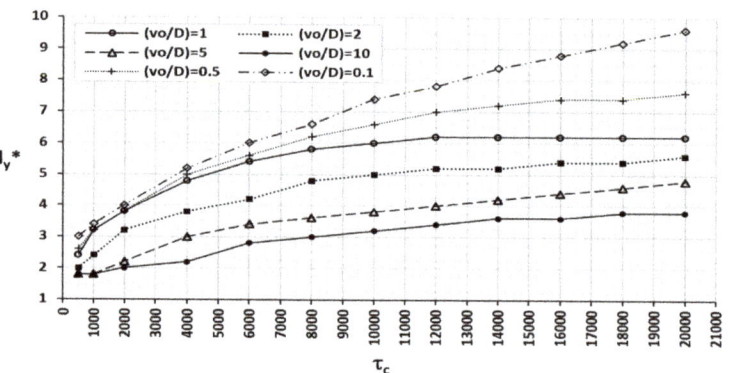

Figure 14. Universal curve $l^*_{y(c'=0.1)}$ as a function of time factor τ_c.

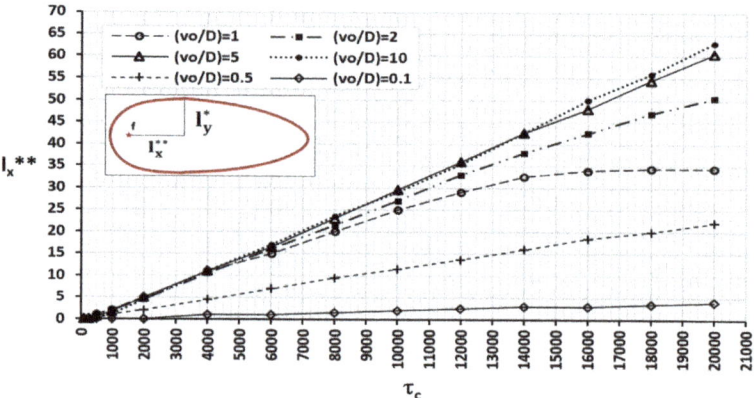

Figure 15. Universal curve $l^{**}_{x(c'=0.1)}$ as a function of time factor τ_c.

To finish, the universal curve corresponding to the location of the stationary contamination fronts to the left of the focus, defined by the extension $l^*_{x(left,c')}$, is shown in Figure 16 for the general case $\frac{v_o}{D} = 1$, in which advection and diffusion process are equally balanced. According to Equation (15), the curve is independent of the regional velocity as long as the ratio $\frac{v_o}{D}$ remains constant. The period of time necessary to achieve this stationary distribution of concentration profiles, estimated from different simulations, is expressed as $t = \frac{1.8}{v_o}$ and varies from 750 to 6000 days, depending on the regional velocity ($v_o = 0.0024$ to 6000 m/d, respectively). Further research may aim to establish new curves as a function of the $\frac{v_o}{D}$ ratio as well as to define the characteristic stabilisation time as a function of this ratio and the regional velocity.

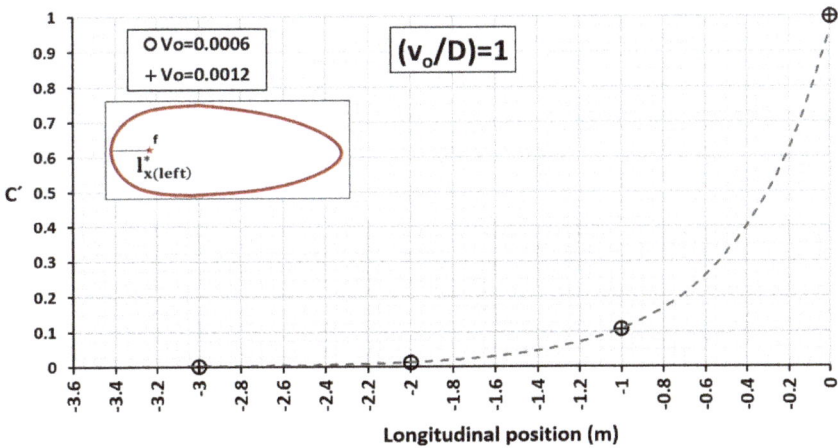

Figure 16. Universal curve c' as a function of longitudinal position for $v_o = 0.0006$ and 0.0012 m/d.

7. Conclusions and Final Comments

The parametric dimensionless characterisation following a discriminated and normalised dimensionless protocol of the governing equations has allowed us to deduce precise information about the dynamics of contaminant plumes in extensive 2D scenarios with a constant concentration focus under the effects of advection and molecular diffusion. The proposed non-dimensionalised procedure as well as the application of the Pi theorem have resulted in accurate expressions of the unknown functions of interest (that is, the horizontal and transversal extensions of the plume) on the physical and geometrical parameters of the problem. In the most complex case, the lengths that define the pattern of the plume extension depend on three parameters: the dimensionless concentration of the isoline, the ratio between molecular diffusivity and regional velocity and a time corrected by the regional Darcy velocity. Based on these results and thanks to a sufficient number of numerical simulations, the functions can be universally represented by means of an easy-to-use abacus that facilitates the monitoring and management of the contamination plume in most real cases.

In the proposed protocol for the search of the dimensionless groups, normalisation makes it possible to approximate the changes or partial derivatives of the dependent variables to the unit when these are averaged over the entire domain of the problem, while discrimination prevents the emergence of monomials of the type of geometric shape factors that unnecessarily increase the number of dimensionless groups.

The coupling of the diffusive and advective flows is not intuitive since the cross values of the variables and their derivatives are combined in the governing equation. Small diffusivities versus advection (to the magnitude of unity, at the most) advance the dragging effect by redistributing the concentration in the posterior or central area of the plume before it affects the diffusive effect. However, if diffusion predominates over advection, it will produce a redistribution of the concentration by diffusion and, over this dampened field of concentrations, advection occurs.

The scenario addressed here is only a sample of the problems of contaminant flow and transport in porous media. First, the geometry of the scenario may be different or finite, including the effects associated with gravity flow. In addition, scenarios with a constant initial concentration (not maintained), with a non-constant concentration or with a contaminated fluid injection well could be tackled. Finally, the study of the dynamics of contaminants under the effects of advection and hydrodynamic dispersion (in general, with negligible molecular diffusivity) is another pending issue to characterise and for which to propose new universal solutions.

Author Contributions: Conceptualisation, I.A. and M.I.; methodology, I.A., G.G.-R. and M.I.; software, I.A.; validation, I.A. and M.I.; formal analysis, I.A. and G.G.-R.; investigation, I.A.; resources, I.A.; data curation, I.A.; writing—original draft preparation, I.A.; writing—review and editing, I.A., G.G.-R. and M.I.; visualisation, I.A. and G.G.-R.; supervision, I.A. All authors have read and agreed to the published version of the manuscript.

Funding: This research received no external funding.

Data Availability Statement: Data openly available in a public repository that issues datasets with DOIs.

Conflicts of Interest: The authors declare no conflict of interest.

References

1. Ghasemizadeh, R.; Hellweger, F.; Butscher, C.; Padilla, I.; Vesper, D.; Field, M.; Alshawabkeh, A. Groundwater flow and transport modeling of karst aquifers, with particular reference to the North Coast Limestone aquifer system of Puerto Rico. *Hydrogeol. J.* **2012**, *20*, 1441–1461. [CrossRef]
2. Huq, M.E.; Fahad, S.; Shao, Z.; Sarven, M.S.; Al-Huqail, A.A.; Siddiqui, M.H.; ur Rahman, M.H.; Khan, I.A.; Alam, M.; Saeed, M.; et al. High arsenic contamination and presence of other trace metals in drinking water of Kushtia district, Bangladesh. *J. Environ. Manag.* **2019**, *242*, 199–209. [CrossRef]
3. Hristov, J.Y.; Planas-Cuchi, E.; Arnaldos, J.; Casal, J. Accidental burning of a fuel layer on a waterbed: A scale analysis of the models predicting the pre-boil over time and tests to published data. *Int. J. Therm. Sci.* **2004**, *43*, 221–239. [CrossRef]
4. Capobianchi, M.; Aziz, A. A scale analysis for natural convective flows over vertical surfaces. *Int. J. Therm. Sci.* **2012**, *54*, 82–88. [CrossRef]
5. Cánovas, M.; Alhama, I.; Alhama, F. Mathematical characterization of Bénard-type geothermal scenarios using discriminated non-dimensionalization of the governing equations. *Int. J. Nonlinear Sci. Numer. Simul.* **2015**, *16*, 23–34. [CrossRef]
6. Canóvas, M.; Alhama, I.; Trigueros, E.; Alhama, F. A review of classical dimensionless numbers for the Yusa problem based on discriminated nondimensionalization of the governing equations. *Hydrol. Process.* **2016**, *30*, 4101–4112. [CrossRef]
7. Alhama, I.; García-Ros, G.; Alhama, F. Universal solution for the characteristic time and the degree of settlement in nonlinear soil consolidation scenarios. A deduction based on nondimensionalization. *Commun. Nonlinear Sci. Numer. Simul.* **2018**, *57*, 186–201. [CrossRef]
8. Sánchez-Pérez, J.F.; Alhama, I. Universal curves for the solution of chlorides penetration in reinforced concrete, water-saturated structures with bound chloride. *Commun. Nonlinear Sci. Numer. Simul.* **2020**, *84*, 105201. [CrossRef]
9. Solder, J.E.; Jurgens, B.; Stackelberg, P.E.; Shope, C.L. Environmental tracer evidence for connection between shallow and bedrock aquifers and high intrinsic susceptibility to contamination of the conterminous US glacial aquifer. *J. Hydrol.* **2020**, *583*, 124505. [CrossRef]
10. Jamin, P.; Brouyère, S. Monitoring transient groundwater fluxes using the finite volume point dilution method. *J. Contam. Hydrol.* **2018**, *218*, 10–18. [CrossRef]
11. Ntanganedzeni, B.; Elumalai, V.; Rajmohan, N. Coastal aquifer contamination and geochemical processes evaluation in Tugela catchment, South Africa—Geochemical and statistical approaches. *Water* **2018**, *10*, 687. [CrossRef]
12. Erostate, M.; Huneau, F.; Garel, E.; Lehmann, M.F.; Kuhn, T.; Aquilina, L.; Vergnaud-Ayraud, V.; Labasque, T.; Santoni, S.; Robert, S.; et al. Delayed nitrate dispersion within a coastal aquifer provides constraints on land-use evolution and nitrate contamination in the past. *Sci. Total Environ.* **2018**, *644*, 928–940. [CrossRef]
13. Bridgman, P.W. *Dimensional Analysis*; Yale University Press: New Haven, CT, USA, 1937; 113p.
14. Langhaar, H.L. *Dimensional Analysis and Theory of Models*; Wiley: New York, NY, USA, 1951; 166p.
15. Sonin, A.A. *The Physical Basis of Dimensional Analysis*; Department of Mechanical Engineering, MIT: Cambridge, MA, USA, 1992; 57p.
16. Gibbings, J.C. *Dimensional Analysis*; Springer: New York, NY, USA, 2011; 284p.
17. Alhama, F.; Madrid, C.N. *Análisis Dimensional Discriminado. Aplicación a Problemas Avanzados de Dinámica de Fluidos y Transferencia de Calor*; Reverté: Barcelona, Spain, 2012; 328p.
18. Zohuri, B. *Dimensional Analysis beyond the Pi Theorem*; Springer: Cham, Switzerland, 2017; 243p.
19. Buckingham, E. On physically similar systems: Illustrations of the use of dimensional equations. *Phys. Rev.* **1914**, *4*, 345–376. [CrossRef]
20. Holzbecher, E. Comment to 'Mixed convection processes below a saline disposal basin' by Simmons, C.T., Narayan, K.A. 1997. *J. Hydrol.* **2000**, *194*, 263–285.
21. Simmons, C.T.; Narayan, K.A.; Sharp, J.M. Response to comment to 'Mixed convection processes below a saline disposal basin' by Simmons, C.T., Narayan, K.A. 1997 in Journal of Hydrology 194, 263-285 by E. Holzbecher. *J. Hydrol.* **2000**. [CrossRef]
22. Huysmans, M.; Dassargues, A. Review of the use of Péclet numbers to determine the relative importance of advection and diffusion in low permeability environments. *Hydrogeol. J.* **2005**, *13*, 895–904. [CrossRef]
23. Sauty, J.P. An analysis of hydrodispersive transfer in aquifers. *Water Resour. Res.* **1980**, *16*, 145–158. [CrossRef]

24. Moench, A.F. Convergent radial dispersion: A Laplace transform solution for aquifer tracer testing. *Water Resour. Res.* **1989**, *25*, 439–447. [CrossRef]
25. Langevin, C.D. *SEAWAT: A Computer Program for Simulation of Variable-Density Groundwater Flow and Multi-Species Solute and Heat Transport*; No. 2009-3047; U.S. Geological Survey: Reston, VA, USA, 2009; 2p.
26. Bakker, M.; Oude Essink, G.H.P.; Langevin, C.D. The rotating movement of three immiscible fluids—A benchmark problem. *J. Hydrol.* **2004**, *287*, 270–278. [CrossRef]
27. Goswami, R.R.; Clement, T.P. Laboratory-scale investigation of saltwater intrusion dynamics. *Water Resour. Res.* **2007**, *43*, W04418. [CrossRef]
28. Bauer, P.; Held, R.J.; Zimmermann, S.; Linn, F.; Kinzelbach, W. Coupled flow and salinity transport modelling in semi-arid environments: The Shashe River Valley, Botswana. *J. Hydrol.* **2006**, *316*, 163–183. [CrossRef]
29. Alhama, I.; Rodriguez Estrella, T.; Alhama, F. Chemical and physical parameters as trace markers of anthropogenic-induced salinity in the Agua Amarga coastal aquifer (southern Spain). *Hydrogeol. J.* **2012**, *20*, 1315–1329. [CrossRef]
30. Bear, J. *Dynamics of Fluid in Porous Media*; Elsevier: Amsterdam, The Netherlands, 1972; 764p.
31. Zheng, C.; Bennett, G.D. *Applied Contaminant Transport. Modeling: Theory and Practice*; Wiley: New York, NY, USA, 1995; 440p.
32. Muskat, M. *The Flow of Homogeneous Fluids through Porous Media*; McGraw-Hill: New York, NY, USA, 1937; 763p.
33. Crank, C. *The Mathematics of Diffusion*; Oxford University Press: Oxford, UK, 1979; 414p.
34. Alhama, I.; Alcaraz, M.; Trigueros, E.; Alhama, F. Dimensionless characterization of salt intrusion benchmark scenarios in anisotropic media. *Appl. Math. Comput.* **2014**, *247*, 1173–1182. [CrossRef]
35. Zimparoz, V.D.; Petkov, V.M. Application of discriminated analysis to low Reynolds numbers swirl flows in circular tubes with twisted-tape inserts. Pressure drops correlations. *Int. Rev. Chem. Eng.* **2009**, *1*, 346–356.
36. Madrid, C.N.; Alhama, F. Discrimination: A fundamental and necessary extension of classical dimensional analysis. *Int. Commun. Heat Mass Transf.* **2006**, *33*, 287–294. [CrossRef]
37. SEAWAT V.4. United States Geological Survey 2012. Available online: https://www.usgs.gov/software/seawat-a-computer-program-simulation-three-dimensional-variable-density-ground-water-flow (accessed on 3 February 2021).

Article

The Dirichlet-to-Neumann Map in a Disk with a One-Step Radial Potential: An Analytical and Numerical Study

Sagrario Lantarón and Susana Merchán *

Departamento de Matemática e Informática Aplicadas a las Ingenierías Civil y Naval, Escuela de Caminos, Canales y Puertos, Universidad Politécnica de Madrid, Calle del Profesor Aranguren, 3, 28040 Madrid, Spain; sagrario.lantaron@upm.es
* Correspondence: susana.merchan@upm.es

Abstract: Herein, we considered the Schrödinger operator with a potential q on a disk and the map that associates to q the corresponding Dirichlet-to-Neumann (DtN) map. We provide some numerical and analytical results on the range of this map and its stability for the particular class of one-step radial potentials.

Keywords: Dirichlet-to-Neumann map; Schrödinger operator; stability

1. Introduction

Let $\Omega \subset \mathbb{R}^2$ be a bounded domain with smooth boundary $\partial \Omega$. For each $q \in L^\infty(\Omega)$, consider the so called Dirichlet-to-Neumann map (DtN) given by:

$$\Lambda_q : H^{1/2}(\partial\Omega) \rightarrow H^{-1/2}(\partial\Omega) \qquad (1)$$
$$f \rightarrow \frac{\partial u}{\partial n}|_{\partial\Omega}.$$

where u is the solution of the following problem:

$$\begin{cases} \Delta u + q(x)u = 0, & x \in \Omega, \\ u = f, & \partial\Omega, \end{cases} \qquad (2)$$

and $\frac{\partial u}{\partial n}|_{\partial\Omega}$ denotes the normal derivative of u on the boundary $\partial\Omega$.

Note that the uniqueness of u as solution of (2) requires that 0 is not a Dirichlet eigenvalue of $\Delta + q$. A sufficient condition to guarantee that Λ_q is well defined is to assume $q(x) < \lambda_1$, the first Dirichlet eigenvalue of the Laplace operator in Ω, since, in this case, the solution in (2) is unique. We assume that this condition holds and lets us define the space

$$L^\infty_{<\lambda_1}(\Omega) = \{q \in L^\infty(\Omega), \text{ s. t. } q(x) < \lambda_1, \text{ a. e. }\}.$$

In this work, we were interested in the following map:

$$\Lambda : L^\infty_{<\lambda_1}(\Omega) \rightarrow \mathcal{L}(H^{1/2}(\partial\Omega); H^{-1/2}(\partial\Omega)) \qquad (3)$$
$$q \rightarrow \Lambda_q.$$

This has an important role in inverse problems, where the aim is to recover the potential q from boundary measurements. In practice, these boundary measurements correspond to the associated DtN map, and therefore, the mathematical statement of the classical inverse problem consists of the inversion of Λ.

It is known that Λ is one-to-one as long as $q \in L^p$ with $p > 2$ (see [1]). Therefore, the inverse map Λ^{-1} can be defined in the range of Λ. There are, however, two related important and difficult questions that are not well understood: A characterization of the range of Λ and its stability, i.e., a quantification of the difference of two potentials, in the L^∞ topology in terms of the distance of their associated DtN maps. Obviously, this stability will affect the efficiency of any inversion or reconstruction algorithm to recover the potential from the DtN map (see [2] and [3]).

The first question, i.e., the characterization of the range of Λ is widely open. To the best of our knowledge, the further result is due to [4], where a characterization is obtained for the adherence, with respect to the weak topology in ℓ^2_{-1}, of the sequence of eigenvalues associated with the orthogonal basis of eigenvectors $\{e^{ink}\}_{k\in\mathbb{Z}}$. Here, ℓ^2_α is the space of sequences $\{c_n\}_{n\in\mathbb{Z}}$, such that $\sum_{n\in\mathbb{Z}} |n|^{2\alpha} |c_n|^2 < \infty$. This topology is not the usual one in $\mathcal{L}(H^{1/2}(\partial\Omega); H^{-1/2}(\partial\Omega))$ and it is not easy to interpret how the adherence enlarges this set. Furthermore, the characterization does not give much practical information on the range, as, for instance, the convexity or accurate bounds on the eigenvalues. In fact, characterizing such properties is one of the main motivations of this work, since we could establish easily a priori if a desired linear map in $\mathcal{L}(H^{1/2}(\partial\Omega); H^{-1/2}(\partial\Omega))$ can be associated with a DtN map. On the contrary, we have to take into account that in practice, the DtN map is estimated from physical measurements, which are subject to errors and may provide only partial information. A precise knowledge of the range of Λ is useful to find the best DtN map that fits the measurements and to design an inversion algorithm in such situations.

Concerning the stability, it is well known that the problem is ill posed and that the most we can expect is logarithmic stability in general (see [5]). There are more explicit results when we assume that the potential q has some smoothness. In particular, if $q \in H^s(\Omega)$ with $s > 0$, the following $\log$-stability condition is known (see [1]):

$$\|q_1 - q_2\|_{L^\infty} \leq V(\|\Lambda_{q_1} - \Lambda_{q_2}\|_{\mathcal{L}(H^{1/2}; H^{-1/2})}), \quad (4)$$

where $V(t) = C \log(1/t)^{-\alpha}$ for some constants $C, \alpha > 0$. Stronger stability conditions are known in some particular cases. For example, in [6], it was shown that when q is piecewise constant and the components where it takes a constant value are fixed and satisfy some technical conditions, the stability is Lipschitz, i.e., there exists a constant $C > 0$, such that:

$$\|q_1 - q_2\|_{L^\infty} \leq C\|\Lambda_{q_1} - \Lambda_{q_2}\|_{\mathcal{L}(H^{1/2}; H^{-1/2})}. \quad (5)$$

In this work, we tried to understand better the situation by considering the simplest case of a disk with one-step radial potentials q. More precisely, we provide some results on the range of Λ and its stability when we restrict to the particular case $\Omega = B(0,1) = \{x \in \mathbb{R}^2 : r = |x| < 1\}$ and $q \in F \subset L^\infty(\Omega)$ given by:

$$F = \{q \in L^\infty(\Omega) : q(r) = \gamma \chi_{(0,b)}(r), \ r = |x|, \ b \in (0,1), \ \gamma \in [0,1]\}, \quad (6)$$

where $\chi_{(0,b)}(r)$ is the characteristic function of the interval $(0,b)$. Note that F is a two-parametric family depending on γ and b.

It is worth mentioning that, as we show below, the solution of (2) is unique for all $b \in (0,1)$ and $\gamma \geq 0$, and therefore, the DtN map is well defined for all of these one-step potentials. In other words, 0 is not an eigenvalue of the operator $\Delta + q$ and, in particular, we do not need to restrict ourselves to the constraint $q(x) < \lambda_1$. However, we still restrict ourselves to the bounded set F to simplify.

Even in this simple case, a complete analytic answer to the previous questions (range of the DtN map and sharp stability conditions) is unknown. Therefore, we also considered a numerical approach based on a discrete sampling of the set F. Given an integer $N > 0$, we define $h = 1/N$ and:

$$F_h = \{q \in L^\infty(\Omega) : q(r) = \gamma \chi_{(0,b)}(r), \ b = hi, \ \gamma = hj,$$
$$i = 1, \ldots, N-1, \ j = 0, \ldots, N\}. \quad (7)$$

Note that F_h has $N(N-1)+1$ functions from F. As $h \to 0$, we can obtain a better description of F and, in particular, we should recover the properties for $q \in F$.

The main contributions of this paper are given below:

1. Concerning the stability of Λ, we show that it fails in the sense that inequality (4) does not hold for any continuous function $V(t)$ with $V(0) = 0$. The proof is an adaptation of the analogous result for the conductivity problem obtained in [7]. In fact, we consider—as potential—the same piecewise constant radial conductivity used in [7]. The stability constant blows up when the support of the inner disk where the value of the potential is constant becomes zero.
2. We obtain estimates for the Lipschitz stability constant in (5), in terms of $b, \gamma \in (0,1)$. However, the stability constant in (5) depends on b^{-4} and therefore blows up as $b \to 0$.
3. We now consider $\gamma \in [0,1]$ fixed and we define the set $G_\gamma \subset F$ as:

$$G_\gamma = \{q \in L^\infty(\Omega) \ : \ q(r) = \gamma \chi_{(0,b)}(r), \ b \in (0,1)\}. \tag{8}$$

We prove that if $b \geq b_0 > 0$, there is stability of the DtN map with respect to the position of the discontinuity b for potentials in G_γ. More precisely, we obtain a stability constant depending on $\gamma^{-1} b^{-3}$, which is uniformly bounded for $b > b_0$ and fixed γ (see Theorem 3 below). Note, however, that this constant blows up as $\gamma \to 0$. This stability result does not give information about the stability with respect to the L^∞ norm of the potentials, but it provides stability with respect to the L^1 norm, which is sensitive to the position of the discontinuities, when $b > b_0 > 0$ and $\gamma > \gamma_0 > 0$. In fact, we show numerical evidence of such stability when considering potentials in F.
4. For the range of Λ, we give a characterization in terms of the first two eigenvalues of the DtN map. We also analyze the region where the stability constant is larger, and, therefore, the potentials for which any recovering algorithm for q from the DtN map will have more difficulties.

We mention that a similar analysis can be conducted for the closely related—and more classical—conductivity problem, where (2) is replaced by:

$$\begin{cases} -\operatorname{div} a(x) \nabla v = 0, & x \in \Omega, \\ v = f, & \partial \Omega, \end{cases} \tag{9}$$

and the Dirichlet-to-Neumann map, or voltage-to-current map, is given by:

$$\begin{aligned} \Lambda_a : H^{1/2}(\partial \Omega) &\to H^{-1/2}(\partial \Omega) \\ f &\to a \tfrac{\partial v}{\partial n}|_{\partial \Omega}. \end{aligned} \tag{10}$$

In this case, the relationship between piecewise constant radial conductivities and the eigenvalues of the DtN map is known [8] through a suitable recurrence formula. However, there is not a direct transformation that relates both problems, and the analysis must be done specifically for this case.

We refer to the review paper [9] and the references therein for theoretical results on the DtN map in this case.

The rest of this paper is divided as follows: In Section 2 below, we characterize the DtN map in terms of its eigenvalues using polar coordinates. In Sections 3 and 4, we analyze the stability and range results, respectively. In Section 5, we briefly describe the main conclusions, and finally, Section 5 contains the proofs of the theorems stated in the previous sections.

2. The Dirichlet-to-Neumann Map

In this section, we characterize the Dirichlet-to-Neumann map in the case of a disk. System (2) in polar coordinates reads:

$$\begin{cases} r^2 \frac{\partial^2 v}{\partial r^2} + r \frac{\partial v}{\partial r} + \frac{\partial^2 v}{\partial \theta^2} + r^2 q(r) v = 0, & (r, \theta) \in (0,1) \times [0, 2\pi), \\ \lim_{r \to 0, r > 0} v(r, \theta) < \infty, \\ v(1, \theta) = g(\theta), & \theta \in [0, 2\pi), \end{cases} \quad (11)$$

where $v(r, \theta) = u(r \cos \theta, r \sin \theta)$ and $g(\theta) = f(\cos \theta, \sin \theta)$ is a periodic function.

An orthonormal basis in $L^2(0, 2\pi)$ is given by $\{e^{in\theta}\}_{n \in \mathbb{Z}}$. Here, we use this complex basis to simplify the notation, but in the analysis below, we only consider the subspace of real valued functions. Therefore, any function $g \in L^2(0, 2\pi)$ can be written as:

$$g(\theta) = \sum_{n \in \mathbb{Z}} g_n e^{in\theta}, \quad g_n = \frac{1}{2\pi} \int_0^{2\pi} g(t) e^{-int} dt, \quad (12)$$

and $\|g\|^2_{L^2(0,2\pi)} = \sum_{n \in \mathbb{Z}} |g_n|^2$. Associated with this basis, we define the usual Hilbert spaces: $H^\alpha_\#$, for $\alpha > 0$, as

$$H^\alpha_\# = \{g \: : \: \|g\|^2_\alpha = \sum_{n \in \mathbb{Z}} (1 + n^2)^\alpha |g_n|^2 < \infty\}.$$

The Dirichlet-to-Neumann map in this case is defined as:

$$\begin{array}{rcl} \Lambda_q : H^{1/2}_\#(0, 2\pi) & \to & H^{-1/2}_\#(0, 2\pi) \\ g & \to & \frac{\partial v}{\partial r}(1, \cdot), \end{array} \quad (13)$$

where v is the unique solution of (11).

In the above basis, the Dirichlet-to-Neumann map turns out to be diagonal. In fact, we have the following result:

Theorem 1. *Let Ω be the unit disk and $q \in F$. Then:*

$$\Lambda_q \left(e^{in\theta} \right) = c_n e^{in\theta}, \quad n \in \mathbb{Z}, \quad (14)$$

where:

$$c_0 = \frac{-b\sqrt{\gamma} J_1(\sqrt{\gamma} b)}{b \log b \sqrt{\gamma} J_1(\sqrt{\gamma} b) + J_0(\sqrt{\gamma} b)}, \quad (15)$$

$$c_n = c_{-n} = n \frac{J_{n-1}(\sqrt{\gamma} b) - b^{2n} J_{n+1}(\sqrt{\gamma} b)}{J_{n-1}(\sqrt{\gamma} b) + b^{2n} J_{n+1}(\sqrt{\gamma} b)}, \quad n \in \mathbb{N}, \quad (16)$$

and $J_n(r)$ are the Bessel functions of the first kind.

Note that the range of Λ, when restricted to F, is characterized by the set of sequences $\{c_n\}_{n \geq 0}$ of the form (15) and (16) for all possible b, γ. In particular, when $q = 0$, we have:

$$c_n = n, \quad n = 0, 1, 2, \ldots, \quad (17)$$

and this sequence must be in the range of Λ.

The norm of Λ_q, when restricted to F, is given by:

$$\|\Lambda_q\|_{\mathcal{L}(H^{1/2}_\#; H^{-1/2}_\#)} = \sup_{n \geq 0} \frac{|c_n|}{1 + n^2}. \quad (18)$$

Proof of Theorem 1. We first compute c_0 in (14). As the boundary data at $r = 1$ in (11) is the constant $g(\theta) = 1$, we assume that $v(r, \theta)$ is radial, i.e., $v(r, \theta) = a_0(r)$. Then, a_0 should satisfy:

$$\begin{cases} r^2 a_0'' + r a_0' + r^2 q(r) a_0 = 0, & 0 < r < 1, \\ a_0(1) = 1, & \lim_{r \to 0, r > 0} a_0(r) < \infty. \end{cases} \quad (19)$$

For $r \in (0, b)$, we solve the ODE with the boundary data at $r = 0$, while for $r \in (b, 1)$, we use the boundary data at $r = 1$. In the first case, the ODE is the Bessel ODE of order 0, and therefore, we have:

$$a_0(r) = \begin{cases} A_0 J_0(\sqrt{\gamma} r), & r \in (0, b), \\ 1 + C_0 \log r, & r \in (b, 1), \end{cases}$$

where J_0 is the Bessel function of the first kind and A_0 and C_0 are constants. These are computed by imposing continuity of a_0 and a_0' at $r = b$. In this way, we obtain:

$$\begin{cases} A_0 J_0(\sqrt{\gamma} b) = 1 + C_0 \log b \\ A_0 \sqrt{\gamma} J_0'(\sqrt{\gamma} b) = C_0 \frac{1}{b}. \end{cases}$$

Solving the system for A_0 and C_0 and taking into account that $\Lambda_q(1) = \frac{\partial v}{\partial r}(1, \theta) = a_0'(1) = C_0$, we easily obtain (14).

Similarly, to compute c_n in (14), we have to consider $g(\theta) = e^{in\theta}$ in (11), and therefore, we assume that the solution $v(r, \theta)$ can be written in separate variables, i.e., $v(r, \theta) = a_n(r) e^{in\theta}$. Then, a_n must satisfy:

$$\begin{cases} r^2 a_n'' + r a_n' + (r^2 q(r) - n^2) a_n = 0, & 0 < r < 1, \\ a_n(1) = 1, & \lim_{r \to 0, r > 0} a_n(r) < \infty, \quad n \geq 1. \end{cases} \quad (20)$$

As in the case of c_0, for $r \in (0, b)$, we solve the ODE with the boundary data at $r = 0$, while for $r \in (b, 1)$, we use the boundary data at $r = 1$. We have:

$$a_n(r) = \begin{cases} A_n J_n(\sqrt{\gamma} r), & r \in (0, b), \\ C_n(r^n - r^{-n}) + r^n, & r \in (b, 1), \end{cases}$$

where A_n and C_n are constants. These are computed by imposing continuity of a_n and a_n' at $r = b$. In this way, we obtain:

$$\begin{cases} A_n J_n(\sqrt{\gamma} b) = C_n(b^n - b^{-n}) + b^n \\ A_n \sqrt{\gamma} J_n'(\sqrt{\gamma} b) = n C_n(b^{n-1} + b^{-n-1}) + n b^{n-1}. \end{cases}$$

Solving the system for A_n and C_n, we obtain, in particular:

$$C_n = \frac{-b^n J_n'(\sqrt{\gamma} b) + n \frac{b^{n-1}}{\sqrt{\gamma}} J_n(\sqrt{\gamma} b)}{-(b^{-n-1} + b^{n-1}) \frac{n}{\sqrt{\gamma}} J_n(\sqrt{\gamma} b) - (b^{-n} - b^n) J_n'(\sqrt{\gamma} b)}.$$

We simplify this expression using the well-known identity:

$$2 J_n'(r) = J_{n-1}(r) - J_{n+1}(r),$$

and we obtain:

$$C_n = \frac{-J_{n+1}(\sqrt{\gamma} b)}{b^{-2n} J_{n-1}(\sqrt{\gamma} b) + J_{n+1}(\sqrt{\gamma} b)}.$$

Now, taking into account that $\Lambda_q(e^{in\theta}) = \frac{\partial v}{\partial r}(1, \theta) = a_n'(1) e^{in\theta} = (2n C_n + n) e^{in\theta}$, we easily obtain (14). □

Remark 1. *In this proof of Theorem 1, we do not use the restriction $\gamma \leq 1$ that satisfies the potentials in F. In fact, the statement of the theorem still holds for any step potential, as in F, but with any arbitrary large $\gamma \geq 0$.*

3. Stability

In this section, we focus on the stability results for the map Λ. Some results are analytical and they are stated as theorems. The proofs are given in Appendix A. We divided this section in three subsections, where we consider the negative stability result for $q \in F$ norm, and some partial results when we consider the subsets F_b defined by:

$$F_b = \left\{ q \in L^\infty(\Omega) \ : \ q(r) = \xi_{(0,b)}(r), \ \gamma \in [0,1] \right\},$$

and G_γ defined in (8).

3.1. Stability for $q \in F$

The first result in this section is the lack of any stability property when $q \in F$. In particular, we prove that inequality (4) fails for any continuous function $V(t)$ with $V(0) = 0$.

Theorem 2. *Given $q_0 \in F$, there exists a sequence $\{q_k\}_{k \geq 1} \subset F$, such that $\|q_0 - q_k\|_{L^\infty} = \gamma > 0$ for all $k \geq 1$, while:*

$$\|\Lambda_{q_0} - \Lambda_{q_k}\|_{\mathcal{L}(H_\#^{1/2}; H_\#^{-1/2})} \to 0, \quad \text{as } k \to \infty. \tag{21}$$

This result contradicts any possible stability result of the DtN map at $q_0 \in F$. Roughly speaking, the idea is that the eigenvalues of Λ, given in Theorem 1 above, depend continuously on b, unlike the L^∞ norm of the potentials. A detailed proof of the Theorem 2 is given in the Appendix A below.

3.2. Partial Stability

We now give two partial stability results when we fix b and γ, respectively.

Theorem 3. *Given $b \in (0,1)$ and $q_1, q_2 \in F_b$, we have:*

$$\|q_1 - q_2\|_{L^\infty} \leq \frac{15.0756}{b^4} \|\Lambda_{q_1} - \Lambda_{q_2}\|_{\mathcal{L}(H_\#^{1/2}; H_\#^{-1/2})}. \tag{22}$$

On the contrary, given $\gamma \in (0,1]$ and $q_1, q_2 \in G_\gamma$, we have:

$$|b_1 - b_2| \leq \frac{3.7489}{\gamma b^3} \|\Lambda_{q_1} - \Lambda_{q_2}\|_{\mathcal{L}(H_\#^{1/2}; H_\#^{-1/2})}, \tag{23}$$

where $b = \min\{b_1, b_2\}$.

The proof of this theorem is in the Appendix A below.

Inequality (22) provides a Lipschitz stability result for Λ when b is fixed. This result is not new, since this situation enters in the framework in [6], as q takes constant values in fixed regions. The contribution here is in the dependence of the Lipschitz constant on the parameter b, which is associated with the size of the region, where q takes a different constant value. An estimate (22) shows also that the lack of Lipschitz stability is related to variations in the position of the discontinuity, which is the main idea in the negative result given in Theorem 2.

A numerical quantification of this Lipschitz stability for b fixed is easily obtained. We fix $b = b_0$ and consider:

$$F_{h,b_0} = \{q \in L^\infty(\Omega) \ : \ q(r) = \gamma \chi_{(0,b)}(r), \ b = b_0, \ \gamma = hj, \ j = 1, ..., 1/h - 1\}.$$

and for $q_0 \in F_{h,b_0}$:

$$C_2(h, q_0, b_0) = \max_{\substack{q \in F_{h,b_0} \\ q \neq q_0}} \frac{\|q_0 - q\|_{L^\infty}}{\|\Lambda_{q_0} - \Lambda_q\|_{\mathcal{L}(H_\#^{1/2}; H_\#^{-1/2})}}, \qquad (24)$$

then, $C_2(h, q_0, b_0)$ remains bounded as $h \to 0$ for all $q_0 \in F_h$. In Figure 1, we show the behavior of $C_2(h, q_0, b_0)$ when $h = 10^{-4}$ for different values of b_0. To illustrate the behavior with respect to $b_0 \to 0$, we plot on the left-hand side of Figure 1 the graphs of the functions:

$$C_{2,\min}(b_0) = \min_{q \in F_{h,b_0}} C_2(10^{-4}, q, b_0), \text{ and } C_{2,\max}(b_0) = \max_{q \in F_{b_0}} C_2(10^{-4}, q, b_0). \qquad (25)$$

We see that both constants become larger for small values of b. We also see that both graphs are close in this logarithmic scale. However, the range of the interval $[C_{2,\min}(b), C_{2,\max}(b)]$ is not small, as shown on the right-hand side of Figure 1.

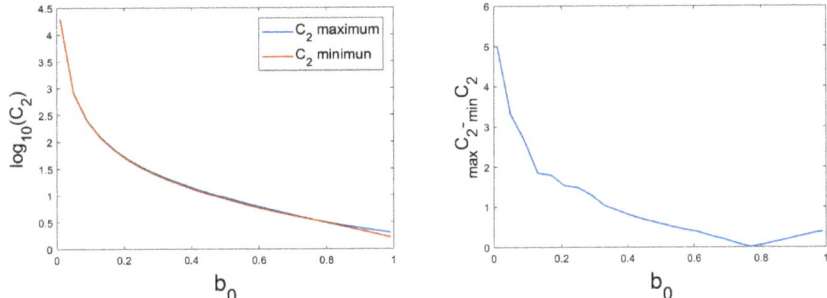

Figure 1. Numerical estimate of the stability constant C_2 in (24) for $h = 10^{-4}$. To illustrate the behavior on b, we plotted the maximum and minimum value when $q \in F_{h,b}$ with respect to b in logarithmic scale (**left**), and its range in normal scale (**right**).

Concerning inequality (23) in Theorem 3, it provides a stability result for Λ with respect to the position of the discontinuity. In particular, this provides Lipschitz stability if we consider a norm for the potentials that is sensitive to the position of the discontinuity. This is not the case for the L^∞ norm, but it is true for the L^p-norm for some $1 \leq p < \infty$. For example, when $p = 1$:

$$\|q_1 - q_2\|_{L^1} = \gamma \pi |b_1^2 - b_2^2| \leq 2\pi \gamma |b_1 - b_2| \leq \frac{7.4978\pi}{b^3} \|\Lambda_{q_0} - \Lambda_q\|_{\mathcal{L}(H_\#^{1/2}; H_\#^{-1/2})}.$$

We can also check this numerically:

$$C_2(h, \gamma_0, b) = \max_{q \in G_{h,\gamma_0}} \frac{\|q_0 - q\|_{L^1}}{\|\Lambda_{q_0} - \Lambda_q\|_{\mathcal{L}(H_\#^{1/2}; H_\#^{-1/2})}}, \qquad (26)$$

is bounded as $h \to 0$ and $b \geq b_0 > 0$, where:

$$G_{h,\gamma_0} = \{q \in L^\infty(\Omega) : q(r) = \gamma \chi_{(0,b)}(r), \gamma = \gamma_0, b = hj, j = 1, ..., 1/h - 1\}.$$

In Figure 2, we show the values when $h = 10^{-4}$. We can observe that the constant blows up as $b \to 0$.

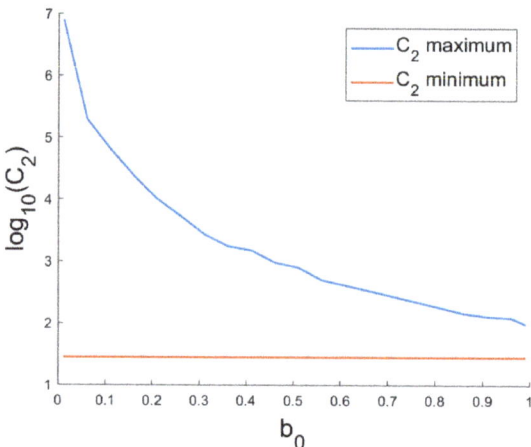

Figure 2. $C_2(h,q)$ for $b > b_0$ when $h = 10^{-4}$.

4. Range of the DtN Map

In this section, we are interested in the range of Λ when $q \in F$, i.e., the set of sequences $\{c_n\}_{n\geq 0}$ of the form (15) and (16) for all possible $b, \gamma \in [0,1] \times [0,1]$.

As F is a bi-parametric family of potentials, it is natural to check if we can characterize the family $\{c_n\}_{n\geq 0}$ with only the first two coefficients c_0 and c_1. In this section, we give numerical evidence of the following facts:

1. The first two coefficients, c_0 and c_1, in (15) and (16) are the most sensitive with respect to (b, γ) and, therefore, are the more relevant ones to identify b and γ from the DtN map.
2. The function:

$$\Lambda^h : F_h \to \mathbb{R}^2 \quad (27)$$
$$q \to (c_0, c_1),$$

is injective. This means, in particular, that the DtN map can be characterized by the coefficients c_0 and c_1, when restricted to functions in F_h. We also illustrate the set of possible coefficients c_0, c_1.
3. The lack of stability for Λ is associated with a higher density of points in the range of Λ^h. This occurs when either b or γ is close to zero.

4.1. Sensitivity of c_n

To analyze the relevance and sensitivity of the coefficients $c_n = c_n(b, \gamma)$ to identify the parameters (b, γ), we computed their range when $(b, \gamma) \in [0,1] \times [0,1]$, and the norm of their gradients. As we can see in Table 1, the range decreases for large n. This means that, for larger values of n, the variability of c_n is smaller and they are likely to be less relevant to identify q.

However, even if the range of c_n becomes smaller for large n, they could be more sensitive to small perturbations in (b, γ) and this would make them useful to distinguish different potentials. However, this is not the case. In Figure 3, we show that for the given values of $\gamma = 0.1, 0.34, 0.67, 0.99$ and $b \in (0,1]$, the gradients of the first two coefficients, with respect to (b, γ), are larger than the others. Therefore, we conclude that the two first coefficients, c_0 and c_1, are the most sensitive and, therefore relevant to identify the potential q.

We also see in Figure 3 that these gradients are very small for $b \ll 1$. This means, in particular, that identifying potentials with small b from the DtN map should be more difficult.

Table 1. Range of the coefficients, i.e., for each c_n, the range is defined as $\max_{q \in F_h} c_n - \min_{q \in F_h} c_n$.

Coefficient	Range
c_0	0.5523
c_1	0.2486
c_2	0.1588
c_3	0.1157
c_4	0.0904
c_5	0.0736

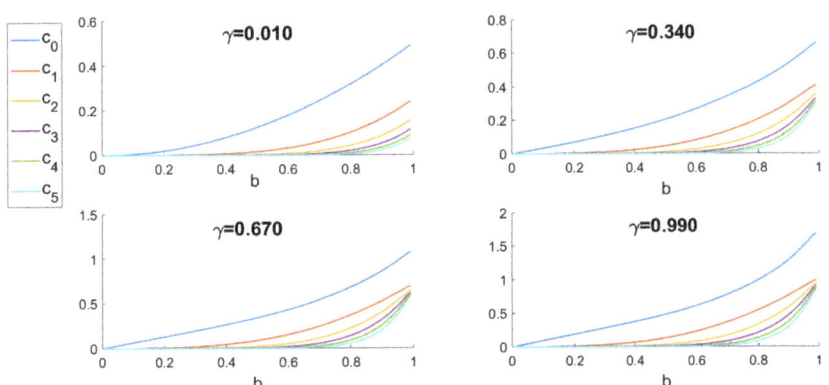

Figure 3. Norm of the gradient of the coefficients $c_n(\gamma, b)$ in terms of $b \in (0, 1)$ for different values of γ. We can see that the gradients of higher coefficients $n \geq 2$ are smaller than those of the first two. We can also observe that these gradients become small for small values of b.

4.2. Range of the DtN in Terms of c_0, c_1

Now, we focus on the range of the DtN in terms of the relevant coefficients (c_0, c_1), i.e., the range of the map Λ^h in (27): $R(\Lambda^h)$. In Figure 4, we show this range.

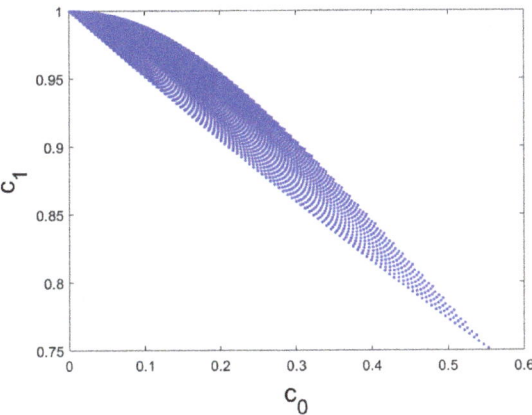

Figure 4. Range of the discrete Dirichlet-to-Neumann (DtN) map in (27) ($h = 10^{-2}$).

Coordinate lines for fixed γ and b are given in Figure 5. We can observe that $R(\Lambda^h)$ is a convex set between the curves:

$$r_{low} : \{(c_0(\gamma,1), c_1(\gamma,1)), \text{ with } \gamma \in [0,1]\},$$
$$r_{up} : \{(c_0(1,b), c_1(1,b)), \text{ with } b \in [0,1]\}.$$

Note also that in the c_0, c_1 plane, the length of the coordinate lines associated with b constant are segments that become smaller as $b \to 0$. Analogously, the length of those associated with constant γ become smaller as $\gamma \to 0$. Thus, the region where either b or γ are small produces a higher density of points in the range of Λ^h. This corresponds to the upper left part of its range (see Figure 4). On the contrary, this Figure provides numerical evidence of the injectivity of Λ^h as well. In fact, any point inside $R(\Lambda^h)$ is the intersection of two coordinate lines associated with some unique b_0 and γ_0.

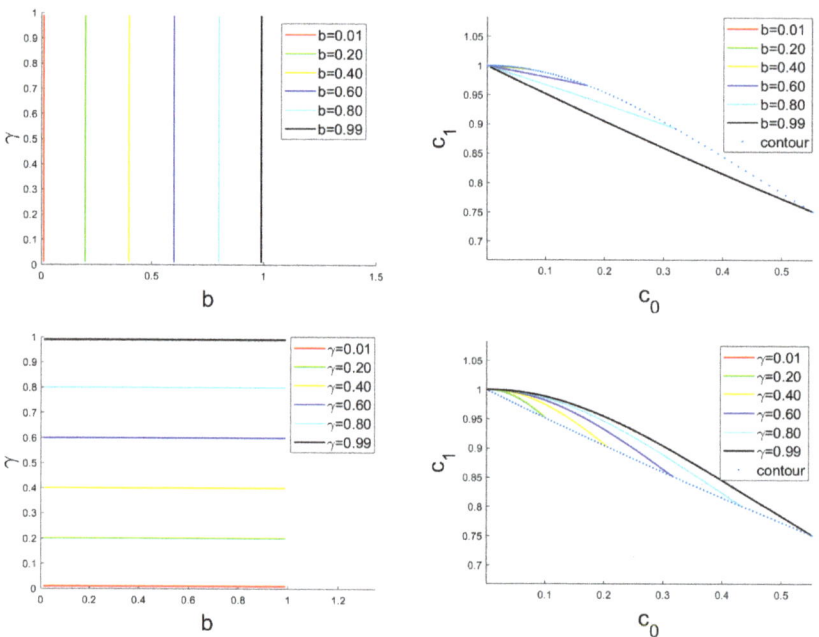

Figure 5. Coordinate lines of the map Λ^h defined in (27) ($h = 10^{-2}$). The upper figure contains the coordinate lines associated with b constant, while the lower one corresponds to γ constant.

The higher density of points in the upper left hand-side of the range of Λ^h should correspond to potentials q with a large stability constant $C_2(h,q)$, defined as:

$$C_2(h,q) = \max_{q \in F} \frac{\|q_0 - q\|_{L^1}}{\|\Lambda_{q_0} - \Lambda_q\|_{\mathcal{L}(H_\#^{1/2}; H_\#^{-1/2})}}.$$

In Figure 6, we show the level sets of $C_2(h,q)$ for $h = 10^{-4}$ and different $q \in F_h$. The region with a larger constant corresponds to small values of b (upper right figure) and larger values of c_1 (upper left and lower figures). On the contrary, the region with a lower stability constant is for b close to $b = 1$, which corresponds to the lower part of the range of Λ^h when c_0 is small.

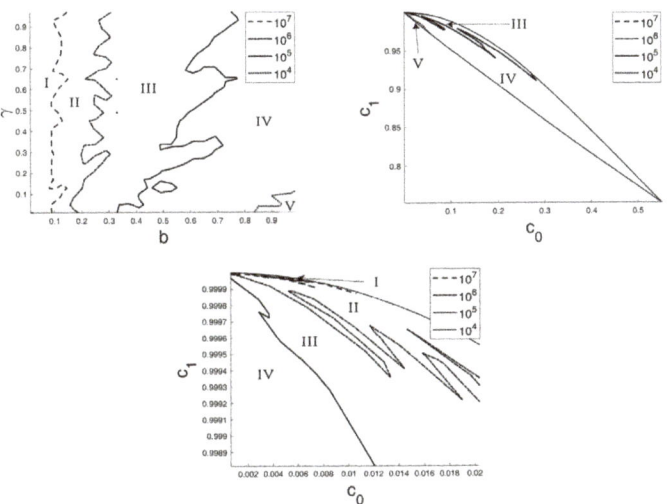

Figure 6. Level sets of the $C_2(b,\gamma)$ for $q \in F_h$ and $h = 10^{-4}$ in terms of (b,γ) (**upper left**) and in terms of (c_0, c_1) (**upper right**), and a close up of the upper left region in this last figure is in the lower figure. Regions separated by level sets are indicated: Region I corresponds to the potentials with a stability constant larger that 10^7, region II corresponds to those with a stability constant lower that 10^7 but larger than 10^6, and so on.

It is interesting to analyze the set of potentials with the same coefficient c_0 or c_1. We provide, in Figure 7, the coordinate lines of the inverse map $(\Lambda^h)^{-1}$. When increasing the value of either c_0 (light lines) or c_1 (dark lines), we obtain lines closer to the left part of the (b,γ) region. We can see that the angle between coordinate lines becomes very small for small b. In this region, close points could be the intersection of the coordinate lines associated with not so close parameters (b,γ). This agrees with the region where the stability constant is larger.

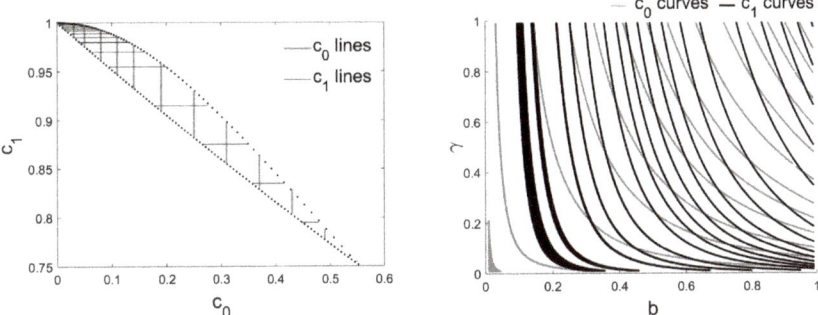

Figure 7. Coordinate lines of the map $(\Lambda^h)^{-1}$ defined in (27).

5. Conclusions

We considered the relationship between the potential in the Schrödinger equation and the associated DtN map in one of the simplest situations, i.e., for a subset of radial one-step potentials in two-dimension. In particular, we focused on two difficult problems: The stability of the map Λ (defined in (3)) and its range. In this case, the map Λ is easily characterized in terms of the Bessel functions and this allows us to give some analytical and numerical results for these problems. We proved the lack of any possible stability result by adapting the argument in [7] [Alessandrini, 1988] for the conductivity problem. We also obtained some partial Lipschitz stability when the position of the discontinuity is fixed in the potential, as well as numerical evidence of the stability with respect to the L^1 norm. Finally, we characterized numerically the range of Λ in terms of the first two eigenvalues of the DtN map and provided some insight into the regions where the stability of Λ is worse. As a future line of work, it could be interesting to consider the problem in a more complicated stage, for instance, one can study not only one-step radial potentials q in the problem, but could add more steps into the definition of the potentials.

Author Contributions: Conceptualization, S.L. and S.M.; Formal analysis, S.L. and S.M.; Investigation, S.L. and S.M.; Methodology, S.L. and S.M.; Visualization, S.L.; Writing—original draft, S.L. and S.M. All authors have read and agreed to the published version of the manuscript.

Funding: This research received of the project PDI2019-110712GB-100 .

Institutional Review Board Statement: Not applicable.

Informed Consent Statement: Not applicable.

Data Availability Statement: Not applicable.

Acknowledgments: The first author was partially supported by project MTM2017-85934-C3-3-P from the MICINN (Spain). The second author was partially supported by project PDI2019-110712GB-100 of the Ministerio de Ciencia e Innovación, Spain. We want to thank to J.A. Barceló and C. Castro their contribution to the research.

Conflicts of Interest: The authors declare no conflict of interest.

Appendix A

To prove Theorems 2 and 3, we need the following technical results regarding the the Bessel functions.

Lemma A1. *Let $J_\mu(r)$ be the Bessel functions of the first kind of order $\mu > -\frac{1}{2}$. It is well known (see [10]) that:*

$$J_\mu(r) = \frac{r^\mu}{2^\mu \Gamma(\mu+1)} + S_\mu(r),$$

where:

$$S_\mu(r) = \frac{r^\mu}{2^\mu \Gamma\left(\mu+\frac{1}{2}\right)\Gamma\left(\frac{1}{2}\right)} \int_{-1}^{1} (\cos rt - 1)\left(1-t^2\right)^{\mu-\frac{1}{2}} dt.$$

For $n = 0, 1, 2, \cdots$ and $r \in (0,1)$, the following holds:

$$-\frac{r^{n+2}}{2^{n+1}\Gamma\left(n+\frac{3}{2}\right)\sqrt{\pi}} \int_0^1 \left(1-t^2\right)^{n+\frac{1}{2}} dt \leq S_n(r) \tag{A1}$$

$$\leq -\frac{r^{n+2} \cos r}{2^{n+1}\Gamma\left(n+\frac{3}{2}\right)\sqrt{\pi}} \int_0^1 \left(1-t^2\right)^{n+\frac{1}{2}} dt,$$

$$0 < \frac{r^n}{2^{n+1} n!} \leq J_n(r) \leq \frac{r^n}{2^n n!}, \tag{A2}$$

and:

$$0 < \frac{r^n}{2^{n+2} n!} \leq J'_{n+1}(r) \leq \frac{r^n}{2^{n+1} n!}. \tag{A3}$$

More explicit estimates for $S_0(r)$ and $S_2(r)$ are given by:

$$-\frac{r^2}{4} \leq S_0(r) \leq -\frac{r^2 \cos r}{4} \leq 0, \tag{A4}$$

$$-\frac{r^4}{15\pi}0.4909 \leq S_2(r) \leq -\frac{r^4\cos r}{15\pi}0.4909. \tag{A5}$$

Proof. To prove (A1), we use:

$$\frac{r^2 t^2}{2}\cos r \leq 1-\cos(rt) \leq \frac{r^2 t^2}{2}, \quad r,t \in (0,1), \tag{A6}$$

and:

$$\int_0^1 t^2\left(1-t^2\right)^{n-\frac{1}{2}}dt = \frac{1}{2\left(n+\frac{1}{2}\right)}\int_0^1 \left(1-t^2\right)^{n+\frac{1}{2}}dt.$$

From (A1) and the well-known identities:

$$\Gamma\left(\tfrac{1}{2}\right) = \sqrt{\pi},$$
$$\Gamma(r+1) = r\Gamma(r), \quad r > 0,$$
$$2J'_{n+1}(r) = J_n(r) - J_{n+2}(r), \quad r > 0,$$

(see [11]), we get (A2), (A3), (A4), and (A5). □

The following lemma is used in the proof of Theorem 3.

Lemma A2. *For $0 < r \leq s < 1$ and $n = 0, 2$, we have:*

$$\int_0^1 (1-\cos(rt))\left(1-t^2\right)^{n-\frac{1}{2}}dt \leq \frac{\pi r^2}{28n+8},$$

and:

$$\int_0^1 (\cos(rt)-\cos(st))\left(1-t^2\right)^{n-\frac{1}{2}}dt \leq \frac{\pi(s^2-r^2)}{28n+8}.$$

Proof. The previous estimates are a consequence of (A6) and the inequality:

$$\cos r - \cos s = 2\sin\frac{s+r}{2}\sin\frac{s-r}{2} \leq \frac{s^2-r^2}{2}.$$

□

Proof of Theorem 2. We take $\gamma = 1$ without loss of generality. For $b_0 \in (0,1)$, we consider the fixed potential:

$$q_0(r,\theta) = \begin{cases} 1, & 0 < r < b_0, \\ 0, & b_0 \leq r < 1, \end{cases}$$

and a positive integer $k(b_0)$ satisfying $b_0 + \frac{1}{k(b_0)} < 1$. We define the potentials:

$$q_k(r,\theta) = \begin{cases} 1, & 0 < r < b_k, \\ 0, & b_k \leq r < 1, \end{cases} \quad k = 1,2,\cdots, \tag{A7}$$

with $b_k = b_0 + \frac{1}{k(b_0)+k}$.

We have $\|q_0 - q_k\|_{L^\infty} = 1$ and to have (21), we have to prove for $g \in H_\#^{1/2}$ that:

$$\|(\Lambda_{q_0} - \Lambda_{q_k})g\|^2_{H_\#^{-1/2}} \leq C|b_0 - b_k|^2 \|g\|^2_{H_\#^{1/2}} \leq \frac{C}{k^2}\|g\|^2_{H_\#^{1/2}}, \tag{A8}$$

where C is a constant independent of k and g.

If $g(\theta) = \sum_{n \in \mathbb{Z}} g_n e^{in\theta}$, by (15) and (16), we have:

$$\|(\Lambda_{q_0} - \Lambda_{q_k})g\|^2_{H^{-1/2}_\#} \leq \left| \frac{b_k J_1(b_k)}{b_k J_1(b_k) \log b_k + J_0(b_k)} - \frac{b_0 J_1(b_0)}{b_0 J_1(b_0) \log b_0 + J_0(b_0)} \right|^2 |g_0|^2$$

$$+ \sum_{n=1}^{\infty} \left| \frac{J_{n-1}(b_k) - b_k^{2n} J_{n+1}(b_k)}{J_{n-1}(b_k) + b_k^{2n} J_{n+1}(b_k)} - \frac{J_{n-1}(b_0) - b_0^{2n} J_{n+1}(b_0)}{J_{n-1}(b_0) + b_0^{2n} J_{n+1}(b_0)} \right|^2 (1+n^2)^{1/2} \left(|g_n|^2 + |g_{-n}|^2 \right)$$

$$= I_0^2 |g_0|^2 + \sum_{n=1}^{\infty} I_n^2 (1+n^2)^{1/2} \left(|g_n|^2 + |g_{-n}|^2 \right).$$

We start by estimating I_0.
From (A2), (A1), and (A4) $J_1(r) \leq \frac{r}{2}$, when $r \in (0,1)$ and:

$$r J_1(r) \log r + J_0(r) \geq \frac{r^2 \log r}{2} + 1 - \frac{r^2}{4}, \quad r \in (0,1).$$

Since $\frac{r^2 \log r}{2} + 1 - \frac{r^2}{4}$ is a decreasing function in $(0,1)$, we have:

$$r J_1(r) \log r + J_0(r) \geq \frac{3}{4}, \quad r \in (0,1). \tag{A9}$$

A simple calculation and this inequality gives us:

$$I_0 \lesssim b_k b_0 J_1(b_k) J_1(b_0) |\log b_k - \log b_0| + J_1(b_k) J_0(b_0) |b_k - b_0|$$

$$+ b_0 J_0(b_k) |J_1(b_k) - J_1(b_0)| + b_0 J_1(b_k) |J_0(b_k) - J_0(b_0)|,$$

where the symbol $\lesssim$ denotes that the left-hand side is bounded by a constant times the right-hand one. Thus, combining the mean value theorem, the identity $J_0'(r) = -J_1(r)$, the fact that $b_k, b_0 \in (0,1)$ and (A2), we easily get:

$$I_0 \lesssim \frac{1}{b_0} |b_k - b_0|. \tag{A10}$$

Now, we deal with I_k, $k = 1, 2, \cdots$. We use the mean value Theorem, $b_k, b_0 \in (0,1)$, $|b_k^{2n} - b_0^{2n}| \lesssim \frac{|b_k - b_0|}{n}$, (A2), and (A3) to obtain:

$$I_n \lesssim \frac{J_{n+1}(b_k) J_{n-1}(b_0) |b_k^{2n} - b_0^{2n}| + b_0^{2n} J_{n-1}(b_0) |J_{n+1}(b_k) - J_{n+1}(b_0)|}{J_{n-1}(b_k) J_{n-1}(b_0)}$$

$$+ \frac{b_k^{2n} J_{n+1}(b_0) |J_{n-1}(b_k) - J_{n-1}(b_0)|}{J_{n-1}(b_k) J_{n-1}(b_0)} \lesssim \frac{b_k - b_0}{n} \leq b_k - b_0.$$

From this estimate and (A10), we have (A8). □

Remark A1. *Theorem 2 can be extended to the case that q_0 is null. In this case, we take in (A7) $k(b_0) = 0$ and from (17):*

$$\|(\Lambda_{q_0} - \Lambda_{q_k})g\|^2_{H^{-1/2}_\#} \leq \left| \frac{b_k J_1(b_k)}{b_k J_1(b_k) \log b_k + J_0(b_k)} \right|^2 |g_0|^2$$

$$+ \sum_{n=1}^{\infty} \left| 1 - \frac{J_{n-1}(b_k) - b_k^{2n} J_{n+1}(b_k)}{J_{n-1}(b_k) + b_k^{2n} J_{n+1}(b_k)} \right|^2 (1+n^2)^{1/2} \left(|g_n|^2 + |g_{-n}|^2 \right),$$

by using $b_k \in (0,1)$, (A9), and (A2):

$$\lesssim b_k^4 |g_0|^2 + \sum_{n=1}^{\infty} \frac{b_k^{4n} J_{n+1}^2(b_k)}{J_{n-1}^2(b_k)}(1+n^2)^{1/2}\left(|g_n|^2 + |g_{-n}|^2\right),$$

$$\lesssim b_k^4 |g_0|^2 + \sum_{n=1}^{\infty} \frac{b_k^{2n+4}}{n(n+1)}(1+n^2)^{1/2}\left(|g_n|^2 + |g_{-n}|^2\right) \lesssim \frac{1}{k^4}\|g\|_{H_\#^{1/2}}^2.$$

Proof of Theorem 3. Let $q_1(x) = \gamma_1 \chi_{B(0,b_1)}(x)$, $q_2(x) = \gamma_2 \chi_{B(0,b_2)}(x)$ in F_b and $g(\theta) = \frac{1}{2^{1/4}} e^{i\theta}$.

$$\|\Lambda_{q_1} - \Lambda_{q_2}\|_{\mathcal{L}(H_\#^{1/2}; H_\#^{-1/2})}^2 \geq \|(\Lambda_{q_1} - \Lambda_{q_2})g\|_{H_\#^{-1/2}}^2$$

$$= \left|\frac{J_0(b_1\sqrt{\gamma_1}) - b_1^2 J_2(b_1\sqrt{\gamma_1})}{J_0(b_1\sqrt{\gamma_1}) + b_1^2 J_2(b_1\sqrt{\gamma_1})} - \frac{J_0(b_2\sqrt{\gamma_2}) - b_2^2 J_2(b_2\sqrt{\gamma_2})}{J_0(b_2\sqrt{\gamma_2}) + b_2^2 J_2(b_2\sqrt{\gamma})}\right|^2 \quad \text{(A11)}$$

$$\geq \frac{4\Pi^2}{\left(1 + \frac{b_1^4 \gamma_1}{8}\right)^2 \left(1 + \frac{b_2^4 \gamma_2}{8}\right)^2},$$

where:

$$\Pi = \left|b_2^2 J_0(b_1\sqrt{\gamma_1}) J_2(b_2\sqrt{\gamma_2}) - b_1^2 J_0(b_2\sqrt{\gamma_2}) J_2(b_1\sqrt{\gamma_1})\right|,$$

and we used (A2) for $n = 0, 2$. On the contrary:

$$\Pi \geq \frac{1}{8}\left|b_2^4 \gamma_2 - b_1^4 \gamma_1\right| - J_1 - J_2 - J_3, \quad \text{(A12)}$$

where:

$$J_1 = \left|b_2^2 S_2(b_2\sqrt{\gamma_2}) - b_1^2 S_2(b_1\sqrt{\gamma_1})\right|, \quad \text{(A13)}$$

$$J_2 = \frac{1}{8}\left|b_2^4 \gamma_2 S_0(b_1\sqrt{\gamma_1}) - b_1^4 \gamma_1 S_0(b_2\sqrt{\gamma_2})\right|, \quad \text{(A14)}$$

and:

$$J_3 = \left|b_2^2 S_0(b_1\sqrt{\gamma_1}) S_2(b_2\sqrt{\gamma_2}) - b_1^2 S_0(b_2\sqrt{\gamma_2}) S_2(b_1\sqrt{\gamma_1})\right|. \quad \text{(A15)}$$

To estimate J_i, $i = 1, 2, 3$, we use (A2), (A4), (A5), and Lemma A2. We get:

$$J_1 \leq \frac{b_2^4 \gamma_2^2 |b_1^2 - b_2^2|}{30\pi} + \frac{b_1^2(b_2^2 \gamma_2 + b_1^2 \gamma_1)|b_2^2 \gamma_2 - b_1^2 \gamma_1|}{96}. \quad \text{(A16)}$$

$$J_2 \leq \frac{b_1^2 \gamma_1 |b_2^4 \gamma_2 - b_1^4 \gamma_1|}{32} + \frac{b_1^4 \gamma_1 |b_2^2 \gamma_2 - b_1^2 \gamma_1|}{32}. \quad \text{(A17)}$$

$$J_3 \leq \frac{b_1^2 b_2^4 \gamma_1 \gamma_2^2 |b_2^2 - b_1^2|}{120\pi} + \frac{b_1^6 \gamma_1 |b_1^2 \gamma_1 - b_2^2 \gamma_2|}{36\pi^{\frac{3}{2}}} + \frac{b_1^4 b_2^2 \gamma_1 \gamma_2 |b_1^2 \gamma_1 - b_2^2 \gamma_2|}{36\pi^{\frac{3}{2}}} \quad \text{(A18)}$$

$$+ \frac{b_1^2 b_2^4 \gamma_1 \gamma_2^2 |b_1^2 \gamma_1 - b_2^2 \gamma_2|}{480\pi^{\frac{3}{2}}}.$$

□

Proof of (22). We suppose that $b_1 = b_2 = b > 0$. We obtain:

$$J_1 \leq \tfrac{b^6}{96}|\gamma_1 - \gamma_2| \leq 0.01041 b^4 \|q_1 - q_2\|_{L^\infty(B(0,1))},$$

$$J_2 \leq \left(\tfrac{b^6}{32} + \tfrac{b^6}{32}\right)|\gamma_1 - \gamma_2| \leq 0.0625 b^4 \|q_1 - q_2\|_{L^\infty(B(0,1))},$$

$$J_3 \leq \left(\tfrac{b^8}{36\pi^{\frac{3}{2}}} + \tfrac{b^{10}}{36\pi^{\frac{3}{2}}} + \tfrac{b^8}{480\pi^{\frac{3}{2}}}\right)|\gamma_1 - \gamma_2| \leq 0.01004 b^4 \|q_1 - q_2\|_{L^\infty(B(0,1))},$$

and from (A11) and the above estimates, we get that:

$$II \geq 0.042 b^4 \|q_1 - q_2\|_{L^\infty}.$$

Since γ_1, γ_2, and b are less than 1, (5.11) and the above estimate gives us:

$$\|\Lambda_{q_1} - \Lambda_{q_2}\|^2_{\mathcal{L}(H_\#^{1/2}; H_\#^{-1/2})} \geq 4\frac{8^4}{9^4}(0,042)^2 b^8 \|q_1 - q_2\|^2_{L^\infty} = 0,0044 b^8 \|q_1 - q_2\|^2_{L^\infty},$$

this implies (22). □

Proof of (23). Now $\gamma_1 = \gamma_2$. Let us define:

$$M(\gamma, b_1, b_2) = \gamma\left(b_1^3 + b_1^2 b_2 + b_1 b_2^2 + b_2^3\right).$$

It is easy to check that:

$$\tfrac{1}{8}|b_2^4 \gamma_2 - b_1^4 \gamma_1| = \tfrac{1}{8} M(\gamma, b_1, b_2)|b_2 - b_1|,$$

$$J_1 \leq \left(\tfrac{1}{30\pi} + \tfrac{1}{9\pi^{\frac{3}{2}}}\right) M(\gamma, b_1, b_2)|b_2 - b_1|,$$

$$J_2 \leq \left(\tfrac{1}{32} + \tfrac{1}{256\pi^{\frac{1}{2}}}\right) M(\gamma, b_1, b_2)|b_2 - b_1|,$$

$$J_3 \leq \left(\tfrac{1}{120} + \tfrac{1}{18\pi^{\frac{3}{2}}} + \tfrac{1}{420\pi^{\frac{3}{2}}}\right) M(\gamma, b_1, b_2)|b_2 - b_1|,$$

therefore:

$$\|\Lambda_{q_1} - \Lambda_{q_2}\|_{\mathcal{L}(H_\#^{1/2}; H_\#^{-1/2})} \geq \frac{2}{\left(1 + \tfrac{1}{8}\right)^2}\left(\tfrac{\gamma}{8}\left|b_1^4 - b_2^4\right| - J_1 - J_2 - J_3\right)$$

$$\geq \frac{2}{\left(1 + \tfrac{1}{8}\right)^2} 0,04216 M(\gamma, b_1, b_2)|b_2 - b_1| \geq 0,2665 \gamma b^3 |b_2 - b_1|,$$

and we obtain (23). □

References

1. Blåsten, E.; Imanuvilov, O.Y.; Yamamoto, M. Stability and uniqueness for a two-dimensional inverse boundary value problem for less regular potentials. *Inverse Probl. Imaging* **2015**, *9*, 709–723.
2. Tejero, J. Reconstruction and stability for piecewise smooth potentials in the plane. *SIAM J. Math. Anal.* **2017**, *49*, 398–420. [CrossRef]
3. Tejero, J. Reconstruction of rough potentials in the plane. *Inverse Probl. Imaging* **2019**, *13*, 863–878. [CrossRef]
4. Ingerman, D.V. Discrete and continuous Dirichlet-to-Neumann maps in the layered case. *SIAM J. Math. Anal.* **2000**, *31*, 1214–1234. [CrossRef]
5. Mandache, N. Exponential instability in an inverse problem for the Schrödinger equation. *Inv. Probl.* **2001**, *17*, 1435–1444. [CrossRef]
6. Beretta, E.; De Hoop, M.V.; Qiu, L. Lipschitz stability of an inverse boundary value problem for a Schrödinger-type equation. *SIAM J. Math. Anal.* **2013**, *45*, 679–699. [CrossRef]
7. Alessandrini, G. Stable determination of conductivity by boundary mea- surements. *Appl. Anal.* **1988**, *27*, 153–172. [CrossRef]

8. Müller, J.; Siltanen, S. *Linear and Nonlinear Inverse Problems with Practical Applications*; SIAM Computational Science and Engineering: Philadelphia, PA, USA, 2012.
9. Uhlmann, G. Inverse problems: Seeing the unseen. *Bull. Math. Sci.* **2014**, *4*, 209–279. [CrossRef]
10. Grafacos, L. *Classical Fourier Analysis*, 2nd ed.; Springer: Berlin/Heidelberg, Germany, 2008.
11. Lebedev, N.N. *Special Functions and Their Applications*; Dover Publications, Inc.: New York, NY, USA, 1972.

Article
Modeling Political Corruption in Spain

Elena de la Poza [1], Lucas Jódar [2] and Paloma Merello [3,*]

[1] Centro de Ingeniería Económica, Universitat Politècnica de València, 46022 Valencia, Spain; elpopla@esp.upv.es
[2] Instituto de Matemática Multidisciplinar, Universitat Politècnica de València, 46022 Valencia, Spain; ljodar@imm.upv.es
[3] Department of Accounting, University of Valencia, 46071 Valencia, Spain
* Correspondence: Paloma.Merello@uv.es

Abstract: Political corruption is a universal phenomenon. Even though it is a cross-country reality, its level of intensity and the manner of its effect vary worldwide. In Spain, the demonstrated political corruption cases that have been echoed by the media in recent years for their economic, judicial and social significance are merely the tip of the iceberg as regards a problem hidden by many interested parties, plus the shortage of the means to fight against it. This study models and quantifies the population at risk of committing political corruption in Spain by identifying and quantifying the drivers that explain political corruption. Having quantified the problem, the model allows changes to be made in parameters, as well as fiscal, economic and legal measures being simulated, to quantify and better understand their impact on Spanish citizenship. Our results suggest increasing women's leadership positions to mitigate this problem, plus changes in the political Parties' Law in Spain and increasing the judiciary system's budget.

Keywords: contagion effect; difference equation; elections; labor condition; mathematical compartmental discrete model; political corruption; revolving doors; sensitivity analysis; simulation

Citation: de la Poza, E.; Jódar, L.; Merello, P. Modeling Political Corruption in Spain. *Mathematics* **2021**, *9*, 952. https://doi.org/10.3390/math9090952

Academic Editor: Mariano Torrisi

Received: 1 March 2021
Accepted: 22 April 2021
Published: 24 April 2021

Publisher's Note: MDPI stays neutral with regard to jurisdictional claims in published maps and institutional affiliations.

Copyright: © 2021 by the authors. Licensee MDPI, Basel, Switzerland. This article is an open access article distributed under the terms and conditions of the Creative Commons Attribution (CC BY) license (https://creativecommons.org/licenses/by/4.0/).

1. Introduction

Political corruption is a universal phenomenon which, even though the times, ideas, laws and cultures of different countries have evolved, has remained unchanged since ancient times [1]. As long as we can remember, political corruption has accompanied the evolution of human kind through its different cultural stages or civilizations. Early referrals to the concept date back to the Pharaonic Egypt period [2], with later evidence indicating that Roman politics hit the bottom due to its corruption in the Republic times of Roman civilization (70 and 50 BC), and as a result the legal code "Twelve Tables" [3] being passed, which imposed the death penalty on judges who accepted bribes and politicians who attempted to influence election results through bribery or other forms of "soft power". This concept lies in the ability to shape others' preferences based on culture and intangible assets, such as the credibility and trustworthy of individuals and institutions [4].

As the political corruption concept is susceptible to ambiguity, we must specify it. Political corruption can be defined as any act or legal or illegal omission of someone who, based on a public office (elected or appointed) embracing political positions, but also on a position in a labor union or business association, favors a particular interest that causes public (not necessarily monetary) harm [5–7]. Hence political corruption can be for private and group enrichment, and also for power preservation, purposes [8,9]. According to [10], these two forms of political corruption are often connected.

The concept given for political corruption highlights two relevant points: first, that which motivates political corruption; that is, the search for self-enrichment, ego and power maintenance (as opposed to a non-corrupt political leader's concern for citizenship's well-being); second, the consequences of political corruption: national impoverishment, institutional decay, arbitrary power, authoritarian tendencies and less freedom and democracy [11–13].

Political corruption decelerates social growth and economic activity [14,15], diverts resources from basic services [16], reduces innovation [17] and, consequently, also the introduction of new products and technologies because innovators and entrepreneurs usually lack political connection [18]. In general, foreign direct investments decrease [19–21], and national firms' value drops [22–25]. Thus, industry must pay more to lenders given the perceived political instability impacting the credit market [26–28]. At those countries where the state institutions are weaker, corruption is often linked to violence, whereas in the so called mature democracies, corruption means the increase of economic and social insecurity and also the opportunity for the privilege to get richer at the expense of everyone else [29].

Hence, literature has analyzed the types of political corruption, its causes or consequences from a theoretical perspective or even empirically focusing on indicators built using historical information. This work means a contribution to the literature and bridges the gap in the literature by: (i) mathematical modeling the political corruption in a free market economy in which democracy does not serve as a warranty of policy making responds to the public interest and (ii) quantifying the total population at risk of committing political corruption. Following, the study identifies the drivers of the problem and highlights the main novelties of the research in terms of methodology employed, data and its contributions.

1.1. Political Corruption in Spain

Even though political corruption is a cross-country reality, its level of intensity and the manner of its effect vary from one nation to another [18]. In this way, [30] since 1995 CPI 2020 has annually issued the corruption perception index to measure from 0 to 100 the perceived level of corruption in the public sector worldwide in accordance with businesspeople and experts. It has become the leading global indicator of public sector corruption. Hence, [30] CPI argues that public sector transparency is the key to ensure public resources being appropriately spent. However, the 2020 annual report shows that more than two thirds of 180 countries score below 50. In particular, Spain ranked 32 worldwide with 63 points, while Western Europe and the European Union scored 66 on average. Denmark had the highest score with 88 points, and Romania the lowest with 44 [30–32].

Thus, identifying the factors that explain political corruption is essential for understanding the trends and differences of this phenomenon among countries. The factors explaining the persistence of political corruption in Spain are discussed below in accordance with previous research [33]. Culture, and particularly religion [34], explain why corruption rates are higher in south than in northern Europe. In Latin America, the culture of the former was inherited from Spanish and Portuguese cultures.

Second, political corruption is explained by the nature of the Spanish political party system and the Parties' Law that guarantees continuity of the establishment [35,36]. In relation to the laws regulating the parties, there is the opacity of financing political parties, which are not obliged to publish their financial information [37]. Indeed, Spain does not enforce political parties disclosing their financial information or candidate funding in their reports, while 93% of OECD countries do [38], even though institutionalized transparency and accountability are the main aspects that promote the integrity and fairness of public decision making [39]. This situation is connected to previous scandals of political corruption that have affected the two longest-standing parties: PP (right-wing) and PSOE (left-wing) [40,41], which resulted in the end of a two-party system [42,43], and in the appearance of new political parties during general and local elections campaigning with vows to get rid of what they brand a "corrupt political elite". However, the most vindictive of these emergent accuser parties has also been prosecuted by irregular financing [44], and thus perpetuates the same phenomenon: political corruption in Spain.

Thus, analyzing the functioning and structure of Spanish political parties would evidence the fulltime politicians whose professional career involves occupying a seat at office for the long term, but lack experience in the wider world [45,46]. This fact fosters patronage as an expression of political corruption; that is, recruiting public sector employees based on political connections rather than on their skills and formal qualifications [47]. Indeed, merit-based bureaucracy, as opposed to one in which politicians appoint employees at will, is expected to reduce corruption [48].

However, on the aforementioned factor, an important connotation is found in gender terms given the systematic differences in how men and women perceive corruption [49,50]. Several studies have empirically evidenced a negative relation between women's presence in politics and its effect on corruption [51,52]. This can be explained by different gender behavior as women are generally more collaborative than men, but also more altruistic and ready to engage in "helping" behavior [53,54]. Moreover, in accordance with [51], political corruption is a deterrent to women's representation because it reinforces "clientelist" networks that privilege men.

In order to commit any crime, two conditions are needed simultaneously [55], sufficient aggressiveness to dare and moral disconnection to bear the thought of our conscience. The evolution of the species [56] has developed a distribution of social roles and habits, which are transmitted genetically, in which males have primarily devoted themselves to tasks related to aggressiveness and are more trained and accustomed, such as counterfeiting, big animals, defense of the territory and war. These social habits make men, in general, more aggressive than women, and for this reason, more prone to commit any crime, and in particular for political corruption. This aggressive training of men makes them have less social shame than women, men care less than women about being discovered in crime. From this, it follows that, at least for a long time, women are less at risk of committing corruption. This implies promoting the presence of women in politics and, consequently, in office as a tool to fight corruption. Moreover, Ref. [57] argues that the longer women remain in office, the lower the corruption levels are, which contrasts with how men in office impact political corruption.

The next factor that aggravates the situation is highly politicized Spanish media [58,59] based on ideological alignments; as Refs. [60,61] argue, presently the media's role does not involve promoting knowledge and defending public interest, but is instead a strategy for political action. In fact, in an attempt to control journalism and the media, political parties run communication and news offices. Indeed, Spanish politicians calling press conferences without allowing questions, and refusing camera operators and reporters admission to election campaigns, have become common practices [62].

The last relevant factor to correlate with political corruption is lack of independence among judiciary, executive and legislative powers. Judiciary Councils are institutions created to protect judges' autonomy. However, according to [63], 36% of Spanish judges perceive their Judiciary Council as not respectful of their independence. This rate was the worst result of the survey carried out by European Network of Judiciary Councils during the 2014–2015 period. This situation is explained by the politicization of the General Council of the Judicial power reported by Nieto [64], which means that the political parties in office control the nomination of the candidates appointed by the parliament.

Thus, imperfect judicial and media independence does not favor the end of the problem [65,66], and even less so when political parties are unable to make decisions against their partisan interests, even when these decisions are for the good of Spanish society and the country's socio-economic future. Likewise, an intoxicating and generalized state of moral relaxation has been established in Spanish society, which excuses the political corruption phenomenon as being inevitable and inherent to the political class and is, therefore, irremediable. Not only does this not slow the problem down, it perpetuates it and amplifies its dimension [55,67], which indiscriminately affects all social chain links [68].

The effect of Spanish political corruption is corrosive because it deteriorates the country's image: economy drops, especially for a country like Spain that is so dependent on tourism [69], with citizens mistrusting the national institution [70], which affects foreign investors and citizens' quality of life, and makes the country's future worse [71,72]. Lack of trust in institutions generates moral disengagement, which makes it easier for citizens to excuse the political class's corruption, who consider it alien, but use it at the same time as an excuse to commit themselves [55]. This kind of contagion is very counterproductive given its social, economic and moral impact on society [73].

The most important factors to explain the current situation are the party system and its laws, where political offices do not respond to citizens, but to the political leader who has appointed them, and where lack of self-criticism, transparency and accountability come into play.

1.2. Novelties of the Study and the Paper's Structure

The demonstrated cases of political corruption that have been echoed by the media in recent years for their economic, judicial and social significance are merely the tip of the iceberg of a problem hidden by many interested parties, and also due to the shortage of the means to fight it. Particularly for Spain, and in accordance with [69,74], political corruption emerged (more than 200 reported cases) during the economic boom between 2000 and 2011, while almost no local corruption was previously registered.

In this work, we quantify the level of risk of committing political corruption for the population living in Spain aged between 16 and 70 years old. Individual behavior is unpredictable, but aggregate behavior can be predicted by mimetic contagious and herding human behavior [75–78]. According to [67], humans are driven by emotions. Unlike previous studies that have centered on political corruption [79,80], we managed a political corruption concept that is not only limited to individuals in the political scenario, but also embraces the rest of the population.

This study is a novel contribution to the literature and bridges the gap in the literature about modeling and quantifying the total population of Spain in accordance with its risk of committing political corruption. It also identifies four levels of risk of committing political corruption. Apart from classifying the population according to their level of risk of committing political corruption, this study also takes into account their employment situation at the time the analysis was carried out.

Our model allowed us to predict the risk of political corruption in Spain given the mimetic nature of humans by constructing a discrete finite epidemiological model [81] that classifies and quantifies citizenship in Spain into subpopulations according to their risk of committing political corruption. Despite previous theoretical approaches to the problem that focus on diagnosing the causes or processes of political corruption cases [82,83], our model is dynamic and classifies the population on an annual basis according to their level of risk of committing political corruption over time during the 2015–2023 period [35,84–86].

Previous studies have employed surveys or historical statistical data to quantify the corruption perceived by different stakeholders at cross-country, national or local levels by building indicators or regression models [6,42,87,88]. In contrast, we study subpopulation trends during the study period by quantifying the annual dynamic transits among subpopulations. These transits are produced by an individual's occupational status at the time the analysis was performed, combined with the following external variables that were quantified for each period: elections, time in office, gender, moral disconnection, economy, religion and the "revolving doors" effect [89]. This effect is the transfer of professionals from the government and public administrations to private companies or social entities, which leads to conflicts of interest and the possibility of corruption [90].

Thus, at each given time point, we quantified the number of individuals by their risk of committing political corruption in Spain. To the best of our knowledge, this is the first study to dynamically score political corruption by levels and sizes in a given country. The relevance of this study lies in reporting the problem to the public authorities responsible for addressing policies to stop this trend.

This article is arranged as follows: Section 2 describes the hypotheses and methodology. Section 3 presents the model construction. Section 4 shows the results and simulations. Section 5 offers the discussion of the results and conclusions.

2. Hypotheses and Methodology

2.1. Subpopulation Definition

The political corruption risk concept is defined in previous sections as the risk of legal or illegal acting or omission by someone based on a public office (elected or appointed) that favors self-interest (or a third party's interest), which causes public damages, which is not necessarily monetary, and should be understood as the suboptimal results obtained from their management. In this definition, the term "public office" embraces political positions, but also includes any management position in labor unions or business associations.

With this political corruption definition, we posed some hypotheses that led to the model's construction.

The target population included residents in Spain aged 16–70 years. This target population was divided into 20 subpopulations by taking into account their level of risk of committing political corruption, and their alternative or complementary professional life to hold public office for year n:

$$P(level\ of\ risk,\ labor\ condition,\ time) \qquad (1)$$

Four levels of risk of committing political corruption were established: zero risk (people who do not hold or are not in contact with public offices); low risk (less than 10%), individuals likely to collaborate with public offices (member of political parties, unions or business associations); medium risk (up to 25%), people who are directly or indirectly elected public representatives, and manage public budgets; high risk (more than 50%), high positions who handle large budgets and/or have relevant decision-making capacity, and have remained in office since previous administrations.

$$Z_j(n) = \text{Zero-risk subpopulation}$$

$$B_j(n) = \text{Low-risk subpopulation}$$

$$M_j(n) = \text{Medium-risk subpopulation}$$

$$A_j(n) = \text{High-risk subpopulation}$$

where j is the occupational status, which can take the values in Table 1.

Table 1. Occupational status classifications.

j	Definition (Age Range)
1	pre-labor (young people aged under 26 years old)
2	unemployed (26, 70)
3	self-employed or employed by a private company (26,70)
4	employed by a public company or public administration (26,70)
5	civil servant (26,70)

2.2. Hypotheses and Initial Subpopulations

The model transits and initial subpopulations at n = 0 (July 2015) were drawn by assuming the following hypotheses:

Hypothesis H1. *Individual behavior is not predictable, but aggregated behavior might be* [73].

Hypothesis H2. *Human behavior is driven by desire and fear* [86].

Hypothesis H3. *The combination of drivers makes subpopulations evolve from one category to another.*

Hypothesis H4. *Mimetic human behavior and herding* [67,75–78,91].

Hypothesis H5. *Subpopulation transits can occur to higher, but also to lower risk categories* [67].

Hypothesis H6. *Retirees (proxy age > 70 years) are assumed to not participate in corruption as only those of working age can accept political positions.*

Hypothesis H7. *The immigrants who reach political management positions are negligible.*

The initial subpopulations are obtained from Spanish statistical data [92,93] according to the following assumptions:

First, some specific labor groups were ruled out from the initial data because of their unavailable access to political management positions. Therefore, the retired population younger than 71 years ($L(0) = 2,239,500$) was not taken into account [93]. The initial data did not consider the house-keeping population in age intervals (16,70) ($H(0) = 2,975,400$) [93]. Finally, pensioners (and widows/widowers with tax-paying pensions) and the disabled subpopulation with fixed incomes younger than 71 years (non tax-paying pensions) [92] were removed from the initial data; $W(0) = 553,800 + 1,961,300 = 2,515,100$.

The zero-risk subpopulation was obtained according to the data of the Spanish National Statistics Institute [93] and the Statistical Bulletin of the Personnel at the Service of Public Administrations, corresponding to January 2016 and published by the Spanish Ministry of Finance and Civil Services [94].

For the low-risk subpopulation, we considered members of political parties and unions to be susceptible, along with those individuals belonging to public or private entities that collaborate or incite corruption, including advisers and trust positions. Therefore, labor subpopulations $Bj(0)$ were obtained as follows (data collected from the Spanish Association of Industrial Participations, and websites of political parties and trade unions):

- $B1(0)$: Youths of political parties and unions, age interval (16,25).
- $B3(0)$: 3/10 Members of left-wing parties and unions + 3/10, and of right-wing and center political parties + 1/3 trade union members, aged (26,70) self-employed or employed by private firms.
- $B4(0)$: 3/10 Members of left-wing parties and unions +3/10 and right-wing and center political parties + 1/3 union members, aged (26,70) employed by public firms or public administrations.
- $B5(0)$: 3/10 Members of left-wing parties and union +3/10 and right-wing and center political parties +1/3 trade union members + 1/10 members only of leftwing parties and are civil servants aged (26,70).

We considered members to be anyone paying a fee.

The medium-risk subpopulation was formed by individuals serving in office (Local Government, Regional or Central Governments) occupying top management posts of public companies and entities (water management entities, hospitals, public TV, universities). Note that according to the subpopulation definition, $M1(0)$ and $M20(0)$ equaled zero. Thus, the labor subpopulations $Mj(0)$ were calculated as:

- $M3(0)$: 94% Union members aged (26,70) working for private companies.
- $M4(0)$: Local governments (City mayor and council) being paid income under 1000 euro/month.
- $M5(0)$: 6% Union members who are civil servant + local governments being paid income under 1000 euros/month who are civil servants.

Finally, individuals in management positions related to public budgets (local, regional or central governments), and the managers of large public organizations (water companies, hospitals, public TV, universities) who have remained in office since previous administrations, are assumed at a high risk of political corruption. Note that according to the subpopulation definition, A1(0) and A2(0) equaled zero. Hence, the labor subpopulations Aj(0) were obtained as:

- A3(0): Manager positions of business associations (CEOE), advisors and board members of private companies.
- A4(0): Local governments (City Mayor/Mayoress + council employees with incomes over 1000 euro/month), members of regional governments, members of the National Parliament, managers of trade unions, managers of public entities, advisors employed by public administrations.
- A5(0): CFO and rectors of public universities.

Note that the total population in Spain aged between 16 and 70 years old at n = 0 is P(0) = 23,985,102. According to the percentages in Table 2, the total risk subpopulations amounted to Σ Zj(0) = 20,701,24, Σ Bj(0) = 2,941,579, Σ Mj(0) = 267,872 and Σ Aj(0) = 74,427.

Table 2. Initial subpopulations in percentages per occupational status, n = 0 (July 2015).

		j = 1	j = 2	j = 3	j = 4	j = 5
Zj(0) =	86.3%	21.4%	24.9%	50.7%	0.3%	2.8%
		4,432,952	5,149,000	10,489,575	54,035	575,662
Bj(0) =	12.3%	2.6%	0.0%	36.5%	28.4%	32.5%
		77,870	-	1,072,500	836,639	954,570
Mj(0) =	1.1%	0.0%	0.0%	88.7%	5.2%	6.0%
		-	-	237,700	14,000	16,172
Aj(0) =	0.3%	0.0%	0.0%	8.4%	91.4%	0.2%
		-	-	6,225	68,020	182
TOTAL	100%	18.8%	21.5%	49.2%	4.1%	6.4%

j = occupational condition: j = 1: pre-labor (young people under 26 years old); j = 2: unemployed; j = 3: self-employed or employed by a private company; j = 4: employed by a public company or public administration; j = 5: civil servant.

3. Model Construction

3.1. Transit Coefficients

This section is divided into subheadings. It should provide a concise and precise description of the experimental results along with their interpretation, and the experimental conclusions that can be drawn. The dynamic population model [78,81,95,96] quantifies the amount of people aged 16–70 years old at risk of committing political corruption in Spain.

Individuals transit to lower or higher levels of probably committing political corruption by the conjunction of factors (Figure 1). The following transit vectors appear: demography, time in office, contagion effect, elections, fear of losing office and the "revolving doors" effect. Other environmental factors, such as gender, culture and religion, economy, lack of political transparency, controlled press and lack of independent justice, can reinforce or encourage dynamics.

In addition, the political chronogram of the study period conditioned the evolution of subpopulations (Figure 2).

Let us define the demographic vector by taking the transit coefficients as constant for the study period (2015–2023). The demographic transit is obtained by adding some incomers (individuals who reached the age of 16 years) and some outgoers (deaths and labor/political retirements).

- A total of I = 427,394 individuals reached the age of 16 years in July 2016 [93]. We distributed incomers between Z_1 and B_1 as $I_1 = \beta_{Z1} I$; $I_{B1} = \beta_{B1} I$, where β_{Z1} = 0.98; β_{M1} = 0.02 are the rates of incomers per level of risk of committing corruption.
- Let us take death rate dj as d_1 = 0.000222; d_j = 0.0034808, $\forall j \neq 1$. Data as of 2015 [93].
- Let us define r_{ij} as the retirement rate from labor and political career at 71 years. Therefore, by considering that a total ΣR_{ij} = 426,626 individuals over 70 years old (January

2015, [93] would leave the model, rates were estimated according to subpopulations' initial weights by taking $r_{i1} = 0; r_{ij} > 0; \forall j \neq 1$.

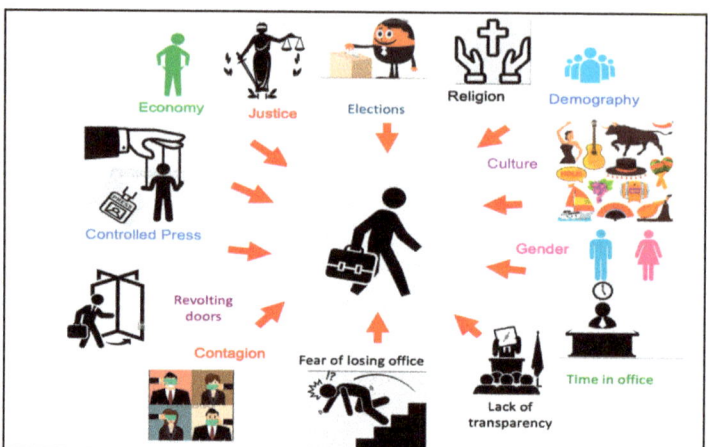

Figure 1. Transits diagram.

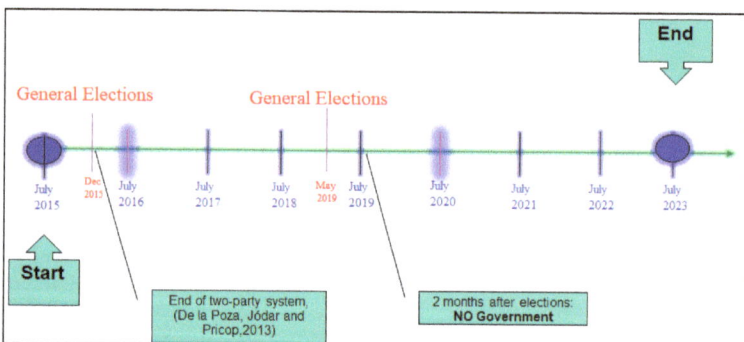

Figure 2. Chronogram.

The economic factor had two opposing effects. On the one hand, when the economy is favorable, transits to higher risk subpopulations increase due to the appeal of accessing positions that manage larger budgets. During the study period, this effect was not included in the model because it did not involve an extensive economic situation in Spain. On the other hand, in bad economic situations, voters' dissatisfaction tends to favor a change of government and alternation of office (hereafter the election effect [97]), as well as loss of party and union members.

As regards the election effect, transits from A_j to B_j occur as a result of change in election results and individuals leaving political seats, while the same amount of transits occurs from B_j to M_j with new political parties appearing and new politic positions being assigned [35]. Therefore, the transit can be assumed as going from A_j to M_j ($A_j \rightarrow B_j \rightarrow M_j$). This transit only occurs with elections (2015, 2019) and takes place at $n + 1$ election year (July 2016 and 2020). Let us call $\mu(n)$ the election effect parameter that takes the value $\mu(1) = \mu(5) = 0.6$ as 40% of positions remain in office after elections [89,98]. Parameter $\mu(n) = 0$ for $n \neq 1, 5$.

The disaffiliation transit is defined in the model as γ. The economic factor leads to loss of members of traditional parties and trade unions because of members' dissatisfaction and/or their inability to pay fees. So it is assumed that in relation to an economic crisis situation, 1% of members of traditional parties and trade unions (i.e., CCOO, UGT, PSOE, PP; 81% of members in 2015) transit from B_j to Z_j. Thus, $\gamma = 0.81 \times 0.01 = 0.0081$ is the population rate from B_j that annually leave.

Let us take the effect of time in office ($\tau_j(n)$) as a detrimental factor because it favors training and access to situations that lead to inappropriate behavior developing when managing public budgets. The effect of this transit takes place at $n + 1$ election year (2016, 2020). By considering that 40% of politicians keep their seat, the politicians who do not keep their seat transit from A_j to B_j, while 50% of politicians who keep their seat transit to a higher category; from M_j to A_j. For those new incomers in office, with political renewal and bipartisanship ending, when a position is renewed and a new individual enters, they transit from M_j to A_j, but in smaller proportions, which increase over time. By considering election years 2015 and 2019 ($n = 0$ and $n = 4$, respectively) and that transits take place at $n + 1$ election year ($n = 1$, $n = 5$), we can assume that new elected officers need at least 1 year to start their corrupt practices. Thus, $\tau_j(1) = 0$, $\forall j$ and $\tau_j(5) = 0$, $\forall j$. The time in office parameter per labor subpopulation takes the following values for the next 3 years of term of office:

- second term of office, $\tau_3(2) = 2.5\%$, $\tau_4(2) = 2.5\%$, $\tau_5(2) = 1.25\%$,
- third term of office, $\tau_3(3) = 5\%$, $\tau_4(3) = 5\%$, $\tau_5(3) = 2.5\%$,
- fourth term of office, $\tau_3(4) = 10\%$, $\tau_4(4) = 10\%$, $\tau_5(4) = 5\%$.

In addition, we have to consider the possibility of unemployment leading to individuals' fear. Civil servants and pre-labor individuals do not face such pressure or fear. However, for those at low- and medium-risks, lack of an alternative professional career ($j = 2,3,4$) and the possibility of losing one's seat in office brings about fear and promotes corrupt behaviors. In line with this, divergence between labor productivity and compensation may increase the tendency to prefer a seat in office [99]. Thus, let us define ρ_{ij} as fear of losing one's seat, which takes the value ρ_{i1}, $\rho_{i5} = 0$ and, given the assumption that the probability of that transit being double for M_j compared to B_j and lower for $j = 4$, the parameter takes the following values: for transits from low- to medium-risk subpopulations $\rho_{B2} = 0.005$, $\rho_{B3} = 0.005$, $\rho_{B4} = 0.0025$; and for transits from medium- to high-risk subpopulations $\rho_{M2} = 0.01$, $\rho_{M3} = 0.01$, $\rho_{M4} = 0.005$.

Human behavior is characterized by an irrational component. Decision making is driven by isomorphism and contagion from individuals in the near environment [55,75–77,100]. This contagion might imply moral disengagement and the normalization of some unethical behaviors in individuals. Indeed religion and ethical codes may cushion the contagion effect [5,6]. We define α_i as the moral disengagement coefficient which affects 90% of the population, excluding the 10% of religious and/or ethical people who are not affected. This factor affects all the subpopulations, but to a lesser extent to the zero-risk individuals. Hence, the parameters are obtained as $\alpha_Z = 0.005 \times 0.9 = 0.0045$ and $\alpha_B = \alpha_M = 3\alpha_Z = 0.135$.

Finally, the "revolving doors" effect is a political factor that needs to be considered. "Revolving doors" are defined as the situation in which an individual leaves his/her political seat and takes a board seat in a large company (e.g., IBEX 35 companies in Spain) [90]. This situation brings about a transit from the high-risk to low-risk subpopulations [98]. Let us define D_j as the "revolving doors" effect parameter, which is calculated by these assumptions: this transit only affects j = 2,3,4; according to the Office of Conflicts of Interest [89], 23 positions per year leave their political seats and take a board seat. However, "revolving doors" affect the politician, and at least one near advisor. In this way, the real individuals affected are at least twice those accounted for. Therefore, $D_j = 0$ for j = 1,5 and $D_j = 15$ for j = 2,3,4.

3.2. Mathematical Model

The study period goes from July 2015 to July 2023. The model considers annual transits, where n is the time parameter in years. Thus, $n = 0$ corresponds to July 2015 and

$n = 8$ to July 2023. Let $P(n)$ be the total population of the individuals in Spain within the contemplated age range (16,70) at risk of committing political corruption.

Thus,
$$P(n) = Z(n) + B(n) + M(n) + A(n), \tag{2}$$

where:
$$Z(n) = Z_1(n) + Z_2(n) + Z_3(n) + Z_4(n) + Z_5(n), \tag{3}$$
$$B(n) = B_1(n) + B_2(n) + B_3(n) + B_4(n) + B_5(n), \tag{4}$$
$$M(n) = M_1(n) + M_2(n) + M_3(n) + M_4(n) + M_5(n), \tag{5}$$
$$A(n) = A_1(n) + A_2(n) + A_3(n) + A_4(n) + A_5(n). \tag{6}$$

The compartmental difference equations model for the risk of committing political corruption dynamics in Spain is presented in the following system for every labor group j,

$$\begin{aligned}
Z_1(n+1) - Z_1(n) &= (I_{z1} - R_{z1}) - d_1 Z_1(n) - \alpha_1 Z_1(n) + \gamma B_1(n), \\
B_1(n+1) - B_1(n) &= (I_{B1} - R_{B1}) - d_1 B_1(n) - \alpha_B B_1(n) + \alpha_Z Z_1(n) - \gamma B_1(n), \\
M_1(n+1) - M_1(n) &= -R_{M1} - d_1 M_1(n) + \alpha_B B_1(n) - \alpha_M M_1(n) + \mu(n) A_1(n), \\
A_1(n+1) - A_1(n) &= -R_{A1} - d_1 A_1(n) - \mu(n) A_1(n) + \alpha_M M_1(n), \\
Z_2(n+1) - Z_2(n) &= (I_{z2} - R_{z2}) - d_2 Z_2(n) - \alpha_Z Z_2(n) + \gamma B_2(n), \\
B_2(n+1) - B_2(n) &= (I_{B2} - R_{B2}) - d_2 B_2(n) - \alpha_B B_2(n) + \alpha_Z Z_2(n) - \rho_{B2} B_2(n) + D_2 A_2(n) - \gamma B_2(n), \\
M_2(n+1) - M_2(n) &= -R_{M2} - d_2 M_2(n) + \alpha_B B_2(n) - \alpha_M M_2(n) - \rho_{M2} M_2(n) + \rho_{B2} B_2(n) + \mu(n) A_2(n), \\
A_2(n+1) - A_2(n) &= -R_{A2} - d_2 A_2(n) - \mu(n) A_2(n) + \alpha_M M_2(n) + \rho_{M2} M_2(n) - D_2 A_2(n), \\
Z_3(n+1) - Z_3(n) &= (I_{z3} - R_{z3}) - d_3 Z_3(n) - \alpha_Z Z_3(n) + \gamma B_3(n), \\
B_3(n+1) - B_3(n) &= (I_{B3} - R_{B3}) - d_3 B_3(n) - \alpha_B B_j(n) + \alpha_Z Z_j(n) - \rho_{B3} B_j(n) + D_3 A_j(n) - \gamma B_3(n), \\
M_3(n+1) - M_3(n) &= -R_{M3} - d_3 M_3(n) + \alpha_B B_3(n) - \alpha_M M_3(n) - \tau_3(n) M_3(n) - \rho_{M3} M_3(n) + \rho_{B3} B_3(n) + \mu(n) A_3(n), \\
A_3(n+1) - A_3(n) &= -R_{A3} - d_3 A_3(n) - \mu(n) A_3(n) + \tau_3(n) M_3(n) + \alpha_M M_3(n) + \rho_{M3} M_3(n) - D_3 A_3(n), \\
Z_4(n+1) - Z_4(n) &= (I_{z4} - R_{z4}) - d_4 Z_4(n) - \alpha_Z Z_4(n) + \gamma B_4(n), \\
B_4(n+1) - B_4(n) &= (I_{B4} - R_{B4}) - d_4 B_4(n) - \alpha_B B_4(n) + \alpha_Z Z_4(n) - \rho_{B4} B_4(n) + D_4 A_4(n) - \gamma B_4(n), \\
M_4(n+1) - M_4(n) &= -R_{M4} - d_4 M_4(n) + \alpha_B B_4(n) - \alpha_M M_4(n) - \tau_4(n) M_4(n) - \rho_{M4} M_4(n) + \rho_{B4} B_4(n) + \mu(n) A_4(n), \\
A_4(n+1) - A_4(n) &= -R_{A4} - d_4 A_4(n) - \mu(n) A_4(n) + \tau_4(n) M_4(n) + \alpha_M M_4(n) + \rho_{M4} M_4(n) - D_4 A_4(n), \\
Z_5(n+1) - Z_5(n) &= (I_{z5} - R_{z5}) - d_5 Z_5(n) - \alpha_Z Z_5(n) + \gamma B_5(n), \\
B_5(n+1) - B_5(n) &= (I_{B5} - R_{B5}) - d_5 B_5(n) - \alpha_B B_5(n) + \alpha_Z Z_5(n) - \gamma B_5(n), \\
M_5(n+1) - M_5(n) &= -R_{M5} - d_5 M_5(n) + \alpha_B B_5(n) - \alpha_M M_5(n) - \tau_5(n) M_5(n) + \mu(n) A_5(n), \\
A_5(n+1) - A_5(n) &= -R_{A5} - d_5 A_5(n) - \mu(n) A_5(n) + \tau_5(n) M_5(n) + \alpha_M M_5(n).
\end{aligned} \tag{7}$$

This can be written in a vector compact form as follows:
$$V(n+1) = G(n) V(n) + C, \tag{8}$$

where $V(n) \in \mathbb{R}^{20 \times 1}$ is the model's unknown vector, including all the subpopulations per labor group at time n, as follows:

$$V(n) = [Z_1(n), B_1(n), M_1(n), A_1(n), Z_2(n), B_2(n), M_2(n), A_2(n), Z_3(n), B_3(n), M_3(n), \\ A_3(n)\, Z_4(n), B_4(n), M_4(n), A_4(n), Z_5(n), B_5(n), M_5(n), A_5(n)]^T$$

Note that matrix $G(n) = (g_{pq}(n)) \in \mathbb{R}^{20 \times 20}$, where

$g_{pq} = 1 - d_x - \alpha_z$, for $p = q = 1, 5, 9, 13, 17$ and $x = 0.75 + p/4$;

$g_{12} = g_{56} = g_{910} = g_{1314} = g_{1718} = \gamma$;

$g_{21} = g_{65} = g_{109} = g_{1413} = g_{1817} = \alpha_Z$;

$g_{34} = g_{78} = g_{1112} = g_{1516} = g_{1920} = \mu(n)$;

$g_{68} = D_2$, $g_{1012} = D_3$, $g_{1416} = D_4$;

$g_{44} = 1 - d_1 - \mu$; $g_{2020} = 1 - d_5 - \mu$; $g_{pq} = 1 - d_x - \mu - D_x$, for $p = q = 8, 12, 16$ and $x = p/4$;

$g_{32} = \alpha_B$; $g_{1918} = \alpha_B + \rho_{Bx}$; for $p = q + 1 = 7, 11, 15$ and $x = 0.5 + q/4$;

$g_{43} = \alpha_M$; $g_{87} = \alpha_M + \rho_{M2}$; $g_{1211} = \tau_3(n) + \alpha_M + \rho_{M3}$; $g_{1615} = \tau_4(n) + \alpha_M + \rho_{M4}$; $g_{2019} = \tau_5(n) + \alpha_M$;

$g_{22} = 1 - d_1 - \alpha_B - \gamma$; $g_{1818} = 1 - d_5 - \alpha_B - \gamma$; $g_{pq} = 1 - d_x - \alpha_B - \rho_{Bx} - \gamma$, for $p = q = 6, 10, 14$ and $x = 0.5 + p/4$;

$g_{33} = 1 - d_1 - \alpha_M$; $g_{77} = 1 - d_2 - \alpha_M - \rho_{M2}$; $g_{1919} = 1 - d_5 - \alpha_M - \tau_5(n)$; $g_{pq} = 1 - d_x - \alpha_M - \tau_x(n) - \rho_{Mx}$, for $p = q = 11, 15$ and $x = 0.25 + p/4$;

and all the other $g_{pq}(n)$ equal zero.

$C \in \mathbb{R}^{20 \times 1}$ is the demographic independent vector given by:

$C = [I_{z1} - R_{z1}, I_{B1} - R_{B1}, -R_{M1}, -R_{A1}, I_{z2} - R_{z2}, I_{B2} - R_{B2}, -R_{M2}, -R_{A2}, I_{z3} - R_{z3}, I_{B3} - R_{B3}, -R_{M3}, -R_{A3}, I_{z4} - R_{z4}, I_{B4} - R_{B4}, -R_{M4}, -R_{A4}, I_{z5} - R_{z5}, I_{B5} - R_{B5}, -R_{M5}, -R_{A5}]^T$;

$C = [418, 846, 8548, 0, 0, -112, 800, 0, 0, 0, -229, 797, -23, 495, -5207, -136, -1184, -18, 328, -307, -1490, -12, 611, -20, 912, -354, -4]^T$.

4. Results

With the mathematical model, we computed solutions from the initial subpopulations (n = 0) for every n until n = 8, (Table 3, Panel A).

Table 3. Numerical results after the 2015 elections (year 2016, n = 1), panel A = individuals, panel B = percentage.

	Panel A					
	TOTAL	j = 1	j = 2	j = 3	j = 4	j = 5
Zj	20,648,347	4,206,758	5,291,150	10,355,955	228,273	566,212
Bj	2,901,249	97,178	28,352	1,067,261	795,750	912,708
Mj	346,885	1051	0	249,704	67,587	28,543
Aj	33,930	0	0	7904	25,739	287
TOTAL	23,930,412	4,304,987	5,319,502	11,680,824	1,117,349	1,507,749
	Panel B					
	TOTAL	j = 1	j = 2	j = 3	j = 4	j = 5
Zj	86.3%	20.4%	25.6%	50.2%	1.1%	2.7%
Bj	12.1%	3.3%	1.0%	3.4%	28.4%	31.0%
Mj	1.4%	0.3%	0.0%	81.4%	8.9%	9.4%
Aj	0.1%	0.0%	0.0%	23.3%	75.9%	0.8%
TOTAL	100.0%	18.0%	2.2%	48.8%	4.7%	6.3%

Thus, we can see how the numerical results show political renovation in 2016 (n = 1) after elections, with a lower percentage for the high-risk subpopulation from 0.3% to 0.1% (Table 3, Panel B). However, the low-risk and medium-risk subpopulations increase from 2015 to 2016.

The results show how the population at high risk of committing political corruption grows for the study period and represents 0.7% of the Spanish population in 2023 (Table 4, Panel B). Even though this percentage may seem low, the socio-economic and moral impact on the Spanish society would be dramatic.

Table 4. Numerical results in n = 2023, panel A = individuals, panel B = percentage.

Panel A						
	TOTAL	j = 1	j = 2	j = 3	j = 4	j = 5
Zj	20,250,555	3,329,507	5,900,978	9,382,973	1,136,106	500,991
Bj	2,684,850	127,717	250,884	1,036,815	599,419	670,015
Mj	441,168	8,039	16,038	233,271	103,937	79,882
Aj	174,554	609	742	110,872	47,910	14,420
TOTAL	23,551,127	3,465,872	6,168,643	10,763,931	1,887,372	1,265,309

Table 4. *Cont.*

Panel B	TOTAL	j = 1	j = 2	j = 3	j = 4	j = 5
Zj	86.0%	16.4%	29.1%	46.3%	5.6%	2.5%
Bj	11.4%	4.8%	9.3%	38.6%	22.3%	25.0%
Mj	1.9%	1.8%	3.6%	52.9%	23.6%	18.1%
Aj	0.7%	0.3%	0.4%	63.5%	27.4%	8.3%
TOTAL	100%	14.7%	26.2%	45.7%	8.0%	5.4%

4.1. Gender Effect Simulation

The literature evidences gender differences in social behavior. Women are more risk-averse [101], more inequality-averse and more cooperative and altruistic than men [49–57].

According to previous evidence, we posit the hypothesis that women are less prone to corruption either because they are subject to more control and expectations or their own attitude toward public service, social engagement and education prevents them from doing so [53,54]. Moreover, in accordance with [52], we considered two different scenarios and computed the results for 2023.

The first simulation considered that women in power are incorruptible. This hypothesis affected the time in office parameter ($\tau_j(n)$). As women occupy 40% of power seats, the time in office parameter would affect only 60% of the population.

As Table 5 shows, the proportion of the population at high risk of committing political corruption drops to almost half its previous value due to the presence of women in top management positions (j = 2 and j = 4)

Table 5. Gender simulation I. Subpopulations for 2023.

	TOTAL	j = 1	j = 2	j = 3	j = 4	j = 5
Zj	86.31%	20.4%	25.6%	50.2%	1.1%	2.7%
Bj	12.09%	3.0%	1.0%	36.9%	27.6%	31.5%
Mj	1.20%	0.4%	0.0%	80.8%	9.0%	9.8%
Aj	0.39%	0.0%	0.0%	27.5%	71.6%	0.9%
TOTAL	100.0%	17.9%	22.2%	48.8%	4.7%	6.3%

We considered a less extreme scenario, in which women in power are less incorruptible. In this way, 50% of the women serving in office would not become corrupt (20% people in power). According to this hypothesis, the time in office parameter would affect 80% of the population. As Table 6 shows, even when not considering an extreme impact of women in office, the Spanish population at risk of committing political corruption would considerably drop.

Table 6. Numerical results as percentages for n = 2023 for gender simulation 2.

	TOTAL	j = 1	j = 2	j = 3	j = 4	j = 5
Zj	86.59%	20.4%	25.6%	50.2%	1.1%	2.7%
Bj	11.82%	2.5%	0.4%	36.6%	28.2%	32.2%
Mj	1.18%	0.4%	0.0%	80.6%	9.1%	9.9%
Aj	0.42%	0.0%	0.0%	30.9%	68.1%	1.0%
TOTAL	100%	17.9%	22.2%	48.8%	4.7%	6.3%

4.2. Sensitivity Analysis

We estimated the sensitivity of all the subpopulations to variation in the α parameter. Figure 3a shows how the differences in the high-risk subpopulation increased over time as the α parameter increased. However, those differences were not so big compared to the evolution of all the subpopulations (Figure 3b).

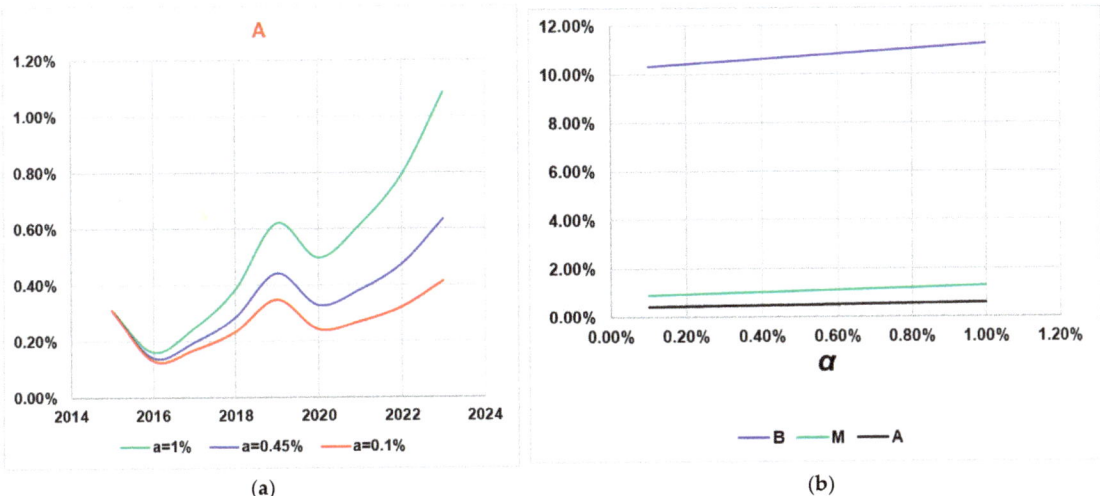

Figure 3. Sensitivity analysis: the α parameter. (**a**) Evolution of the high-risk subpopulations with different α parameter values; (**b**) subpopulations in 2023 according to the α parameter.

5. Discussion and Conclusions

This study quantifies the population at risk of committing political corruption in Spain by identifying and quantifying the drivers that explain political corruption.

Having quantified the problem, the model allowed the implementation of changes in parameters, as well as the simulation of fiscal, economic and legal measures to be simulated in order to quantify and better understand their impact on Spanish citizenship.

One of the potential advantages of the model is its applicability to other geographical areas using local data. However, its application to other areas requires the careful reworking and adaptation to each region's idiosyncrasies.

Stopping this social problem requires policy makers' action. Specifically, changing the Spanish electoral law of parties is advisable to increase politicians' transparency and accountability. This could be much better controlled by hiring "advisers" in office, but by also regulating local administration's wages (small town councils). In addition, the "revolving doors" effect needs to end [102].

Political corruption risk consequences are numerous, starting with economic ones as they involve more reluctance from investors to flow capitals in unstable political environments, but are also negative for industry in general and social development [50], and are particularly relevant for the Spanish economy because tourism is one of the main economic engines of Southern Europe that is negatively affected by political corruption [103]. Combating public corruption can not only directly improve Spanish business performance, but can also facilitate it via access to credit [27]. It is also necessary to distance political and managerial spheres.

Our model shows the importance of women's empowerment as their presence in leadership roles and their representation in government are useful for mitigating the political corruption phenomenon, which falls in line with [51,104]. Thus, women's capacity to deliver a more relational leadership style gives better results than the self-critical style linked with men [104].

Finally, increasing funding for open government initiatives [105], building capacity toward strategic planning and performance evaluation [106], rather than investing in entities of doubtful nature like non-profit organizations and/or public companies [107], plus devoting more funds to the judicial system (district attorneys and judges) [108], and information technology for open innovation [109] are recommended.

As authors in [110] claim, the need to encourage a type of press that favors interpretative contribution is urgent because it would allow citizenship to understand and comprehend corruption.

Further research deserves attention to claiming for suitable changes in order to limit or reduce both the possibilities of committing political corruption and also reducing the impact. These measures are local, depend on culture and cannot be implemented without willpower, or at least with citizens' pressure through their votes. The case of Spain would require:

(i) Changes in the electoral law to allow the transparency and accountability of elected political representatives.

In fact, the present Spanish Political Parties and Electoral Laws are a closed system with a block list of candidates provided for each political party. This means that citizens must accept all the listed people, but they cannot select some of them. This procedure eliminates the accountability of political actions taken and decisions made by representatives. They are simply accountable to the political party's leader, but not to citizens. This fact is a double source of corruption: one is lack of the representative's transparency and independency because they have no freedom to voice their opinion against that of the party's leader. The other is the party's leader is potentially, and at least, a commander or political boss, and even a dictator.

(ii) Not increasing bureaucracy measures.

These measures have a paralyzing effect on the Administration, reduce labor motivation and are used to produce new sources of corruption.

(iii) Selecting independent inspectors of parties' accounts.

(iv) Introducing new laws that forbid financial support from the national budget for private communication companies.

(v) Measures that address cutting public expenses spent on political advisors based on confidence criteria.

There is no way to discriminate between political favor and other forms of selection that make corruption possible.

(vi) Bearing in mind that the higher the public GDP, the more possible corruption is because public money is not administrated like private money.

(vii) Albeit difficult, implementing a measure because it is not acceptable for inexperienced persons who administer personal budgets to have the capacity to administer public budgets. Implementing some minimum level of capacity to manage public budgets.

Author Contributions: E.d.l.P., L.J. and P.M. contributed equally to this work. All authors have read and agreed to the published version of the manuscript.

Funding: Not applicable.

Data Availability Statement: Not applicable.

Conflicts of Interest: The authors declare no conflict of interest.

References

1. Brioschi, C.A.; Shugaar, A. *Corruption: A Short History*; Brookings Institution Press: Washington, DC, USA, 2017. Available online: http://www.jstor.org/stable/10.7864/j.ctt1hfr1sk (accessed on 11 February 2021).
2. Eyre, C.J. Patronage, power, and corruption in Pharaonic Egypt. *Int. J. Public Adm.* **2011**, *34*, 701–711. [CrossRef]
3. Conant, E. The Laws of the Twelve Tables. Washington University Law Review. 1928, Volume 13. Available online: https://openscholarship.wustl.edu/law_lawreview/vol13/iss4/ (accessed on 6 February 2021).
4. Nye, J.S., Jr. *The Powers to Lead*; Oxford University Press: New York, NY, USA, 2008.
5. La Porta, R.; López de Silanes, F.; Shleifer, A.; Visnhy, R. Trust in Large Organizations. *Am. Econ. Rev.* **1997**, *87*, 333–338.
6. Lederman, D.; Loayza, N.; Soares, R. Accountability and Corruption: Political Institutions Matter. *Econ. Politics* **2005**, *17*, 1–35. [CrossRef]
7. Nisnevich, Y.A. What is political corruption. *Mirovaya Ekonomika I Mezhdunarodnye Otnosheniya* **2020**, *64*, 133–138. [CrossRef]
8. Chiappinelli, O. Political corruption in the execution of public contracts. *J. Econ. Behav. Organ.* **2020**, *179*, 116–140. [CrossRef]
9. Cruz, C.O.; Sarmento, J.M. Public Management and Cost Overruns in Public Projects. *Int. Public Manag. J.* **2020**. [CrossRef]

10. Amundsen, I. Political Corruption. Anticorruption Resource Center. 2006. Available online: https://www.cmi.no/publications/2565-political-corruption (accessed on 15 January 2021).
11. Caplan, B. *The Myth of the Rationale Voter*; Princeton University Press: Princeton, NJ, USA, 2007.
12. Stockemer, D.; LaMontagne, B.; Scruggs, L. Bribes and ballots: The impact of corruption on voter turnout in democracies. *Int. Polit. Sci. Rev.* **2013**, *34*, 74–90. [CrossRef]
13. Kubbe, I. Elites and Corruption in European Democracies. In *Parties, Governments and Elites: The Comparative Study of Democracy*; Springer: Berlin, Germany, 2017; pp. 249–279. [CrossRef]
14. Rose-Ackerman, S. *International Handbook on the Economics of Corruption*; Edgar Elgar: Cheltenham, UK, 2006.
15. Serritzlew, S.; Sonderskov, K.M.; Svendsen, G.T. Do corruption and social trust affect economic growth? A review. *J. Comp. Pol. Anal. Res. Pract.* **2014**, *16*, 121–139. [CrossRef]
16. Johnston, M. *Syndromes of Corruption: Wealth, Power, and Democracy*; Cambridge University Press: Cambridge, UK, 2005. [CrossRef]
17. Huang, Q. Does Political Corruption Impede Firm Innovation? Evidence from the United States. *J. Financ. Quant. Anal.* **2021**, *56*, 213–248. [CrossRef]
18. Warf, B. Global geographies of corruption. *Geo J.* **2016**, *81*, 657–669. [CrossRef]
19. Habib, M.; Zurawicki, L. Corruption and foreign direct investment. *J. Int. Bus. Stud.* **2002**, *32*, 291–307. [CrossRef]
20. Springis, M. The Impact of the Host Country Corruption On Inward Fdi. *Curr. Issues Manag. Bus. Soc. Dev.* **2011**, *2011*, 680–689.
21. Peres, M.; Ameer, W.; Xu, H. The impact of institutional quality on foreign direct investment inflows: Evidence for developed and developing countries. *Econ. Res. Ekonomska Istraživanja* **2018**, *31*, 626–644. [CrossRef]
22. Dass, N.; Nanda, V.; Xiao, S.C. Public corruption in the United States: Implications for local firms. *Rev. Corp. Financ. Stud.* **2016**, *5*, 102–138. [CrossRef]
23. Le, A.T.; Dohan, A.T. Corruption and financial fragility of small and medium enterprises: International evidence. *J. Multinatl. Financ. Manag.* **2020**, *57–58*, 100660. [CrossRef]
24. Brown, N.C.; Smith, J.D.; White, R.M.; Zutter, C.J. *Greased Wheels or Just Greased Palms: Political Corruption and Firm Value (August 2015)*; SSRN: Rochester, NY, USA, 2011. [CrossRef]
25. Brown, N.C.; Smith, J.D.; White, R.M.; Zutter, C.J. Political Corruption and Firm Value in the US: Do Rents and Monitoring Matter? *J. Bus. Ethics* **2021**, *168*, 335–351. [CrossRef]
26. Hossain, A.T.; Kryzanowski, L. Political corruption shielding and corporate acquisitions. *Financ. Rev.* **2021**, *56*, 55–83. [CrossRef]
27. Bermpei, T.; Kalyvas, A.N.; Leonida, L. Local Public Corruption and Bank Lending Activity in the United States. *J. Bus Ethics* **2020**. [CrossRef]
28. Hossain, A.T.; Kryzanowski, L.; Ma, X.B. U.S. political corruption and loan pricing. *J. Financ. Res.* **2020**, *43*, 459–489. [CrossRef]
29. Johnston, M. More than Necessary, Less than Sufficient: Democratization and the Control of Corruption. *Soc. Res.* **2013**, *80*, 1237–1258. Available online: http://www.jstor.org/stable/24385658 (accessed on 19 April 2021).
30. Transparency International. Corruption Perceptions Index 2020. Available online: https://images.transparencycdn.org/images/CPI2020_Report_EN_0802-WEB-1_2021-02-08-103053.pdf (accessed on 15 January 2021).
31. Muro, D. Political mistrust in southern Europe since the Great Recession. *Mediterr. Politics* **2017**, *22*, 197–217. [CrossRef]
32. Rajh, E.; Budak, J. Determinants of corruption pressures on local government in the EU. *Econ. Res. Ekonomska istraživanja* **2020**, *33*, 3492–3508. [CrossRef]
33. Preston, P. *A People Betrayed: A History of Corruption, Political Incompetence, and Social Division in Modern Spain*; HarperCollins Publishers: Scotland, UK, 2021.
34. Melián, I. Inharmonious corruption. *Revista Investigaciones Politicas Sociologicas* **2018**, *17*, 181–206. [CrossRef]
35. De la Poza, E.; Jódar, L.; Pricop, A. Mathematical modeling of the propagation of democratic support of extreme ideologies in Spain:Causes, Effects and Recommendations for its stop. *Abstr. Appl. Anal.* **2013**. [CrossRef]
36. Gomez Fortes, B. Political Corruption and the End of two-party system after the May 2015 Spanish Regional Elections. *Reg. Fed. Stud.* **2015**, *25*, 379–389. [CrossRef]
37. Moreno Zacarés, J. The Iron Triangle of Urban Entrepreneurialism: The Political Economy of Urban Corruption in Spain. *Antipode* **2020**, *52*, 1351–1372. [CrossRef]
38. OECD. Increasing Transparency and Accountability through Disclosure of Political Party and Election-Campaign Funding. In *Financing Democracy: Funding of Political Parties and Election Campaigns and the Risk of Policy Capture*; OECD Publishing: Paris, France, 2016. [CrossRef]
39. UNODC (United Nations Office on Drugs and Crime). United Nations Convention against Corruption. 2003. Available online: www.unodc.org/unodc/en/treaties/CAC/ (accessed on 19 January 2021).
40. The Guardian. Spain's Ruling Party Ran Secret Fund for 18 Years, Investigating Judge Finds. Available online: https://www.theguardian.com/world/2015/mar/23/spain-ruling-peoples-party-secret-fund-18-years-investigating-judge (accessed on 23 January 2021).
41. El País. Spain's Socialists Try to Contain Fallout from ERE Corruption Case Ruling. Available online: https://english.elpais.com/elpais/2019/11/20/inenglish/1574238009_170261.html (accessed on 17 February 2021).
42. Jimenez, F. Building Boom and Political Corruption in Spain. *South Eur. Soc. Politics* **2009**, *14*, 255–272. [CrossRef]
43. De La Poza, E.; Jódar Sánchez, L.A.; Pricop, A.G. Modelling and analysing voting behaviour: The case of the Spanish general elections. *Appl. Econ.* **2017**, *49*, 1287–1297. [CrossRef]

44. El País. Judge Calls Members of Podemos Leadership to Testify after Accusations of Irregular Financing. Available online: https://english.elpais.com/politics/2020-08-11/judge-calls-members-of-podemos-leadership-to-testify-after-accusations-of-irregular-financing.html (accessed on 17 February 2021).
45. Allen, N.; Magni, G.; Searing, D.; Warncke, P. What is a career politician? Theories, concepts, and measures. *Eur. Political Sci. Rev.* **2020**, *12*, 199–217. [CrossRef]
46. El Español. Podemos, el Partido con Mayor Porcentaje de Diputados que Nunca ha Trabajado. August 2017. Available online: https://www.elespanol.com/espana/politica/20170812/238476359_0.html (accessed on 17 February 2021). (In Spanish)
47. Campbell, J.W. Buying the honor of thieves? Performance pay, political patronage, and corruption. *Int. J. Law Crime Justice* **2020**, *63*, 100439. [CrossRef]
48. Charron, N.; Dahlström, C.; Fazekas, M.; Lapuente, V. Careers, Connections, and Corruption Risks: Investigating the Impact of Bureaucratic Meritocracy on Public Procurement Processes. *J. Politics* **2016**, *79*, 89–104. [CrossRef]
49. Alexander, A.C. Gender, gender equality and corruption: A review of theory and evidence. In *Oxford Handbook of Quality of Government*; Bagenholm, A., Bauhr, M., Grimes, M., Rothstein, B., Eds.; Oxford University Press: Oxford, UK, 2021.
50. Martín, J.C.; Román, C.; Viñán, C. An institutional trust indicator based on fuzzy logic and ideal solutions. *Mathematics* **2020**, *8*, 807. [CrossRef]
51. Jha, C.K.; Sarangi, S. Women and corruption: What positions must they hold to make a difference? *J. Econ. Behav. Organ.* **2018**, *151*, 219–233. [CrossRef]
52. Esarey, J.; Schwindt-Bayer, L.A. Estimating Causal Relationships between Women's Representation in Government and Corruption. *Comp. Political Stud.* **2019**, *52*, 1713–1741. [CrossRef]
53. Boehm, F. Are Men and Women Equally Corrupt? Chr. Michelsen Institute. U4 Brief. No. 6. 2015. Available online: https://www.u4.no/publications/are-men-and-women-equally-corrupt.pdf (accessed on 19 January 2021).
54. Bauhr, M.; Charron, N. Do Men and Women Perceive Corruption Differently? Gender Differences in Perception of Need and Greed Corruption. *Politics Gov.* **2020**, *8*, 92–102. [CrossRef]
55. Bandura, A. Moral disengagement in the perpetration of inhumanities. *Personal. Soc. Psychol. Rev.* **1999**, *3*, 193–209. [CrossRef]
56. Darwin, C. *The Origin of Species*; Collier Books: London, UK, 1962.
57. Afridi, F.; Iversen, V.; Sharan, M.R. Women political leaders, corruption, and learning: Evidence from a large public program in India. *Econ. Dev. Cult. Chang.* **2017**, *66*, 1–30. [CrossRef]
58. González Rodríguez, J.J.; Rodríguez Díaz, R.; Rodriguez Castromil, A. A Case of Polarized Pluralism in a Mediterranean country. The Media and Politics in Spain. *Global Media J.* **2010**, *5*, 1–9.
59. Feenstra, R.A.; Tormey, S.; Casero-Ripollés, A.; Keane, J. *Refiguring Democracy: The Spanish Political Laboratory*; Routledge: New York, NY, USA, 2017.
60. Mazzoleni, G.; Winfried, S. Mediatization of Politics: A Challenge for Democracy? *Political Commun.* **1999**, *16*, 247–261. [CrossRef]
61. Strömbäck, J. Four Phases of Mediatization: An Analysis of the Mediatization of Politics. *Int. J. Press Politics* **2008**, *13*, 228–246. [CrossRef]
62. Humanes, M.L.; Martínez, M.; Saperas, E. Political Journalism in Spain. Practices, Roles and Attitudes. *Estudios Sobre Mensaje Periodístico* **2013**, *19*, 715–731. [CrossRef]
63. Castillo Ortiz, P.J. Councils of the Judiciary and Judges' Perceptions of Respect to their Independence in Europe. *Hague J. Rule Law* **2017**, *9*, 315–336. [CrossRef]
64. Nieto, A. *El Malestar de Los Jueces*; Trotta: Madrid, Spain, 2010.
65. Zurutuza-Muñoz, C.; Verón-Lassa, J.J. The Spanish press's loss of confidence, in a context of political corruption: The 'Barcenas scandal' in El Pais and El Mundo. *Análisi* **2015**, 113–127. [CrossRef]
66. Camison Yague, J.A. 2016 and 2017 Compliance Reports on the Recommendations Proposed by GRECO for the Prevention of Judicial Corruption in Spain: Chronicle of Non-Compliances. Teoría y Realidad Constitucional Publicación Semestral del Departamento de Derecho Político. *Universidad Nacional Educación Distancia* **2018**, *41*, 337–356. (In Spanish)
67. Damasio, A. *The Strange Order of Things. Life, Feelings, and the Making of Cultures*; Pantheon Books: New York, NY, USA, 2018.
68. The Guardian. Spain's Former King Juan Carlos Faces New Corruption Allegations. Available online: https://www.theguardian.com/world/2020/nov/03/spains-former-king-juan-carlos-faces-new-corruption-allegations (accessed on 17 February 2021).
69. Jiménez, J.L.; Nombela, G.; Suárez-Alemán, A. Tourist municipalities and local political corruption. *Int. J. Tour. Res.* **2017**, *19*, 515–523. [CrossRef]
70. Rubí, G.; Toledano, L.F.; Riquer i Permanyer, B. Political corruption in Catalonia within contemporary Spain. *Revista Catalana Dret Públic* **2020**, *60*, 1–19. [CrossRef]
71. Rius-Ulldemolins, J.; Flor Moreno, V.; Hernandez i Marti, G.M. The dark side of cultural policy: Economic and political instrumentalisation, white elephants, and corruption in Valencian cultural institutions. *Int. J. Cult. Policy* **2019**, *25*, 282–297. [CrossRef]
72. Rueda, M.R. Political Agency, Election Quality, and Corruption. *J. Politics* **2020**, *82*, 1256–1270. [CrossRef]
73. Standing, G. *The Precariat: The New Dangerous Class*; Bloomsbury: London, UK, 2011.
74. Nogués-Pedregal, A.M. The instrumental time of memory: Local politics and urban aesthetics in a tourism context. *J. Tour. Anal.* **2019**, *26*, 2–24. [CrossRef]

75. Christakis, N.A.; Fowler, J.H. *Connected: The Surprising Power of Our Social Networks and How They Shape Our Lives*; Little Brown and Company: Boston, MA, USA, 2009.
76. Girard, R. *Things Hidden since the Foundation of the World (Des Choses Cachées Depuis la Fondation du Monde)*; Stanford University Press: Stanford, UK, 2002.
77. Girard, R. *Mimesis and Theory: Essays on Literature and Criticism, 1953–2005*; Stanford University Press: Palo Alto, CA, USA, 2008.
78. Raafat, R.M.; Chater, N.; Frith, C. Herding in Humans. *Trends Cogn. Sci.* **2009**, *13*, 420–428. [CrossRef] [PubMed]
79. Philp, M. Defining Political Corruption. *Political Stud.* **1997**, *45*, 436–462. [CrossRef]
80. Schumacher, I. Political stability, corruption and trust in politicians. *Econ. Model.* **2013**, *31*, 359–369. [CrossRef]
81. Haddad, W.M.; Chellaboina, V.; Nersesov, S.G. Hybrid nonnegative and compartmental dynamical systems. *Math. Probl. Eng.* **2002**, *8*, 493–515. [CrossRef]
82. Treisman, D. The causes of corruption: A cross-national study. *J. Public Econ.* **2000**, *76*, 399–457. [CrossRef]
83. Aidt, T.S. Corruption, institutions, and economic development. *Oxf. Rev. Econ. Policy* **2009**, *25*, 271–291. [CrossRef]
84. De la Poza, E.; Jodar, L.; Merello, P.; Todoli-Signes, A. Explaining the rising precariat in Spain. *Technol. Econ. Dev. Econ.* **2020**, *26*, 165–185. [CrossRef]
85. Merello, P.; De la Poza, E.; Jódar, L. Explaining shopping behavior in a market economy country: A short-term mathematical model applied to the case of Spain. *Math. Methods Appl. Sci.* **2020**, *43*, 8089–8104. [CrossRef]
86. De la Poza, E.; Jódar, L.; Douklia, G. Modeling the Spread of Suicide in Greece. *Complex Syst.* **2019**, *28*, 475–489. [CrossRef]
87. Dell'Anno, R. Corruption around the world: An analysis by partial least squares—Structural equation modeling. *Public Choice* **2020**, *184*, 327–350. [CrossRef]
88. Viloria, M.; Jiménez, F. La corrupción en España (2004–2010): Datos, percepción y efectos. *Revista Española Investigaciones Sociológicas* **2012**, *138*, 109–134. [CrossRef]
89. Montero, L.M. *El Club de Las Puertas Giratorias*; La Esfera de los libros: Madrid, Spain, 2016.
90. Cerrillo-Martinez, A. Beyond Revolving Doors: The Prevention of Conflicts of Interests Through Regulation. *Public Integr.* **2017**, *19*, 357–373. [CrossRef]
91. Spinoza, B. *Tratado Politico*; Alianza: Madrid, Spain, 1986.
92. Spanish Ministry of Labour and Social Security. Pensiones de la Seguridad Social. 2015. Available online: http://www.seg-social.es/wps/portal/wss/internet/EstadisticasPresupuestosEstudios/Estadisticas/EST23/2575/3030 (accessed on 20 February 2021).
93. Spanish National Statistics Institute. 2015. Available online: http://www.ine.es (accessed on 20 February 2021).
94. Spanish Ministry of Finance and Civil Service. Available online: https://www.hacienda.gob.es/en-gb/paginas/home.aspx (accessed on 20 February 2021).
95. MacCluer, C.R. *Industrial Mathematics, Modeling in Industry, Science and Government*; Prentice-Hall: Upper Saddle River, NJ, USA, 2000.
96. Goldthorpe, J.H. *Sociology as a Population Science*; Cambridge University Press: Cambridge, UK, 2016. [CrossRef]
97. Kriesi, H. The Political Consequences of the Financial and Economic Crisis in Europe: Electoral Punishment and Popular Protest. *Swiss Political Sci. Rev.* **2012**, *18*, 518–522. [CrossRef]
98. El Mundo. Consulta los Resultados de las Elecciones Generales y Comprueba Quién Podría Gobernar con Nuestro Pactómetro. Available online: https://www.elmundo.es/espana/2016/06/27/5770ccfbe2704e457c8b4610.html (accessed on 20 February 2021). (In Spanish)
99. Gil-Alana, L.A.; Škare, M.; Claudio-Quiroga, G. Innovation and knowledge as drivers of the 'great decoupling' in China: Using long memory methods. *J. Innov. Knowl.* **2020**, *5*, 266–278. [CrossRef]
100. Veblen, T. *The Theory of the Leisure Class*; Macmillan: New York, NY, USA, 1899.
101. Tran, C.D.; Phung, M.T.; Yang, F.J.; Wang, Y.H. The role of gender diversity in downside risk: Empirical evidence from Vietnamese listed firms. *Mathematics* **2020**, *8*, 933. [CrossRef]
102. El Diario.es. La Oficina de Conflicto de Intereses Permite 525 Puertas Giratorias de Altos Cargos del Gobierno y solo veta 11 en Más de una Década. 11 marzo 2019. Available online: https://www.eldiario.es/economia/oficina-puertas-giratorias-conflicto-intereses_1_1162477.html (accessed on 20 February 2021). (In Spanish)
103. Jiménez, A.; Duran, J.J.; de la Fuente, J.M. Political Risk as a Determinant of Investment by Spanish Multinational Firms in Europe. *Appl. Econ. Lett.* **2011**, *18*, 789–793. [CrossRef]
104. Johnson, S.K. Is a Tipping Point for Female Leaders. Bloomberg Opinion. 2021. Available online: https://www.bloomberg.com/opinion/articles/2021-01-31/women-leaders-are-doing-better-during-the-pandemic (accessed on 20 February 2021).
105. Perez-Arellano, L.A.; Blanco-Mesa, F.; Leon-Castro, E.; Alfaro-Garcia, V. Bonferroni prioritized aggregation operators applied to government transparency. *Mathematics* **2021**, *9*, 24. [CrossRef]
106. Drápalová, E.; Di Mascio, F. Islands of Good Government: Explaining Successful Corruption Control in Two Spanish Cities. *Politics Gov.* **2020**, *8*, 128–139. [CrossRef]
107. El País. La Autoridad Fiscal Denuncia la Falta de Evaluación de 14.000 Millones en Subvenciones Públicas. Available online: https://elpais.com/economia/2019/06/03/actualidad/1559558438_034795.html (accessed on 20 February 2021).
108. Ramió, C. *La Renovación de la Función Pública. Estrategias para Frenar la Corrupción Política en España*; Catarata: Madrid, Spain, 2016. (In Spanish)

109. Adamides, E.; Karacapilidis, N. Information technology for supporting the development and maintenance of open innovation capabilities. *J. Innov. Knowl.* **2020**, *5*, 29–38. [CrossRef]
110. Sola-Morales, S. Media Discourse and News Frames on Political Corruption in Spain. *Convergencia* **2019**, 1–24. [CrossRef]

Article

Modeling of Fundus Laser Exposure for Estimating Safe Laser Coagulation Parameters in the Treatment of Diabetic Retinopathy

Aleksandr Shirokanev [1,2], Nataly Ilyasova [12], Nikita Andriyanov [3], Evgeniy Zamytskiy [4], Andrey Zolotarev [5] and Dmitriy Kirsh [1,2,*]

[1] Department of Technical Cybernetics, Samara National Research University, 443086 Samara, Russia; shirokanev@ipsiras.ru (A.S.); ilyasova@ipsiras.ru (N.I.)
[2] Image Processing Systems Institute of the RAS—Branch of the FSRC "Crystallography and Photonics" RAS, 443001 Samara, Russia
[3] Department of Data Analysis and Machine Learning, Financial University under the Government Russian Federation, 125993 Moscow, Russia; naandriyanov@fa.ru
[4] Department of Ophthalmology, Samara State Medical University, 443099 Samara, Russia; undue_@mail.ru
[5] Samara Regional Clinical Ophthalmological Hospital named after T.I. Eroshevsky, 443068 Samara, Russia; avz@zrenie-samara.ru
* Correspondence: kirsh@ssau.ru; Tel.: +7-917-162-8411

Citation: Shirokanev, A.; Ilyasova, N.; Andriyanov, N.; Zamytskiy, E.; Zolotarev, A.; Kirsh, D. Modeling of Fundus Laser Exposure for Estimating Safe Laser Coagulation Parameters in the Treatment of Diabetic Retinopathy. *Mathematics* 2021, 9, 967. https://doi.org/10.3390/math9090967

Academic Editor: Lucas Jódar

Received: 2 April 2021
Accepted: 24 April 2021
Published: 26 April 2021

Publisher's Note: MDPI stays neutral with regard to jurisdictional claims in published maps and institutional affiliations.

Copyright: © 2021 by the authors. Licensee MDPI, Basel, Switzerland. This article is an open access article distributed under the terms and conditions of the Creative Commons Attribution (CC BY) license (https://creativecommons.org/licenses/by/4.0/).

Abstract: A personalized medical approach can make diabetic retinopathy treatment more effective. To select effective methods of treatment, deep analysis and diagnostic data of a patient's fundus are required. For this purpose, flat optical coherence tomography images are used to restore the three-dimensional structure of the fundus. Heat propagation through this structure is simulated via numerical methods. The article proposes algorithms for smooth segmentation of the retina for 3D model reconstruction and mathematical modeling of laser exposure while considering various parameters. The experiment was based on a two-fold improvement in the number of intervals and the calculation of the root mean square deviation between the modeled temperature values and the corresponding coordinates shown for the convergence of the integro-interpolation method (balance method). By doubling the number of intervals for a specific spatial or temporal coordinate, a decrease in the root mean square deviation takes place between the simulated temperature values by a factor of 1.7–5.9. This modeling allows us to estimate the basic parameters required for the actual practice of diabetic retinopathy treatment while optimizing for efficiency and safety. Mathematical modeling is used to estimate retina heating caused by the spread of heat from the vascular layer, where the temperature rose to 45 °C in 0.2 ms. It was identified that the formation of two coagulates is possible when they are located at least 180 μm from each other. Moreover, the distance can be reduced to 160 μm with a 15 ms delay between imaging.

Keywords: mathematical modeling; numerical methods; integro-interpolation method; splitting method; convergence of models; standard deviation of the error; diabetic retinopathy; ocular fundus; laser coagulation; optical coherence tomography; image processing; segmentation; safe treatment

1. Introduction

The assessment of both the result of a service rendered to a person and the way it is provided is becoming extremely relevant; however, this approach seems far from being well developed, though modern computing systems have allowed the customization of unique patient treatment parameters by taking into account patient-specific features. This helps to increase treatment efficiency and patient opinion regarding the course of treatment. The application of personalized methods and means in treatment, both in diagnosis and monitoring, ensures the maximum treatment effect. In a broad sense, the concept of personalized medicine can be applied to specific diseases. In particular, taking into account the structure of a patient's fundus and the parameters of laser exposure will help to increase the number of successful operations in the future.

Laser coagulation of the retina in the treatment of a diabetic macular edema is widespread today [1]. It involves multiple exposures to the edema to form coagulates [2]; however, coagulate deposition may have negative effects. When choosing unsuitable laser exposure parameters and locations for coagulates, the coagulation can lead to retinal damage and complete blindness [3]. This is because the use of a laser with increasing intensity leads to strong heating of a large area beyond the boundaries of the diabetic macular edema area, which, in turn, leads to the formation of a coagulum in the healthy retina during its critical heating. The maximum therapeutic effect becomes significantly complicated with an increase in the distance between coagulates, as well as under the conditions of lasers acting on blood vessels, retinal hemorrhages, and "solid" exudates [4]. These drawbacks in treatment, one way or another, often lead to irreversible decreases in vision. The best treatment can be obtained when coagulates are located at the same distance from each other and do not extend beyond the edema area, where the anatomical structures of the retina are not affected by the laser exposure.

Three modes provide the formation of coagulants under laser action on the fundus [5]: manual, semiautomatic, and navigation. Manual mode implies pointed laser shots which form respective coagulates in units. Semi-automatic mode denotes a series of shots leading to the formation of a group of coagulates in accordance with preset template areas. Thus, the locations of coagulates significantly affect treatment effectiveness; however, it is clear that a universal plan for their arrangement does not exist. The most effective one to date depends on the specific fundus structure and the location of the diabetic macular edema. The effectiveness of the placement is estimated by the locations of the centers of the coagulates formed. The most advanced means for arranging coagulates today is the Navilas unit, which provides high therapeutic efficiency. Operation in navigation mode requires the use of a preliminary coagulate arrangement plan based on a fundus image [6].

This approach does not solve at least two problems. The first is how to build a preliminary plan for coagulates, and the second is defining the optimal power and time of exposure for the ocular fundus. To solve the first problem, algorithms must be developed for the analysis of optical coherence tomography (OCT) images to isolate sensitive areas in the fundus, detect diabetic macular edema areas, and then choose the locations for coagulates [7]. For this purpose, various algorithms for pattern recognition, object detection, and image segmentation [8–11] can be used. To cope with the second problem, mathematical modeling has been proposed to specify the laser exposure parameters [12]. Indeed, on the one hand, it is necessary to achieve a predetermined temperature in the area of a diabetic macular edema for the formation of a coagulum. On the other hand, it is required to ensure a permissible temperature in the healthy area of the retina and other elements of the fundus to avoid negative effects. Consequently, the use of a mathematical model for heat propagation along the fundus should lead to an ideal temperature distribution in the fundus in a certain period, depending on the locations of coagulates, the power and duration of laser exposure, the duration of pauses between shots, etc.

Today, special attention in research is directed to the development of algorithms for processing fundus images [13–17]. Most importantly, the analysis of patient images using computer vision techniques in the treatment of diabetic retinopathy makes the process personalized and therefore more efficient and safer. We briefly consider the current state of research in the field of laser therapy.

In Reference [18], the estimation of parameters for laser coagulation was considered, but for the treatment of varicose veins. The authors noted unsafe treatments at high power radiation in the range of 8–20 W with a wavelength of 810–980 nm and 5–15 W for electromagnetic waves with a 1470–1550 nm wavelength. The study simulated the laser action of a solid-state laser. The advantages of the approach in comparison with real experiments was noted. Based on the obtained temperature values, the range of the permissible exposure power was identified. The lower and upper power levels were set accordingly. At the low level, the required effect was achieved, i.e., thrombus formation. The upper level had power peaks, which caused irreversible damage to veins. The disadvantage of this work is that the model describes the operation of only one laser of a specific type and brand.

A detailed review of various lasers was presented in Reference [19]. The authors carried out an in-depth analysis of studies of various characteristics of lasers around the world. They mainly compared argon or diode lasers, as well as lasers based on panretinal photocoagulation (PRP) technology. The analysis showed that argon lasers, which have been considered the standard in the treatment of diabetic retinopathy for a long time, today no longer always provide the best effects.

The work in Reference [20] was conducted to assess anatomical and functional outcomes for patients undergoing treatment for diabetic retinopathy with a 532 nm laser PRP method (Supra Scan® Quantel Medical) and laser pinpoint coagulation at a wavelength of 532 nm (PASCAL® Topcon). The study was carried out with 48 patients and aimed to identify the correlation of their comfort with the laser exposure parameters, thereby showing the advantage of a multipoint laser.

A study of the thickness of the layer of retinal nerve fibers in the context of laser action was reported in Reference [21]. The article discusses lasers based on PRP technology. The effectiveness of red laser radiation treatment was compared to that of green laser treatment. The red laser showed an increase in the thickness of the retinal nerve fiber layer by 3–10 microns, which is 1.5 times higher than the green laser facilitated. The study also showed that the increase in thickness usually does not coincide with age or the number of burns.

In addition, Reference [22] showed that when comparing pain scores for patients with diabetic retinopathy, new systems, such as the novel navigated laser (NNL) system, have an advantage over conventional PRP-based systems.

The work of Reference [23] concerns laser action efficiency in terms of promoting the therapeutic treatment of diabetic retinopathy. Single-spot laser (SSL) and pre-stabilized laser (PSL) methods were compared, where the latter showed the best performance in terms of pain and effectiveness of treatment.

The analysis of the literature has shown that a limited number of works have been devoted to the problem of the mathematical modeling of fundus laser exposure. Nevertheless, this problem can now be solved via the use of numerical methods. In Reference [24], numerical methods based on an implicit difference scheme were considered. Such methods are often used to obtain fuzzy wave equations. The main idea of the work was the formulation of a new space of coordinates with respect to time using implicit schemes and the theory of fuzzy sets. The use of fuzzy sets allowed the authors to perform fuzzy analysis of the resulting equations and ensured function convergence. The use of numerical methods allowed the authors to reduce the time required for constructing wave equations with minimal losses in accuracy when describing real processes. The confirmation of the convergence of the methods made it possible to assess the possibility of using such mathematical models with a given error.

Indirectly, thermal energy distribution and transfer mechanics were studied in Reference [25], where a model of the flow of dusty nanofluid $Cu-Al_2O_3$/water was developed. Using numerical methods, the authors obtained ordinary differential equations (ODEs) that described the physical process of such a flow. In this case, double solutions were obtained, which were investigated for convergence and after which only one solution was chosen as a reliable one. On that basis, various parameters of the process were evaluated. For example, it has been found that nanosized particles (Cu) have a significant effect on heat transfer.

Finally, Reference [26] also studied heat propagation; however, the authors investigated the dynamics of micropolar fluids using numerical methods. Two edge conditions were investigated: an isothermal wall and isothermal flow. On the basis of numerical methods, the authors considered systems of nonlinear ODEs which were solved using the sequential relaxation method and Simpson's rule. They investigated different mesh dimensions using numerical methods. The increase in the dimensions allowed the authors to gradually obtain more accurate solutions.

Despite the widespread use of numerical methods, including the integro-interpolation method, to solve various applied problems, their application to 3D modeling of fundus laser exposure involves significant computational difficulties. The development of algorithms that efficiently use modern High Performance Computing resources allowed us to make significant progress in resolving this problem.

In the article, we discuss algorithms and methods for analyzing OCT images of the fundus and propose algorithms for the mathematical modeling of laser exposure to the fundus to estimate the appropriate parameters for safe and effective treatment. Moreover, this article is the first to study the convergence of the integro-interpolation method for the problem of mathematical modeling of laser exposure to the fundus. Thus, we developed a new 3D fundus structure model based on processing of OCT images and using approximating functions to describe the boundaries of the retina. One of the most important factors determining the safety of treatment is the distance between coagulates and the delay between laser shots. In this case, one of the options for evaluating safe parameters is mathematical modeling. Indeed, simulating heat propagation at various values with the noted parameters allowed us to estimate the basic parameters required for the actual practice of diabetic retinopathy treatment while optimizing for efficiency and safety.

2. Materials and Methods

2.1. Research Material

A diabetic macular edema is one of the most unfavorable consequences of diabetic retinopathy, and can cause blindness. Figure 1 shows diagnostic images of a healthy fundus and a fundus damaged by macular edema.

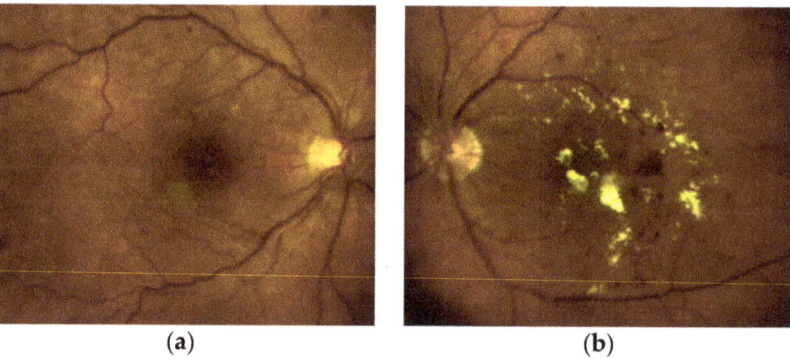

Figure 1. Fundus images: (**a**) healthy fundus; (**b**) fundus damaged by diabetic macular edema.

The fundus has a three-dimensional structure. For this reason, flat image analysis may not always be effective. Currently, the OCT scanning of the fundus provides sections of the retinal image obtained in the oXZ plane. As a result, 85 cross-sectional images are issued for different positions of the Y coordinate. Thus, the three-dimensional structure of the fundus can be reconstructed using the data of a sequence of images, given the exact Y value in each OCT image. Figure 2 shows one of the sections obtained by OCT registration and also schematically shows the laser contacting the fundus.

The edge condition is an area where the laser action has no effect on the vascular layer. The presence and positioning of such an area is extremely important for laser exposure modeling. As the analysis of the subject area has shown, the identification of new methods of effective treatment for diabetic retinopathy is possible when a sufficiently large number of fundus images can be analyzed. Additional features can be extracted from the analysis of OCT images. This requires a lengthy and thorough analysis of a sample of images with the involvement of medical experts.

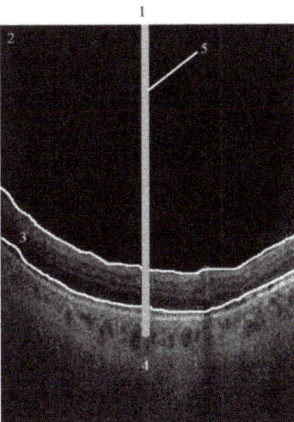

Figure 2. The structure of the retina as it pertains to laser exposure: 1—edge conditions, 2—vitreous, 3—retina, 4—vascular layer, 5—laser radiation.

The collection of a large database for the assessment of laser exposure effects on the fundus seems to be a great challenge. At present, the only way to assess the effective parameters of laser exposure is to carry out laser coagulation and fix the values under various conditions. This approach requires an excessively large number of patients as samples, since the physician cannot test different laser exposure parameters on real patients to study the effectiveness of the laser coagulation. A small sample of treated patients may not contain enough information to identify safe parameters. One approach is recommended that would require a small patient base containing heterogeneous fundus structures. As a result, the optimal material for research is a mathematical model that describes the spread of heat along the fundus in three-dimensional space, depending on the laser power, the duration of the exposure, etc. Having defined the temperature inside the fundus under normal conditions and the temperature of the formation of coagulates, the optimal parameters of laser exposure can be determined, including the target area. This makes it possible to explore various changes in the action of an object by replacing it with a mathematical model.

2.2. OCT Image Analysis and Reconstruction of 3D Fundus Structure Model

To assess the parameters of laser exposure, the preliminary analysis of OCT images may be required. Greater accuracy in determining the main layers on the cross-sections of the fundus images will result in a more suitable reconstructed model of the fundus surface. The reconstructed 3D model is needed to estimate the heat propagation.

At present, there are no universal methods for registering a fundus that allow obtaining undistorted images of parts of the eye. For various reasons, the recorded areas of the eye's surface in the original image may contain additional interference (Figure 3a).

Consequently, the first task is to highlight different layers in the noisy fundus image. The preprocessing stage may include the filtering procedures. For example, as mentioned earlier, median filtering ensures the elimination of strong impulse noise by brightness replacement with the median value in a certain neighborhood. If the noise is additively distributed over all pixels in the form of additive Gaussian white noise (AGWN), then it is possible to use recurrent Kalman filtering [27].

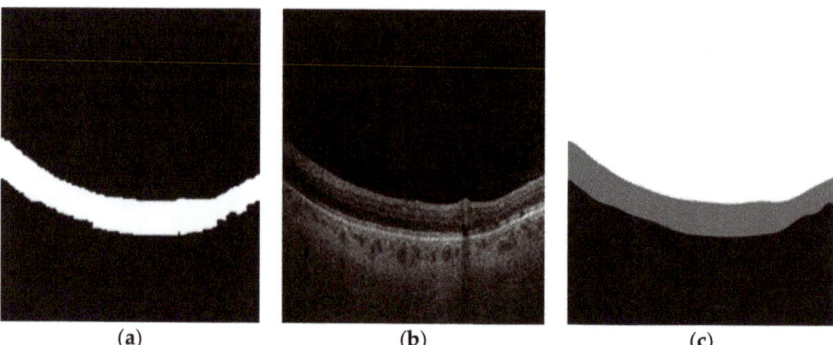

Figure 3. The result of the optical coherence tomography (OCT) fundus image processing: (**a**) original OCT image against the background of noise; (**b**) preliminary segmentation of the retinal layer; (**c**) fundus image model based on parametric functions.

Thus, preprocessing does not solve a specific applied problem but improves the quality of the solution obtained in the course of the main image processing. Based on the image shown in Figure 3a, the segmentation of three layers is required: the vitreous, retinal, and epithelial (vascular) layers; however, using the a priori knowledge that the retina always separates the other two layers, the task is simplified to binary segmentation, where the retinal layer will be highlighted. With segmentation, it is important to take into account that the retinal layer must be evenly filled and distributed over the entire width of the image. In the image, every pixel inside the retina should have zero brightness, and every pixel outside the retina should have maximum brightness. At the first stage, the manual selection of borders and filling of the corresponding area can be performed. Figure 3b shows the result of such processing as a binary image.

The analysis of Figure 3b shows a rather complex structure of the boundaries of the retinal layer, which can lead to significant difficulties in the reconstruction of a three-dimensional model. Therefore, the next processing step is the generation of a binary image, for which the boundary lines of the retinal layer can be approximated using mathematical functions and have a smooth appearance. Continuous parametric functions are used to describe the upper and lower boundaries of the retinal layer. Figure 3c shows the resulting image after approximation.

As can be seen from Figure 3c, in addition to smoothing, three areas of the fundus are segmented separately in the final modeled image.

Thus, after preliminary processing, the considered algorithm will consist of the following five steps [28]:

Step 1. Constructing a halftone image.
Step 2. Detection of the contour lines of the retina.
Step 3. Selection of a group of points located on contour lines.
Step 4. Approximation of contour lines through selected points.
Step 5. Construction of the image with selected layers based on smoothed contour lines.

The sequential processing of all OCT images for the fundus allows one to obtain a set of approximating functions and use them to restore the three-dimensional structure of the fundus from OCT images. Figure 4 shows an example of such reconstruction. The processing of OCT images and reconstruction of a 3D model was performed using an implementation of the proposed algorithm in MATLAB.

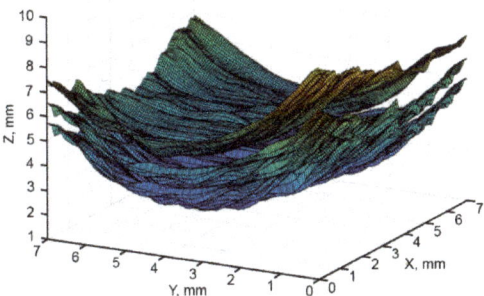

Figure 4. Fundus model reconstructed from OCT images.

Using 3D fundus models, based on the modeling of laser exposure, it is possible to analyze to what temperature this or that point of the fundus will be heated during laser exposure with certain parameters. This allows the prediction of the formation of coagulates in accordance with the chosen plan and to assess the likelihood of unwanted side effects, including heating the retinal layer to a critical temperature.

The analysis of Figure 4 shows that the use of approximating functions for two-dimensional OCT images makes it possible to reconstruct three-dimensional models of the fundus.

2.3. Heat Propagation Modeling

In the mathematical modeling of laser exposure, it is necessary to take into account the fact that energy is converted in such a model, i.e., a laser light pulse (light energy) leads to heating of the fundus (thermal energy). Indeed, the electromagnetic energy of the laser action is converted into heat when interacting with the vascular layer [29,30].

The intensity of light energy is described by Equation (1):

$$I(r) = \frac{P}{\pi a^2} e^{-(\frac{r}{a})^2}, \qquad (1)$$

where r is the distance from the light source, a is the spot radius, and P is the light source's power.

Using Equation (2), the temperature distribution in three-dimensional space can be determined after the end of the laser exposure:

$$\psi(x,y,z) = \frac{e^{-\int_0^z \beta(x,y,\xi)d\xi} \beta I(r) \Delta t}{C_v} + T_c, \qquad (2)$$

where T_c is the temperature at the time of the laser shot, $\beta = \beta(x,y,z)$ is the environment absorption function, $C_v = C_v(x,y,z)$ is a function of the environment volumetric heat capacity at a fixed timestamp, $r = \sqrt{(x-x_0)^2 + (y-y_0)^2 + (z-z_0)^2}$ is the distance to the point (x_0, y_0, z_0) where the laser coagulation was initialized, and Δt is the laser exposure duration.

Considering very small values of Δt, it is possible to neglect the diffraction of light [27]. Since this is indeed the case in practice, the model of the heat propagation after laser exposure can be rewritten in the form of Equation (3):

$$\begin{cases} C_v \frac{\partial T}{\partial t} = div[k \cdot grad_{xyz}(T)], \\ T|_{t=0} = \psi(x,y,z), \\ T|_E = T_0, \end{cases} \quad (3)$$

where $T = T(x,y,z,t)$ is temperature distribution depending on the spatial and time coordinates, $k = k(x,y,z,T)$ is a function of thermal conductivity of the environment in space and time, $C_v = C_v(x,y,z,T)$ is a function of the environment volumetric heat capacity which changes during heating or cooling, E is the edge laser exposure influence on temperature, div is a vector field divergence operator, $grad_{xyz}$ is an operator for calculating the gradient of a function by coordinates x, y, z, and T_0 is the temperature at the edge region.

When using Equation (3), it is necessary that the region of determination of the thermal field in space is large enough to ensure that, under laser action, the propagation of heat only occurs up to the edge through which the laser passes. Two methods can be used to meet this condition.

First, the function $\psi(x,y,z)$ must be specified at the bearing edge, i.e., one that changes over time. In addition, it is necessary to perform a special linear replacement in order to reduce the task to fixed (or zero) edge conditions. Thus, the thermal conductivity will be described by an inhomogeneous differential equation. Solution retrieval is possible via expansion into a solution for the corresponding homogeneous equation with the initial edge and initial conditions. Next, a solution to the inhomogeneous equation is found for which a zero edge and initial conditions are provided. Consequently, the final solution will be found from an equation in the form of Equation (3).

Nevertheless, several difficulties usually arise when searching for such a replacement and solving an inhomogeneous equation. Therefore, in this work, it was decided to use the second method. This method is based on the fact that it is necessary to artificially expand the domain of definition, and then, based on the resulting extension, divide it into informative and non-informative areas. In this case, for the first region, the condition of a relatively negligible pulse duration is satisfied, which makes it possible to describe the temperature distribution at the initial moment of time in accordance with Equation (2). Determining the spread of heat in a non-informative area will not be difficult, since for solving this problem it will be sufficient to use a symmetric display; however, the already fixed temperature value will correspond to the border in the uninformative area. At the same time, when analyzing a mathematical model, it will be possible to use only data in the informative area which further simplifies the task.

Unfortunately, due to the dependence of the volumetric heat capacity and temperature-conductivity on temperature, the task will be nonlinear. Nevertheless, it is possible to assess the change in the shape of the retina based on the temperature values of its layers, i.e., by assessing the given layer and temperature. Equation (4) is similar to Equation (3) in which the functions of the volumetric heat capacity C_v and thermal diffusivity k depend only on spatial coordinates.

The analysis of Equation (4) shows that it is not difficult to obtain zero edge conditions. Indeed, by replacing $T = \tilde{T} + T_0$, where $\tilde{T}$ describes the simulated temperature in the form of the direct effect of laser exposure (i.e., how much a given point of the fundus has heated as a result of laser exposure). Absolute temperatures on the fundus surface T are defined as the sum of the heating temperature and the initial temperature T_0.

$$\begin{cases} C_v(x,y,z) \frac{\partial T}{\partial t} = div[k(x,y,z) \cdot grad_{xyz}(T)], \\ T|_{t=0} = \frac{e^{-\beta(x,y,z)z} \beta(x,y,z) I(r) \Delta t}{C_v(x,y,z)} + T_c, \\ T|_E = T_0. \end{cases} \quad (4)$$

A normal tissue temperature (~36.5 °C) can be used as the starting temperature. The solution of Equation (4) makes it possible to obtain a model of the temperature change after the application of a laser with a given power at the point (x_0, y_0, z_0) throughout the structure of the fundus. Then, by combining the resulting model with a three-dimensional fundus model, it is possible to predict the absolute temperature at each point of the reconstructed model. The temperature on the retina is estimated based on this alignment. If, at some point in the retinal layer, the value exceeds the critical temperature ($T_{sr} = 38 - 40\,°C$), then treatment with the given parameters may be unsafe. Otherwise, the temperature in the area of the diabetic macular edema is estimated. If this temperature exceeds 39 °C, then a coagulum will appear and the treatment can be considered effective.

The solution of the task in an analytical form is not possible; however, the temperature distribution can be estimated using numerical methods [31]. For example, based on the splitting scheme, it is possible to go from a multidimensional task to a one-dimensional one. This allows the use of an implicit scheme without solving the linear relationship system. The application of a sweep method significantly speeds up the solution retrieval for this task.

To use the splitting method, it is necessary to rewrite Equation (4) in the form of Equation (5).

$$\begin{cases} s_v(x,y,z)\frac{\partial T}{\partial t} = \frac{\partial}{\partial x}(k(x,y,z)\frac{\partial T}{\partial x}) + \\ + \frac{\partial}{\partial y}(k(x,y,z)\frac{\partial T}{\partial y}) + \frac{\partial}{\partial z}(k(x,y,z)\frac{\partial T}{\partial z}), \\ T|_{t=0} = \psi(x,y,z), \\ T|_E = T_0. \end{cases} \quad (5)$$

The numerical solution of Equation (5) using the splitting method is described in a more convenient form below. The idea behind the method is that sampling with the same step must first be performed for a given time segment. After that, the solution is reduced to solving iterative Equations (6) and (7). At the first stage, the coordinate is split off along the oY axis, since this is the main axis of the OCT images.

$$\begin{cases} c_v(x,y,z)\frac{\partial W}{\partial t} = \frac{\partial}{\partial y}(k(x,y,z)\frac{\partial W}{\partial y}), \\ W|_{t=t_k} = T|_{t=t_k}, \\ W|_E = 0 \end{cases} \quad (6)$$

$$\begin{cases} s_v(x,y,z)\frac{\partial V}{\partial t} = \frac{\partial}{\partial x}(k(x,y,z)\frac{\partial V}{\partial x}) + \\ + \frac{\partial}{\partial z}(k(x,y,z)\frac{\partial V}{\partial z}), \\ V|_{t=t_k} = W|_{t=t_k}, \\ V|_E = 0. \end{cases} \quad (7)$$

The iterative solutions for these tasks are given as follows: First, a solution to Equation (6) is sought on the time interval $[t_k, t_{k+1}]$. In this case, the initial condition of Equation (6) at moment t_k coincides with the solution to Equation (5) at the same time. Further, for Equation (7), a solution is sought under the condition that the initial condition at the moment t_k is nothing more than the result of solving Equation (6) at the time t_{k+1} or $W_{t_{k+1}}$. It should be noted that for the splitting method the solution to the main problem, namely, temperature distribution at time t_{k+1}, can be represented as a solution to task (7), i.e., $T|_{t=t_{k+1}} \approx V|_{t=t_{k+1}}$.

It is necessary to understand that Equation (6) describes a set of tasks taking into account the dependence of the sought functions on all spatial coordinates. Similar dependencies are observed for Equation (7). The solution to Equation (6) can be obtained using the finite difference method based on an implicit difference scheme.

To summarize, the three-dimensional problem is reduced to solving Equations (6) and (7). In this case, the resulting two-dimensional task can also be decomposed into one-dimensional tasks or solved using numerical methods.

To carry out computational experiments of laser exposure to the fundus, the proposed algorithm for heat propagation modeling was implemented in C++.

3. Results

Several important results were obtained as a result of solving the laser exposure mathematical modeling task on the basis of numerical methods. First, the modeling of the laser action of the fundus was carried out while taking into account its three-dimensional structure. Figure 5 shows the result of this simulation. In addition, the simulation results can be used to estimate the parameters of laser exposure. By superimposing the obtained three-dimensional model of heat distribution at a given moment in time on the fundus model under the condition of normal tissue temperature, it is possible to estimate the absolute values of temperature at a given moment in time. Moreover, it is possible to estimate how quickly the temperature is restored to normal after the termination of the laser action. In Figure 5, some heat dissipation can be observed in the vicinity of the center point of the shot.

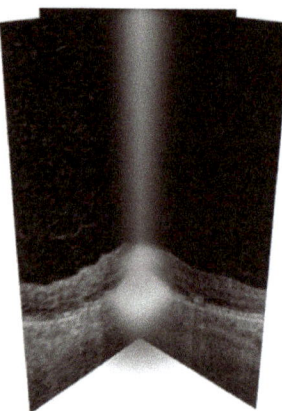

Figure 5. Three-dimensional model of laser exposure.

The use of numerical methods, generally speaking, can lead to different simulation results for different mesh sizes. Moreover, the larger the mesh size is, the longer the simulation will take. This gives rise to the problem of investigating the convergence of solutions.

The presence of discontinuities in the coefficients of thermal conductivity and volumetric heat capacity leads to the absence of the existence of derivatives at points of discontinuity. The integro-interpolation method (IIM) [31,32] aims at eliminating this problem, and consists of calculating integrals over the limits of the neighborhoods of the grid nodes for the main equation of the problem and the initial and boundary conditions. As a result of calculating the integrals, a difference scheme is formed which is considered to be resistant to discontinuities. It is very difficult to analytically show the effectiveness of the IIM in comparison with the finite difference method, as existing works have only been aimed at analyzing particular problems related to heat conduction [24–26]; however, there is a way to experimentally evaluate the convergence of the method. This method involves the numerical simulation of the coagulation process for two grids in which the first grid consist of N intervals and the second one consist of $2N$ intervals.

Since the modeling is performed with three spatial coordinates and one time coordinate, a convergence study was carried out on the basis of modeling at different sampling steps along one of the axes. At the same time, for all nodes available in two neighboring models, the temperature discrepancies were estimated and the standard deviation or root mean square (RMS) values were calculated. The results of the convergence study are shown in Tables 1–4. Here I, J, K are the numbers of intervals in spatial coordinates x, y, z, respectively, and S is number of intervals in time coordinate t.

Table 1. Convergence for different mesh sizes along the oX axis.

I	J	K	S	RMS
60	200	500	1000	0.015130427
120	200	500	1000	0.002634052
240	200	500	1000	0.00050212
480	200	500	1000	0.000172598

Table 2. Convergence for different mesh sizes along the oY axis.

I	J	K	S	RMS
200	60	500	1000	0.015772823
200	120	500	1000	0.003441242
200	240	500	1000	0.001191759
200	480	500	1000	0.000531296

Table 3. Convergence for different mesh sizes along the oZ axis.

I	J	K	S	RMS
200	200	60	1000	0.025006277
200	200	120	1000	0.004207547
200	200	240	1000	0.002474279
200	200	480	1000	0.001344305
200	200	960	1000	0.000694162

Table 4. Convergence for different mesh sizes along the time axis.

I	J	K	S	RMS
200	200	500	200	0.000323827
200	200	500	400	0.000186017
200	200	500	800	0.000103741
200	200	500	1600	0.00005.58805
200	200	500	3200	0.00002.92848

Analysis of the results presented in Tables 1–4 shows that with an increase in the number of grid dimensions, the results of mathematical modeling converge; however, different convergence rates are set for different coordinates. For example, for the oX axis, an increase in the number of intervals from 60 to 480 (8 times) leads to a decrease in the standard deviation by 87 times—but 29 times for the oY axis and less than 36 times for the oZ axis. A halving in the number of steps along the spatial coordinates leads on average to decreases in the standard deviation by 2.9–5.7 times along the oX axis, 2.2–4.5 times along the oY axis, and 1.7–5.9 times along the oZ axis. A stable decrease in the standard deviation indicates the convergence of the IIM. Most discontinuities exist in the oZ direction and the convergence is slow there. In the presence of discontinuities, the standard deviation usually decreases by no more than 50%. There are discontinuities in the oX and oY directions which are not as pronounced as in the Z direction. In these directions, the standard deviation can decrease by more than 50%, which is shown in the tables above. When studying the influence of the dimensions when passing from 200 intervals to 3200 intervals, a decrease in the deviation by a factor of 11 was observed; however, in terms of absolute values, the results showed far lower values for spatial coordinates with a value of approximately 5 °C. Reducing the sampling time step by a factor of two led to a decrease in the standard deviation of no more than 50%. On average, as can be seen in Table 4, the standard deviation decreased by 1.8 times. Thus, the algorithm has good convergence; that is, an increase in the grid dimension size will lead to an improvement in the accuracy of the solution.

Heat propagation with various laser displacements was also simulated. In particular, the coordinates along the oY and oZ axes were fixed as y_0 and z_0, and research was carried out for different x_0 values. Figure 6 shows the results of such simulations for three different x_0 scenarios. It should be noted that the assessment was carried out with absolute parameters characterizing the displacement distance x_0 from the left edge, where $x_0 = 0$.

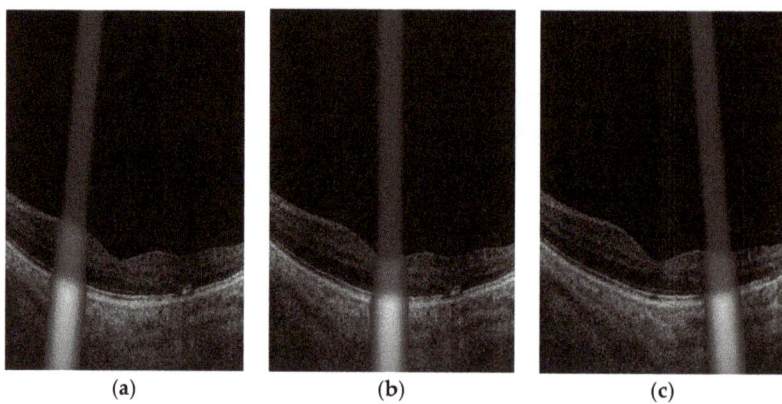

Figure 6. Simulation of thermal propagation at various coordinates for the point of influence along the oX axis: (**a**) 200 μm; (**b**) 350 μm; (**c**) 500 μm.

Analysis of the simulation results presented in Figure 6 shows that the structure (shape) of the retina had little effect on the spread of heat. For the considered example, the thickness of the retinal layer was approximately the same for all displacements; however, there were slight differences in the spread of heat along the fundus under the influence of different displacements. This difference should be taken into account when forming several coagulates side by side in order to prevent reaching a critical temperature on the retinal layer. Figure 7 shows the results of modeling the temperature distribution on the retinal layer after laser exposure at different points in time.

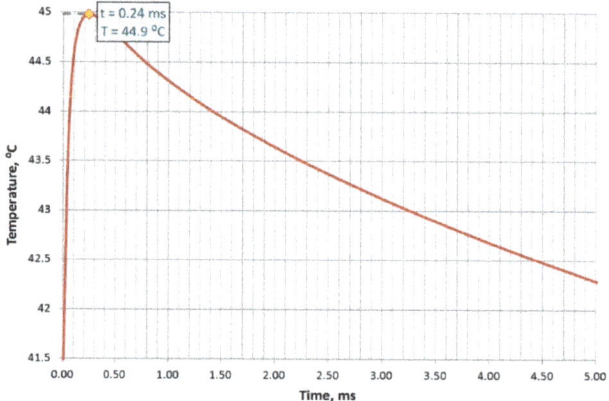

Figure 7. Simulation of temperature versus time.

Figure 7 shows that the retina heats up rather quickly. It should be noted that the value at time zero is associated with laser action on the adjacent layer.

It can be seen that within 0.24 ms, due to the spread of heat from the vascular layer, the temperature on the retina could reach 44.9 °C, after which there would be a slow decrease in temperature with a limit trending to the normal temperature of the tissue. It is clear that if one more shot is fired during these 0.24 ms, then the peak temperature may even exceed 44.9 °C. Based on this model, the required delay time between shots can be estimated.

The resulting model allowed us to simulate two shots in the area of the epithelial layer. Figure 8 shows an example of the implementation of the model. The coordinate along the oY axe was fixed, and the points of the shot were specified with different offsets along the oX and oZ axes. It should be noted that the different colors correspond to heating temperatures, so the figure can be interpreted in the same way in K and °C.

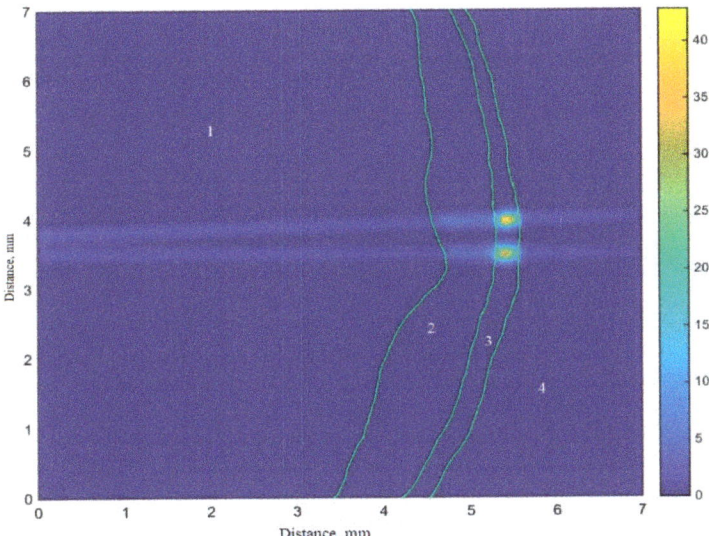

Figure 8. Modeling of laser action for the formation of two coagulates: 1—vitreous layer, 2—retinal layer, 3—epithelial layer, 4—vascular layer.

Analysis of Figure 8 shows that as a result of each shot, heat spread from the epithelial layer to the retinal layer. A short time after exposure, the temperature of the retina can increase by 5 °C due to the spread of heat from one shot. The second shot, taking into account the proximity of the distance between the coagulates, will also contribute to the spread of heat in the local neighborhood. During treatment, it is impossible to allow a critical temperature on the retinal layer to be reached.

In this regard, in order to identify safe treatment parameters, heat propagation was simulated for various distances between the centers of coagulates and delay times between shots. In particular, the situations of laser exposure with two shots with different displacements and delay times were simulated. Figure 9 shows the results of such a simulation at different coordinates along the oX axis. It should be noted that the temperature is calculated as the maximum temperature over the entire area as a result of simulating two laser shots.

Analysis of Figure 9 shows that a safe distance between coagulates is about 180 μm. Another option to ensure safe treatment is to increase the delay time between shots to 15 ms; the distance can be reduced by increasing the delay between shots. A distance of around 160 μm may be used with a delay greater than 15 ms.

Thus, the main results here are the mathematical models of laser exposure which make it possible to estimate the parameters of safe treatment. Notably, these models feature the property of convergence.

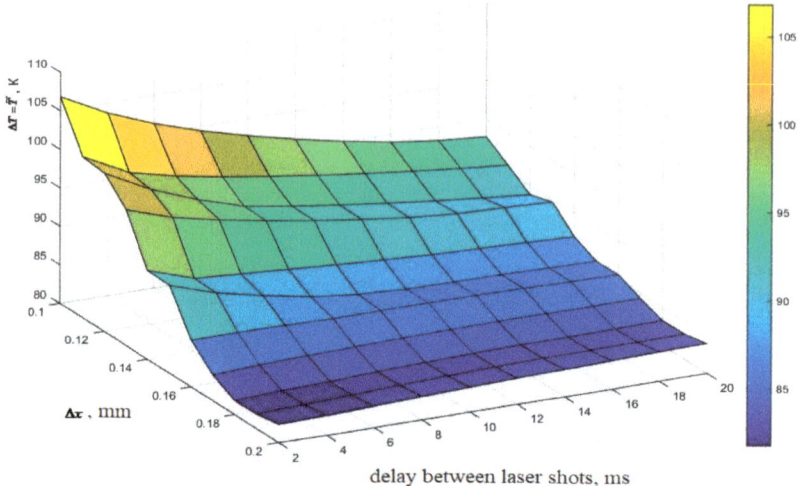

Figure 9. Dependence of the maximum temperature on the epithelial layer during the implementation of two shots on the delay between shots and the coordinates of the shots.

4. Discussion

The results obtained within the framework of the study can be taken into account for the treatment of diabetic retinopathy, and the models developed can be used in treatment to select parameters of laser exposure that will provide the maximum therapeutic effect. The use of the proposed fundus models and laser exposure parameters is another step towards personalized medical care. Indeed, the OCT images make it possible to restore the three-dimensional structure of the fundus where the laser exposure is simulated. The modeling of heating the retinal, vascular, and epithelial layers allows conclusions to be formed regarding safe shifts between laser shots and pause durations.

Many quantitative results have been obtained. In particular, at a given laser power, a safe pause of 15 ms was revealed with a distance between coagulates of 160 μm. In addition, a minimum safe distance between shots was 180 μm, which did not depend on the delay between shots. Convergence of the numerical methods used in modeling has been established with a decrease in the sampling step size by a factor of two, where the standard deviation stably decreases. A decrease in the step size by 50% along the spatial coordinates leads, on average, to decreases in the standard deviation of 2.9–5.7 times along the oX axis, 2.2–4.5 times along the oY axis, and 1.7–5.9 times along the oZ axis. A stable decrease in the standard deviation indicates the convergence of the integro-interpolation method. This becomes especially important when arranging coagulates not as a single formation, but as a group. Moreover, on the basis of mathematical modeling, it is potentially possible to adjust the plans of coagulates in order to maximize the therapeutic effects. In the future, we plan to study the placement of a group of coagulates, as well as the performance of the proposed algorithms.

Author Contributions: Methodology, A.S., E.Z. and A.Z.; writing—original draft, N.A., A.S. and N.I.; writing—review and editing, N.A. and D.K.; supervision, N.I.; funding acquisition, N.I. All authors have read and agreed to the published version of the manuscript.

Funding: This research was funded by RFBR (grant number 19-29-01135); Ministry of Science and Higher Education of the Russian Federation within (Government project).

Institutional Review Board Statement: Not applicable.

Informed Consent Statement: Not applicable.

Data Availability Statement: Not applicable.

Conflicts of Interest: The authors declare no conflict of interest.

References

1. Lipatov, D.V.; Smirnova, N.B.; Aleksandrova, V.K. Modern algorithm for laser coagulation of the retina in diabetic retinopathy. *Diabetes Mellit.* **2007**, *3*, 45–46. [CrossRef]
2. Gafurov, S.D.; Katakhonov, S.M.; Holmonov, M.M. Features of the use of lasers in medicine. *Eur. Sci.* **2019**, *3*, 92–95.
3. Zamytsky, E.A.; Zolotarev, A.V.; Karlova, E.V.; Zamytsky, P.A. Analysis of the intensity of coagulates in laser treatment of diabetic macular edema using a robotic laser Navilas. *Saratov. J. Med. Sci. Res.* **2017**, *13*, 375–378.
4. Kotsur, T.V.; Izmailov, A.S. The effectiveness of laser coagulation in the macula and high-density microphotocoagulation in the treatment of diabetic maculopathy. *Ophthalmol. Stat.* **2016**, *9*, 43–45. [CrossRef]
5. Kozak, I.; Luttrull, J.K. Modern retinal laser therapy. *Saudi J. Ophthalmol.* **2015**, *29*, 137–146. [CrossRef]
6. Chhablani, J.; Kozak, I.; Barteselli, G.; El-Emam, S. A novel navigated laser system brings new efficacy to the treatment of retinovascular disorders. *Oman J. Ophthalmol.* **2013**, *6*, 18–22. [CrossRef]
7. Shirokanev, A.S.; Kirsh, D.V.; Ilyasova, N.Y.; Kupriyanov, A.V. Investigation of algorithms for placing coagulates on the fundus image. *Comput. Opt.* **2018**, *42*, 712–721. [CrossRef]
8. Ilyasova, N.Y.; Shirokanev, A.S.; Kupriynov, A.V.; Paringer, R.A. Technology of intellectual feature selection for a system of automatic formation of a coagulate plan on retina. *Comput. Opt.* **2019**, *43*, 304–315. [CrossRef]
9. Mukhin, A.; Kilbas, I.; Paringer, R.; Ilyasova, N. Application of the gradient descent for data balancing in diagnostic image analysis problems. In Proceedings of the IEEE Xplore, 2020 International Conference on Information Technology and Nanotechnology (ITNT), Samara, Russia, 26–29 May 2020; pp. 1–4. [CrossRef]
10. Ilyasova, N.Yu.; Demin, N.S.; Shirokanev, A.S.; Kupriyanov, A.V.; Zamytskiy, E.A. Method for selection macular edema region using optical coherence tomography data. *Comput. Opt.* **2020**, *44*, 250–258. [CrossRef]
11. Andriyanov, N.A.; Dementiev, V.E. Developing and studying the algorithm for segmentation of simple images using detectors based on doubly stochastic random fields. *Pattern Recognit. Image Anal.* **2019**, *29*, 1–9. [CrossRef]
12. Shirokanev, A.S.; Kibitkina, A.S.; Ilyasova, N.Y.; Degtyarev, A.A. Methods of mathematical modeling of fundus laser exposure for therapeutic effect evaluation. *Comput. Opt.* **2020**, *44*, 809–820. [CrossRef]
13. Fiandono, I.; Firdausy, K. Median filtering for optic disc segmentation in retinal image. *Kinetik* **2018**, *3*, 75–82. [CrossRef]
14. Joon, H.L.; Joonseok, L.; Sooah, C.; Ji, E.S.; Minyoung, L.; Sung, H.K.; Jin, Y.L.; Dae, H.S.; Joon, M.K.; Jung, H.B.; et al. Development of decision support software for deep learning-based automated retinal disease screening using relatively limited fundus photograph data. *Electronics* **2021**, *10*, 163. [CrossRef]
15. Arfan, G.; Chan, H.S.; Vaisakh, S.; Jahanzeb, A.; Raed, A.A. Accelerating retinal fundus image classification using artificial neural networks (ANNs) and reconfigurable hardware (FPGA). *Electronics* **2019**, *8*, 1522. [CrossRef]
16. Ling, L.; Dingyu, X.; Xinglong, F. Automatic diabetic retinopathy grading via self-knowledge distillation. *Electronics* **2020**, *9*, 1337. [CrossRef]
17. Jyostna, D.B.; Veeranjaneyulu, N.; Shaik, N.Sh.; Saqib, H.; Muhammad, B.; Praveen, K.; Reddy, M.; Ohyun, J. Blended multi-modal deep ConvNet features for diabetic retinopathy severity prediction. *Electronics* **2020**, *9*, 914. [CrossRef]
18. Artemov, S.; Belyaev, A.; Bushukina, O.; Khrushchalina, S.; Kostin, S.; Lyapin, A.; Ryabochkina, P.; Taratynova, A. Endovenous laser coagulation using two-micron laser radiation: Mathematical modeling and in vivo experiments. In Proceedings of the International Conference on Advanced Laser Technologies (ALT), Prague, Czech Republic, 15–20 September 2019; Volume 19, pp. 14–21.
19. Moutray, T.; Evans, J.R.; Lois, N.; Armstrong, D.J.; Peto, T.; Azuara-Blanco, A. Different lasers and techniques for proliferative diabetic retinopathy. *Cochrane Database Syst. Rev.* **2018**, *3*, 1–87. [CrossRef] [PubMed]
20. Belucio-Neto, J.; De Oliveira, X.C.; De Oliveira, D.R.; Passos, R.M.; Maia, M.; Maia, A. Functional and anatomical outcomes in patients submitted to panretinal photocoagulation using 577nm multispot vs 532nm single-spot laser: A clinical trial. *Investig. Ophthalmol. Vis. Sci.* **2016**, *57*, 1–18.
21. Ghassemi, F.; Ebrahimiadib, N.; Roohipoor, R.; Moghimi, S.; Alipour, F. Nerve fiber layer thickness in eyes treated with red versus green laser in proliferative diabetic retinopathy: Short-term results. *Ophthalmologica* **2013**, *230*, 195–200. [CrossRef]
22. Inan, U.U.; Polat, O.; Inan, S.; Yigit, S.; Baysal, Z. Comparison of pain scores between patients undergoing panretinal photocoagulation using navigated or pattern scan laser systems. *Arq. Bras. O1almol.* **2016**, *79*, 8–15. [CrossRef]
23. Zhang, S.; Cao, G.F.; Xu, X.Z.; Wang, C.H. Pattern scan laser versus single spot laser in panretinal photocoagulation treatment for proliferative diabetic retinopathy. *Int. Eye Sci.* **2017**, *17*, 205–208.
24. Almutairi, M.; Zureigat, H.; Izani Ismail, A.; Fareed Jameel, A. Fuzzy numerical solution via finite difference scheme of wave equation in double parametrical fuzzy number form. *Mathematics* **2021**, *9*, 667. [CrossRef]
25. Anuar, N.S.; Bachok, N.; Pop, I. Numerical computation of dusty hybrid nanofluid flow and heat transfer over a deformable sheet with slip fffect. *Mathematics* **2021**, *9*, 643. [CrossRef]
26. Ahmad, F.; Almatroud, A.O.; Hussain, S.; Farooq, S.E.; Ullah, R. Numerical solution of nonlinear diff. equations for heat transfer in micropolar fluids over a stretching domain. *Mathematics* **2020**, *8*, 854. [CrossRef]
27. Pak, J.M. Switching extended Kalman filter bank for indoor localization using wireless sensor networks. *Electronics* **2021**, *10*, 718. [CrossRef]
28. Shirikanev, A.; Ilyasova, A.; Demin, N.S.; Zamyckij, E. Extracting a DME area based on graph-based image segmentation and collation of OCT retinal images. *J. Phys.* **2021**, 1–9. [CrossRef]

29. Kistenev, Y.; Buligin, A.; Sandykova, E.; Sim, E.; Vrazhnov, D. Modeling of IR laser radiation propagation in bio-tissues. In *Proceedings of SPIE 2019*; SPIE Press: Bellingham, WA, USA, 2019; Volume 11208, pp. 1–4.
30. Samarsky, A.A. Schemes of the increased order of accuracy for the multidimensional heat conduction equation. *J. Comput. Math. Math. Phys.* **1963**, *3*, 812–840.
31. Anufriev, I.E.; Osipov, P.A. *Mathematical Methods for Modeling Physical Processes. Finite Difference Method. With Solutions of Typical Problems*; St. Petersburg State Polytechnic University, Institute of Applied Mathematics and Mechanics: Saint Petersburg, Russia, 2014; pp. 1–282.
32. Azima, Y.I. Use of the integro-interpolation method for construction of difference equations for determination of thermal properties and unsteady-state heat fluxes. *Eng. Phys. Thermophys.* **1998**, *71*, 795–802. [CrossRef]

Article
Mechanical Models for Hermite Interpolation on the Unit Circle

Elías Berriochoa [1,*,†], Alicia Cachafeiro [1,*,†], Héctor García Rábade [2,†] and José Manuel García-Amor [3,†]

1. Departamento de Matemática Aplicada I, Universidad de Vigo, 36310 Vigo, Spain
2. Departamento de Matemática Aplicada II, Universidad de Vigo, 32004 Ourense, Spain; hector.garcia.rabade@uvigo.es
3. Departamento de Matemáticas, Instituto E. S. Valle Inclán, 36001 Pontevedra, Spain; garciaamor@edu.xunta.gal
* Correspondence: esnaola@uvigo.es (E.B.); acachafe@uvigo.es (A.C.); Tel.: +34-988-387216 (E.B.); +34-986-812138 (A.C.)
† These authors contributed equally to this work.

Abstract: In the present paper, we delve into the study of nodal systems on the unit circle that meet certain separation properties. Our aim was to study the Hermite interpolation process on the unit circle by using these nodal arrays. The target was to develop the corresponding interpolation theory in order to make practical use of these nodal systems linked to certain mechanical models that fit these distributions.

Keywords: Hermite interpolation; nodal systems; unit circle; convergence

1. Introduction

Hermite polynomial interpolation problems on the real line and on the bounded interval have been widely studied by many researchers. Most of the main contributions have been obtained by using as nodal points the zeros of the classical orthogonal polynomials and some of their generalizations (see [1–3]). When the nodal points are zeros of general orthogonal polynomials, the papers of Freud [4,5] (see [6]) deserve to be mentioned. The first attempt to use nodal systems not linked to measures and therefore to orthogonal polynomials was carried out by Fejér who introduced the so-called normal systems. In the same direction, we must mention the contributions of Grünwald who introduced in [7] the strongly normal nodal systems. Unfortunately, the use of these systems has not been consolidated and has not been continued. More recently, some nodal systems called well-spaced ones have been used for studying Lagrange interpolation problems. Although they are not connected with measures, any reasonable choice of interpolation nodes fulfills the conditions of being well spaced (see [8]). Thus, if one wants to work on the bounded interval with nodal systems that are not connected with measures, it seems convenient that the nodes have a distribution that is not far from the Chebyshev distribution. This closeness can be established in terms of suitable separation properties. The advantages of using these types of nodal systems is that they can be obtained through a random uniform distribution (see the examples given in [9]).

During the last few decades, several problems related to Hermite polynomial interpolation on the unit circle have been studied in depth. In some cases, the study has been connected to the Hermite interpolation on the interval and to the trigonometric interpolation. The nodal systems usually employed were the equispaced ones, that is those constituted by the n roots of complex numbers with modulus one. It is well known that there are methods to compute the Laurent polynomials of Hermite interpolation in an efficient way, when considering equally spaced nodal points on the unit circumference. By using these nodal systems, the convergence of the Hermite–Fejér interpolation polynomials related to continuous functions was studied in [10], obtaining a Fejér-type theorem. The

convergence of the Hermite interpolation polynomials taking nonvanishing conditions for the derivatives was studied in [11] giving several versions of the second Fejér-type theorem.

Other more general nodal systems that have been used are those formed by the zeros of para-orthogonal polynomials, associated with good measures on the unit circle. We recall that a polynomial $Q_n(z)$ of degree n is para-orthogonal if it satisfies that $Q_n(z) = z^n \overline{Q_n}(1/z)$ and it is orthogonal to $\{z^k\}_{k=1}^{n-1}$ with respect to a measure on the unit circle. It deserves to mention measures in the Baxter class or analytic measures and, in particular, measures in the Szegő class with the Szegő function having an analytic extension outside the unit disk (see [12,13]). In these cases, the nodal points are characterized by satisfying suitable separation properties, and it is possible to obtain the properties of the nodal polynomials without the need to compute their zeros explicitly. Most of these properties play an important role in the results for convergence and related problems that were studied.

The use of another type of nodal system adds complexity to the problem due mainly to the difficulty of computing their zeros. In this sense, some advances have been made by using nodal polynomials that are close to the equispaced ones in the unit circumference and that can be obtained through a perturbation of the uniform random distribution. The starting point is to ask the nodes for some separation properties that allow obtaining the main properties of the nodal polynomials playing a fundamental role in the interpolation processes.

If the arguments of the nodal points are θ_j, $j = 1, \cdots, n$, we recall that in the Baxter class, the separation property satisfied by the nodal points is $\theta_{j+1} - \theta_j = \frac{2\pi}{n} + \mathcal{O}(\frac{1}{n})$ (see [14]). When the nodal points are the zeros of para-orthogonal polynomials with respect to analytic weights on the unit circle or with respect to measures in the Szegő class having an analytic extension outside the unit disk, then the separation property is $\theta_{j+1} - \theta_j = \frac{2\pi}{n} + \mathcal{O}(\frac{1}{n^2})$ (see [15]). Thus, the nodal points behave like perturbations of the roots of complex unimodular numbers.

In [9], we changed the focus of the interpolation problem, and our starting point was to use nodal systems that were not related to measures and that were only characterized to fulfil certain separation properties between the nodes. Thus, we studied the Lagrange interpolation problem on the unit circle by using nodal systems that are not connected with any measure, and they are only characterized by satisfying a separation property of the type: $\theta_{j+1} - \theta_j = \frac{2\pi}{n} + \mathcal{O}(\frac{1}{n^2})$. In the aforementioned paper, a detailed study of their properties was done. Moreover, the Lebesgue constant of the process was obtained, as well as some results for the convergence and the rate of convergence for different smooth continuous functions.

In the present paper, we studied nodal systems like those used in [9] that meet the separation properties mentioned above, and we present mechanical models that fit these distributions. The aim of this work was to study the Hermite interpolation process on the unit circle by using these nodal systems characterized by a separation property between the nodes. The target was to develop the corresponding interpolation theory that allows us to make practical use of these nodal systems linked to certain mechanical models. Thus, in the first part of the paper (see Sections 2 and 3), we recall the properties given in [9] satisfied by the nodal system, and we obtain some new ones that play an important role in the Hermite process. Our main result was to prove a new version of Fejér's theorem for continuous functions on the unit circle by using these nodal systems. We also studied the complete problem, that is the Hermite interpolation problem with nonvanishing conditions for the derivatives, giving a sufficient condition on the derivatives in order to assure convergence. These results are gathered in Section 4, with the study of the convergence of the Hermite interpolation polynomial related to analytic and smooth functions (see [16]), and finally, in Section 5 we obtain the corresponding results on the bounded interval. Indeed, we transformed the problem into a new one on the bounded interval studying the Hermite interpolation problem related to continuous functions on $[-1, 1]$ and by using nodal points characterized by some separation properties. In Section 6, we present some mechanical models generating the nodal systems according to our distribution, as well as some numerical examples by applying our results. Finally, Section 7 gathers the main notation concerning the different classes of polynomials used throughout the paper; Section 8 briefs the materials and methods used and Section 9 offers a discussion of the problem.

2. Preliminaries

The aim of this paper was to study Hermite interpolation problems on the unit circle $\mathbb{T} = \{z \in \mathbb{C} : |z| = 1\}$ by using nodal systems satisfying some suitable separation properties, which are not connected with orthogonality nor para-orthogonality with respect to any measure.

We denote the nodal polynomials by $W_n(z)$ and their zeros by $\{\alpha_{j,n}\}_{j=1}^n$, that is, we assumed that $W_n(z) = \Pi_{j=1}^n(z - \alpha_{j,n})$, where $|\alpha_{j,n}| = 1$ for $j = 1, \cdots, n$, and $\alpha_{j,n} \neq \alpha_{k,n}$ for $j \neq k$. For simplicity, we omit the subscript n and write α_j instead of $\alpha_{j,n}$ for $j = 1, \cdots, n$. First, we recall some well-known definitions related to interpolation problems on the unit circle. We work in the space of Laurent polynomials and, in particular, in the subspaces $\Lambda_{p,q}[z] = span\{z^k : p \leq k \leq q\}$, with p and q integers $p \leq q$.

If $\{u_j\}_{j=1}^n$ and $\{v_j\}_{j=1}^n$ are arbitrary complex numbers and $W_n(z)$ is the nodal polynomial, then the Hermite interpolation problem consists of determining the Laurent polynomial $\mathcal{H}_{-n,n-1}(z)$ satisfying the interpolation conditions:

$$\mathcal{H}_{-n,n-1}(\alpha_j) = u_j, \quad \mathcal{H}'_{-n,n-1}(\alpha_j) = v_j, \quad \text{for } j = 1, \cdots, n.$$

This polynomial can be rewritten as $\mathcal{H}_{-n,n-1}(z) = \mathcal{HF}_{-n,n-1}(z) + \mathcal{HD}_{-n,n-1}(z)$, where the Hermite–Fejér interpolation polynomial satisfies that:

$$\mathcal{HF}_{-n,n-1}(\alpha_j) = u_j, \quad \mathcal{HF}'_{-n,n-1}(\alpha_j) = 0, \quad \text{for } j = 1, \cdots, n, \qquad (1)$$

and $\mathcal{HD}_{-n,n-1}(z)$ satisfies that

$$\mathcal{HD}_{-n,n-1}(\alpha_j) = 0, \quad \mathcal{HD}'_{-n,n-1}(\alpha_j) = v_j, \quad \text{for } j = 1, \cdots, n. \qquad (2)$$

If f is a function, $u_j = f(\alpha_j)$ and $v_j = 0$, we denote the corresponding Laurent polynomial $\mathcal{HF}_{-n,n-1}(f,z)$. In the same way, if $u_j = 0, v_j = f'(\alpha_j)$, we denote the corresponding Laurent polynomial by $\mathcal{HD}_{-n,n-1}(f,z)$ and also denote by $\mathcal{H}_{-n,n-1}(f,z) = \mathcal{HF}_{-n,n-1}(f,z) + \mathcal{HD}_{-n,n-1}(f,z)$. In the case, when $u_j = f(\alpha_j)$ and v_j is arbitrary, if $\gamma_n = (v_j)_{j=1}^n$, we denote by $\mathcal{H}_{-n,n-1}(f,\gamma_n,z) = \mathcal{HF}_{-n,n-1}(f,z) + \mathcal{HD}_{-n,n-1}(z)$.

Let us recall that the preceding polynomials can be computed by using the following expressions (see [17]).

$$\mathcal{HF}_{-n,n-1}(z) = \sum_{k=1}^n \mathfrak{h}_{k,n}(z) u_k \qquad (3)$$

and

$$\mathcal{HD}_{-n,n-1}(z) = \sum_{k=1}^n \mathfrak{k}_{k,n}(z) v_k, \qquad (4)$$

where $\mathfrak{h}_{k,n}$ and $\mathfrak{k}_{k,n}$ are the fundamental polynomials of Hermite interpolation given by:

$$\mathfrak{h}_{k,n}(z) = \frac{(W_n(z))^2}{z^n} \frac{1}{(W_n'(\alpha_k))^2} \left[\frac{\alpha_k^n}{(z-\alpha_k)^2} + \frac{\alpha_k^{n-1}}{(z-\alpha_k)} \left(n - \frac{\alpha_k W_n''(\alpha_k)}{W_n'(\alpha_k)} \right) \right], \qquad (5)$$

and:

$$\mathfrak{k}_{k,n}(z) = \frac{(W_n(z))^2}{z^n} \frac{\alpha_k^n}{(W_n'(\alpha_k))^2} \frac{1}{(z-\alpha_k)}. \qquad (6)$$

We can obtain more suitable expressions of these polynomials by using the following relations:

If $z, \alpha_k \in \mathbb{T}$, then $(z - \alpha_k)^2 = -z\alpha_k|z - \alpha_k|^2$, and therefore, if $W_n(z) = \prod_{k=1}^{n}(z - \alpha_k)$, then we get $(W_n(z))^2 = (-1)^n z^n (\prod_{k=1}^{n} \alpha_k)|W_n(z)|^2$. Moreover, since $W'_n(\alpha_k) = \prod_{j=1, j \neq k}^{n}(\alpha_k - \alpha_j)$, then $(W'_n(\alpha_k))^2 = (-1)^{n-1} \alpha_k^{n-2}(\prod_{j=1}^{n} \alpha_j)|W'_n(\alpha_k)|^2$.

Hence, by substituting these relations in (5) and (6), we obtain:

$$\mathfrak{h}_{k,n}(z) = \frac{|W_n(z)|^2}{|W'_n(\alpha_k)|^2}\left[\frac{\alpha_k}{z|z-\alpha_k|^2} - \frac{\alpha_k}{(z-\alpha_k)}\left(n - \frac{\alpha_k W''_n(\alpha_k)}{W'_n(\alpha_k)}\right)\right], \quad (7)$$

and:

$$\mathfrak{k}_{k,n}(z) = -\frac{|W_n(z)|^2}{|W'_n(\alpha_k)|^2}\frac{\alpha_k^2}{(z-\alpha_k)}, \quad (8)$$

and therefore, we have the following desired expressions for the Hermite interpolation polynomials:

$$\mathcal{HF}_{-n,n-1}(z) = \sum_{k=1}^{n}\frac{|W_n(z)|^2}{|W'_n(\alpha_k)|^2}\left[\frac{\alpha_k}{z|z-\alpha_k|^2} - \frac{\alpha_k}{(z-\alpha_k)}\left(n - \frac{\alpha_k W''_n(\alpha_k)}{W'_n(\alpha_k)}\right)\right]u_k, \quad (9)$$

and

$$\mathcal{HD}_{-n,n-1}(z) = -\sum_{k=1}^{n}\frac{|W_n(z)|^2}{|W'_n(\alpha_k)|^2}\frac{\alpha_k^2}{(z-\alpha_k)}v_k. \quad (10)$$

In practice, it is more convenient to use the barycentric expressions for computing the interpolation polynomials. Thus, in what follows, we obtain these types of formulas.

By taking into account that $\mathcal{H}_{-n,n-1}(z) = \mathcal{HF}_{-n,n-1}(z) + \mathcal{HD}_{-n,n-1}(z)$ and $1 = \mathcal{HF}_{-n,n-1}(1,z)$, then we get:

$$\mathcal{H}_{-n,n-1}(z) = \frac{\mathcal{HF}_{-n,n-1}(z) + \mathcal{HD}_{-n,n-1}(z)}{\mathcal{HF}_{-n,n-1}(1,z)}.$$

Thus, if we use (3) and (4), we have:

$$\mathcal{H}_{-n,n-1}(z) = \frac{\sum_{k=1}^{n}\mathfrak{h}_{k,n}(z)u_k + \sum_{k=1}^{n}\mathfrak{k}_{k,n}(z)v_k}{\sum_{k=1}^{n}\mathfrak{h}_{k,n}(z)}.$$

On the one hand, if we apply (5) and (6), we obtain the following barycentric expression:

$$\mathcal{H}_{-n,n-1}(z) = \frac{\sum_{k=1}^{n}\frac{\alpha_k^{n-1}}{(W'_n(\alpha_k))^2}\left[\left[\frac{\alpha_k}{(z-\alpha_k)^2} + \frac{1}{(z-\alpha_k)}\left(n - \frac{\alpha_k W''_n(\alpha_k)}{W'_n(\alpha_k)}\right)\right]u_k + \frac{\alpha_k}{(z-\alpha_k)}v_k\right]}{\sum_{k=1}^{n}\frac{\alpha_k^{n-1}}{(W'_n(\alpha_k))^2}\left[\frac{\alpha_k}{(z-\alpha_k)^2} + \frac{1}{(z-\alpha_k)}\left(n - \frac{\alpha_k W''_n(\alpha_k)}{W'_n(\alpha_k)}\right)\right]}.$$

On the other hand, if we use Equations (7) and (8) instead of (5) and (6), we obtain the equivalent barycentric expression:

$$\mathcal{H}_{-n,n-1}(z) = \frac{\sum_{k=1}^{n}\frac{1}{|W'_n(\alpha_k)|^2}\left[\left[\frac{\alpha_k}{z|z-\alpha_k|^2} - \frac{\alpha_k}{(z-\alpha_k)}\left(n - \frac{\alpha_k W''_n(\alpha_k)}{W'_n(\alpha_k)}\right)\right]u_k - \frac{\alpha_k^2}{(z-\alpha_k)}v_k\right]}{\sum_{k=1}^{n}\frac{1}{|W'_n(\alpha_k)|^2}\left[\frac{\alpha_k}{z|z-\alpha_k|^2} - \frac{\alpha_k}{(z-\alpha_k)}\left(n - \frac{\alpha_k W''_n(\alpha_k)}{W'_n(\alpha_k)}\right)\right]}. \quad (11)$$

Proceeding in a similar way, we can obtain the two following barycentric expressions for the Hermite–Fejér interpolation polynomial:

$$\mathcal{HF}_{-n,n-1}(z) = \frac{\sum_{k=1}^{n} \frac{\alpha_k^{n-1}}{(W_n'(\alpha_k))^2} \left[\frac{\alpha_k}{(z-\alpha_k)^2} + \frac{1}{(z-\alpha_k)}\left(n - \frac{\alpha_k W_n''(\alpha_k)}{W_n'(\alpha_k)}\right) \right] u_k}{\sum_{k=1}^{n} \frac{\alpha_k^{n-1}}{(W_n'(\alpha_k))^2} \left[\frac{\alpha_k}{(z-\alpha_k)^2} + \frac{1}{(z-\alpha_k)}\left(n - \frac{\alpha_k W_n''(\alpha_k)}{W_n'(\alpha_k)}\right) \right]},$$

$$\mathcal{HF}_{-n,n-1}(z) = \frac{\sum_{k=1}^{n} \frac{1}{|W_n'(\alpha_k)|^2} \left[\frac{\alpha_k}{z|z-\alpha_k|^2} - \frac{\alpha_k}{(z-\alpha_k)}\left(n - \frac{\alpha_k W_n''(\alpha_k)}{W_n'(\alpha_k)}\right) \right] u_k}{\sum_{k=1}^{n} \frac{1}{|W_n'(\alpha_k)|^2} \left[\frac{\alpha_k}{z|z-\alpha_k|^2} - \frac{\alpha_k}{(z-\alpha_k)}\left(n - \frac{\alpha_k W_n''(\alpha_k)}{W_n'(\alpha_k)}\right) \right]}. \quad (12)$$

3. The Focus: The Nodal System

Throughout the paper, we assume that the zeros of the nodal polynomials $W_n(z)$ satisfy the following separation property: there exists a positive constant A, $A < \pi$ such that the length of the shortest arc between two consecutive nodes α_j and α_{j+1} satisfies:

$$\widehat{\alpha_j - \alpha_{j+1}} = \frac{2\pi}{n} + \frac{A(j)}{n^2} \text{ with } |A(j)| \leq A, \ \forall j = 1, \cdots, n, \quad (13)$$

where $\alpha_{n+1} = \alpha_1$. As we said before, we denote the length of the shortest arc between any two points of the unit circle, z_1 and z_2, by $\widehat{z_1 - z_2}$.

If we use Landau's notation for complex sequences, denoting by $a_n = \mathcal{O}(b_n)$ if $|\frac{a_n}{b_n}|$ is bounded, then the preceding property can be formulated as follows:

$$\widehat{\alpha_j - \alpha_{j+1}} = \frac{2\pi}{n} + \mathcal{O}(\frac{1}{n^2}).$$

We use the same $\mathcal{O}$ to denote different sequences. Unless we mention it explicitly, the limits we obtained from (13) were uniform.

We also considered other nodal polynomials, $\widetilde{W}_{n,j}(z), j = 1, \cdots, n$, well connected with $W_n(z)$. We define $\widetilde{W}_{n,j}(z) = z^n - \lambda_j$, with $\lambda_j = \alpha_j^n$, and we denote their zeros by $\beta_{k,j} = \sqrt[n]{\lambda_j}$, $k = 1, \cdots, n$, and it holds $\beta_{1,j} = \alpha_j$.

In what follows, we take for simplicity $j = 1$, that is we work with $\widetilde{W}_{n,1}(z) = z^n - \lambda_1$, and in order to simplify the notation, we denote $\widetilde{W}_{n,1}(z)$ by $\widetilde{W}_n(z)$ and its zeros $\beta_{k,1}$ by β_k for $k = 1, \cdots, n$, with $\beta_{1,1} = \beta_1 = \alpha_1$.

Hence, it is clear that the separation property satisfied by the zeros $\{\beta_j\}$ of $\widetilde{W}_n(z)$ is:

$$\widehat{\beta_j - \beta_{j+1}} = \frac{2\pi}{n}, \ \forall j = 1, \cdots, n, \quad (14)$$

where $\beta_{n+1} = \beta_1$.

In this section, we present the properties satisfied by these nodal polynomials $W_n(z)$ that play an important role in our interpolatory scheme.

First, we recall the following well-known relation between arcs and strings that we use to obtain the nodal properties based on the convex character of the arcsin function: If z_1 and z_2 belong to $\mathbb{T}$, then:

$$\frac{2}{\pi}(\widehat{z_1 - z_2}) \leq |z_1 - z_2| \leq (\widehat{z_1 - z_2}). \quad (15)$$

Secondly, we recall a separation property, given in [9], between both nodal systems $W_n(z)$ and $\widetilde{W}_n(z)$.

If $\{\alpha_j\}_{j=1}^n$ and $\{\beta_j\}_{j=1}^n$, with $\alpha_1 = \beta_1$, are the nodal points satisfying the separation properties (13) and (14), respectively, and we assume they are numbered in the clockwise sense, then:

$$\widehat{\alpha_j - \beta_j} \leq (j-1)\frac{A}{n^2}, \text{ for } j \geq 1, \text{ and } \widehat{\alpha_{n-j} - \beta_{n-j}} \leq (j+1)\frac{A}{n^2}, \text{ for } j \geq 0.$$

Notice that we can write the preceding relations as follows:

$$\widehat{\alpha_j - \beta_j} = (j-1)\mathcal{O}(\frac{1}{n^2}), \text{ for } j \geq 1, \text{ and } \widehat{\alpha_{n-j} - \beta_{n-j}} = (j+1)\mathcal{O}(\frac{1}{n^2}), \text{ for } j \geq 0. \quad (16)$$

In the next propositions, we present the main properties of the nodal polynomials $W_n(z)$ involved in the interpolatory schemes of Lagrange and Hermite.

Proposition 1.
(i) For every n, it holds:

$$|W_n(z)| < 2e^A \text{ and } \frac{|W_n'(z)|}{n} < 2e^A, \quad \forall z \in \mathbb{T}. \quad (17)$$

(ii) There exists a positive constant $C > 0$ such that for n large enough:

$$\frac{|W_n'(\alpha_j)|}{n} > C, \quad \forall j = 1, \cdots, n. \quad (18)$$

(iii) There exists a positive constant $D > 0$ such that for every n:

$$\frac{|W_n(z)|^2}{n^2} \sum_{j=1}^n \frac{1}{|z-\alpha_j|^2} < D, \quad \forall z \in \mathbb{T}. \quad (19)$$

Proof. See [9]. □

We finish the section with the next property, which turns out to be the key property to study the convergence of the Hermite interpolation polynomials.

Proposition 2. *There exists a positive constant $E > 0$ such that for every n:*

$$|nW_n'(\alpha_j) - \alpha_j W_n''(\alpha_j)| < En\log n, \quad \forall j = 1, \cdots, n. \quad (20)$$

Proof. Since $|nW_n'(\alpha_j) - \alpha_j W_n''(\alpha_j)| = |W_n'(\alpha_j)||n - \frac{\alpha_j W_n''(\alpha_j)}{W_n'(\alpha_j)}|$, by applying (17), we obtain that $|nW_n'(\alpha_j) - \alpha_j W_n''(\alpha_j)| \leq 2ne^A|n - \frac{\alpha_j W_n''(\alpha_j)}{W_n'(\alpha_j)}|$.

Thus, we prove that there exists a positive constant F such that:

$$|n - \frac{\alpha_j W_n''(\alpha_j)}{W_n'(\alpha_j)}| \leq F\log n,$$

and for simplicity and without loss of generality, we took, in what follows, $j = 1$. If we write $W_n(z) = (z-\alpha_1)P_{n-1}(z)$, with $P_{n-1}(z) = \prod_{j=2}^n (z-\alpha_j)$, then $W_n'(\alpha_1) = P_{n-1}(\alpha_1) = \prod_{j=2}^n (\alpha_1 - \alpha_j)$ and $W_n''(\alpha_1) = 2P_{n-1}'(\alpha_1) = \sum_{k=2}^n \prod_{j=2, j\neq k}^n (\alpha_1 - \alpha_j)$. Hence:

$$\frac{\alpha_1 W_n''(\alpha_1)}{W_n'(\alpha_1)} = \frac{2\alpha_1 P_{n-1}'(\alpha_1)}{P_{n-1}(\alpha_1)} = 2\sum_{j=2}^{n} \frac{1}{1-\left(\frac{\alpha_j}{\alpha_1}\right)}. \tag{21}$$

Since $\alpha_j = \beta_j e^{i(\widehat{\beta_j - \alpha_j})}$, then $\frac{\alpha_j}{\alpha_1} = \frac{\beta_j}{\beta_1} e^{i(\widehat{\beta_j - \alpha_j})} = \delta_j e^{i(\widehat{\beta_j - \alpha_j})}, j = 2, \cdots, n$, where $\{\delta_j\}_{j=2}^{n}$ are the n-th roots of the unity different from one. By taking into account (16), we have that:

$$1 - \frac{\alpha_j}{\alpha_1} = 1 - \delta_j e^{i(j-1)\mathcal{O}(\frac{1}{n^2})} = 1 - \delta_j - (j-1)\mathcal{O}(\frac{1}{n^2}),$$

where $\mathcal{O}(\frac{1}{n^2}) = \frac{A_n}{n^2}$, with $|A_n| \leq A$. Hence, (21) can be rewritten as:

$$\frac{\alpha_1 W_n''(\alpha_1)}{W_n'(\alpha_1)} = 2\sum_{j=2}^{n} \frac{1}{1-\delta_j - (j-1)\mathcal{O}(\frac{1}{n^2})}. \tag{22}$$

In order to simplify the calculus in the preceding sum, we used the identity:

$$\frac{\mathcal{A}}{\mathcal{B}+\mathcal{D}} = \frac{\mathcal{A}}{\mathcal{B}} - \frac{\mathcal{A}\mathcal{D}}{\mathcal{B}(\mathcal{B}+\mathcal{D})} \tag{23}$$

taking $\mathcal{A} = 1$, $\mathcal{B} = 1-\delta_j$, and $\mathcal{D} = -(j-1)\mathcal{O}(\frac{1}{n^2})$. To analyze $\mathcal{B} = 1-\delta_j$, we took into account that:

$$\frac{2}{\pi}\widehat{(1-\delta_j)} \leq |1-\delta_j| \leq \widehat{(1-\delta_j)},$$

that is,

$$\frac{4}{n}(j-1) \leq |1-\delta_j| \leq \frac{2\pi}{n}(j-1).$$

Hence, $1 - \delta_j = (j-1)\mathcal{O}_1(\frac{1}{n})$, where $\mathcal{O}_1(\frac{1}{n}) = \frac{B_n}{n}$, with $4 < |B_n| < 2\pi$. Now, by applying (23) in Equation (22), we can rewrite (21) as follows:

$$\frac{\alpha_1 W_n''(\alpha_1)}{W_n'(\alpha_1)} = 2\sum_{j=2}^{n}\left(\frac{1}{1-\delta_j} + \frac{(j-1)\mathcal{O}(\frac{1}{n^2})}{(j-1)\mathcal{O}_1(\frac{1}{n})[(j-1)\mathcal{O}_1(\frac{1}{n}) - (j-1)\mathcal{O}(\frac{1}{n^2})]}\right)$$

Hence, if we use that $2\sum_{j=2}^{n} \frac{1}{1-\delta_j} = n-1$, then:

$$\frac{\alpha_1 W_n''(\alpha_1)}{W_n'(\alpha_1)} - n = -1 + 2\sum_{j=2}^{n} \frac{\mathcal{O}(\frac{1}{n^2})}{(j-1)\mathcal{O}_1(\frac{1}{n})[\mathcal{O}_1(\frac{1}{n}) - \mathcal{O}(\frac{1}{n^2})]}$$

and therefore:

$$\left|\frac{\alpha_1 W_n''(\alpha_1)}{W_n'(\alpha_1)} - n\right| = \left|-1 + 2\sum_{j=2}^{n} \frac{\frac{A_n}{n^2}}{(j-1)\frac{B_n}{n}\left(\frac{B_n}{n} - \frac{A_n}{n^2}\right)}\right| \leq 1 + 2\sum_{j=2}^{n} \frac{|A_n|}{(j-1)|B_n(B_n - \frac{A_n}{n})|} <$$
$$1 + \frac{A}{2}\sum_{j=2}^{n} \frac{1}{(j-1)|B_n - \frac{A_n}{n}|}.$$

Since $|B_n - \frac{A_n}{n}| \geq |B_n| - \frac{|A_n|}{n} > 4 - \frac{A}{n} > 4 - \pi$, then:

$$\left|\frac{\alpha_1 W_n''(\alpha_1)}{W_n'(\alpha_1)} - n\right| < 1 + \frac{A}{2(4-\pi)}\sum_{j=2}^{n} \frac{1}{j-1} = 1 + \frac{A}{2(4-\pi)}(\log(n-1) + C + \varepsilon_{n-1}),$$

with $\varepsilon_{n-1} = o(1)$. Therefore, it is immediate that the last expression is $\mathcal{O}(\log n)$, and thus, the proposition is proven. Notice that for another nodal point α_j with $j \neq 1$, one can proceed in a similar way. □

Remark 1. In [17], we considered as nodal polynomials the para-orthogonal polynomials related to measures in the Szegő class with the Szegő function having an analytic extension outside the unit disk (see [12,13]). In this situation, Condition (13) is satisfied (see [15]) and Properties (17), (18), and (19) also hold, but Property (20) was different. Now, in the present paper, we had that $|nW'_n(\alpha_j) - \alpha_j W''_n(\alpha_j)| < En \log n$, while in [17], the relation was $|nW'_n(\alpha_j) - \alpha_j W''_n(\alpha_j)| < En$.

4. Hermite–Fejér and Hermite Processes

4.1. Convergence of Hermite–Fejér Interpolation in the Case of Continuous Functions

Next, by following the ideas given in [10] for extending Fejér's theorem for continuous functions (see [18]) to the unit circle, we obtained our main result. Reference [10] was the first extension in the case when the nodal points are the n roots of a complex number unimodular. In [17], Fejér's theorem was proven when the nodal polynomial was para-orthogonal with respect to appropriate measures or when it satisfied certain properties. Now, we give a new version of Fejér's theorem for continuous functions on the unit circle with nodal systems satisfying (13).

Theorem 1. *If F is a continuous function on $\mathbb{T}$, then $\mathcal{HF}_{-n,n-1}(F,z)$ converges to F uniformly on $\mathbb{T}$.*

Proof. Since $\mathcal{HF}_{-n,n-1}(F,z) = \sum_{k=1}^{n} \mathfrak{h}_{k,n}(z) F(\alpha_k)$ and $F(z) = F(z) \sum_{k=1}^{n} \mathfrak{h}_{k,n}(z)$, then we have:

$$|F(z) - \mathcal{HF}_{-n,n-1}(F,z)| \leq \sum_{k=1}^{n} |\mathfrak{h}_{k,n}(z)| |F(z) - F(\alpha_k)|. \qquad (24)$$

By taking into account that for $\varepsilon > 0$, there exists $\delta > 0$ such that if $|z-y| < \delta$, then $|F(z) - F(y)| < \varepsilon$, let us take n such that $\frac{1}{\sqrt[4]{n}} < \delta$.

Thus, we rewrite the last term of (24) as follows:

$$\sum_{k=1}^{n} |\mathfrak{h}_{k,n}(z)||F(z) - F(\alpha_k)| = \sum_{k \in I_{1,n}} |\mathfrak{h}_{k,n}(z)||F(z) - F(\alpha_k)| + \sum_{k \in I_{2,n}} |\mathfrak{h}_{k,n}(z)||F(z) - F(\alpha_k)|,$$

where $I_{1,n} = \{k \in \{1, \cdots, n\} : |z - \alpha_k| < \frac{1}{\sqrt[4]{n}}\}$ and $I_{2,n} = \{k \in \{1, \cdots, n\} : |z - \alpha_k| \geq \frac{1}{\sqrt[4]{n}}\}$.

On the one hand, $\sum_{k \in I_{2,n}} |\mathfrak{h}_{k,n}(z)||F(z) - F(\alpha_k)| \leq 2 \|F\|_\infty \sum_{k \in I_{2,n}} |\mathfrak{h}_{k,n}(z)|$, and by using (7), (18), (20), and (17), respectively, we get:

$$\sum_{k \in I_{2,n}} |\mathfrak{h}_{k,n}(z)| \leq$$
$$|W_n(z)|^2 \sum_{k \in I_{2,n}} \frac{1}{|W'_n(\alpha_k)|^2 |z - \alpha_k|^2} + |W_n(z)|^2 \sum_{k \in I_{2,n}} \frac{1}{|z - \alpha_k|} \frac{|nW'_n(\alpha_k) - \alpha_k W''_n(\alpha_k)|}{|W'_n(\alpha_k)|^3} \leq$$
$$|W_n(z)|^2 \sum_{k \in I_{2,n}} \frac{1}{n^2 C^2} \frac{1}{|z - \alpha_k|^2} + |W_n(z)|^2 \sum_{k \in I_{2,n}} \frac{1}{|z - \alpha_k|} \frac{E \log n}{n^2 C^3} \leq \qquad (25)$$
$$4e^{2A} \sum_{k \in I_{2,n}} \frac{1}{C^2 n^{\frac{3}{2}}} + 4e^{2A} \sum_{k \in I_{2,n}} \frac{E \log n}{C^3 n^{\frac{7}{4}}} \leq 4e^{2A} \left(\frac{1}{C^2 n^{\frac{1}{2}}} + \frac{E \log n}{C^3 n^{\frac{3}{4}}} \right),$$

which goes to zero for n large enough. On the other hand, we also obtain $\sum_{k \in I_{1,n}} |\mathfrak{h}_{k,n}(z)||F(z) - F(\alpha_k)| \leq \varepsilon \sum_{k \in I_{1,n}} |\mathfrak{h}_{k,n}(z)|$, and by applying (7), (18), (20), (17), and (19), respectively, we get:

$$\sum_{k \in I_{1,n}} |\mathfrak{h}_{k,n}(z)| = \sum_{k \in I_{1,n}} \frac{|W_n(z)|^2}{|W_n'(\alpha_k)|^2} \left| \frac{\alpha_k}{z|z-\alpha_k|^2} - \frac{\alpha_k}{(z-\alpha_k)} \left(n - \frac{\alpha_k W_n''(\alpha_k)}{W_n'(\alpha_k)} \right) \right| \leq$$

$$\sum_{k \in I_{1,n}} \frac{|W_n(z)|^2}{|z-\alpha_k|^2} \left(\frac{1}{|W_n'(\alpha_k)|^2} + \frac{|z-\alpha_k||nW_n'(\alpha_k) - \alpha_k W_n''(\alpha_k)|}{|W_n'(\alpha_k)|^3} \right) \leq \quad (26)$$

$$\sum_{k \in I_{1,n}} \frac{|W_n(z)|^2}{|z-\alpha_k|^2} \left(\frac{1}{C^2 n^2} + \frac{1}{\sqrt[4]{n}} \frac{E n \log n}{C^3 n^3} \right) < D \left(\frac{1}{C^2} + \frac{E \log n}{C^3 \sqrt[4]{n}} \right).$$

Therefore, $\sum_{k \in I_{1,n}} |\mathfrak{h}_{k,n}(z)||F(z) - F(\alpha_k)| \leq \varepsilon T$, for some positive constant T and n large enough. Hence the Hermite–Fejér-type theorem is proven. □

Corollary 1. *There exists a positive constant $L > 0$ such that for every F bounded on $\mathbb{T}$, it holds that:*

$$|\mathcal{H}\mathcal{F}_{-n,n-1}(F,z)| \leq L \| F \|_{\infty},$$

for every $z \in \mathbb{T}$.

Next, we studied the complete problem, that is the Hermite interpolation problem with nonvanishing conditions for the derivatives. In [17], under suitable conditions for the nodal systems, we gave a sufficient condition on the derivatives, which cannot be improved, in order to obtain convergence for continuous functions. Now, we prove that the same condition works.

Proposition 3. *Let F be a continuous function on $\mathbb{T}$, and assume that $\gamma_n = (v_1, \cdots, v_n)$ satisfies that $\lim_{n \to \infty} \frac{\| \gamma_n \|_2}{n} = 0$, then $\mathcal{H}_{-n,n-1}(F, \gamma_n, z)$ converges to F uniformly on $\mathbb{T}$.*

Proof. If we apply Expression (10), we get:

$$|\mathcal{H}\mathcal{D}_{-n,n-1}(z)| = \left| \sum_{k=1}^{n} -\frac{|W_n(z)|^2}{|W_n'(\alpha_k)|^2} \frac{\alpha_k^2}{(z-\alpha_k)} v_k \right| \leq |W_n(z)|^2 \sum_{k=1}^{n} \frac{1}{|W_n'(\alpha_k)|^2} \frac{|v_k|}{|z-\alpha_k|} \leq$$

$$\frac{|W_n(z)|^2}{C^2 n^2} \sum_{k=1}^{n} \frac{|v_k|}{|z-\alpha_k|} \leq \frac{|W_n(z)|^2}{C^2 n^2} \left(\sum_{k=1}^{n} \frac{1}{|z-\alpha_k|^2} \right)^{\frac{1}{2}} \left(\sum_{k=1}^{n} |v_k|^2 \right)^{\frac{1}{2}} \leq \quad (27)$$

$$\leq \frac{|W_n(z)|}{C^2 n} \left(\sum_{k=1}^{n} \frac{|W_n(z)|^2}{n^2 |z-\alpha_k|^2} \right)^{\frac{1}{2}} \| \gamma_n \|_2 < \frac{2e^A}{C^2 n} \sqrt{D} \| \gamma_n \|_2,$$

which goes to zero uniformly on $\mathbb{T}$. Hence, if we take into account $\mathcal{H}_{-n,n-1}(F, \gamma_n, z) = \mathcal{H}\mathcal{F}_{-n,n-1}(F,z) + \mathcal{H}\mathcal{D}_{-n,n-1}(z)$ joint with Theorem 1, then the result is proven. □

Notice that following the ideas given in [11], it is possible to give other sufficient conditions.

4.2. Convergence of Hermite Interpolation in the Case of Smooth Functions

This section is devoted to studying the convergence of the Hermite interpolation polynomials related to analytic functions and certain types of smooth functions.

Proposition 4. *If F is an analytic function in an open annulus containing $\mathbb{T}$, then $\mathcal{H}_{-n,n-1}(F,z)$ uniformly converges to F on $\mathbb{T}$, and the order of convergence is geometric.*

Proof. F can be written $F(z) = \sum_{j=-\infty}^{\infty} A_j z^j$, with $|A_j| \leq Pr^{|j|}$ for some $P > 0$ and $0 < r < 1$. Thus, if we decompose $F(z) = F_{1,n}(z) + F_{2,n}(z) + F_{3,n}(z)$, with $F_{1,n}(z) = \sum_{j=-n}^{n-1} A_j z^j$, $F_{2,n}(z) = \sum_{j=n}^{\infty} A_j z^j$, and $F_{3,n}(z) = \sum_{j=-\infty}^{-n-1} A_j z^j$, then it is clear that $\mathcal{H}_{-n,n-1}(F_{1,n}, z) = F_{1,n}(z)$.
Now, to obtain our result, first we studied the behavior of the difference $|\mathcal{H}_{-n,n-1}(F_{2,n}, z) - F_{2,n}(z)|$. Indeed:

$$|\mathcal{H}_{-n,n-1}(F_{2,n}, z) - F_{2,n}(z)| = |\sum_{j=n}^{\infty} A_j z^j - \sum_{j=n}^{\infty} A_j \mathcal{H}_{-n,n-1}(z^j, z)| =$$

$$|\sum_{j=n}^{\infty} A_j z^j - \sum_{j=n}^{\infty} A_j \sum_{k=1}^{n} \mathfrak{h}_{k,n}(z) \alpha_k^j - \sum_{j=n}^{\infty} A_j \sum_{k=1}^{n} \mathfrak{k}_{k,n}(z) j \alpha_k^{j-1}| =$$

$$|\sum_{j=n}^{\infty} A_j \left(z^j - \sum_{k=1}^{n} \mathfrak{h}_{k,n}(z) \alpha_k^j - \sum_{k=1}^{n} \mathfrak{k}_{k,n}(z) j \alpha_k^{j-1} \right)| \leq$$

$$\sum_{j=n}^{\infty} |A_j| \left(1 + \sum_{k=1}^{n} |\mathfrak{h}_{k,n}(z)| + \sum_{k=1}^{n} j|\mathfrak{k}_{k,n}(z)| \right).$$

By taking into account (25) and (26), we get that:

$$\sum_{k=1}^{n} |\mathfrak{h}_{k,n}(z)| \leq 4e^{2A} \left(\frac{1}{C^2 n^{\frac{1}{2}}} + \frac{E \log n}{C^3 n^{\frac{3}{4}}} \right) + D \left(\frac{1}{C^2} + \frac{E \log n}{C^3 \sqrt[4]{n}} \right) \leq M + Q \frac{\log n}{\sqrt{n}}, \quad (28)$$

for some positive constants M and Q.
In the same way, by using (6) and (27) with $\gamma_n = (1, \cdots, 1)$, we have:

$$\sum_{k=1}^{n} |\mathfrak{k}_{k,n}(z)| = |W_n(z)|^2 \sum_{k=1}^{n} \frac{1}{|W_n'(\alpha_k)|^2} \frac{1}{|z - \alpha_k|} \leq \frac{2e^A}{C^2 n} \sqrt{D} \sqrt{n} \leq \frac{R}{\sqrt{n}}, \quad (29)$$

for some positive constant R.
Hence:

$$|\mathcal{H}_{-n,n-1}(F_{2,n}, z) - F_{2,n}(z)| \leq$$

$$\sum_{j=n}^{\infty} |A_j| \left(1 + M + Q \frac{\log n}{\sqrt{n}} + j \frac{R}{\sqrt{n}} \right) \leq \sum_{j=n}^{\infty} P r^j \left(1 + M + Q \frac{\log n}{\sqrt{n}} + j \frac{R}{\sqrt{n}} \right) \leq$$

$$P \frac{r^n}{1-r} \left(1 + M + Q \frac{\log n}{\sqrt{n}} + \frac{(n(1-r)+r)}{(1-r)} \frac{R}{\sqrt{n}} \right) \leq S r_1^n,$$

for some positive constant r_1 such that $r < r_1 < 1$. Then, it is clear that it goes to zero uniformly on $\mathbb{T}$, and the order of convergence is geometric.
Finally, one can obtain an analogous result for the difference $|\mathcal{H}_{-n,n-1}(F_{3,n}, z) - F_{3,n}(z)|$, and hence, the result is proven. □

Proposition 5. *If $F(z) = \sum_{j=-\infty}^{\infty} A_j z^j$ is a function defined on $\mathbb{T}$ with $|A_j| \leq K \frac{1}{|j|^c}$, for some positive constant K, $j \neq 0$, $c > 2$, then $\mathcal{H}_{-n,n-1}(F, z)$ converges to $F(z)$ uniformly on $\mathbb{T}$. Moreover, the order of convergence is $\mathcal{O}(\frac{1}{(n-1)^{c-2}})$.*

Proof. If we decompose $F(z) = F_{1,n}(z) + F_{2,n}(z) + F_{3,n}(z)$, with $F_{1,n}(z) = \sum_{j=-n}^{n-1} A_j z^j$,

$F_{2,n}(z) = \sum_{j=n}^{\infty} A_j z^j$, and $F_{3,n}(z) = \sum_{j=-\infty}^{-n-1} A_j z^j$, then:

$$\mathcal{H}_{-n,n-1}(F,z) = \mathcal{H}_{-n,n-1}(F_{1,n},z) + \mathcal{H}_{-n,n-1}(F_{2,n},z) + \mathcal{H}_{-n,n-1}(F_{3,n},z).$$

Since $\mathcal{H}_{-n,n-1}(F_{1,n},z) = F_{1,n}(z)$, then:

$$|\mathcal{H}_{-n,n-1}(F,z) - F(z)| \leq |\mathcal{H}_{-n,n-1}(F_{2,n},z) - F_{2,n}(z)| + |\mathcal{H}_{-n,n-1}(F_{3,n},z) - F_{3,n}(z)|,$$

and we have to study the behavior of the differences $|\mathcal{H}_{-n,n-1}(F_{i,n},z) - F_{i,n}(z)|$, $i = 2, 3$, to obtain the uniform convergence of $\mathcal{H}_{-n,n-1}(F,z)$ to F.
Indeed, if $z \in \mathbb{T}$:

$$|\mathcal{H}_{-n,n-1}(F_{2,n},z) - F_{2,n}(z)| = |\mathcal{H}_{-n,n-1}(\sum_{j=n}^{\infty} A_j z^j, z) - \sum_{j=n}^{\infty} A_j z^j| =$$

$$|\sum_{j=n}^{\infty} A_j \mathcal{H}_{-n,n-1}(z^j,z) - \sum_{j=n}^{\infty} A_j z^j| = |\sum_{j=n}^{\infty} A_j(z^j - \mathcal{H}_{-n,n-1}(z^j,z))| =$$

$$|\sum_{j=n}^{\infty} A_j(z^j - \sum_{k=1}^{n} \mathfrak{h}_{k,n}(z)\alpha_k^j - \sum_{k=1}^{n} \mathfrak{k}_{k,n}(z) j \alpha_k^{j-1})| \leq \sum_{j=n}^{\infty} |A_j|(1 + \sum_{k=1}^{n} |\mathfrak{h}_{k,n}(z)| + j \sum_{k=1}^{n} |\mathfrak{k}_{k,n}(z)|).$$

If we apply (28) and (29) in the preceding proposition, we get

$$|\mathcal{H}_{-n,n-1}(F_{2,n},z) - F_{2,n}(z)| \leq \sum_{j=n}^{\infty} |A_j|(1 + M + Q\frac{\log n}{\sqrt{n}} + j\frac{R}{\sqrt{n}}) \leq$$

$$\sum_{j=n}^{\infty} |A_j|S + \sum_{j=n}^{\infty} j|A_j|\frac{R}{\sqrt{n}} \leq SK \sum_{j=n}^{\infty} \frac{1}{j^c} + \frac{RK}{\sqrt{n}} \sum_{j=n}^{\infty} \frac{1}{j^{c-1}} =$$

$$SK(H(c) - H_{n-1,c}) + \frac{RK}{\sqrt{n}}(H(c-1) - H_{n-1,c-1}) \leq$$

$$\frac{SK}{(c-1)(n-1)^{c-1}} + \frac{RK}{\sqrt{n}(c-2)(n-1)^{c-2}},$$

for some positive constant S, with $H(c)$ being the sum of the harmonic series $\sum_{j=1}^{\infty} \frac{1}{j^c}$ and $H_{n-1,c}$ its $(n-1)$-partial sum.

Hence, $|\mathcal{H}_{-n,n-1}(F_{2,n},z) - F_{2,n}(z)|$ goes to zero, and the order of convergence is $\mathcal{O}(\frac{1}{(n-1)^{c-2}})$. Notice that the same result is valid for $|\mathcal{H}_{-n,n-1}(F_{3,n},z) - F_{3,n}(z)|$, and hence, the result is proven. □

Remark 2. *Notice that for smooth functions, we obtained closed results to those given in [16] related to the Chebyshev process on the bounded interval. In particular, for the functions studied in the previous proposition, we obtained an accuracy of $\mathcal{O}(\frac{1}{(n-1)^{c-2}})$, while the Chebyshev-related result was $\mathcal{O}(\frac{1}{(n-1)^{c-1}})$, which is a better result. However, for analytic functions, we obtained a quite similar result.*

5. The Case of the Bounded Interval

In this section, we consider nodal systems in the interval $[-1, 1]$, which are closely connected with those considered on the unit circle in Section 3. Indeed let us consider the nodal array $\{x_j\}_{j=1}^n \subset [-1, 1]$ numbered as follows $-1 \leq x_n < x_{n-1} < \cdots < x_2 < x_1 \leq 1$. We distinguished four cases depending on $x_1 = 1$ or $x_1 \neq 1$ and $x_n = -1$ or $x_n \neq -1$. In all the cases, we assumed that there exists a positive constant $A < \pi$ such that one of the following separation properties holds:

(i) $x_1 = 1, x_n = -1$, and $\arccos x_{j+1} - \arccos x_j = \frac{\pi}{n} + \frac{a(j)}{n^2}$, with $|a(j)| \leq A$, $\forall j = 1, \cdots, n-1$.

(ii) $x_1 < 1, x_n = -1$, $\arccos x_{j+1} - \arccos x_j = \frac{\pi}{n} + \frac{a(j)}{n^2}$, with $|a(j)| \leq A$, $\forall j = 1, \cdots, n-1$, and $2\arccos x_1 = \frac{\pi}{n} + \frac{a(0)}{n^2}$, with $|a(0)| \leq A$.

(iii) $x_1 = 1, x_n > -1$, $\arccos x_{j+1} - \arccos x_j = \frac{\pi}{n} + \frac{a(j)}{n^2}$, with $|a(j)| \leq A$, $\forall j = 1, \cdots, n-1$, and $2(\pi - \arccos x_n) = \frac{\pi}{n} + \frac{a(n)}{n^2}$, with $|a(n)| \leq A$.

(iv) $x_1 < 1, x_n > -1$, $\arccos x_{j+1} - \arccos x_j = \frac{\pi}{n} + \frac{a(j)}{n^2}$, with $|a(j)| \leq A$, $\forall j = 1, \cdots, n-1$, $2\arccos x_1 = \frac{\pi}{n} + \frac{a(0)}{n^2}$, with $|a(0)| \leq A$, and $2(\pi - \arccos x_n) = \frac{\pi}{n} + \frac{a(n)}{n^2}$, with $|a(n)| \leq A$.

Our aim was to study Hermite interpolation problems on the interval with these nodal systems. First, we considered a real function f continuous on $[-1, 1]$, and we studied the convergence of the Hermite interpolation polynomial $h_{2n-1}(f, x)$ satisfying the conditions:

$$h_{2n-1}(f, x_j) = f(x_j),\ h'_{2n-1}(f, x_j) = v_j,\ \text{with } v_j \in \mathbb{R},\ j = 1, \cdots, n. \tag{30}$$

Proposition 6. *Let $\{x_j\}_{j=1}^n$ be a nodal system on $[-1, 1]$ satisfying one of the separation properties (i), (ii), (iii), or (iv) given before. Let f be a real continuous function on $[-1, 1]$, and assume that $\{v_j\}_{j=1}^n$ satisfies:*

$$\lim_{n \to \infty} \frac{\| (v_1 \sqrt{1 - x_1^2}, \cdots, v_n \sqrt{1 - x_n^2}) \|_2}{n} = 0.$$

Then, the Hermite interpolation polynomials fulfilling (30) converge uniformly to f on $[-1, 1]$.

Proof. To fix the ideas, we assumed that $\{x_j\}_{j=1}^n$ satisfies Case (iv).

To prove the result, we transformed our interpolation problem in $[-1, 1]$ into the interpolation problem on the unit circle studied in the previous sections. First of all, we transformed the nodal systems through the Szegő transformation, obtaining the transformed system $\{\alpha_j\}_{j=1}^n \cup \{\overline{\alpha_j}\}_{j=1}^n$ such that $x_j = \frac{\alpha_j + \overline{\alpha_j}}{2}$, $j = 1, \cdots, n$. If $w_n(x) = \prod_{j=1}^n (x - x_j)$, then $w_n(\frac{z + \frac{1}{z}}{2}) = \prod_{j=1}^n (\frac{z + \frac{1}{z}}{2} - \frac{\alpha_j + \frac{1}{\alpha_j}}{2}) = \frac{1}{2^n} \frac{1}{z^n} \prod_{j=1}^n (z - \alpha_j)(z - \overline{\alpha_j})$, which implies that the transformed nodal polynomial is $W_{2n}(z) = \prod_{j=1}^n (z - \alpha_j)(z - \overline{\alpha_j}) = 2^n z^n w_n(\frac{z + \frac{1}{z}}{2})$, with $z \in \mathbb{T}$. If we renumber the zeros of $W_{2n}(z)$ as follows $\alpha_{n+1} = \overline{\alpha_n}, \cdots, \alpha_{2n} = \overline{\alpha_1}$, then $W_{2n}(z) = \prod_{j=1}^{2n} (z - \alpha_j)$. Now, it is immediate that the nodal points $\alpha_j = e^{i\theta_j}$, $j = 1, \cdots, 2n$, with $\theta_{n+1} = -\theta_n, \cdots, \theta_{2n} = -\theta_1$ satisfy:

$$\theta_{j+1} - \theta_j = \frac{2\pi}{2n} + \frac{A(j)}{n^2},$$

with $|A(j)| \leq A$, $j = 1, \cdots, 2n$ and $\theta_{2n+1} = \theta_1$, that is their arguments satisfy Property (13). We define a continuous function F on $\mathbb{T}$ by $F(z) = F(\overline{z}) = f(x)$ for $z \in \mathbb{T}$, being $x = \frac{z + \frac{1}{z}}{2} \in [-1, 1]$, and we pose the Hermite interpolation problem of finding the Laurent polynomial $\mathcal{H}_{-2n, 2n-1}(F, z)$ satisfying the interpolation conditions:

$$\mathcal{H}_{-2n, 2n-1}(F, \alpha_j) = \mathcal{H}_{-2n, 2n-1}(F, \overline{\alpha_j}) = f(x_j),\ j = 1, \cdots, n,\ \text{and}$$

$$\mathcal{H}'_{-2n,2n-1}(F,\alpha_j) = v_j\sqrt{1-x_j^2}\,\overline{\alpha_j}i, \ \mathcal{H}'_{-2n,2n-1}(F,\overline{\alpha_j}) = -v_j\sqrt{1-x_j^2}\,\alpha_j i, \ j=1,\cdots,n.$$

Now, we can apply Proposition 3 by taking:

$$\gamma_{2n} = (v_1\sqrt{1-x_1^2}\,\overline{\alpha_1}i,\cdots,v_n\sqrt{1-x_n^2}\,\overline{\alpha_n}i,-v_n\sqrt{1-x_n^2}\,\alpha_n i,\cdots,-v_1\sqrt{1-x_1^2}\,\alpha_1 i).$$

Since $\lim_{n\to\infty}\frac{\|\gamma_{2n}\|_2}{2n} = 0$, we obtain that $\mathcal{H}_{-2n,2n-1}(F,z)$ uniformly converges to F on $\mathbb{T}$. If we define:

$$h_{2n-1}(f,x) = \frac{\mathcal{H}_{-2n,2n-1}(F,z) + \mathcal{H}_{-2n,2n-1}(F,\frac{1}{z})}{2},$$

it is clear that $h_{2n-1}(f,x)$ satisfies (30) and converges to f uniformly on $[-1,1]$.
The other cases, (i), (ii), and (iii), can be obtained in a similar form. □

Remark 3. *Conditions (i), (ii), (iii), and (iv) on the nodal systems in $[-1,1]$, given before, can be substituted by other equivalent conditions.*

Indeed, $\arccos x_{j+1} - \arccos x_j = \frac{\pi}{n} + \frac{a(j)}{n^2}$, with $|a(j)| \leq A$, $\forall j=1,\cdots,n-1$, is equivalent to $x_j\sqrt{1-x_{j+1}^2} - x_{j+1}\sqrt{1-x_j^2} = \sin(\frac{\pi}{n}) + \mathcal{O}(\frac{1}{n^2})$, or alternatively to $x_j\sqrt{1-x_{j+1}^2} - x_{j+1}\sqrt{1-x_j^2} = \frac{\pi}{n} + \mathcal{O}(\frac{1}{n^2})$, $\forall j=1,\cdots,n-1$.

In the same way, $2\arccos x_1 = \frac{\pi}{n} + \frac{a(0)}{n^2}$, with $|a(0)| \leq A$, is equivalent to $\sqrt{1-x_1^2} = \sin(\frac{\pi}{2n}) + \mathcal{O}(\frac{1}{n^2})$, or alternatively to $\sqrt{1-x_1^2} = \frac{\pi}{2n} + \mathcal{O}(\frac{1}{n^2})$.

Finally, $2(\pi - \arccos x_n) = \frac{\pi}{n} + \frac{a(n)}{n^2}$, with $|a(n)| \leq A$, is equivalent to $\sqrt{1-x_n^2} = \sin(\frac{\pi}{2n}) + \mathcal{O}(\frac{1}{n^2})$, or alternatively to $\sqrt{1-x_n^2} = \frac{\pi}{2n} + \mathcal{O}(\frac{1}{n^2})$.

Proceeding as in the previous Proposition 6, we can study Hermite interpolation problems by using these nodal systems, obtaining results for the convergence, as well as the order of convergence when we deal with analytic functions.

Proposition 7. *Let $\{x_j\}_{j=1}^n$ be a nodal system on $[-1,1]$ satisfying one of the separation properties (i), (ii), (iii), or (iv) given at the beginning of this section, and let f be a real function analytic on $[-1,1]$. Then, the Hermite interpolation polynomials $h_{2n-1}(f,x)$ fulfilling:*

$$h_{2n-1}(f,x_j) = f(x_j), \ h'_{2n-1}(f,x_j) = f'(x_j), \ \text{for } j=1,\cdots,n. \tag{31}$$

converge uniformly to f on $[-1,1]$, and the order of convergence is geometric.

Proof. We proceed as in Proposition 6, transforming the nodal system $\{x_j\}_{j=1}^n$ fulfilling (iv) into the nodal system $\{\alpha_j\}_{j=1}^n \cup \{\overline{\alpha_j}\}_{j=1}^n$ satisfying Property (13).

We applied to f the following result that can be seen in [13]: if f is analytic on $[-1,1]$, then its expansion in Fourier–Chebyshev series $f(x) \sim \sum_{k=0}^\infty a_k T_k(x)$ converges to f in the interior of the greatest ellipse with foci ± 1, in which f is regular. The expansion diverges in the exterior of this ellipse, and the sum R of the semi-axes of the ellipse is $R = \liminf \frac{1}{\sqrt[n]{|a_n|}}$.

Then, if we define $F(z) = \sum_{k=0}^\infty a_k z^k$, it holds that F is analytic on the open disk $D(0,R)$, and therefore, the function $G(z) = \frac{1}{2}(F(z) + F(\frac{1}{z}))$ is analytic in the annulus $\frac{1}{R} < |z| < R$. Clearly, for $z \in \mathbb{T}$ and $x = \frac{z+\frac{1}{z}}{2} \in [-1,1]$, it holds that $G(z) = f(x)$, and thus, $G(\alpha_j) = G(\overline{\alpha_j}) = f(x_j)$ for $j=1,\cdots,n$. In the same way:

$$f'(x) = \sum_{k=1}^\infty ka_k U_{k-1}(x) = \sum_{k=1}^\infty ka_k \left(\frac{z^k - z^{-k}}{z - z^{-1}}\right)$$

and therefore, $\frac{(z^2-1)}{2z^2}f'(x) = G'(z)$ for $z \in \mathbb{T}$ and $x = \frac{z+\frac{1}{z}}{2}$. Hence, $G'(\alpha_j) = \frac{1}{2}(1 - \overline{\alpha_j}^2)f'(x_j)$ and $G'(\overline{\alpha_j}) = \frac{1}{2}(1 - \alpha_j^2)f'(x_j)$ for $j = 1, \cdots, n$.

Let us consider the Hermite interpolation problem of finding a Laurent polynomial $\mathcal{H}_{-2n,2n-1}(G, z)$ satisfying the interpolation conditions:

$$\mathcal{H}_{-2n,2n-1}(G, \alpha_j) = \mathcal{H}_{-2n,2n-1}(G, \overline{\alpha_j}) = f(x_j), \text{ for } j = 1, \cdots, n, \text{ and}$$

$$\mathcal{H}'_{-2n,2n-1}(G, \alpha_j) = \frac{1}{2}(1 - \overline{\alpha_j}^2)f'(x_j), \quad \mathcal{H}'_{-2n,2n-1}(G, \overline{\alpha_j}) = \frac{1}{2}(1 - \alpha_j^2)f'(x_j), \text{ for } j = 1, \cdots, n.$$

By applying Proposition 4, we have that $\mathcal{H}_{-2n,2n-1}(G, z)$ uniformly converges to G on $\mathbb{T}$ with the geometrical order of convergence. If we define the polynomial:

$$h_{2n-1}(f, x) = \frac{\mathcal{H}_{-2n,2n-1}(G, z) + \mathcal{H}_{-2n,2n-1}(G, \frac{1}{z})}{2},$$

it is clear that $h_{2n-1}(f, x)$ satisfies (31) and converges to f uniformly on $[-1, 1]$ with the geometrical order of convergence.

Notice that for the other cases (i), (ii), and (iii), one can proceed in a similar way. □

6. Mechanical Models and Numerical Examples

In the previous section, our results were transferred to the real case, that is to the bounded interval, but the natural process is to arrive at the trigonometric case, which we do not detail due to its simplicity. Certainly, it is in this last situation where we can find many examples using our nodal systems.

Example 1. *A certain periodic phenomenon of period $T = 2\pi$ is observed using a certain measuring device. An essential element of this device is a pair of facing disks, one of which (Disk 1) rotates at an angular constant speed of $\frac{2\pi(n+1)}{T}$. The other disk (Disk 2) is driven by a Cardan transmission with a β angle ($0 \leq \beta < \frac{\pi}{2}$) whose driving shaft rotates with an angular speed of $\frac{2\pi}{T}$. Originally, the idea was to take measurements at evenly spaced times. For last-minute indications, the measurements are made when the light inside the device is at its maximum. Considering the moments in which this happens allows for additional video recording, and it occurs when the holes located on the discs are facing each other (event), as initially. Both discs have a hole at the same distance from the center. Therefore, the question is: Could we use the new scheme with the same properties of convergence?*

The Cardan transmission is a well-known device, and the equations that govern the angular displacement for the driving and driven shafts can be found in books on mechanics, such as [19] (Section 2.6) and [20]. With $\Theta_2(t)$ denoting the angular displacement of Disk 2 and supposing that $\Theta_2(0) = 0$, we can state that:

$$\Theta_2(t) = \arctan\left(\frac{\sin(\frac{2\pi t}{T})}{\cos(\frac{2\pi t}{T})\cos\beta}\right) \text{ and } w_2(t) = \frac{d\Theta_2(t)}{dt} = \frac{2\pi}{T}\frac{\cos\beta}{(1 - \sin^2\beta)\cos^2(\frac{2\pi t}{T})}.$$

The last equation above shows one characteristic of the considered Cardan joint: it is not homokinetic. Indeed, $\frac{d\Theta_2(t)}{dt}$ is not constant, and there exist positive numbers A and B, with $B < 1$, such that:

$$(1 - B)\frac{2\pi}{T} = \cos\beta\frac{2\pi}{T} \leq \frac{2\pi}{T}(1 - C(t)) = w_2(t) \leq \frac{1}{\cos\beta}\frac{2\pi}{T} = (1 + A)\frac{2\pi}{T} \quad (32)$$

with $1 - C(t) > 0$ and bounded.

On the other hand, we denote the angular displacement of Disk 1 by $\Theta_1(t)$, and if we suppose that $\Theta_1(0) = 0$, we have $\Theta_1(t) = \frac{2\pi(n+1)}{T}t$. After $t = T$, the hole of Disk 1 reaches that of Disk 2 n times. Moreover, we can bound the time between two of these events. For Δt, the angle of displacement for Disk 1 is $\frac{2\pi(n+1)}{T}\Delta t$ and for Disk 2 is $w_2(t_1)\Delta t$ (t_1 is an intermediate value between the t values for both events). Since the first displacement is 2π radians greater than the second one, we have:

$$2\pi = \frac{2\pi(n+1)}{T}\Delta t - w_2(t_1)\Delta t \Rightarrow \Delta t = \frac{2\pi}{\frac{2\pi(n+1)}{T} - (1-C(t_1))\frac{2\pi}{T}} = \frac{T}{n+C(t_1)} \Rightarrow$$

$$\Delta t = \frac{T}{n} - \frac{C(t_1)T}{n(n+C(t_1))}. \quad (33)$$

In other words, we have:

$$\Delta t = \frac{T}{n} + \mathcal{O}\left(\frac{1}{n^2}\right) = \frac{2\pi}{n} + \mathcal{O}\left(\frac{1}{n^2}\right).$$

When the period is $T \neq 2\pi$, we can change the independent variable t by the new variable $t' = \frac{2\pi}{T}t$ and use the previous analysis. As can be seen, holes would only face each other for evenly spaced times for a null angle, that is for homokinetic joints. Notice that the nodal system determined by the events satisfies the separation relation (13).

We must point out that we can use the interpolation scheme on the unit circle to perform trigonometrical interpolation on the interval $[0, 2\pi]$ for periodic functions. The idea is quite similar to that developed in Section 5. Here, we must consider the change $z = e^{i\theta}$; in particular, the change $z = e^{it}$. In this case, the nodal system satisfies (13), and we can confidently use the interpolation methods. To compute the interpolation polynomials, we used the barycentric representation given by (11).

We obtained a nodal system employing for the previous scheme $n = 40$ and $\beta = \frac{\pi}{7}$. As a test function, we used $\sum_{k=1}^{\infty} \frac{2}{k^6}\cos(k\theta)$ which led to $f(z) = \sum_{k=1}^{\infty} \frac{1}{k^6}(z^k + z^{-k})$. The left-hand side in Figure 1 shows the real part of the interpolator along with the interpolated function. We observed that they were indistinguishable. On the right-hand side, we represent the real part of the error. The imaginary parts in both cases are irrelevant.

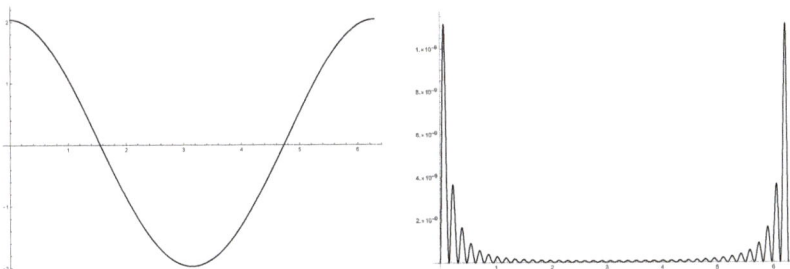

Figure 1. Left: representation of $f(z)$ and $\Re(\mathcal{H}_{-n,n-1}(f,z))$. Right: representation of $\Re(\mathcal{H}_{-n,n-1}(f,z)) - f(z)$, with $f(z) = \sum_{k=1}^{\infty} \frac{1}{k^6}(z^k + z^{-k})$, $z = e^{i\theta}$, $\theta \in [0,2\pi]$, and $n = 40$.

Example 2. *A certain periodic magnitude of period $T = 2\pi$ is observed using a certain measuring device with which n measurements are taken. Each one is intended to be obtained just after a countdown of $\frac{2\pi}{n}$, this time needed for the correct performance of the device. It includes an automatic repair of four possible errors. Every error has a probability of p, and they are independent. Each repair requires a time of $\frac{1}{n^2}$, which can lead to the fact that between two measures, the time is greater than or equal to $\frac{2\pi}{n}$.*

Thus, in this case, we cannot ensure evenly spaced measurements. Between two of them, we have $\Delta t = \frac{2\pi}{n} + \frac{A}{n^2}$ where A is random. Indeed, it is a binomial variable $B(4,p)$ and $0 \leq A \leq 4$ in the way that $\Delta t = \frac{2\pi}{n} + \mathcal{O}\left(\frac{1}{n^2}\right)$. Again, the nodal system satisfies the separation property (13).

Naturally, the rest of the ideas of the previous example can be applied. Consequently, we can be confident when using Hermite or Hermite–Fejér interpolation based on the corresponding random nodal systems.

We obtained a particular nodal system using for the previous scheme $n = 400$ and $p = 0.6$. As a test function, we used the continuous non-derivable function $\cos t \sin(\frac{1}{\cos t})$, which led to $f(z) = \frac{z+\frac{1}{z}}{2} \sin\left(\frac{2}{z+\frac{1}{z}}\right)$, and we performed Hermite–Fejér interpolation. To obtain the corresponding interpolator, we used the barycentric formula (12). The left-hand side in Figure 2 shows the real part of the interpolator together with the interpolated function on $[\frac{\pi}{2}, \frac{3\pi}{4}]$. On the right-hand side, we represent the same pair in a quite small interval. We point out the characteristic shape of the Hermite–Fejér interpolation on the nodes. The imaginary parts in both cases are irrelevant.

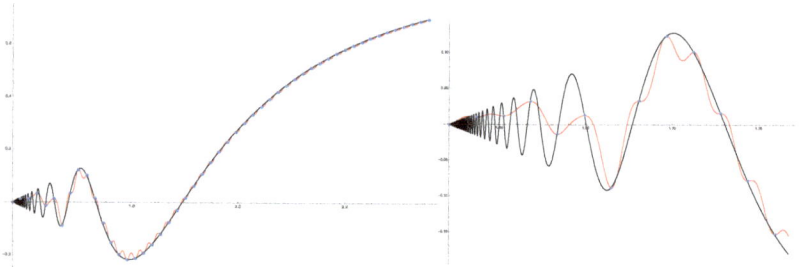

Figure 2. $f(z)$ and $\Re(\mathcal{HF}_{-n,n-1}(f,z))$ with the nodes marked on the lines, $f(z) = \frac{z+\frac{1}{z}}{2} \sin\left(\frac{2}{z+\frac{1}{z}}\right)$, $n = 400$, and $z = e^{i\theta}$. Left: $\theta \in [\frac{\pi}{2}, \frac{3\pi}{4}]$. Right: $\theta \in [\frac{\pi}{2}, \frac{9\pi}{16}]$.

Example 3. *An artificial satellite takes measurements of a periodic phenomenon of period T_1 that has its origin at the point at infinity of the perpendicular to its elliptical orbit of period T around a star. The satellite can rotate on itself, and it can vary its speed of rotation, although not its orbit. The orbit was intended to be circular, but ended up being elliptical. Observations have to be taken when one particular point of the satellite is in its solar noon, that is when the center of the star, the center of the satellite, and the point are aligned (event).*

In this scenario, we can successfully apply the ideas in Section 4. Let us suppose that at Time 0, the satellite is at its aphelion and at its solar noon. By using the laws of the two-body problem, we can know its true anomaly ϕ at time T_1 and adjust the rotation of the satellite so that the point is at its (n) solar noon, having rotated $(n+1)2\pi + \phi$. Logically, the time lapse between two solar noons would be equal if the orbit were circular. However, as a consequence of Kepler's second law, the rate of variation of the angular velocity of the satellite (in orbit) w_1 is a variable that oscillates between the values that it takes at aphelion and perihelion. Note that if $T_1 = T$, the satellite would rotate at an angular speed of $\frac{2\pi(n+1)}{T}$ leading to Equation (29). Furthermore, $C(t)$ would be bounded, the mean w_1 value being $\frac{2\pi}{T}$. Thus, it is possible to ensure that the times between the satellite noons are variable, but with the form $\frac{T_1}{n} + \mathcal{O}(\frac{1}{n^2})$, that is the nodal system fulfills Relation (13). Using the new variable t' given by $t' = \frac{2\pi}{T_1}t$, we have a Hermite or Hermite–Fejér interpolation problem based on a nodal system satisfying (13). Hence, we can use the mechanism.

When determining the noon of the satellite, we must use some elements of the mechanics of elliptical orbits. The ideas are quite similar to those of Example 1. However, in this case, we must use equations related to elliptical orbits. The true anomaly is determined by t, T and the eccentricity of the orbit e. The algorithm can be found in [21] (Chapter 3, Section 4).

We obtained a particular nodal system using the previous scheme with $n = 1000$, $e = 0.25$, and for simplicity, $T_1 = T = 8,640,000$. We used $\frac{1}{[1+(1.02)^2 - 2.04\cos\theta][1+(1.2)^2 + 2.4\sin\theta]}$ as the test function, which led to $f(z) = \frac{1}{(z-1.02)(\frac{1}{z}-1.02)(z+1.2i)(\frac{1}{z}-1.2i)}$. The left-hand side in Figure 3 shows the real part of the interpolator with the interpolated function near $z = 1$, the most problematic area. We obtained the interpolator by using the formula (11). We observed that they were indistinguishable.

On the right-hand side, we represent the real part of the error. The imaginary parts in both cases are irrelevant.

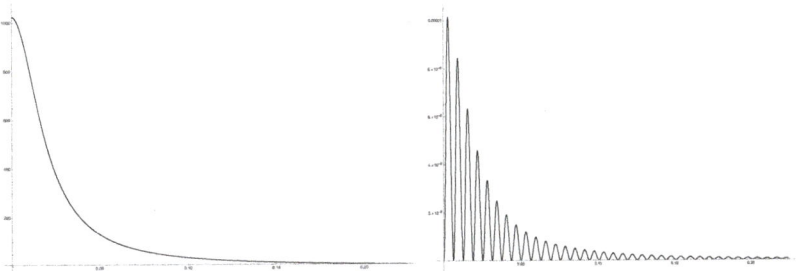

Figure 3. Left: representation of $f(z)$ and $\Re(\mathcal{H}_{-n,n-1}(f,z))$. Right: representation of $\Re(\mathcal{H}_{-n,n-1}(f,z)) - f(z)$, with $f(z) = \frac{1}{(z-1.02)(\frac{1}{z}-1.02)(z+1.2i)(\frac{1}{z}-1.2i)}$, $z = e^{i\theta}$, and $n = 1000$.

7. Notation

Next, we summarize the notation related to the polynomials used throughout the paper:
$W_n(z) = \prod_{j=1}^{n}(z - \alpha_j)$ is the nodal polynomial.
The Laurent polynomials of Hermite interpolation are denoted by:
$\mathcal{H}_{-n,n-1}(z)$ for the values $\{u_j\}_{j=1}^{n}$ and $\{v_j\}_{j=1}^{n}$.
$\mathcal{HF}_{-n,n-1}(z)$ for the values $\{u_j\}_{j=1}^{n}$ and $\{0\}_{j=1}^{n}$.
$\mathcal{HD}_{-n,n-1}(z)$ for the values $\{0\}_{j=1}^{n}$ and $\{v_j\}_{j=1}^{n}$.
Clearly, $\mathcal{H}_{-n,n-1}(z) = \mathcal{HF}_{-n,n-1}(z) + \mathcal{HD}_{-n,n-1}(z)$.
$\mathcal{H}_{-n,n-1}(f,z)$ for the values $\{f(\alpha_j)\}_{j=1}^{n}$ and $\{f'(\alpha_j)\}_{j=1}^{n}$.
$\mathcal{HF}_{-n,n-1}(f,z)$ for the values $\{f(\alpha_j)\}_{j=1}^{n}$ and $\{0\}_{j=1}^{n}$.
$\mathcal{HD}_{-n,n-1}(f,z)$ for the values $\{0\}_{j=1}^{n}$ and $\{f'(\alpha_j)\}_{j=1}^{n}$.
Clearly, $\mathcal{H}_{-n,n-1}(f,z) = \mathcal{HF}_{-n,n-1}(f,z) + \mathcal{HD}_{-n,n-1}(f,z)$.
$\mathcal{H}_{-n,n-1}(f,\gamma_n,z)$ for the values $\{f(\alpha_j)\}_{j=1}^{n}$ and $\{v_j\}_{j=1}^{n}$, where $\gamma_n = \{v_j\}_{j=1}^{n}$.

8. Materials and Methods

To perform the three numerical experiments (arrays, interpolators, and plots) included in the previous section, we used the formulae included in the paper, and we elaborated three programs that could be obtained, with public access, at the url https://www.dropbox.com/sh/0cx9chq3jfzov2w/AAA_SvL2i7HlC7ChMGpuG-Ata?dl=0 (accessed on 22 March 2021). Actually, these programs are notebooks (files with the names cardan, kepler, and random, which have as the extension .nb) elaborated with Mathematica (Mathematica is a trademark property of Wolfram Research). Mathematica is a quite standard software for mathematical computing. In particular, we used Version 12 Release 2. We do not hesitate to state that the programs (notebooks) run correctly in recent previous versions and in future versions because we used quite simple commands. Moreover, we did not use compiled routines nor other software.

9. Discussion

The nodal systems usually used for interpolation problems on the real line and the unit circle are related to measures. Normally, the zeros of orthogonal or para-orthogonal polynomials with respect to measures on the real line or the circumference, respectively, are considered as nodal points.

In the previous paper [9], new nodal systems were introduced, which in the case of the circumference proceeded from a perturbation of the roots of unity. Moreover, the Lagrange interpolation theory based on these nodal points was developed. In [9], the study of other types of interpolation was also suggested by using these nodal arrays as a future new research line. By following this idea, in the present paper, we developed the corresponding Hermite polynomial interpolation theory on the circle, as well as on the bounded interval.

The new nodal systems, whose study was completed in this paper, are very suitable to solve Hermite interpolation problems that appear in practice, most of which are linked to interesting mechanical models.

These nodal systems have not been used before in the Hermite interpolation due to the lack of a convergence theory to support their use. Thus, in this article, we dedicated a section to the study of the convergence of Hermite–Fejér interpolants in the case of continuous functions and the convergence of Hermite interpolants for smooth functions. With the theory developed in this paper, we are in conditions to use these nodal distributions for those models that fit to them. Taking into account these ideas, we presented three mechanical models for which the application of our results is very suitable.

The nodal system of the first example appeared as a consequence of a Cardan movement. We used the Hermite scheme given in Section 4.2 to recover an analytic function. The nodal system of the second example was a consequence of a random process. In this case, we used the Hermite–Fejér interpolator to approximate a quite variable continuous function. The convergence of this process is guaranteed by the results in Section 4.1. The last example is linked to a planetarium movement and to recovering a smooth non-analytic function following Section 4.2. Clearly, the examples only pretend to visualize the theoretical results previously obtained.

A possible future research line connected with this work could be the study of the corresponding Gibbs–Wilbraham phenomena.

Author Contributions: Conceptualization, E.B., A.C., H.G.R. and J.M.G.-A.; investigation, E.B., A.C., H.G.R. and J.M.G.-A.; software, E.B., A.C., H.G.R. and J.M.G.-A.; writing—original draft, E.B., A.C., H.G.R. and J.M.G.-A. All authors read and agreed to the published version of the manuscript.

Funding: This research received no external funding.

Institutional Review Board Statement: Not applicable.

Informed Consent Statement: Not applicable.

Data Availability Statement: Not applicable.

Conflicts of Interest: The authors declare no conflict of interest.

References

1. Nevai, P.; Vértesi, P. Convergence of Hermite–Fejér interpolation at zeros of generalized Jacobi polynomials. *Acta Sci. Math. (Szeged)* **1989**, *53*, 77–98.
2. Szabados, J. On Hermite–Fejér interpolation for the Jacobi abscissas. *Acta Math. Acad. Sci. Hung.* **1972**, *23*, 449–464. [CrossRef]
3. Vértesi, P. Notes on the Hermite–Fejér interpolation based on the Jacobi abscissas. *Acta Math. Acad. Sci. Hung.* **1973**, *24*, 233–239. [CrossRef]
4. Freud, G. On Hermite–Fejér interpolation processes. *Stud. Sci. Math. Hung.* **1972**, *7*, 307–316.
5. Freud, G. On Hermite–Fejér interpolation sequences. *Acta Math. Acad. Sci. Hung.* **1972**, *23*, 175–178. [CrossRef]
6. Nevai, P. Géza Freud, orthogonal polynomials and Christoffel functions. A case study. *J. Approx. Theory* **1986**, *48*, 3–167. [CrossRef]
7. Grünwald, G. On the theory of interpolation. *Acta Math.* **1943**, *75*, 219–245. [CrossRef]
8. Bos, L.; De Marchi, S.; Hormann, K.; Sidon, J. Bounding the Lebesgue constant for Berrut's rational interpolant at general nodes. *J. Approx. Theory* **2013**, *169*, 7–22. [CrossRef]
9. Berriochoa, E.; Cachafeiro, A.; Castejón, A.; García-Amor, J.M. Classical Lagrange interpolation based on general nodal systems at perturbed roots of unity. *Mathematics* **2020**, *8*, 498. [CrossRef]
10. Darius, L.; González-Vera, P. A note on Hermite–Fejér interpolation for the unit circle. *Appl. Math. Lett.* **2001**, *14*, 997–1003. [CrossRef]
11. Berriochoa, E.; Cachafeiro, A.; García-Amor, J.M. An extension of Fejér's condition for Hermite interpolation. *Complex Anal. Oper. Theory* **2012**, *6*, 651–664. [CrossRef]
12. Simon, B. *Orthogonal Polynomials on the Unit Circle: Part 1 and Part 2*; American Mathematical Society Colloquium Publications: Providence, RI, USA, 2005; Volume 54.
13. Szegő, G. *Orthogonal Polynomials*, 4th ed.; American Mathematical Society Colloquium Publications: Providence, RI, USA, 1975; Volume 23.

14. Simon, B. Fine structure of the zeros of orthogonal polynomials, I. A tale of two pictures. *Electron. Trans. Numer. Anal.* **2006**, *25*, 328–368.
15. Berriochoa, E.; Cachafeiro, A.; Marcellán, F. Szegő transformation and zeros of analytic perturbations of Chebyshev weights. *J. Math. Anal. Appl.* **2019**, *470*, 571–583. [CrossRef]
16. Trefethen, L.N. *Approximation Theory and Approximation Practice*; Society for Industrial and Applied Mathematics (SIAM): Philadelphia, PA, USA, 2013.
17. Berriochoa, E.; Cachafeiro, A.; Martínez, E. About measures and nodal systems for which the Hermite interpolants uniformly converge to continuous functions on the circle and interval. *Appl. Math. Comput.* **2012**, *218*, 4813–4824. [CrossRef]
18. Davis, P.J. *Interpolation and Approximation*; Dover Publications: New York, NY, USA, 1975.
19. Gupta, B.V.R. *Theory of Machines: Kinematics and Dynamics*; I. K. International Publishing House Pvt. Ltd.: New Delhi, India, 2011.
20. McGill, D.J.; King, W.W. *Engineering Mechanics: Statics and an Introduction to Dynamics*, 3rd ed.; Cengage Learning, Inc.: London, UK, 1995.
21. Curtis, H.D. *Orbital Mechanics for Engineering Students*; Elsevier Butterworth-Heinemann: Amsterdam, The Netherlands, 2005.

Article

Steady Fluid–Structure Coupling Interface of Circular Membrane under Liquid Weight Loading: Closed-Form Solution for Differential-Integral Equations

Xue Li [1], Jun-Yi Sun [1,2,*], Xiao-Chen Lu [1], Zhi-Xin Yang [1] and Xiao-Ting He [1,2]

[1] School of Civil Engineering, Chongqing University, Chongqing 400045, China; 20161602025t@cqu.edu.cn (X.L.); luxiaochen9653@126.com (X.-C.L.); 20141602063@cqu.edu.cn (Z.-X.Y.); hexiaoting@cqu.edu.cn (X.-T.H.)
[2] Key Laboratory of New Technology for Construction of Cities in Mountain Area, Chongqing University, Ministry of Education, Chongqing 400045, China
* Correspondence: sunjunyi@cqu.edu.cn; Tel.: +86-(0)23-65120720

Citation: Li, X.; Sun, J.-Y.; Lu, X.-C.; Yang, Z.-X.; He, X.-T. Steady Fluid–Structure Coupling Interface of Circular Membrane under Liquid Weight Loading: Closed-Form Solution for Differential-Integral Equations. *Mathematics* **2021**, *9*, 1105. https://doi.org/10.3390/math9101105

Academic Editors: Lucas Jódar and Rafael Company

Received: 9 March 2021
Accepted: 10 May 2021
Published: 13 May 2021

Publisher's Note: MDPI stays neutral with regard to jurisdictional claims in published maps and institutional affiliations.

Copyright: © 2021 by the authors. Licensee MDPI, Basel, Switzerland. This article is an open access article distributed under the terms and conditions of the Creative Commons Attribution (CC BY) license (https://creativecommons.org/licenses/by/4.0/).

Abstract: In this paper, the problem of fluid–structure interaction of a circular membrane under liquid weight loading is formulated and is solved analytically. The circular membrane is initially flat and works as the bottom of a cylindrical cup or bucket. The initially flat circular membrane will undergo axisymmetric deformation and deflection after a certain amount of liquid is poured into the cylindrical cup. The amount of the liquid poured determines the deformation and deflection of the circular membrane, while in turn, the deformation and deflection of the circular membrane changes the shape and distribution of the liquid poured on the deformed and deflected circular membrane, resulting in the so-called fluid-structure interaction between liquid and membrane. For a given amount of liquid, the fluid-structure interaction will eventually reach a static equilibrium and the fluid-structure coupling interface is steady, resulting in a static problem of axisymmetric deformation and deflection of the circular membrane under the weight of given liquid. The established governing equations for the static problem contain both differential operation and integral operation and the power series method plays an irreplaceable role in solving the differential-integral equations. Finally, the closed-form solutions for stress and deflection are presented and are confirmed to be convergent by the numerical examples conducted.

Keywords: circular membrane; fluid-structure interaction; differential-integral equations; power series method; closed-form solution

1. Introduction

Elastic membrane structures or structural components have applications in various fields [1–5]. These applications have provided an impetus for scholars to investigate the phenomena of large deflection of membrane [6–8]. Such investigations usually give rise to nonlinear equations with differential and even integral operation. These somewhat intractable nonlinear equations may present serious analytical difficulties when applied to boundary value problems [9–13].

The usually so-called circular membrane problem refers to the problem of axisymmetric deformation and deflection of an initially flat, peripherally fixed circular membrane subjected to transverse loads. Three main loading forms of transverse loads are involved in the existing studies: ① the uniformly distributed loads applied to the entire circular membrane [14–22], ② the uniformly distributed loads applied to the central portion of the circular membrane [23], and ③ the concentrated force applied to the center of the circular membrane [24–28]. Hencky was the first scholar to deal with the circular membrane problem concerning the first loading form of transverse loads and presented a closed-form solution in the form of power series [14]. A computational error in reference [14] was corrected by Chien [15] and Alekseev [16], respectively. The problem originally dealt with by Hencky is usually called the well-known Hencky problem; i.e., the problem of axisymmetric deformation and deflection of an initially flat, peripherally fixed circular membrane under the action of a uniformly distributed transverse loads, where the weight of the circular membrane is usually ignored because it is usually very small in comparison with the transverse loads. The solution of the well-known Hencky problem is usually called the well-known Hencky solution, which is the first solution of the circular membrane problem and is often cited in some studies of related issues [15–22]. Chien et al. [23] analytically dealt with the symmetrical deformation of circular membrane under the action of uniformly distributed loads in its central portion, i.e., the circular membrane problem concerning the second loading form of transverse load. As for the third loading form, the concentrated force applied to the center of the circular membrane, it is, in fact, the limit case of the second loading form of transverse loads.

If an initially flat circular membrane is used as the bottom of a cylindrical cup or bucket and a certain amount of liquid is poured into the cylindrical cup or bucket, then the initially flat circular membrane will undergo axisymmetric deformation and deflection. The amount of the liquid poured determines the deformation and deflection of the circular membrane, while in turn, the deformation and deflection of the circular membrane changes the shape or distribution of the liquid over the deformed and deflected circular membrane. This results in an interaction between the liquid and the membrane, which is often referred to as a fluid–structure interaction. Obviously, for a given amount of liquid, the interaction between the liquid and the membrane will eventually reach a static equilibrium and a steady fluid–solid coupling interface will appear. Our main interest here is the static problem of axisymmetric deformation and deflection of the circular membrane under the given liquid weight loading. The closed-form solution of this static problem is expected to be used in the development of a new rain gauge [29–31]. However, such a fluid–structure coupling problem will give rise to governing equations containing both differential operation and integral operation. The power series method plays a unique and key role in solving these kinds of differential-integral equations analytically, as will be seen later.

The paper is organized as follows: in Section 2, the governing equations are established and the closed-form solutions for stress and deflection are presented. In Section 3, the numerical examples are conducted to show the differences between the presented solution and the well-known Hencky solution, and the convergence of the power-series solution for deflection and stress is verified. The concluding remarks are shown in Section 4.

2. Membrane Equation and Its Solution

An initially flat circular unstretched membrane with Young's modulus of elasticity E, Poisson's ratio ν, thickness h and radius a is fixed at the lower end of a vertically placed rigid round tube of finite length to form a cylindrical cup or bucket of inner radius a having a closed soft bottom with elastic deformation capability, and then a colored liquid with density ρ is slowly poured into the cup until the height of liquid reaches H, as shown in Figure 1, where H is the distance from the liquid level to the plane in which the initially flat circular membrane is located, w_m denotes the maximum deflection of the deflected circular membrane at static equilibrium. Based on the anticipated use of this study for rain gauge, only the case of $H \geq 0$ is considered here.

Let us take out a free body of a piece of circular membrane with radius r ($0 \leq r \leq a$) from the central portion of the whole deformed circular membrane, to study the static problem of equilibrium of this free body under the joint actions of the external force $F(r)$ produced by the transverse distributed loads $q(r)$ within r and the total force $2\pi r \sigma_r h$ produced by the membrane force $\sigma_r h$ acting on the boundary r, as shown in Figure 2; where a cylindrical coordinate system (r, φ, w) is introduced, the polar coordinate plane (r, φ) is located in the plane in which the geometric middle plane of the initially flat circular membrane is located; o denotes the origin of the cylindrical coordinate system (r, φ, w), which is placed in the centroid of the geometric intermediate plane, r denotes the radial coordinate, w denotes the axial coordinate of the cylindrical coordinate system (r, φ, w) as well as the transverse displacement of a point on the deflected circular membrane, θ denotes the slope angle of the deflecting membrane, and σ_r denotes the radial stress, while the angle coordinate φ is not represented in Figure 2.

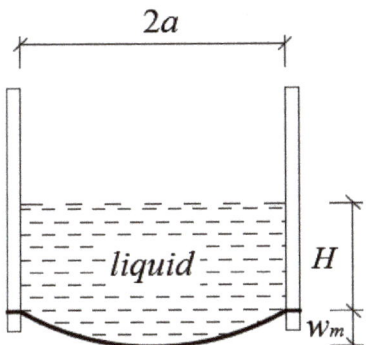

Figure 1. Geometry of the circular membrane under prescribed liquid along a diameter.

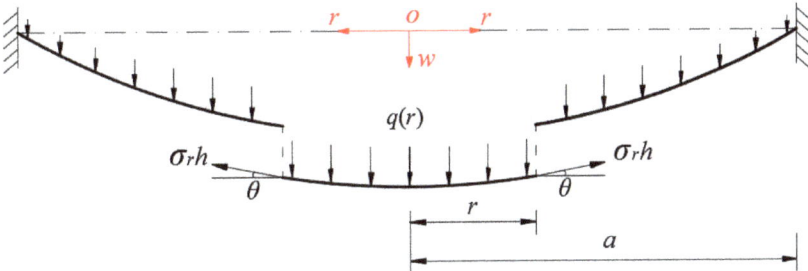

Figure 2. Free body diagram of the deformed circular membrane with radius $0 \leq r \leq a$.

Obviously, the external force $F(r)$ produced by $q(r)$ within radius r is equal to the weight of the liquid within radius r, and is given by

$$F(r) = \rho g \int_0^r [w(r) + H] \cdot 2\pi r dr = 2\pi \rho g \int_0^r w(r) r dr + \rho g \pi r^2 H, \quad (1)$$

where g is the acceleration of gravity and $w(r)$ is the transverse displacement at r. Equation (1) is the usually so-called fluid-structure coupling equation at static equilibrium. The direction of the external force $F(r)$ is always perpendicular to the plane in which the initially flat circular membrane is located and vertically downward. Right here, the vertical upward force is $2\pi r \sigma_r h \sin \theta$, that is, the vertical component of the total membrane force $2\pi r \sigma_r h$ at r. Therefore, after ignoring the weight of the circular membrane, the equilibrium condition in the vertical direction, i.e., the so-called out-of-plane equilibrium equation, is given by

$$2\pi r \sigma_r h \sin \theta = F(r) = 2\pi \rho g \int_0^r w(r) r dr + \rho g \pi r^2 H, \quad (2)$$

where

$$\sin \theta \cong \tan \theta = -\frac{dw}{dr}. \quad (3)$$

Substituting Equation (3) into Equation (2) yields

$$2r\sigma_r h \frac{dw}{dr} + 2\rho g \int_0^r w(r)rdr + \rho g r^2 H = 0. \tag{4}$$

In the horizontal direction, there are two horizontal forces, the circumferential membrane force $\sigma_t h$ and the horizontal component of the radial membrane force $\sigma_r h$, where σ_t denotes the circumferential stress. Then, the equilibrium condition in the horizontal direction (i.e., the so-called in-plane equilibrium equation) is [23]

$$\frac{d}{dr}(r\sigma_r h) - \sigma_t h = 0. \tag{5}$$

Suppose that the radial strain is e_r, the circumferential strain is e_t and the radial displacement at r is $u(r)$. Then, the relations of the strain and displacement, the so-called geometric equations, may be written as [23]

$$e_r = \frac{du}{dr} + \frac{1}{2}(\frac{dw}{dr})^2 \tag{6}$$

and

$$e_t = \frac{u}{r}. \tag{7}$$

The relations of the stress and strain (i.e., the so-called physical equations) are [23]

$$\sigma_r = \frac{E}{1 - \nu^2}(e_r + \nu e_t) \tag{8}$$

and

$$\sigma_t = \frac{E}{1 - \nu^2}(e_t + \nu e_r). \tag{9}$$

Substituting Equations (6) and (7) into Equations (8) and (9) (to eliminate e_r and e_t in Equations (8) and (9)) yields

$$\sigma_r = \frac{E}{1 - \nu^2}[\frac{du}{dr} + \frac{1}{2}(\frac{dw}{dr})^2 + \nu\frac{u}{r}] \tag{10}$$

and

$$\sigma_t = \frac{E}{1 - \nu^2}[\frac{u}{r} + \nu\frac{du}{dr} + \nu\frac{1}{2}(\frac{dw}{dr})^2]. \tag{11}$$

By means of Equations (10), (11) and (5), one has

$$\frac{u}{r} = \frac{1}{Eh}(\sigma_t h - \nu\sigma_r h) = \frac{1}{Eh}[\frac{d}{dr}(r\sigma_r h) - \nu\sigma_r h]. \tag{12}$$

Eliminating u from Equations (10) and (12) yields

$$r\frac{d}{dr}[\frac{1}{r}\frac{d}{dr}(r^2\sigma_r h)] + \frac{Eh}{2}(\frac{dw}{dr})^2 = 0. \tag{13}$$

Equation (13) is usually called a consistency equation. Equations (4) and (13) are two equations for the solutions of σ_r and w.

The boundary conditions, under which Equations (4) and (13) may be solved, are

$$\frac{dw}{dr} = 0 \text{ at } r = 0, \tag{14}$$

$$\frac{u}{r} = \frac{1}{Eh}[\frac{d}{dr}(r\sigma_r h) - \nu\sigma_r h] = 0 \text{ at } r = a \tag{15}$$

and
$$w = 0 \text{ at } r = a. \tag{16}$$

Let us introduce the following nondimensionalization:

$$W = \frac{w}{a}, \ S_r = \frac{\sigma_r}{E}, \ S_t = \frac{\sigma_t}{E}, \ x = \frac{r}{a}, \ H_0 = \frac{H}{a}, \ G = \frac{\rho g a^2}{Eh}, \tag{17}$$

and transform Equations (4), (13), (5), (14), (15) and (16) into

$$2xS_r \frac{dW}{dx} + 2G \int_0^x W(x) x \, dx + x^2 G H_0 = 0, \tag{18}$$

$$x^2 \frac{d^2 S_r}{dx^2} + 3x \frac{dS_r}{dx} + \frac{1}{2}\left(\frac{dW}{dx}\right)^2 = 0, \tag{19}$$

$$S_t = S_r + x \frac{dS_r}{dx}, \tag{20}$$

$$\frac{dW}{dx} = 0 \text{ at } x = 0, \tag{21}$$

$$\frac{u}{r} = (1-\nu)S_r + x \frac{dS_r}{dx} = 0 \text{ at } x = 1 \tag{22}$$

and
$$W = 0 \text{ at } x = 1. \tag{23}$$

In view of the physical phenomenon that the values of stress and deflection are both finite at $x = 0$, S_r and W can be expanded into the power series of x; i.e., let

$$S_r = \sum_{i=0}^{n} c_i x^i \tag{24}$$

and
$$W = \sum_{i=0}^{n} d_i x^i. \tag{25}$$

After substituting Equations (24) and (25) into Equations (18) and (19), it is found, by using the mathematical software Maple 2018, that, $c_i \equiv 0$ and $d_i \equiv 0$ when $i = 1, 3, 5, \ldots$, and when $i = 2, 4, 6, \ldots$, the coefficients c_i and d_i can be expressed into the polynomial with regard to the coefficients c_0 and d_0 (see Appendices A and B).

The remaining two coefficients c_0 and d_0 are called undetermined constants, which can be determined by using the boundary conditions Equations (22) and (23) as follows. From Equation (24), Equation (22) gives

$$(1-\nu) \sum_{i=0}^{n} c_i + \sum_{i=1}^{n} i c_i = 0, \tag{26}$$

and from Equation (25), Equation (23) gives

$$\sum_{i=0}^{n} d_i = 0. \tag{27}$$

For a concrete problem, the values of a, h, E, ν, ρ and H are known in advance. Therefore, after substituting all expressions of c_i and d_i (which are expressed by c_0 and d_0, see Appendices A and B) into Equations (26) and (27), we can obtain a system of equations containing only c_0 and d_0. The undetermined constants c_0 and d_0 can be determined by solving this system of equations. Furthermore, with the known c_0 and d_0, the other coefficients c_i and d_i ($i = 2, 4, 6, \ldots$) can easily be determined and the expressions of S_r and W can also be determined. The problems dealt with here are thus solved.

3. Results and Discussions

It is obvious that the boundary condition Equation (14), i.e., $dw/dr = 0$ at $r = 0$, has not been used yet. Now, let us see whether the closed-form solution obtained in Section 2 meets this boundary condition. From Equations (17) and (25) the dimensional form of the deflection w can be written as

$$w = \sum_{i=0}^{\infty} \frac{d_i}{a^{i-1}} r^i, \qquad (28)$$

and the first derivative of Equation (28) is

$$\frac{dw}{dr} = \sum_{i=1}^{\infty} \frac{i d_i}{a^{i-1}} r^{i-1}. \qquad (29)$$

Equation (29) gives $dw/dr = d_1$ at $r = 0$. Therefore $dw/dr \equiv 0$ at $r = 0$, due to $d_1 \equiv 0$ (see the description after Equation (25)). It indicates that Equation (14) can be automatically satisfied, which, to some extent, proves the validity of the closed-form solution obtained in Section 2.

3.1. Comparison with the Well-Known Hencky Solution

It is well known that the well-known Hencky solution applies only to the case where the transverse loads applied to the whole deflected circular membrane must, regardless of the deflection of the membrane, be uniformly distributed [14]. Obviously, the more uneven the distribution of the transverse loads is, the greater the error caused by using the well-known Hencky solution. It is not hard to imagine from Figure 1 that, for a given amount of liquid (keep the liquid level H constant), the thinner or softer the membrane is, the greater the deflection of the membrane is, while the greater the deflection of the membrane is, the more uneven the distribution of the liquid over the whole deflected circular membrane is. On the other hand, for a given circular membrane, the uniformity of liquid distribution will also change with the increase of the liquid level H. Now, let us consider a numerical example to examine the difference between using the well-known Hencky solution and the solution obtained in Section 2. When the well-known Hencky solution is used, its uniformly distributed transverse loads q are given here by $q = \rho g H$.

Suppose that a circular rubber membrane with radius $a = 20$ mm, thickness $h = 0.1$ mm, Young's modulus of elasticity and Poisson's ratio $\nu = 0.47$ is subjected to a liquid with a density of $\rho = 1 \times 10^{-6}$ kg/mm^3. After the fluid–structure interaction reaches a static equilibrium, the liquid level H is assumed to be equal to 0.5 mm, 50 mm and 200 mm, respectively. The acceleration of gravity is assumed to be $g = 10$ m/s^2. The deflection and radial stress curves along radius are shown in Figures 3 and 4, respectively, where the solid lines represent the results calculated by the solution obtained in Section 2 and the dotted lines by the well-known Hencky solution. The concrete values of the maximum deflection and radial stress are listed in Table 1, where the "errors" are given by the absolute value of the results by the well-known Hencky solution minus the results by the solution presented in this paper and then divided by the results by the solution presented in this paper.

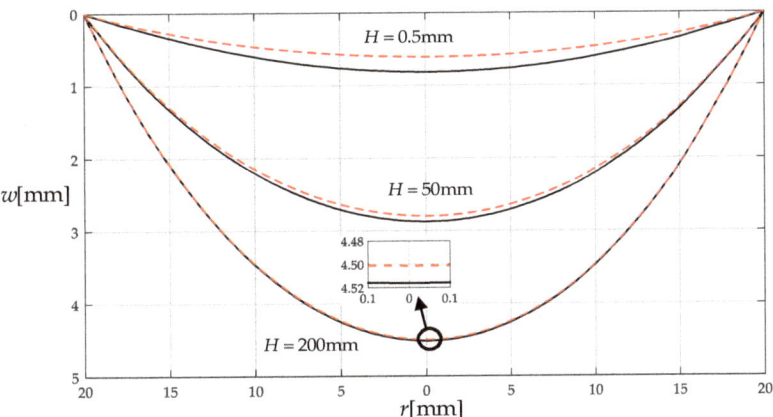

Figure 3. Deflection w along radius r when H takes 0.5 mm, 50 mm and 200 mm, respectively, where the solid lines by the solution presented in this paper and the dotted lines by the well-known Hencky solution.

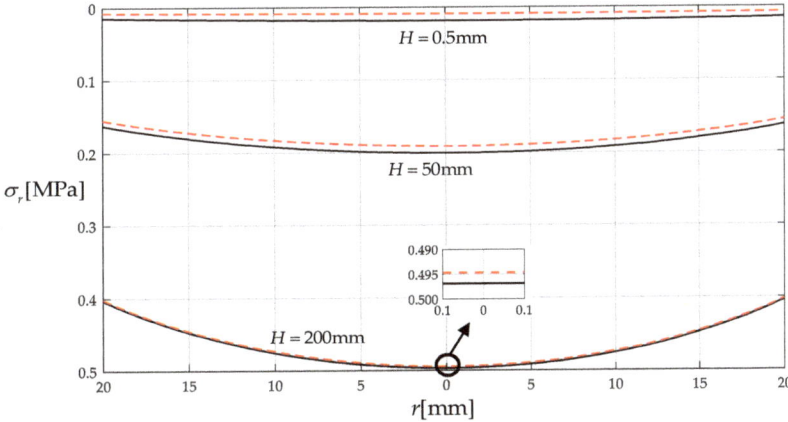

Figure 4. Radial stress σ_r along radius r when H takes 0.5 mm, 50 mm and 200 mm, respectively, where the solid lines represent the solution presented in this paper and the dotted lines represent the well-known Hencky solution.

Table 1. Maximum deflection and radial stress values at different H calculated by the solution presented in this paper and the well-known Hencky solution.

H	Maximum Deflection [mm]			Maximum Radial Stress [MPa]		
	Presented Solution	Hencky Solution	Errors	Presented Solution	Hencky Solution	Errors
0.5	0.8169	0.6089	25.5%	0.0184	0.0091	50.8%
50	2.8733	2.7977	2.6%	0.2005	0.1911	4.7%
200	4.5161	4.5013	0.3%	0.4969	0.4947	0.4%

From Figures 3 and 4, it can be easily seen that the distance between the dotted line and the solid line decreases as the liquid level H increases. When H = 0.5 mm, the distance between the dotted line and the solid line is the largest and the error between the results calculated by the solution presented in this paper and the well-known Hencky solution are also the largest (see Table 1), while H = 200 mm, both the distance and the error are very small. This means that when $H = 0.5$ mm, the distribution of the liquid over the whole deflected circular membrane is the most uneven, and consequently the difference between using the well-known Hencky solution and the solution presented in this paper is the most obvious (the maximum value of relative error is about 25.5% for deflection and 50.8% for radial stress; see the first row in Table 1). The main reason behind this is that the uniformly distributed transverse loads q used for the well-known Hencky solution are given by $q = \rho g H$ (where H takes 0.5 mm); while $H = 0.5$ mm, the actual height of the liquid over the whole deflected circular membrane is 0.5 mm at the edge of the circular membrane and is about 1.3169 (0.5 + 0.8169) mm (see the first column in Table 1) at the center of the circular membrane (the relative error is about $(1.3169 - 0.5)/0.5 = 163.38\%$). Therefore, the distribution of the liquid over the whole deflected circular membrane is actually very uneven. Just as stated above, the more uneven the distribution of the transverse loads is, the greater the error caused by using the well-known Hencky solution. On the other hand, when $H = 200$ mm, the actual height of the liquid over the whole deflected circular membrane is 200 mm at the edge of the circular membrane and is about 204.5161 (200 + 4.5161) mm at the center of the circular membrane (the relative error is about $(204.5161 - 200)/200 = 2.26\%$). Therefore, in this case, the distribution of the liquid over the whole deflected circular membrane is actually very uniform. In other words, in this case, the external force $F(a)$ produced by $q(r)$ within radius a, which is applied to the whole deflected circular membrane, is largely determined by $\rho g \pi a^2 H$, and the contribution of the fluid–structure interaction $2\pi \rho g \int_0^a w(r) r dr$ can be ignored, see Equation (1). In addition, the phenomenon that the results calculated by the solution presented in this paper gradually converge to the results by the well-known Hencky solution as the liquid level H increases also proves to some extent that the closed-form solution obtained in Section 2 are basically reliable, as far as the well-known Hencky solution is considered to be a reliable solution.

3.2. Verification of Convergence of the Power Series Solution

In this section, the convergence of the power series solution obtained in Section 2 will be discussed. In general, it is better to discuss the convergence of the general solution rather than that of the special solution, because the special solution will converge if the general solution converges. However, we here have to discuss the convergence of the special solution, because the discussion on the convergence of the general solution cannot be conducted due to the complexity of the coefficients c_i and d_i ($i = 2, 4, 6, \ldots$) expressed by the undetermined constants c_0 and d_0 (see Appendices A and B). From the derivation in Section 2, we know that the undetermined constants c_0 and d_0 can be determined by simultaneous solutions of Equations (26) and (27); the special solutions for $S_r(x)$ and $W(x)$ can be easily obtained as long as the undetermined constants c_0 and d_0 can be determined. When calculating the undetermined constants c_0 and d_0, we have to substitute the partial sum of former n terms of Equations (24) and (25), rather than the infinite series of Equations (24) and (25), into Equations (26) and (27), otherwise the resulting Equations (26) and (27) will contain two infinite series and are thus difficult to be solved. Therefore, it seems that the terms n will determine the values of the undetermined constants c_0 and d_0, and different n will determine the different values of c_0 and d_0. Hence, the discussion on the convergence of the special solution should focus on giving the variations of c_0 and d_0 with terms n. If the undetermined constants c_0 and d_0 converge as the terms n increase, then the special solution can be concluded to converge as well.

We will continue with the numerical example given in Section 3.1. A circular rubber membrane with radius $a = 20$ mm, thickness $h = 0.1$ mm, Young's modulus of elasticity $E = 7.84$ MPa and Poisson's ratio $\nu = 0.47$ is subjected to the liquid weight loading with the liquid level $H = 50$ mm. We start the numerical calculations of c_0 and d_0 from $n = 4$; that is, start from the partial sum of the former four terms of Equations (24) and (25). The variations of c_0 and d_0 with terms n are shown in Figures 5 and 6, respectively. From Figures 5 and 6, we may see that with the increase of the terms n, the values of c_0 and d_0 are gradually close to some certain values (i.e., their exact values), and are almost no longer changed when the terms n reach around 14, which indicates that the undetermined constants c_0 and d_0 converge reasonably well. Therefore, we here show only the results of the coefficients c_i and d_i when $n \leq 24$, which are listed in Tables 2 and 3, and when $n = 24$, the variations of the coefficients c_i and d_i with i ($i = 0, 2, 4, \ldots, 24$) are shown in Figures 7 and 8. From Figures 7 and 8, it can be seen that c_{24} and d_{24} are already very close to 0, which means that the values of S_r and W are already very close to their exact values when $n = 24$.

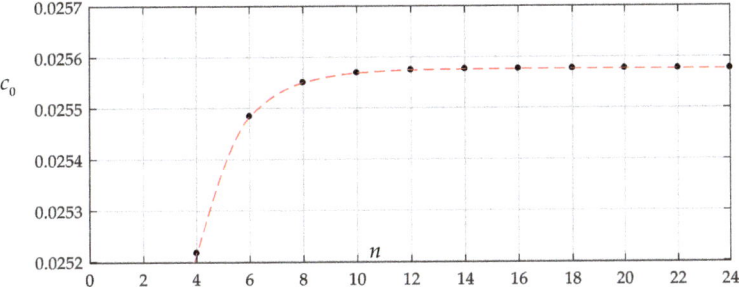

Figure 5. Variation of c_0 with n.

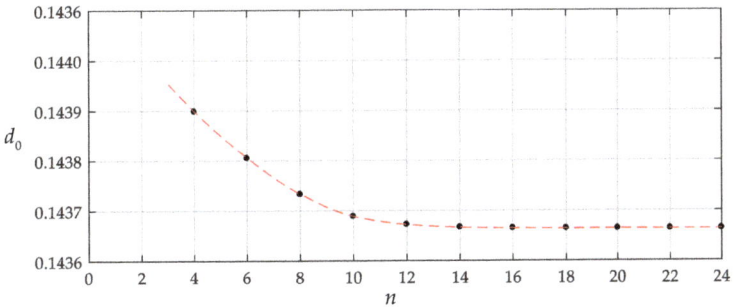

Figure 6. Variation of d_0 with n.

Table 2. (a) The values of c_i at different n; (b) The values of c_i at different n; (c) The values of c_i at different n; (d) The value of c_{24} at $n = 24$.

(a) The values of c_i at different n				
n	c_0	c_2	c_4	c_6
4	2.5216650×10^{-2}	$-4.4711868 \times 10^{-3}$	$-4.5313948 \times 10^{-4}$	-
6	2.5484214×10^{-2}	$-4.3775807 \times 10^{-3}$	$-4.2827528 \times 10^{-4}$	$-7.5232361 \times 10^{-5}$
8	2.5550954×10^{-2}	$-4.3544054 \times 10^{-3}$	$-4.2226218 \times 10^{-4}$	$-7.3567411 \times 10^{-5}$
10	2.5569445×10^{-2}	$-4.3479638 \times 10^{-3}$	$-4.2060234 \times 10^{-4}$	$-7.3110446 \times 10^{-5}$
12	2.5574842×10^{-2}	$-4.3460754 \times 10^{-3}$	$-4.2011702 \times 10^{-4}$	$-7.2977069 \times 10^{-5}$
14	2.5576464×10^{-2}	$-4.3455055 \times 10^{-3}$	$-4.1997075 \times 10^{-4}$	$-7.2936900 \times 10^{-5}$
16	2.5576960×10^{-2}	$-4.3453305 \times 10^{-3}$	$-4.1992585 \times 10^{-4}$	$-7.2924572 \times 10^{-5}$
18	2.5577114×10^{-2}	$-4.3452760 \times 10^{-3}$	$-4.1991189 \times 10^{-4}$	$-7.2920739 \times 10^{-5}$
20	2.5577163×10^{-2}	$-4.3452589 \times 10^{-3}$	$-4.1990751 \times 10^{-4}$	$-7.2919536 \times 10^{-5}$
22	2.5577178×10^{-2}	$-4.3452535 \times 10^{-3}$	$-4.1990612 \times 10^{-4}$	$-7.2919156 \times 10^{-5}$
24	2.5577183×10^{-2}	$-4.3452518 \times 10^{-3}$	$-4.1990568 \times 10^{-4}$	$-7.2919035 \times 10^{-5}$

Table 2. *Cont.*

	(b) The values of c_i at different n			
n	c_8	c_{10}	c_{12}	c_{14}
8	$-1.5488504 \times 10^{-5}$	-	-	-
10	$-1.5357045 \times 10^{-5}$	$-3.5823498 \times 10^{-6}$	-	-
12	$-1.5318738 \times 10^{-5}$	$-3.5710160 \times 10^{-6}$	$-8.9013785 \times 10^{-7}$	-
14	$-1.5307207 \times 10^{-5}$	$-3.5676061 \times 10^{-6}$	$-8.8910802 \times 10^{-7}$	$-2.3216086 \times 10^{-7}$
16	$-1.5303669 \times 10^{-5}$	$-3.5665600 \times 10^{-6}$	$-8.8879213 \times 10^{-7}$	$-2.3206398 \times 10^{-7}$
18	$-1.5302569 \times 10^{-5}$	$-3.5662349 \times 10^{-6}$	$-8.8869395 \times 10^{-7}$	$-2.3203386 \times 10^{-7}$
20	$-1.5302224 \times 10^{-5}$	$-3.5661328 \times 10^{-6}$	$-8.8866314 \times 10^{-7}$	$-2.3202441 \times 10^{-7}$
22	$-1.5302115 \times 10^{-5}$	$-3.5661005 \times 10^{-6}$	$-8.8865339 \times 10^{-7}$	$-2.3202143 \times 10^{-7}$
24	$-1.5302080 \times 10^{-5}$	$-3.5660903 \times 10^{-6}$	$-8.8865029 \times 10^{-7}$	$-2.3202047 \times 10^{-8}$
	(c) The values of c_i at different n			
n	c_{16}	c_{18}	c_{20}	c_{22}
16	$-6.2711514 \times 10^{-8}$	-	-	-
18	$-6.2702167 \times 10^{-8}$	$-1.7397329 \times 10^{-8}$	-	-
20	$-6.2699234 \times 10^{-8}$	$-1.7396410 \times 10^{-8}$	$-4.9288360 \times 10^{-9}$	-
22	$-6.2698306 \times 10^{-8}$	$-1.7396119 \times 10^{-8}$	$-4.9287442 \times 10^{-9}$	$-1.4203606 \times 10^{-9}$
24	$-6.2698011 \times 10^{-8}$	$-1.7396027 \times 10^{-8}$	$-4.9287150 \times 10^{-9}$	$-1.4203513 \times 10^{-9}$
	(d) The value of c_{24} at $n = 24$			
n	c_{24}			
24	$-4.1511468 \times 10^{-10}$	-	-	-

Table 3. (a) The values of d_i at different n; (b) The values of d_i at different n; (c) The values of d_i at different n; (d) The value of d_{24} at $n = 24$.

	(a) The values of d_i at different n			
n	d_0	d_2	d_4	d_6
4	1.4389897×10^{-1}	$-1.3373387 \times 10^{-1}$	$-1.0165102 \times 10^{-2}$	-
6	1.4380527×10^{-1}	$-1.3232658 \times 10^{-1}$	$-9.7095072 \times 10^{-3}$	$-1.7991832 \times 10^{-3}$
8	1.4373309×10^{-1}	$-1.3197584 \times 10^{-1}$	$-9.5986248 \times 10^{-3}$	$-1.7643172 \times 10^{-3}$
10	1.4368872×10^{-1}	$-1.3187818 \times 10^{-1}$	$-9.5679741 \times 10^{-3}$	$-1.7547332 \times 10^{-3}$
12	1.4367240×10^{-1}	$-1.3184954 \times 10^{-1}$	$-9.5590098 \times 10^{-3}$	$-1.7519351 \times 10^{-3}$
14	1.4366673×10^{-1}	$-1.3184090 \times 10^{-1}$	$-9.5563083 \times 10^{-3}$	$-1.7510923 \times 10^{-3}$
16	1.4366481×10^{-1}	$-1.3183824 \times 10^{-1}$	$-9.5554791 \times 10^{-3}$	$-1.7508337 \times 10^{-3}$
18	1.4366417×10^{-1}	$-1.3183741 \times 10^{-1}$	$-9.5552212 \times 10^{-3}$	$-1.7507533 \times 10^{-3}$
20	1.4366395×10^{-1}	$-1.3183716 \times 10^{-1}$	$-9.5551403 \times 10^{-3}$	$-1.7507281 \times 10^{-3}$
22	1.4366388×10^{-1}	$-1.3183707 \times 10^{-1}$	$-9.5551147 \times 10^{-3}$	$-1.7507201 \times 10^{-3}$
24	1.4366386×10^{-1}	$-1.3183705 \times 10^{-1}$	$-9.5551066 \times 10^{-3}$	$-1.7507175 \times 10^{-3}$
	(b) The values of d_i at different n			
n	d_8	d_{10}	d_{12}	d_{14}
8	$-3.9431453 \times 10^{-4}$	-	-	-
10	$-3.9128051 \times 10^{-4}$	$-9.6550465 \times 10^{-5}$	-	-
12	$-3.9039607 \times 10^{-4}$	$-9.6267818 \times 10^{-5}$	$-2.5251202 \times 10^{-5}$	-
14	$-3.9012984 \times 10^{-4}$	$-9.6182775 \times 10^{-5}$	$-2.5223804 \times 10^{-5}$	$-6.8962634 \times 10^{-6}$
16	$-3.9004815 \times 10^{-4}$	$-9.6156686 \times 10^{-5}$	$-2.5215401 \times 10^{-5}$	$-6.8935385 \times 10^{-6}$
18	$-3.9002276 \times 10^{-4}$	$-9.6148576 \times 10^{-5}$	$-2.5212789 \times 10^{-5}$	$-6.8926917 \times 10^{-6}$
20	$-3.9001479 \times 10^{-4}$	$-9.6146031 \times 10^{-5}$	$-2.5211969 \times 10^{-5}$	$-6.8924259 \times 10^{-6}$
22	$-3.9001227 \times 10^{-4}$	$-9.6145227 \times 10^{-5}$	$-2.5211710 \times 10^{-5}$	$-6.8923419 \times 10^{-6}$
24	$-3.9001147 \times 10^{-4}$	$-9.6144970 \times 10^{-5}$	$-2.5211627 \times 10^{-5}$	$-6.8923151 \times 10^{-6}$

Table 3. *Cont.*

	(c) The values of d_i at different n			
n	d_{16}	d_{18}	d_{20}	d_{22}
16	$-1.9420161 \times 10^{-6}$	-	-	-
18	$-1.9417401 \times 10^{-6}$	$-5.5954589 \times 10^{-7}$	-	-
20	$-1.9416535 \times 10^{-6}$	$-5.5951754 \times 10^{-7}$	$-1.6411662 \times 10^{-7}$	-
22	$-1.9416261 \times 10^{-6}$	$-5.5950858 \times 10^{-7}$	$-1.6411368 \times 10^{-7}$	$-4.8827855 \times 10^{-8}$
24	$-1.9416174 \times 10^{-6}$	$-5.5950573 \times 10^{-7}$	$-1.6411274 \times 10^{-7}$	$-4.8827547 \times 10^{-8}$
	(d) The value of d_{24} at $n = 24$			
n	d_{24}	-	-	-
24	$-1.4698279 \times 10^{-8}$	-	-	-

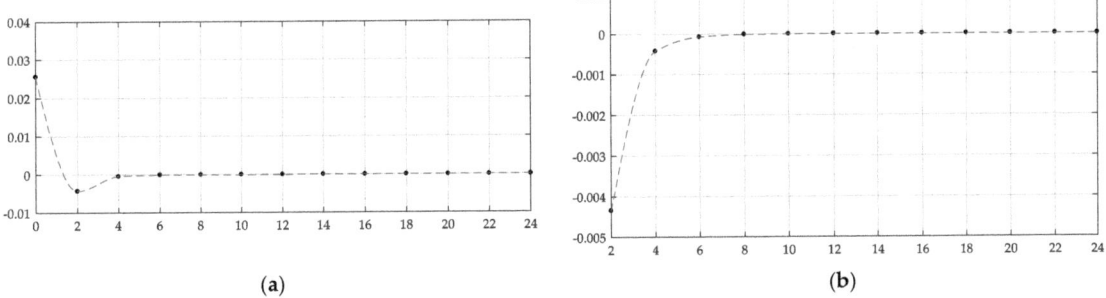

Figure 7. The values of c_i when $n = 24$: (**a**) for $i = 0, 2, 4, 6, \ldots, 24$; (**b**) for $i = 2, 4, 6, \ldots, 24$.

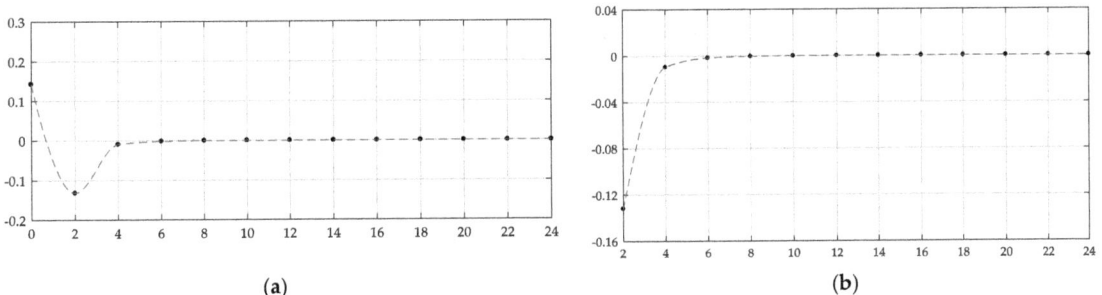

Figure 8. The values of d_i when $n = 24$: (**a**) for $i = 0, 2, 4, 6, \ldots, 24$; (**b**) for $i = 2, 4, 6, \ldots, 24$.

4. Concluding Remarks

In this paper, we analytically solved the problem of axisymmetric deformation and deflection of a circular membrane under liquid weight loading and presented the closed-form solution for stress and deflection. The following conclusions can be drawn from this study:

i. The power series method is effective for the analytical solution to differential-integral equations.

ii. When the amount of liquid applied onto the circular membrane is large enough, the difference between the solution presented in this paper and the well-known Hencky solution will become relatively small. If the requirement for calculation accuracy is not too high, the problem of axisymmetric deformation of the circular membrane under liquid self-weight loading may be treated as the well-known Hencky problem; the fluid–structure interaction may be neglected.

iii. When the amount of liquid applied onto the circular membrane is relatively small, the solution presented in this paper will be quite different from the well-known Hencky solution. For a higher calculation accuracy, the fluid–structure interaction should be taken into account.

iv. The numerical example conducted shows that the closed-form solution obtained in this paper has good convergence.

The work presented here could further be combined with the research and development of new rain gauges.

Author Contributions: Conceptualization, X.L. and J.-Y.S.; methodology, X.L. and J.-Y.S.; validation, X.-T.H.; writing—original draft preparation, X.L. and X.-C.L.; writing—review and editing, X.L. and X.-T.H.; visualization, X.L. and Z.-X.Y.; funding acquisition, J.-Y.S. All authors have read and agreed to the published version of the manuscript.

Funding: This research was funded by the National Natural Science Foundation of China (Grant No. 11772072).

Institutional Review Board Statement: Not applicable.

Informed Consent Statement: Not applicable.

Data Availability Statement: Not applicable.

Conflicts of Interest: The authors declare no conflict of interest.

Nomenclatures

a	Radius of the circular membrane
h	Thickness of the circular membrane
E	Young's modulus of elasticity
v	Poisson's ratio
H	Height of the liquid poured into the cup
ρ	Density of the poured liquid
g	Acceleration of gravity
r	Radial coordinate of the cylindrical coordinate system
φ	Circumferential coordinate of the cylindrical coordinate system
w	Axial coordinate of the cylindrical coordinate system as well as transverse displacement of a point on the deflected circular membrane
u	Radial displacement of a point on the deflected circular membrane
w_m	Maximum deflection of the deflected circular membrane
$q(r)$	Transverse distributed loads over the circular membrane produced by the gravity of the liquid within radius r
$F(r)$	External force produced by $q(r)$ within radius r
σ_r	Radial stress
σ_t	Circumferential stress
e_r	Radial strain
e_t	Circumferential strain
θ	Slope angle of the deflected membrane
π	Pi (ratio of circumference to diameter)
W	Dimensionless transverse displacement (w/a)
S_r	Dimensionless radial stress (σ_r/E)
S_t	Dimensionless circumferential stress (σ_t/E)
H_0	Dimensionless height of liquid (H/a)
G	Dimensionless quantity ($\rho g a^2/Eh$)
x	Dimensionless radial coordinate (r/a)
c_i	Coefficients of the power series for S_r
d_i	Coefficients of the power series for W

Appendix A

$$c_2 = -\frac{G^2(H_0+d_0)^2}{64c_0^2},$$

$$c_4 = -\frac{G^3(H_0+d_0)^2}{6144c_0^5}(GH_0^2 + 2GH_0d_0 + Gd_0^2 - 8c_0^2),$$

$$c_6 = -\frac{G^4(H_0+d_0)^2}{4718592c_0^8}(GH_0^2 + 2GH_0d_0 + Gd_0^2 - 8c_0^2)(13GH_0^2 + 26GH_0d_0 + 13Gd_0^2 - 40c_0^2),$$

$$c_8 = -\frac{G^5(H_0+d_0)^2}{1509949440c_0^{11}}(GH_0^2 + 2GH_0d_0 + Gd_0^2 - 8c_0^2)(85G^2H_0^4 + 340G^2H_0^3d_0 + 85G^2d_0^4$$
$$+510G^2H_0^2d_0^2 + 340G^2H_0d_0^3 - 510GH_0^2c_0^2 - 1024GH_0c_0^2d_0 - 512Gc_0^2d_0^2 + 448c_0^4),$$

$$c_{10} = -\frac{G^6(H_0+d_0)^2}{724775731200c_0^{14}}(GH_0^2 + 2GH_0d_0 + Gd_0^2 - 8c_0^2)(925G^3H_0^6 + 5550G^3H_0^5d_0$$
$$+13875G^3H_0^4d_0^2 + 18500G^3H_0^3d_0^3 + 13875G^3H_0^2d_0^4 + 5550G^3H_0d_0^5 + 925G^3d_0^6$$
$$-8252G^2H_0^4c_0^2 - 33008G^2H_0^3c_0^2d_0 - 49512G^2H_0^2c_0^2d_0^2 - 33008G^2H_0c_0^2d_0^3$$
$$-8252G^2c_0^2d_0^4 + 17344GH_0^2c_0^4 + 34688GH_0c_0^4d_0 + 17344Gc_0^4d_0^2 - 5376c_0^6),$$

$$c_{12} = -\frac{G^7(H_0+d_0)^2}{974098582732800c_0^{17}}(GH_0{}^2 + 2GH_0d_0 + Gd_0{}^2 - 8c_0{}^2)(30125G^4H_0{}^8$$
$$+241000G^4H_0{}^7d_0 + 843500G^4H_0{}^6d_0{}^2 + 1687000G^4H_0{}^5d_0{}^3 + 2108750G^4H_0{}^4d_0{}^4$$
$$+1687000G^4H_0{}^3d_0{}^5 + 843500G^4H_0{}^2d_0{}^6 + 241000G^4H_0d_0{}^7 + 30125G^4d_0{}^8$$
$$-355344G^3H_0{}^6c_0{}^2 - 2132064G^3H_0{}^5c_0{}^2d_0 - 5330160G^3H_0{}^4c_0{}^2d_0{}^2$$
$$-7106880G^3H_0{}^3c_0{}^2d_0{}^3 - 5330160G^3H_0{}^2c_0{}^2d_0{}^4 - 2132064G^3H_0c_0{}^2d_0{}^5$$
$$-355344G^3c_0{}^2d_0{}^6 + 1217664G^2H_0{}^4c_0{}^4 + 4870656G^2H_0{}^3c_0{}^4d_0$$
$$+7305984G^2H_0{}^2c_0{}^4d_0{}^2 + 4870656G^2H_0c_0{}^4d_0{}^3 + 1217664G^2c_0{}^4d_0{}^4$$
$$-1178624GH_0{}^2c_0{}^6 - 2357248GH_0c_0{}^6d_0 - 1178624Gc_0{}^6d_0{}^2 + 135168c_0{}^8),$$

$$c_{14} = -\frac{G^8(H_0+d_0)^2}{6982338641028710400c_0^{20}}(GH_0{}^2 + 2GH_0d_0 + Gd_0{}^2 - 8c_0{}^2)(5481025G^5H_0{}^{10}$$
$$+54810250G^5H_0{}^9d_0 + 246646125G^5H_0{}^8d_0{}^2 + 657723000G^5H_0{}^7d_0{}^3$$
$$+1151015250G^5H_0{}^6d_0{}^4 + 1381218300G^5H_0{}^5d_0{}^5 + 1151015250G^5H_0{}^4d_0{}^6$$
$$+657723000G^5H_0{}^3d_0{}^7 + 246646125G^5H_0{}^2d_0{}^8 + 54810250G^5H_0d_0{}^9 + 5481025G^5d_0{}^{10}$$
$$-80340520G^4H_0{}^8c_0{}^2 - 642724160G^4H_0{}^7c_0{}^2d_0 - 2249534560G^4H_0{}^6c_0{}^2d_0{}^2$$
$$-4499069120G^4H_0{}^5c_0{}^2d_0{}^3 - 5623836400G^4H_0{}^4c_0{}^2d_0{}^4 - 4499069120G^4H_0{}^3c_0{}^2d_0{}^5$$
$$-2249534560G^4H_0{}^2c_0{}^2d_0{}^6 - 642724160G^4H_0c_0{}^2d_0{}^7 - 80340520G^4c_0{}^2d_0{}^8$$
$$+384915840G^3H_0{}^6c_0{}^4 + 2309495040G^3H_0{}^5c_0{}^4d_0 + 5773737600G^3H_0{}^4c_0{}^4d_0{}^2$$
$$+7698316800G^3H_0{}^3c_0{}^4d_0{}^3 + 5773737600G^3H_0{}^2c_0{}^4d_0{}^4 + 2309495040G^3H_0c_0{}^4d_0{}^5$$
$$+384915840G^3c_0{}^4d_0{}^6 - 674397184G^2H_0{}^4c_0{}^6 - 2697588736G^2H_0{}^3c_0{}^6d_0$$
$$-4046383104G^2H_0{}^2c_0{}^6d_0{}^2 - 2697588736G^2H_0c_0{}^6d_0{}^3 - 674397184G^2c_0{}^6d_0{}^4$$
$$+334532608GH_0{}^2c_0{}^8 + 669065216GH_0c_0{}^8d_0 + 334532608Gc_0{}^8d_0{}^2 - 14057472c_0{}^{10}),$$

$$c_{16} = -\frac{G^9(H_0+d_0)^2}{804365411446507438 0800c_0^{23}}(GH_0{}^2 + 2GH_0d_0 + Gd_0{}^2 - 8c_0{}^2)$$
$$(165851725G^6H_0{}^{12} + 1990220700G^6H_0{}^{11}d_0 + 10946213850G^6H_0{}^{10}d_0{}^2$$
$$+36487379500G^6H_0{}^9d_0{}^3 + 82096603875G^6H_0{}^8d_0{}^4 + 131354566200G^6H_0{}^7d_0{}^5$$
$$+153246993900G^6H_0{}^6d_0{}^6 + 131354566200G^6H_0{}^5d_0{}^7 + 82096603875G^6H_0{}^4d_0{}^8$$
$$+36487379500G^6H_0{}^3d_0{}^9 + 10946213850G^6H_0{}^2d_0{}^{10} + 1990220700G^6H_0d_0{}^{11}$$
$$+165851725G^6d_0{}^{12} - 2904572016G^5H_0{}^{10}c_0{}^2 - 29045720160G^5H_0{}^9c_0{}^2d_0$$
$$-130705740720G^5H_0{}^8c_0{}^2d_0{}^2 - 348548641920G^5H_0{}^7c_0{}^2d_0{}^3$$
$$-609960123360G^5H_0{}^6c_0{}^2d_0{}^4 - 731952148032G^5H_0{}^5c_0{}^2d_0{}^5$$
$$-609960123360G^5H_0{}^4c_0{}^2d_0{}^6 - 348548641920G^5H_0{}^3c_0{}^2d_0{}^7$$
$$-130705740720G^5H_0{}^2c_0{}^2d_0{}^8 - 29045720160G^5H_0c_0{}^2d_0{}^9$$
$$-2904572016G^5c_0{}^2d_0{}^{10} + 17934897216G^4H_0{}^8c_0{}^4$$
$$+143479177728G^4H_0{}^7c_0{}^4d_0 + 502177122048G^4H_0{}^6c_0{}^4d_0{}^2$$
$$+1004354244096G^4H_0{}^5c_0{}^4d_0{}^3 + 1255442805120G^4H_0{}^4c_0{}^4d_0{}^4$$
$$+1004354244096G^4H_0{}^3c_0{}^4d_0{}^5 + 502177122048G^4H_0{}^2c_0{}^4d_0{}^6$$
$$+143479177728G^4H_0c_0{}^4d_0{}^7 + 17934897216G^4c_0{}^4d_0{}^8$$

$$-46544832512G^3H_0^6c_0^6 - 279268995072G^3H_0^5c_0^6d_0$$
$$-698172487680G^3H_0^4c_0^6d_0^2 - 930896650240G^3H_0^3c_0^6d_0^3$$
$$-698172487680G^3H_0^2c_0^6d_0^4 - 279268995072G^3H_0c_0^6d_0^5$$
$$-46544832512G^3c_0^6d_0^6 + 46777135104G^2H_0^4c_0^8$$
$$+187108540416G^2H_0^3c_0^8d_0 + 280662810624G^2H_0^2c_0^8d_0^2$$
$$+187108540416G^2H_0c_0^8d_0^3 + 46777135104G^2c_0^8d_0^4$$
$$-12516261888GH_0^2c_0^{10} - 25032523776GH_0c_0^{10}d_0$$
$$-12516261888Gc_0^{10}d_0^2 + 187432960c_0^{12})$$

$$c_{18} = -\frac{G^{10}(H_0+d_0)^2}{115828619248297071083520 00c_0^{26}}(GH_0^2 + 2GH_0d_0 + Gd_0^2 - 8c_0^2)$$

$$(6440470375G^7H_0^{14} + 90166585250G^7H_0^{13}d_0 + 586082804125G^7H_0^{12}d_0^2$$
$$+2344331216500G^7H_0^{11}d_0^3 + 6446910845375G^7H_0^{10}d_0^4$$
$$+12893821690750G^7H_0^9d_0^5 + 19340732536125G^7H_0^8d_0^6$$
$$+22103694327000G^7H_0^7d_0^7 + 19340732536125G^7H_0^6d_0^8$$
$$+12893821690750G^7H_0^5d_0^9 + 6446910845375G^7H_0^4d_0^{10}$$
$$+2344331216500G^7H_0^3d_0^{11} + 586082804125G^7H_0^2d_0^{12}$$
$$+90166585250G^7H_0d_0^{13} + 6440470375G^7d_0^{14}$$
$$-131152225780G^6H_0^{12}c_0^2 + 992711744640G^5c_0^4d_0^{10}$$
$$-1573826709360G^6H_0^{11}c_0^2d_0 - 8656046901480G^6H_0^{10}c_0^2d_0^2$$
$$-28853489671600G^6H_0^9c_0^2d_0^3 - 64920351761100G^6H_0^8c_0^2d_0^4$$
$$-103872562817760G^6H_0^7c_0^2d_0^5 - 121184566620720G^6H_0^6c_0^2d_0^6$$
$$-103872562817760G^6H_0^5c_0^2d_0^7 - 64920351761100G^6H_0^4c_0^2d_0^8$$
$$-28853489671600G^6H_0^3c_0^2d_0^9 - 8656046901480G^6H_0^2c_0^2d_0^{10}$$
$$-1573826709360G^6H_0c_0^2d_0^{11} - 131152225780G^6c_0^2d_0^{12}$$
$$+992711744640G^5H_0^{10}c_0^4 + 9927117446400G^5H_0^9c_0^4d_0$$
$$+44672028508800G^5H_0^8c_0^4d_0^2 + 119125409356800G^5H_0^7c_0^4d_0^3$$
$$+208469466374400G^5H_0^6c_0^4d_0^4 + 250163359649280G^5H_0^5c_0^4d_0^5$$
$$+208469466374400G^5H_0^4c_0^4d_0^6 + 119125409356800G^5H_0^3c_0^4d_0^7$$
$$+44672028508800G^5H_0^2c_0^4d_0^8 + 9927117446400G^5H_0c_0^4d_0^9$$
$$-3445151675648G^4H_0^8c_0^6 - 27561213405184G^4H_0^7c_0^6d_0$$
$$-96464246918144G^4H_0^6c_0^6d_0^2 + 5444708495360G^3H_0^6c_0^8$$
$$-192928493836288G^4H_0^5c_0^6d_0^3 - 241160617295360G^4H_0^4c_0^6d_0^4$$
$$-192928493836288G^4H_0^3c_0^6d_0^5 - 96464246918144G^4H_0^2c_0^6d_0^6$$
$$-27561213405184G^4H_0c_0^6d_0^7 - 3445151675648G^4c_0^6d_0^8$$
$$+32668250972160G^3H_0^5c_0^8d_0 + 81670627430400G^3H_0^4c_0^8d_0^2$$
$$+108894169907200G^3H_0^3c_0^8d_0^3 + 81670627430400G^3H_0^2c_0^8d_0^4$$
$$+32668250972160G^3H_0c_0^8d_0^5 + 5444708495360G^3c_0^8d_0^6$$
$$-3333820760064G^2H_0^4c_0^{10} + 493015269376GH_0^2c_0^{12}$$
$$-13335283040256G^2H_0^3c_0^{10}d_0 - 20002924560384G^2H_0^2c_0^{10}d_0^2$$
$$-13335283040256G^2H_0c_0^{10}d_0^3 - 3333820760064G^2c_0^{10}d_0^4$$
$$+986030538752GH_0c_0^{12}d_0 + 493015269376Gc_0^{12}d_0^2 - 2549088256c_0^{14})$$

$$c_{20} = -\frac{G^{11}(H_0+d_0)^2}{4077167397540056902139904 0000c_0^{29}}(GH_0^2 + 2GH_0d_0 + Gd_0^2 - 8c_0^2)$$

$$(624247690625G^8H_0^{16} + 9987963050000G^8H_0^{15}d_0 + 74909722875000G^8H_0^{14}d_0^2$$
$$+349578706750000G^8H_0^{13}d_0^3 + 1136130796937500G^8H_0^{12}d_0^4$$
$$+2726713912650000G^8H_0^{11}d_0^5 + 8034067778343750G^8H_0^8d_0^8$$
$$+4998975506525000G^8H_0^{10}d_0^6 + 7141393580750000G^8H_0^9d_0^7$$
$$+7141393580750000G^8H_0^7d_0^9 + 4998975506525000G^8H_0^6d_0^{10}$$
$$+2726713912650000G^8H_0^5d_0^{11} + 74909722875000G^8H_0^2d_0^{14}$$
$$+1136130796937500G^8H_0^4d_0^{12} + 349578706750000G^8H_0^3d_0^{13}$$
$$+9987963050000G^8H_0d_0^{15} + 624247690625G^8d_0^{16} - 14489792053000G^7H_0^{14}c_0^2$$

$$
\begin{aligned}
&-202857088742000 G^7 H_0{}^{13} c_0{}^2 d_0 - 1318571076823000 G^7 H_0{}^{12} c_0{}^2 d_0{}^2 \\
&-5274284307292000 G^7 H_0{}^{11} c_0{}^2 d_0{}^3 - 14504281845053000 G^7 H_0{}^{10} c_0{}^2 d_0{}^4 \\
&-29008563690106000 G^7 H_0{}^9 c_0{}^2 d_0{}^5 - 43512845535159000 G^7 H_0{}^8 c_0{}^2 d_0{}^6 \\
&-49728966325896000 G^7 H_0{}^7 c_0{}^2 d_0{}^7 - 43512845535159000 G^7 H_0{}^6 c_0{}^2 d_0{}^8 \\
&-29008563690106000 G^7 H_0{}^5 c_0{}^2 d_0{}^9 - 14504281845053000 G^7 H_0{}^4 c_0{}^2 d_0{}^{10} \\
&-5274284307292000 G^7 H_0{}^3 c_0{}^2 d_0{}^{11} - 1318571076823000 G^7 H_0{}^2 c_0{}^2 d_0{}^{12} \\
&-202857088742000 G^7 H_0 c_0{}^2 d_0{}^{13} - 14489792053000 G^7 c_0{}^2 d_0{}^{14} \\
&+129976428743936 G^6 H_0{}^{12} c_0{}^4 + 1559717144927232 G^6 H_0{}^{11} c_0{}^4 d_0 \\
&+8578444297099776 G^6 H_0{}^{10} c_0{}^4 d_0{}^2 + 28594814323665920 G^6 H_0{}^9 c_0{}^4 d_0{}^3 \\
&+64338332228248320 G^6 H_0{}^8 c_0{}^4 d_0{}^4 + 102941331565197312 G^6 H_0{}^7 c_0{}^4 d_0{}^5 \\
&+120098220159396864 G^6 H_0{}^6 c_0{}^4 d_0{}^6 + 102941331565197312 G^6 H_0{}^5 c_0{}^4 d_0{}^7 \\
&+64338332228248320 G^6 H_0{}^4 c_0{}^4 d_0{}^8 + 28594814323665920 G^6 H_0{}^3 c_0{}^4 d_0{}^9 \\
&+8578444297099776 G^6 H_0{}^2 c_0{}^4 d_0{}^{10} + 1559717144927232 G^6 H_0 c_0{}^4 d_0{}^{11} \\
&+129976428743936 G^6 c_0{}^4 d_0{}^{12} - 567342708341248 G^5 H_0{}^{10} c_0{}^6 \\
&-5673427083412480 G^5 H_0{}^9 c_0{}^6 d_0 - 25530421875356160 G^5 H_0{}^8 c_0{}^6 d_0{}^2 \\
&-68081125000949760 G^5 H_0{}^7 c_0{}^6 d_0{}^3 - 119141968751662080 G^5 H_0{}^6 c_0{}^6 d_0{}^4 \\
&-142970362501994496 G^5 H_0{}^5 c_0{}^6 d_0{}^5 - 119141968751662080 G^5 H_0{}^4 c_0{}^6 d_0{}^6 \\
&-68081125000949760 G^5 H_0{}^3 c_0{}^6 d_0{}^7 - 25530421875356160 G^5 H_0{}^2 c_0{}^6 d_0{}^8 \\
&-5673427083412480 G^5 H_0 c_0{}^6 d_0{}^9 - 567342708341248 G^5 c_0{}^6 d_0{}^{10} \\
&+1244939899486208 G^4 H_0{}^8 c_0{}^8 + 9959519195889664 G^4 H_0{}^7 c_0{}^8 d_0 \\
&+34858317185613824 G^4 H_0{}^6 c_0{}^8 d_0{}^2 + 69716634371227648 G^4 H_0{}^5 c_0{}^8 d_0{}^3 \\
&+87145792964034560 G^4 H_0{}^4 c_0{}^8 d_0{}^4 + 69716634371227648 G^4 H_0{}^3 c_0{}^8 d_0{}^5 \\
&+34858317185613824 G^4 H_0{}^2 c_0{}^8 d_0{}^6 + 9959519195889664 G^4 H_0 c_0{}^8 d_0{}^7 \\
&+1244939899486208 G^4 c_0{}^8 d_0{}^8 - 1276001808646144 G^3 H_0{}^6 c_0{}^{10} \\
&-7656010851876864 G^3 H_0{}^5 c_0{}^{10} d_0 - 19140027129692160 G^3 H_0{}^4 c_0{}^{10} d_0{}^2 \\
&-25520036172922880 G^3 H_0{}^3 c_0{}^{10} d_0{}^3 - 19140027129692160 G^3 H_0{}^2 c_0{}^{10} d_0{}^4 \\
&-7656010851876864 G^3 H_0 c_0{}^{10} d_0{}^5 - 1276001808646144 G^3 c_0{}^{10} d_0{}^6 \\
&+493067700273152 G^2 H_0{}^4 c_0{}^{12} + 1972270801092608 G^2 H_0{}^3 c_0{}^{12} d_0 \\
&+2958406201638912 G^2 H_0{}^2 c_0{}^{12} d_0{}^2 + 1972270801092608 G^2 H_0 c_0{}^{12} d_0{}^3 \\
&+493067700273152 G^2 c_0{}^{12} d_0{}^4 - 40655185248256 G H_0{}^2 c_0{}^{14} \\
&-81310370496512 G H_0 c_0{}^{14} d_0 - 40655185248256 G c_0{}^{14} d_0{}^2 + 70447529984 c_0{}^{16})
\end{aligned}
$$

$$
c_{22} = -\frac{G^{12}(H_0+d_0)^2}{1722195508720920035463895449600000 c_0{}^{32}}(GH_0{}^2 + 2GH_0 d_0 + G d_0{}^2 - 8c_0{}^2)
$$

$$
\begin{aligned}
&(73847994008125 G^9 H_0{}^{18} + 1329263892146250 G^9 H_0{}^{17} d_0 \\
&+11298743083243125 G^9 H_0{}^{16} d_0{}^2 - 1924297443407700 G^8 H_0{}^{16} c_0{}^2 \\
&+60259963110630000 G^9 H_0{}^{15} d_0{}^3 + 225974861664862500 G^9 H_0{}^{14} d_0{}^4 \\
&+632729612661615000 G^9 H_0{}^{13} d_0{}^5 + 1370914160768832500 G^9 H_0{}^{12} d_0{}^6 \\
&+2350138561314570000 G^9 H_0{}^{11} d_0{}^7 + 3231440521807533750 G^9 H_0{}^{10} d_0{}^8 \\
&+3590489468675037500 G^9 H_0{}^9 d_0{}^9 + 3231440521807533750 G^9 H_0{}^8 d_0{}^{10} \\
&+2350138561314570000 G^9 H_0{}^7 d_0{}^{11} + 1370914160768832500 G^9 H_0{}^6 d_0{}^{12} \\
&+632729612661615000 G^9 H_0{}^5 d_0{}^{13} + 225974861664862500 G^9 H_0{}^4 d_0{}^{14} \\
&+60259963110630000 G^9 H_0{}^3 d_0{}^{15} + 11298743083243125 G^9 H_0{}^2 d_0{}^{16} \\
&+1329263892146250 G^9 H_0 d_0{}^{17} + 73847994008125 G^9 d_0{}^{18} \\
&-30788759094523200 G^8 H_0{}^{15} c_0{}^2 d_0 - 230915693208924000 G^8 H_0{}^{14} c_0{}^2 d_0{}^2 \\
&-1077606568308312000 G^8 H_0{}^{13} c_0{}^2 d_0{}^3 - 3502221347002014000 G^8 H_0{}^{12} c_0{}^2 d_0{}^4 \\
&-8405331232804833600 G^8 H_0{}^{11} c_0{}^2 d_0{}^5 - 15409773926808861600 G^8 H_0{}^{10} c_0{}^2 d_0{}^6 \\
&-22013962752584088000 G^8 H_0{}^9 c_0{}^2 d_0{}^7 - 24765708096657099000 G^8 H_0{}^8 c_0{}^2 d_0{}^8 \\
&-22013962752584088000 G^8 H_0{}^7 c_0{}^2 d_0{}^9 - 15409773926808861600 G^8 H_0{}^6 c_0{}^2 d_0{}^{10} \\
&-8405331232804833600 G^8 H_0{}^5 c_0{}^2 d_0{}^{11} - 3502221347002014000 G^8 H_0{}^4 c_0{}^2 d_0{}^{12} \\
&-1077606568308312000 G^8 H_0{}^3 c_0{}^2 d_0{}^{13} - 230915693208924000 G^8 H_0{}^2 c_0{}^2 d_0{}^{14} \\
&-30788759094523200 G^8 H_0 c_0{}^2 d_0{}^{15} - 1924297443407700 G^8 c_0{}^2 d_0{}^{16}
\end{aligned}
$$

$$\begin{aligned}
&+19965729788090944 G^7 H_0{}^{14} c_0{}^4 + 279520217033273216 G^7 H_0{}^{13} c_0{}^4 d_0\\
&+1816881410716275904 G^7 H_0{}^{12} c_0{}^4 d_0{}^2 + 7267525642865103616 G^7 H_0{}^{11} c_0{}^4 d_0{}^3\\
&+19985695517879034944 G^7 H_0{}^{10} c_0{}^4 d_0{}^4 + 39971391035758069888 G^7 H_0{}^9 c_0{}^4 d_0{}^5\\
&+59957086553637104832 G^7 H_0{}^8 c_0{}^4 d_0{}^6 + 68522384632728119808 G^7 H_0{}^7 c_0{}^4 d_0{}^7\\
&+59957086553637104832 G^7 H_0{}^6 c_0{}^4 d_0{}^8 + 39971391035758069888 G^7 H_0{}^5 c_0{}^4 d_0{}^9\\
&+19985695517879034944 G^7 H_0{}^4 c_0{}^4 d_0{}^{10} + 7267525642865103616 G^7 H_0{}^3 c_0{}^4 d_0{}^{11}\\
&+1816881410716275904 G^7 H_0{}^2 c_0{}^4 d_0{}^{12} + 279520217033273216 G^7 H_0 c_0{}^4 d_0{}^{13}\\
&+19965729788090944 G^7 c_0{}^4 d_0{}^{14} - 105258286088052736 G^6 H_0{}^{12} c_0{}^6\\
&-1263099433056632832 G^6 H_0{}^{11} c_0{}^6 d_0 - 6947046881811480576 G^6 H_0{}^{10} c_0{}^6 d_0{}^2\\
&-23156822939371601920 G^6 H_0{}^9 c_0{}^6 d_0{}^3 - 52102851613586104320 G^6 H_0{}^8 c_0{}^6 d_0{}^4\\
&-83364562581737766912 G^6 H_0{}^7 c_0{}^6 d_0{}^5 - 97258656345360728064 G^6 H_0{}^6 c_0{}^6 d_0{}^6\\
&-83364562581737766912 G^6 H_0{}^5 c_0{}^6 d_0{}^7 - 52102851613586104320 G^6 H_0{}^4 c_0{}^6 d_0{}^8\\
&-23156822939371601920 G^6 H_0{}^3 c_0{}^6 d_0{}^9 - 6947046881811480576 G^6 H_0{}^2 c_0{}^6 d_0{}^{10}\\
&-1263099433056632832 G^6 H_0 c_0{}^6 d_0{}^{11} - 105258286088052736 G^6 c_0{}^6 d_0{}^{12}\\
&+298141919987527680 G^5 H_0{}^{10} c_0{}^8 + 2981419199875276800 G^5 H_0{}^9 c_0{}^8 d_0\\
&+13416386399438745600 G^5 H_0{}^8 c_0{}^8 d_0{}^2 + 35777030398503321600 G^5 H_0{}^7 c_0{}^8 d_0{}^3\\
&+62609803197380812800 G^5 H_0{}^6 c_0{}^8 d_0{}^4 + 75131763836856975360 G^5 H_0{}^5 c_0{}^8 d_0{}^5\\
&+62609803197380812800 G^5 H_0{}^4 c_0{}^8 d_0{}^6 + 35777030398503321600 G^5 H_0{}^3 c_0{}^8 d_0{}^7\\
&+13416386399438745600 G^5 H_0{}^2 c_0{}^8 d_0{}^8 + 2981419199875276800 G^5 H_0 c_0{}^8 d_0{}^9\\
&+298141919987527680 G^5 c_0{}^8 d_0{}^{10} - 440975944806858752 G^4 H_0{}^8 c_0{}^{10}\\
&-3527807558454870016 G^4 H_0{}^7 c_0{}^{10} d_0 - 12347326454592045056 G^4 H_0{}^6 c_0{}^{10} d_0{}^2\\
&-24694652909184090112 G^4 H_0{}^5 c_0{}^{10} d_0{}^3 - 30868316136480112640 G^4 H_0{}^4 c_0{}^{10} d_0{}^4\\
&-24694652909184090112 G^4 H_0{}^3 c_0{}^{10} d_0{}^5 - 12347326454592045056 G^4 H_0{}^2 c_0{}^{10} d_0{}^6\\
&-3527807558454870016 G^4 H_0 c_0{}^{10} d_0{}^7 - 440975944806858752 G^4 c_0{}^{10} d_0{}^8\\
&+304616906657457152 G^3 H_0{}^6 c_0{}^{12} + 1827701763944742912 G^3 H_0{}^5 c_0{}^{12} d_0\\
&+4569254409861857280 G^3 H_0{}^4 c_0{}^{12} d_0{}^2 + 6092339213149143040 G^3 H_0{}^3 c_0{}^{12} d_0{}^3\\
&+4569254409861857280 G^3 H_0{}^2 c_0{}^{12} d_0{}^4 + 1827701763944742912 G^3 H_0 c_0{}^{12} d_0{}^5\\
&+304616906657457152 G^3 c_0{}^{12} d_0{}^6 - 75883562624090112 G^2 H_0{}^4 c_0{}^{14}\\
&-303534250496360448 G^2 H_0{}^3 c_0{}^{14} d_0 - 455301375744540672 G^2 H_0{}^2 c_0{}^{14} d_0{}^2\\
&-303534250496360448 G^2 H_0 c_0{}^{14} d_0{}^3 - 75883562624090112 G^2 c_0{}^{14} d_0{}^4\\
&+3486326154330112 G H_0{}^2 c_0{}^{16} + 6972652308660224 G H_0 c_0{}^{16} d_0\\
&+3486326154330112 G c_0{}^{16} d_0{}^2 - 1972530839552 c_0{}^{18})
\end{aligned}$$

$$c_{24} = -\frac{G^{13}(H_0+d_0)^2}{85971999795348328170357660844032 0000 c_0{}^{35}}(GH_0^2 + 2GH_0 d_0 + G d_0^2 - 8 c_0^2)$$

$$\begin{aligned}
&(10470550396840625 G^{10} H_0{}^{20} + 209411007936812500 G^{10} H_0{}^{19} d_0\\
&+1989404575399718750 G^{10} H_0{}^{18} d_0{}^2 + 11936427452398312500 G^{10} H_0{}^{17} d_0{}^3\\
&+50729816672692828125 G^{10} H_0{}^{16} d_0{}^4 + 162335413352617050000 G^{10} H_0{}^{15} d_0{}^5\\
&+405838533381542625000 G^{10} H_0{}^{14} d_0{}^6 + 811677066763085250000 G^{10} H_0{}^{13} d_0{}^7\\
&+1318975233490013531250 G^{10} H_0{}^{12} d_0{}^8 + 1758633644653351375000 G^{10} H_0{}^{11} d_0{}^9\\
&+1934497009118686512500 G^{10} H_0{}^{10} d_0{}^{10} + 1758633644653351375000 G^{10} H_0{}^9 d_0{}^{11}\\
&+1318975233490013531250 G^{10} H_0{}^8 d_0{}^{12} + 811677066763085250000 G^{10} H_0{}^7 d_0{}^{13}\\
&+405838533381542625000 G^{10} H_0{}^6 d_0{}^{14} + 162335413352617050000 G^{10} H_0{}^5 d_0{}^{15}\\
&+50729816672692828125 G^{10} H_0{}^4 d_0{}^{16} + 11936427452398312500 G^{10} H_0{}^3 d_0{}^{17}\\
&+1989404575399718750 G^{10} H_0{}^2 d_0{}^{18} + 209411007936812500 G^{10} H_0 d_0{}^{19}\\
&+10470550396840625 G^{10} d_0{}^{20} - 302621805505979000 G^9 H_0{}^{18} c_0{}^2\\
&-5447192499107622000 G^9 H_0{}^{17} c_0{}^2 d_0 - 46301136242414787000 G^9 H_0{}^{16} c_0{}^2 d_0{}^2\\
&-246939393292878864000 G^9 H_0{}^{15} c_0{}^2 d_0{}^3 - 926022724848295740000 G^9 H_0{}^{14} c_0{}^2 d_0{}^4\\
&-2592863629575228072000 G^9 H_0{}^{13} c_0{}^2 d_0{}^5 - 5617871197412994156000 G^9 H_0{}^{12} c_0{}^2 d_0{}^6\\
&-9630636338422275696000 G^9 H_0{}^{11} c_0{}^2 d_0{}^7 - 13242124965330629082000 G^9 H_0{}^{10} c_0{}^2 d_0{}^8\\
&-14713472183700698980000 G^9 H_0{}^9 c_0{}^2 d_0{}^9 - 13242124965330629082000 G^9 H_0{}^8 c_0{}^2 d_0{}^{10}\\
&-9630636338422275696000 G^9 H_0{}^7 c_0{}^2 d_0{}^{11} - 5617871197412994156000 G^9 H_0{}^6 c_0{}^2 d_0{}^{12}
\end{aligned}$$

$$\begin{aligned}
&-2592863629575228072000 G^9 H_0{}^5 c_0{}^2 d_0{}^{13} - 9260227248482957400000 G^9 H_0{}^4 c_0{}^2 d_0{}^{14}\\
&-246939392928788640000 G^9 H_0{}^3 c_0{}^2 d_0{}^{15} - 46301136242414787000 G^9 H_0{}^2 c_0{}^2 d_0{}^{16}\\
&-5447192499107622000 G^9 H_0 c_0{}^2 d_0{}^{17} - 302621805505979000 G^9 c_0{}^2 d_0{}^{18}\\
&+3566092425128384960 G^8 H_0{}^{16} c_0{}^4 + 57057478802054159360 G^8 H_0{}^{15} c_0{}^4 d_0\\
&+427931091015406195200 G^8 H_0{}^{14} c_0{}^4 d_0{}^2 + 1997011758071895577600 G^8 H_0{}^{13} c_0{}^4 d_0{}^3\\
&+6490288213733660627200 G^8 H_0{}^{12} c_0{}^4 d_0{}^4 + 15576691712960785505280 G^8 H_0{}^{11} c_0{}^4 d_0{}^5\\
&+28557268140428106759680 G^8 H_0{}^{10} c_0{}^4 d_0{}^6 + 40796097343468723942400 G^8 H_0{}^9 c_0{}^4 d_0{}^7\\
&+45895609511402314435200 G^8 H_0{}^8 c_0{}^4 d_0{}^8 + 40796097343468723942400 G^8 H_0{}^7 c_0{}^4 d_0{}^9\\
&+28557268140428106759680 G^8 H_0{}^6 c_0{}^4 d_0{}^{10} + 15576691712960785505280 G^8 H_0{}^5 c_0{}^4 d_0{}^{11}\\
&+6490288213733660627200 G^8 H_0{}^4 c_0{}^4 d_0{}^{12} + 1997011758071895577600 G^8 H_0{}^3 c_0{}^4 d_0{}^{13}\\
&+427931091015406195200 G^8 H_0{}^2 c_0{}^4 d_0{}^{14} + 57057478802054159360 G^8 H_0 c_0{}^4 d_0{}^{15}\\
&+3566092425128384960 G^8 c_0{}^4 d_0{}^{16} - 22064961114736215040 G^7 H_0{}^{14} c_0{}^6\\
&-308909455606307010560 G^7 H_0{}^{13} c_0{}^6 d_0 - 2007911461440995568640 G^7 H_0{}^{12} c_0{}^6 d_0{}^2\\
&-8031645845763982274560 G^7 H_0{}^{11} c_0{}^6 d_0{}^3 - 22087026075850951255040 G^7 H_0{}^{10} c_0{}^6 d_0{}^4\\
&-44174052151701902510080 G^7 H_0{}^9 c_0{}^6 d_0{}^5 - 66261078227552853765120 G^7 H_0{}^8 c_0{}^6 d_0{}^6\\
&-75726946545774690017280 G^7 H_0{}^7 c_0{}^6 d_0{}^7 - 66261078227552853765120 G^7 H_0{}^6 c_0{}^6 d_0{}^8\\
&-44174052151701902510080 G^7 H_0{}^5 c_0{}^6 d_0{}^9 - 22087026075850951255040 G^7 H_0{}^4 c_0{}^6 d_0{}^{10}\\
&-8031645845763982274560 G^7 H_0{}^3 c_0{}^6 d_0{}^{11} - 2007911461440995568640 G^7 H_0{}^2 c_0{}^6 d_0{}^{12}\\
&-308909455606307010560 G^7 H_0 c_0{}^6 d_0{}^{13} - 22064961114736215040 G^7 c_0{}^6 d_0{}^{14}\\
&+76935695715733790720 G^6 H_0{}^{12} c_0{}^8 + 923228348588805488640 G^6 H_0{}^{11} c_0{}^8 d_0\\
&+5077559172384301875200 G^6 H_0{}^{10} c_0{}^8 d_0{}^2 + 16925853057461433958400 G^6 H_0{}^9 c_0{}^8 d_0{}^3\\
&+38083169379288226406400 G^6 H_0{}^8 c_0{}^8 d_0{}^4 + 60933071006861162250240 G^6 H_0{}^7 c_0{}^8 d_0{}^5\\
&+71088582841338022625280 G^6 H_0{}^6 c_0{}^8 d_0{}^6 + 60933071006861162250240 G^6 H_0{}^5 c_0{}^8 d_0{}^7\\
&+38083169379288226406400 G^6 H_0{}^4 c_0{}^8 d_0{}^8 + 16925853057461433958400 G^6 H_0{}^3 c_0{}^8 d_0{}^9\\
&+5077559172384301875200 G^6 H_0{}^2 c_0{}^8 d_0{}^{10} + 923228348588805488640 G^6 H_0 c_0{}^8 d_0{}^{11}\\
&+76935695715733790720 G^6 c_0{}^8 d_0{}^{12} - 150811020027895349248 G^5 H_0{}^{10} c_0{}^{10}\\
&-1508110200278953492480 G^5 H_0{}^9 c_0{}^{10} d_0 - 6786495901255290716160 G^5 H_0{}^8 c_0{}^{10} d_0{}^2\\
&-18097322403347441909760 G^5 H_0{}^7 c_0{}^{10} d_0{}^3 - 31670314205858023342080 G^5 H_0{}^6 c_0{}^{10} d_0{}^4\\
&-38004377047029628010496 G^5 H_0{}^5 c_0{}^{10} d_0{}^5 - 31670314205858023342080 G^5 H_0{}^4 c_0{}^{10} d_0{}^6\\
&-18097322403347441909760 G^5 H_0{}^3 c_0{}^{10} d_0{}^7 - 6786495901255290716160 G^5 H_0{}^2 c_0{}^{10} d_0{}^8\\
&-1508110200278953492480 G^5 H_0 c_0{}^{10} d_0{}^9 - 150811020027895349248 G^5 c_0{}^{10} d_0{}^{10}\\
&+156446903803372371968 G^4 H_0{}^8 c_0{}^{12} + 1251575230426978975744 G^4 H_0{}^7 c_0{}^{12} d_0\\
&+4380513306494426415104 G^4 H_0{}^6 c_0{}^{12} d_0{}^2 + 8761026612988852830208 G^4 H_0{}^5 c_0{}^{12} d_0{}^3\\
&+10951283266236066037760 G^4 H_0{}^4 c_0{}^{12} d_0{}^4 + 8761026612988852830208 G^4 H_0{}^3 c_0{}^{12} d_0{}^5\\
&+4380513306494426415104 G^4 H_0{}^2 c_0{}^{12} d_0{}^6 + 1251575230426978975744 G^4 H_0 c_0{}^{12} d_0{}^7\\
&+156446903803372371968 G^4 c_0{}^{12} d_0{}^8 - 74687799413960605696 G^3 H_0{}^6 c_0{}^{14}\\
&-448126796483763634176 G^3 H_0{}^5 c_0{}^{14} d_0 - 1120316991209409085440 G^3 H_0{}^4 c_0{}^{14} d_0{}^2\\
&-1493755988279212113920 G^3 H_0{}^3 c_0{}^{14} d_0{}^3 - 1120316991209409085440 G^3 H_0{}^2 c_0{}^{14} d_0{}^4\\
&-448126796483763634176 G^3 H_0 c_0{}^{14} d_0{}^5 - 74687799413960605696 G^3 c_0{}^{14} d_0{}^6\\
&+12147792181792342016 G^2 H_0{}^4 c_0{}^{16} + 48591168727169368064 G^2 H_0{}^3 c_0{}^{16} d_0\\
&+72886753090754052096 G^2 H_0{}^2 c_0{}^{16} d_0{}^2 + 48591168727169368064 G^2 H_0 c_0{}^{16} d_0{}^3\\
&+12147792181792342016 G^2 c_0{}^{16} d_0{}^4 - 309052965610586112 G H_0{}^2 c_0{}^{18}\\
&-618105931221172224 G H_0 c_0{}^{18} d_0 - 309052965610586112 G c_0{}^{18} d_0{}^2\\
&+55837796073472 c_0{}^{20})
\end{aligned}$$

Appendix B

$$d_2 = -\frac{G(H_0+d_0)}{4c_0},$$

$$d_4 = -\frac{G^2(H_0+d_0)}{512c_0^4}(GH_0^2 + 2GH_0d_0 + Gd_0^2 - 8c_0^2),$$

$$d_6 = -\frac{G^3(H_0+d_0)}{147456c_0^7}(GH_0^2 + 2GH_0d_0 + Gd_0^2 - 8c_0^2)(5GH_0^2 + 10GH_0d_0 + 5Gd_0^2 - 8c_0^2),$$

$$d_8 = -\frac{G^4(H_0+d_0)}{75497472c_0^{10}}(GH_0^2 + 2GH_0d_0 + Gd_0^2 - 8c_0^2)(55G^2H_0^4 + 220G^2H_0^3d_0 + 55G^2d_0^4$$
$$+330G^2H_0^2d_0^2 + 220G^2H_0d_0^3 - 224GH_0^2c_0^2 - 448GH_0c_0^2d_0 - 224Gc_0^2d_0^2 + 64c_0^4)$$

$$d_{10} = -\frac{G^5(H_0+d_0)}{30198988800c_0^{13}}(GH_0^2 + 2GH_0d_0 + Gd_0^2 - 8c_0^2)(525G^3H_0^6 + 3150G^3H_0^5d_0$$
$$+7875G^3H_0^4d_0^2 + 10500G^3H_0^3d_0^3 + 7875G^3H_0^2d_0^4 + 3150G^3H_0d_0^5 - 14128G^2H_0c_0^2d_0^3$$
$$+525G^3d_0^6 - 3532G^2H_0^4c_0^2 - 14128G^2H_0^3c_0^2d_0 - 21192G^2H_0^2c_0^2d_0^2 - 3532G^2c_0^2d_0^4$$
$$+4544GH_0^2c_0^4 + 9088GH_0c_0^4d_0 + 4544Gc_0^4d_0^2 - 256c_0^6)$$

$$d_{12} = -\frac{G^6(H_0+d_0)}{34789235097600c_0^{16}}(GH_0^2 + 2GH_0d_0 + Gd_0^2 - 8c_0^2)(15375G^4H_0^8$$
$$+123000G^4H_0^7d_0 + 430500G^4H_0^6d_0^2 + 861000G^4H_0^5d_0^3 + 1076250G^4H_0^4d_0^4$$
$$+861000G^4H_0^3d_0^5 + 430500G^4H_0^2d_0^6 + 123000G^4H_0d_0^7 + 15375G^4d_0^8$$
$$-145364G^3H_0^6c_0^2 - 872184G^3H_0^5c_0^2d_0 - 872184G^3H_0c_0^2d_0^5$$
$$-2180460G^3H_0^4c_0^2d_0^2 - 2907280G^3H_0^3c_0^2d_0^3 - 2180460G^3H_0^2c_0^2d_0^4$$
$$-145364G^3c_0^2d_0^6 + 360864G^2H_0^4c_0^4 + 1443456G^2H_0^3c_0^4d_0$$
$$+2165184G^2H_0^2c_0^4d_0^2 + 1443456G^2H_0c_0^4d_0^3 + 360864G^2c_0^4d_0^4$$
$$-194304GH_0^2c_0^6 - 388608GH_0c_0^6d_0 - 194304Gc_0^6d_0^2 + 2048c_0^8)$$

$$d_{14} = -\frac{G^7(H_0+d_0)}{109099041266073600c_0^{19}}(GH_0^2 + 2GH_0d_0 + Gd_0^2 - 8c_0^2)(1278825G^5H_0^{10}$$
$$+12788250G^5H_0^9d_0 + 57547125G^5H_0^8d_0^2 + 153459000G^5H_0^7d_0^3$$
$$+268553250G^5H_0^6d_0^4 + 322263900G^5H_0^5d_0^5 + 268553250G^5H_0^4d_0^6$$
$$+153459000G^5H_0^3d_0^7 + 57547125G^5H_0^2d_0^8 + 12788250G^5H_0d_0^9$$
$$+1278825G^5d_0^{10} - 15624872G^4H_0^8c_0^2 - 124998976G^4H_0^7c_0^2d_0$$
$$-437496416G^4H_0^6c_0^2d_0^2 - 874992832G^4H_0^5c_0^2d_0^3$$
$$-1093741040G^4H_0^4c_0^2d_0^4 - 874992832G^4H_0^3c_0^2d_0^5$$
$$-437496416G^4H_0^2c_0^2d_0^6 - 124998976G^4H_0c_0^2d_0^7$$
$$-15624872G^4c_0^2d_0^8 + 58775168G^3H_0^6c_0^4 + 352651008G^3H_0^5c_0^4d_0$$
$$+881627520G^3H_0^4c_0^4d_0^2 + 1175503360G^3H_0^3c_0^4d_0^3$$
$$+881627520G^3H_0^2c_0^4d_0^4 + 352651008G^3H_0c_0^4d_0^5$$
$$+58775168G^3c_0^4d_0^6 - 71685120G^2H_0^4c_0^6$$
$$-286740480G^2H_0^3c_0^6d_0 - 430110720G^2H_0^2c_0^6d_0^2$$
$$-286740480G^2H_0c_0^6d_0^3 - 71685120G^2c_0^6d_0^4 + 17813504GH_0^2c_0^8$$
$$+35627008GH_0c_0^8d_0 + 17813504Gc_0^8d_0^2 - 32768c_0^{10})$$

$$d_{16} = -\frac{G^8(H_0+d_0)}{22343483651291873280 0c_0^{22}}(GH_0^2 + 2GH_0d_0 + Gd_0^2 - 8c_0^2)$$
$$(71612125G^6H_0^{12} + 859345500G^6H_0^{11}d_0 + 4726400250G^6H_0^{10}d_0^2$$
$$+15754667500G^6H_0^9d_0^3 + 35448001875G^6H_0^8d_0^4 + 56716803000G^6H_0^7d_0^5$$
$$+66169603500G^6H_0^6d_0^6 + 56716803000G^6H_0^5d_0^7 + 35448001875G^6H_0^4d_0^8$$
$$+15754667500G^6H_0^3d_0^9 + 4726400250G^6H_0^2d_0^{10} + 859345500G^6H_0d_0^{11}$$
$$+71612125G^6d_0^{12} - 1074406560G^5H_0^{10}c_0^2 - 10744065600G^5H_0^9c_0^2d_0$$
$$-48348295200G^5H_0^8c_0^2d_0^2 - 128928787200G^5H_0^7c_0^2d_0^3$$
$$-225625377600G^5H_0^6c_0^2d_0^4 - 270750453120G^5H_0^5c_0^2d_0^5$$
$$-225625377600G^5H_0^4c_0^2d_0^6 - 128928787200G^5H_0^3c_0^2d_0^7$$
$$-48348295200G^5H_0^2c_0^2d_0^8 + 305919828480G^4H_0^5c_0^4d_0^3$$
$$-10744065600G^5H_0c_0^2d_0^9 - 1074406560G^5c_0^2d_0^{10} + 5462854080G^4H_0^8c_0^4$$
$$+43702832640G^4H_0^7c_0^4d_0 + 152959914240G^4H_0^6c_0^4d_0^2$$
$$+382399785600G^4H_0^4c_0^4d_0^4 + 305919828480G^4H_0^3c_0^4d_0^5$$
$$+152959914240G^4H_0^2c_0^4d_0^6 - 217917767680G^3H_0^3c_0^6d_0^3$$
$$+43702832640G^4H_0c_0^4d_0^7 + 5462854080G^4c_0^4d_0^8 - 10895888384G^3H_0^6c_0^6$$

$$-65375330304G^3H_0^5c_0^6d_0 - 163438325760G^3H_0^4c_0^6d_0^2$$
$$-163438325760G^3H_0^2c_0^6d_0^4 - 65375330304G^3H_0c_0^6d_0^5$$
$$-10895888384G^3c_0^6d_0^6 + 43822227456G^2H_0^2c_0^8d_0^2$$
$$+7303704576G^2H_0^4c_0^8 + 29214818304G^2H_0^3c_0^8d_0$$
$$+29214818304G^2H_0c_0^8d_0^3 + 7303704576G^2c_0^8d_0^4 - 866254848GH_0^2c_0^{10}$$
$$-1732509696GH_0c_0^{10}d_0 - 866254848Gc_0^{10}d_0^2 + 262144c_0^{12})$$

$$d_{18} = -\frac{G^9(H_0+d_0)}{289571548120742677708800c_0^{25}}(GH_0^2 + 2GH_0d_0 + Gd_0^2 - 8c_0^2)$$
$$(2596581625G^7H_0^{14} + 36352142750G^7H_0^{13}d_0 + 236288927875G^7H_0^{12}d_0^2$$
$$+945155711500G^7H_0^{11}d_0^3 + 2599178206625G^7H_0^{10}d_0^4$$
$$+5198356413250G^7H_0^9d_0^5 + 7797534619875G^7H_0^6d_0^8$$
$$+7797534619875G^7H_0^8d_0^6 + 8911468137000G^7H_0^7d_0^7$$
$$+5198356413250G^7H_0^5d_0^9 + 2599178206625G^7H_0^4d_0^{10}$$
$$+236288927875G^7H_0^2d_0^{12} + 36352142750G^7H_0d_0^{13}$$
$$-46223906500G^6H_0^{12}c_0^2 - 554686878000G^6H_0^{11}c_0^2d_0$$
$$-3050777829000G^6H_0^{10}c_0^2d_0^2 + 2596581625G^7d_0^{14} + 945155711500G^7H_0^3d_0^{11}$$
$$-10169259430000G^6H_0^9c_0^2d_0^3 - 22880833717500G^6H_0^8c_0^2d_0^4$$
$$-36609333948000G^6H_0^7c_0^2d_0^5 - 42710889606000G^6H_0^6c_0^2d_0^6$$
$$-36609333948000G^6H_0^5c_0^2d_0^7 - 22880833717500G^6H_0^4c_0^2d_0^8$$
$$-10169259430000G^6H_0^3c_0^2d_0^9 - 3050777829000G^6H_0^2c_0^2d_0^{10}$$
$$-554686878000G^6H_0c_0^2d_0^{11} - 46223906500G^6c_0^2d_0^{12}$$
$$+297379154880G^5H_0^{10}c_0^4 + 2973791548800G^5H_0^9c_0^4d_0$$
$$+13382061969600G^5H_0^8c_0^4d_0^2 + 35685498585600G^5H_0^7c_0^4d_0^3$$
$$+62449622524800G^5H_0^6c_0^4d_0^4 + 74939547029760G^5H_0^5c_0^4d_0^5$$
$$+62449622524800G^5H_0^4c_0^4d_0^6 + 35685498585600G^5H_0^3c_0^4d_0^7$$
$$+13382061969600G^5H_0^2c_0^4d_0^8 + 2973791548800G^5H_0c_0^4d_0^9$$
$$+297379154880G^5c_0^4d_0^{10} - 838832062208G^4H_0^8c_0^6$$
$$-6710656497664G^4H_0^7c_0^6d_0 - 23487297741824G^4H_0^6c_0^6d_0^2$$
$$-46974595483648G^4H_0^5c_0^6d_0^3 - 58718244354560G^4H_0^4c_0^6d_0^4$$
$$-46974595483648G^4H_0^3c_0^6d_0^5 - 23487297741824G^4H_0^2c_0^6d_0^6$$
$$-6710656497664G^4H_0c_0^6d_0^7 - 838832062208G^4c_0^6d_0^8$$
$$+995841536000G^3H_0^6c_0^8 + 5975049216000G^3H_0^5c_0^8d_0$$
$$+14937623040000G^3H_0^4c_0^8d_0^2 + 19916830720000G^3H_0^3c_0^8d_0^3$$
$$+14937623040000G^3H_0^2c_0^8d_0^4 + 5975049216000G^3H_0c_0^8d_0^5$$
$$+995841536000G^3c_0^8d_0^6 - 388074553344G^2H_0^4c_0^{10}$$
$$-1552298213376G^2H_0^3c_0^{10}d_0 - 2328447320064G^2H_0^2c_0^{10}d_0^2$$
$$-1552298213376G^2H_0c_0^{10}d_0^3 - 388074553344G^2c_0^{10}d_0^4$$
$$+22064398336GH_0^2c_0^{12} + 44128796672GH_0c_0^{12}d_0$$
$$+22064398336Gc_0^{12}d_0^2 - 1048576c_0^{14})$$

$$d_{20} = -\frac{G^{10}(H_0+d_0)}{46331447699318828433408000c_0^{28}}(GH_0^2 + 2GH_0d_0 + Gd_0^2 - 8c_0^2)$$
$$(118334925625G^8H_0^{16} + 1893358810000G^8H_0^{15}d_0$$
$$+14200191075000G^8H_0^{14}d_0^2 + 66267558350000G^8H_0^{13}d_0^3$$
$$+215369564637500G^8H_0^{12}d_0^4 + 516886955130000G^8H_0^{11}d_0^5$$
$$+947626084405000G^8H_0^{10}d_0^6 + 1353751549150000G^8H_0^9d_0^7$$
$$+1522970492793750G^8H_0^8d_0^8 + 1353751549150000G^8H_0^7d_0^9$$
$$+947626084405000G^8H_0^6d_0^{10} + 516886955130000G^8H_0^5d_0^{11}$$
$$+215369564637500G^8H_0^4d_0^{12} + 1893358810000G^8H_0d_0^{15}$$
$$+66267558350000G^8H_0^3d_0^{13} + 14200191075000G^8H_0^2d_0^{14}$$
$$+118334925625G^8d_0^{16} - 2438867435750G^7H_0^{14}c_0^2$$
$$-34144144100500G^7H_0^{13}c_0^2d_0 - 221936936653250G^7H_0^{12}c_0^2d_0^2$$

$$\begin{aligned}
&-887747746613000 G^7 H_0^{11} c_0^2 d_0^3 - 2438867435750 G^7 c_0^2 d_0^{14}\\
&-2441306303185750 G^7 H_0^{10} c_0^2 d_0^4 - 4882612606371500 G^7 H_0^9 c_0^2 d_0^5\\
&-7323918909557250 G^7 H_0^8 c_0^2 d_0^6 - 8370193039494000 G^7 H_0^7 c_0^2 d_0^7\\
&-7323918909557250 G^7 H_0^6 c_0^2 d_0^8 - 4882612606371500 G^7 H_0^5 c_0^2 d_0^9\\
&-2441306303185750 G^7 H_0^4 c_0^2 d_0^{10} - 887747746613000 G^7 H_0^3 c_0^2 d_0^{11}\\
&-2219369366653250 G^7 H_0^2 c_0^2 d_0^{12} - 34144144100500 G^7 H_0 c_0^2 d_0^{13}\\
&+19021198932112 G^6 H_0^{12} c_0^4 + 228254387185344 G^6 H_0^{11} c_0^4 d_0\\
&+1255399129519392 G^6 H_0^{10} c_0^4 d_0^2 - 69963325870208 G^5 H_0^{10} c_0^6\\
&+4184663765064640 G^6 H_0^9 c_0^4 d_0^3 + 9415493471395440 G^6 H_0^8 c_0^4 d_0^4\\
&+15064789554232704 G^6 H_0^7 c_0^4 d_0^5 + 17575587813271488 G^6 H_0^6 c_0^4 d_0^6\\
&+15064789554232704 G^6 H_0^5 c_0^4 d_0^7 + 9415493471395440 G^6 H_0^4 c_0^4 d_0^8\\
&+4184663765064640 G^6 H_0^3 c_0^4 d_0^9 + 1255399129519392 G^6 H_0^2 c_0^4 d_0^{10}\\
&+228254387185344 G^6 H_0 c_0^4 d_0^{11} + 19021198932112 G^6 c_0^4 d_0^{12}\\
&-699633258702080 G^5 H_0^9 c_0^6 d_0 - 3148349664159360 G^5 H_0^8 c_0^6 d_0^2\\
&-8395599104424960 G^5 H_0^7 c_0^6 d_0^3 - 14692298432743680 G^5 H_0^6 c_0^6 d_0^4\\
&-17630758119292416 G^5 H_0^5 c_0^6 d_0^5 - 14692298432743680 G^5 H_0^4 c_0^6 d_0^6\\
&-8395599104424960 G^5 H_0^3 c_0^6 d_0^7 - 3148349664159360 G^5 H_0^2 c_0^6 d_0^8\\
&-699633258702080 G^5 H_0 c_0^6 d_0^9 - 69963325870208 G^5 c_0^6 d_0^{10}\\
&+123048169851904 G^4 H_0^8 c_0^8 - 92437077204992 G^3 H_0^6 c_0^{10}\\
&+984385358815232 G^4 H_0^7 c_0^8 d_0 + 3445348755853312 G^4 H_0^6 c_0^8 d_0^2\\
&+6890697511706624 G^4 H_0^5 c_0^8 d_0^3 + 8613371889633280 G^4 H_0^4 c_0^8 d_0^4\\
&+6890697511706624 G^4 H_0^3 c_0^8 d_0^5 + 3445348755853312 G^4 H_0^2 c_0^8 d_0^6\\
&+984385358815232 G^4 H_0 c_0^8 d_0^7 + 123048169851904 G^4 c_0^8 d_0^8\\
&-554622463229952 G^3 H_0^5 c_0^{10} d_0 - 1386556158074880 G^3 H_0^4 c_0^{10} d_0^2\\
&-1848741544099840 G^3 H_0^3 c_0^{10} d_0^3 - 1386556158074880 G^3 H_0^2 c_0^{10} d_0^4\\
&-554622463229952 G^3 H_0 c_0^{10} d_0^5 - 92437077204992 G^3 c_0^{10} d_0^6\\
&+21592071405568 G^2 H_0^4 c_0^{12} - 582969982976 G H_0^2 c_0^{14}\\
&+86368285622272 G^2 H_0^3 c_0^{12} d_0 + 129552428433408 G^2 H_0^2 c_0^{12} d_0^2\\
&+86368285622272 G^2 H_0 c_0^{12} d_0^3 + 21592071405568 G^2 c_0^{12} d_0^4\\
&-1165939965952 G H_0 c_0^{14} d_0 - 582969982976 G c_0^{14} d_0^2 + 4194304 c_0^{16})
\end{aligned}$$

$$d_{22} = -\frac{G^{11}(H_0 + d_0)}{358790730983525007388311552000 c_0^{31}}(G H_0^2 + 2 G H_0 d_0 + G d_0^2 - 8 c_0^2)$$

$$\begin{aligned}
&(26479758670625 G^9 H_0^{18} + 476635656071250 G^9 H_0^{17} d_0\\
&+4051403076605625 G^9 H_0^{16} d_0^2 - 620277594363700 G^8 H_0^{16} c_0^2\\
&+21607483075230000 G^9 H_0^{15} d_0^3 + 81028061532112500 G^9 H_0^{14} d_0^4\\
&+226878572289915000 G^9 H_0^{13} d_0^5 + 491570239961482500 G^9 H_0^{12} d_0^6\\
&+842691839933970000 G^9 H_0^{11} d_0^7 + 1158701279909208750 G^9 H_0^{10} d_0^8\\
&+1287445866565787500 G^9 H_0^9 d_0^9 + 1158701279909208750 G^9 H_0^8 d_0^{10}\\
&+842691839933970000 G^9 H_0^7 d_0^{11} + 491570239961482500 G^9 H_0^6 d_0^{12}\\
&+226878572289915000 G^9 H_0^5 d_0^{13} + 81028061532112500 G^9 H_0^4 d_0^{14}\\
&+21607483075230000 G^9 H_0^3 d_0^{15} + 4051403076605625 G^9 H_0^2 d_0^{16}\\
&+476635656071250 G^9 H_0 d_0^{17} + 26479758670625 G^9 d_0^{18}\\
&-9924441509819200 G^8 H_0^{15} c_0^2 d_0 - 74433311323644000 G^8 H_0^{14} c_0^2 d_0^2\\
&-347355452843672000 G^8 H_0^{13} c_0^2 d_0^3 - 1128905221741934000 G^8 H_0^{12} c_0^2 d_0^4\\
&-2709372532180641600 G^8 H_0^{11} c_0^2 d_0^5 - 4967182975664509600 G^8 H_0^{10} c_0^2 d_0^6\\
&-7059975679520728000 G^8 H_0^9 c_0^2 d_0^7 - 7982972639460819000 G^8 H_0^8 c_0^2 d_0^8\\
&-7059975679520728000 G^8 H_0^7 c_0^2 d_0^9 - 4967182975664509600 G^8 H_0^6 c_0^2 d_0^{10}\\
&-2709372532180641600 G^8 H_0^5 c_0^2 d_0^{11} - 1128905221741934000 G^8 H_0^4 c_0^2 d_0^{12}\\
&-347355452843672000 G^8 H_0^3 c_0^2 d_0^{13} - 74433311323644000 G^8 H_0^2 c_0^2 d_0^{14}\\
&-9924441509819200 G^8 H_0 c_0^2 d_0^{15} - 620277594363700 G^8 c_0^2 d_0^{16}\\
&+5691966058468800 G^7 H_0^{14} c_0^4 + 79687524818563200 G^7 H_0^{13} c_0^4 d_0\\
&+517968911320660800 G^7 H_0^{12} c_0^4 d_0^2 + 2071875645282643200 G^7 H_0^{11} c_0^4 d_0^3
\end{aligned}$$

$$\begin{aligned}
&+5697658024527268800 G^7 H_0{}^{10} c_0{}^4 d_0{}^4 + 11395316049054537600 G^7 H_0{}^9 c_0{}^4 d_0{}^5 \\
&+17092974073581806400 G^7 H_0{}^8 c_0{}^4 d_0{}^6 + 19534827512664921600 G^7 H_0{}^7 c_0{}^4 d_0{}^7 \\
&+17092974073581806400 G^7 H_0{}^6 c_0{}^4 d_0{}^8 + 11395316049054537600 G^7 H_0{}^5 c_0{}^4 d_0{}^9 \\
&+5697658024527268800 G^7 H_0{}^4 c_0{}^4 d_0{}^{10} + 2071875645282643200 G^7 H_0{}^3 c_0{}^4 d_0{}^{11} \\
&+517968911320660800 G^7 H_0{}^2 c_0{}^4 d_0{}^{12} + 79687524818563200 G^7 H_0 c_0{}^4 d_0{}^{13} \\
&+5691966058468800 G^7 c_0{}^4 d_0{}^{14} - 25932933890948096 G^6 H_0{}^{12} c_0{}^6 \\
&-311195206691377152 G^6 H_0{}^{11} c_0{}^6 d_0 - 1711573636802574336 G^6 H_0{}^{10} c_0{}^6 d_0{}^2 \\
&-5705245456008581120 G^6 H_0{}^9 c_0{}^6 d_0{}^3 - 12836802276019307520 G^6 H_0{}^8 c_0{}^6 d_0{}^4 \\
&-20538883641630892032 G^6 H_0{}^7 c_0{}^6 d_0{}^5 - 23962030915236040704 G^6 H_0{}^6 c_0{}^6 d_0{}^6 \\
&-20538883641630892032 G^6 H_0{}^5 c_0{}^6 d_0{}^7 - 12836802276019307520 G^6 H_0{}^4 c_0{}^6 d_0{}^8 \\
&-5705245456008581120 G^6 H_0{}^3 c_0{}^6 d_0{}^9 - 1711573636802574336 G^6 H_0{}^2 c_0{}^6 d_0{}^{10} \\
&-311195206691377152 G^6 H_0 c_0{}^6 d_0{}^{11} - 25932933890948096 G^6 c_0{}^6 d_0{}^{12} \\
&+61322321008062464 G^5 H_0{}^{10} c_0{}^8 + 613223210080624640 G^5 H_0{}^9 c_0{}^8 d_0 \\
&+2759504445362810880 G^5 H_0{}^8 c_0{}^8 d_0{}^2 + 7358678520967495680 G^5 H_0{}^7 c_0{}^8 d_0{}^3 \\
&+12877687411693117440 G^5 H_0{}^6 c_0{}^8 d_0{}^4 + 15453224894031740928 G^5 H_0{}^5 c_0{}^8 d_0{}^5 \\
&+12877687411693117440 G^5 H_0{}^4 c_0{}^8 d_0{}^6 + 7358678520967495680 G^5 H_0{}^3 c_0{}^8 d_0{}^7 \\
&+2759504445362810880 G^5 H_0{}^2 c_0{}^8 d_0{}^8 + 613223210080624640 G^5 H_0 c_0{}^8 d_0{}^9 \\
&+61322321008062464 G^5 c_0{}^8 d_0{}^{10} - 71614989018365952 G^4 H_0{}^8 c_0{}^{10} \\
&-572919912146927616 G^4 H_0{}^7 c_0{}^{10} d_0 - 2005219692514246656 G^4 H_0{}^6 c_0{}^{10} d_0{}^2 \\
&-4010439385028493312 G^4 H_0{}^5 c_0{}^{10} d_0{}^3 - 5013049231285616640 G^4 H_0{}^4 c_0{}^{10} d_0{}^4 \\
&-4010439385028493312 G^4 H_0{}^3 c_0{}^{10} d_0{}^5 - 2005219692514246656 G^4 H_0{}^2 c_0{}^{10} d_0{}^6 \\
&-572919912146927616 G^4 H_0 c_0{}^{10} d_0{}^7 - 71614989018365952 G^4 c_0{}^{10} d_0{}^8 \\
&+35312095237242880 G^3 H_0{}^6 c_0{}^{12} + 211872571423457280 G^3 H_0{}^5 c_0{}^{12} d_0 \\
&+529681428558643200 G^3 H_0{}^4 c_0{}^{12} d_0{}^2 + 706241904744857600 G^3 H_0{}^3 c_0{}^{12} d_0{}^3 \\
&+529681428558643200 G^3 H_0{}^2 c_0{}^{12} d_0{}^4 + 211872571423457280 G^3 H_0 c_0{}^{12} d_0{}^5 \\
&+35312095237242880 G^3 c_0{}^{12} d_0{}^6 - 5025844518977536 G^2 H_0{}^4 c_0{}^{14} \\
&-20103378075910144 G^2 H_0{}^3 c_0{}^{14} d_0 - 30155067113865216 G^2 H_0{}^2 c_0{}^{14} d_0{}^2 \\
&-20103378075910144 G^2 H_0 c_0{}^{14} d_0{}^3 - 5025844518977536 G^2 c_0{}^{14} d_0{}^4 \\
&+63454920572928 G H_0{}^2 c_0{}^{16} + 126909841145856 G H_0 c_0{}^{16} d_0 \\
&+63454920572928 G c_0{}^{16} d_0{}^2 - 67108864 c_0{}^{18})
\end{aligned}$$

$$d_{24} = -\frac{G^{12}(H_0 + d_0)}{41332692209302080851133 49079040000 c_0{}^{34}}(G H_0{}^2 + 2 G H_0 d_0 + G d_0{}^2 - 8 c_0{}^2)$$

$$\begin{aligned}
&(891919562511250 G^{10} H_0{}^{20} + 17838391250225000 G^{10} H_0{}^{19} d_0 \\
&+169464716877137500 G^{10} H_0{}^{18} d_0{}^2 + 1016788301262825000 G^{10} H_0{}^{17} d_0{}^3 \\
&+4321350280367006250 G^{10} H_0{}^{16} d_0{}^4 + 13828320897174420000 G^{10} H_0{}^{15} d_0{}^5 \\
&+34570802242936050000 G^{10} H_0{}^{14} d_0{}^6 + 69141604485872100000 G^{10} H_0{}^{13} d_0{}^7 \\
&+112355107289542162500 G^{10} H_0{}^{12} d_0{}^8 + 149806809719389550000 G^{10} H_0{}^{11} d_0{}^9 \\
&+164787490691328505000 G^{10} H_0{}^{10} d_0{}^{10} + 149806809719389550000 G^{10} H_0{}^9 d_0{}^{11} \\
&+112355107289542162500 G^{10} H_0{}^8 d_0{}^{12} + 69141604485872100000 G^{10} H_0{}^7 d_0{}^{13} \\
&+34570802242936050000 G^{10} H_0{}^6 d_0{}^{14} + 13828320897174420000 G^{10} H_0{}^5 d_0{}^{15} \\
&+4321350280367006250 G^{10} H_0{}^4 d_0{}^{16} + 1016788301262825000 G^{10} H_0{}^3 d_0{}^{17} \\
&+169464716877137500 G^{10} H_0{}^2 d_0{}^{18} + 17838391250225000 G^{10} H_0 d_0{}^{19} \\
&+891919562511250 G^{10} d_0{}^{20} - 23407785060336525 G^9 H_0{}^{18} c_0{}^2 \\
&-421340131086057450 G^9 H_0{}^{17} c_0{}^2 d_0 - 3581391114231488325 G^9 H_0{}^{16} c_0{}^2 d_0{}^2 \\
&-19100752609234604400 G^9 H_0{}^{15} c_0{}^2 d_0{}^3 - 71627822284629766500 G^9 H_0{}^{14} c_0{}^2 d_0{}^4 \\
&-200557902396963346200 G^9 H_0{}^{13} c_0{}^2 d_0{}^5 - 434542121860087250100 G^9 H_0{}^{12} c_0{}^2 d_0{}^6 \\
&-744929351760149571600 G^9 H_0{}^{11} c_0{}^2 d_0{}^7 - 1024277858670205660950 G^9 H_0{}^{10} c_0{}^2 d_0{}^8 \\
&-1138086509633561845500 G^9 H_0{}^9 c_0{}^2 d_0{}^9 - 1024277858670205660950 G^9 H_0{}^8 c_0{}^2 d_0{}^{10} \\
&-744929351760149571600 G^9 H_0{}^7 c_0{}^2 d_0{}^{11} - 434542121860087250100 G^9 H_0{}^6 c_0{}^2 d_0{}^{12} \\
&-200557902396963346200 G^9 H_0{}^5 c_0{}^2 d_0{}^{13} - 71627822284629766500 G^9 H_0{}^4 c_0{}^2 d_0{}^{14} \\
&-19100752609234604400 G^9 H_0{}^3 c_0{}^2 d_0{}^{15} - 3581391114231488325 G^9 H_0{}^2 c_0{}^2 d_0{}^{16} \\
&-421340131086057450 G^9 H_0 c_0{}^2 d_0{}^{17} - 23407785060336525 G^9 c_0{}^2 d_0{}^{18}
\end{aligned}$$

$$\begin{aligned}
&+247234276374827672 G^8 H_0^{16} c_0^4 + 3955748421997242752 G^8 H_0^{15} c_0^4 d_0 \\
&+29668113164979320640 G^8 H_0^{14} c_0^4 d_0^2 + 138451194769903496320 G^8 H_0^{13} c_0^4 d_0^3 \\
&+449966383002186363040 G^8 H_0^{12} c_0^4 d_0^4 + 1079919319205247271296 G^8 H_0^{11} c_0^4 d_0^5 \\
&+1979852085209619997376 G^8 H_0^{10} c_0^4 d_0^6 + 2828360121728028567680 G^8 H_0^9 c_0^4 d_0^7 \\
&+3181905136944032138640 G^8 H_0^8 c_0^4 d_0^8 + 2828360121728028567680 G^8 H_0^7 c_0^4 d_0^9 \\
&+1979852085209619997376 G^8 H_0^6 c_0^4 d_0^{10} + 1079919319205247271296 G^8 H_0^5 c_0^4 d_0^{11} \\
&+449966383002186363040 G^8 H_0^4 c_0^4 d_0^{12} + 138451194769903496320 G^8 H_0^3 c_0^4 d_0^{13} \\
&+29668113164979320640 G^8 H_0^2 c_0^4 d_0^{14} + 3955748421997242752 G^8 H_0 c_0^4 d_0^{15} \\
&+247234276374827672 G^8 c_0^4 d_0^{16} - 1346957387631448192 G^7 H_0^{14} c_0^6 \\
&-18857403426840274688 G^7 H_0^{13} c_0^6 d_0 - 122573122274461785472 G^7 H_0^{12} c_0^6 d_0^2 \\
&-490292489097847141888 G^7 H_0^{11} c_0^6 d_0^3 - 1348304345019079640192 G^7 H_0^{10} c_0^6 d_0^4 \\
&+1979852085209619997376 G^8 H_0^{10} c_0^4 d_0^6 + 2828360121728028567680 G^8 H_0^9 c_0^4 d_0^7 \\
&+3181905136944032138640 G^8 H_0^8 c_0^4 d_0^8 + 2828360121728028567680 G^8 H_0^7 c_0^4 d_0^9 \\
&+1979852085209619997376 G^8 H_0^6 c_0^4 d_0^{10} + 1079919319205247271296 G^8 H_0^5 c_0^4 d_0^{11} \\
&+449966383002186363040 G^8 H_0^4 c_0^4 d_0^{12} + 138451194769903496320 G^8 H_0^3 c_0^4 d_0^{13} \\
&+29668113164979320640 G^8 H_0^2 c_0^4 d_0^{14} + 3955748421997242752 G^8 H_0 c_0^4 d_0^{15} \\
&+247234276374827672 G^8 c_0^4 d_0^{16} - 1346957387631448192 G^7 H_0^{14} c_0^6 \\
&-18857403426840274688 G^7 H_0^{13} c_0^6 d_0 - 122573122274461785472 G^7 H_0^{12} c_0^6 d_0^2 \\
&-490292489097847141888 G^7 H_0^{11} c_0^6 d_0^3 - 1348304345019079640192 G^7 H_0^{10} c_0^6 d_0^4 \\
&-2696608690038159280384 G^7 H_0^9 c_0^6 d_0^5 - 4044913035057238920576 G^7 H_0^8 c_0^6 d_0^6 \\
&-4622757754351130194944 G^7 H_0^7 c_0^6 d_0^7 - 4044913035057238920576 G^7 H_0^6 c_0^6 d_0^8 \\
&-2696608690038159280384 G^7 H_0^5 c_0^6 d_0^9 - 1348304345019079640192 G^7 H_0^4 c_0^6 d_0^{10} \\
&-490292489097847141888 G^7 H_0^3 c_0^6 d_0^{11} - 122573122274461785472 G^7 H_0^2 c_0^6 d_0^{12} \\
&-18857403426840274688 G^7 H_0 c_0^6 d_0^{13} - 1346957387631448192 G^7 c_0^6 d_0^{14} \\
&+4032118651150257152 G^6 H_0^{12} c_0^8 + 48385423813803085824 G^6 H_0^{11} c_0^8 d_0 \\
&+261198309759169720 32 G^6 H_0^{10} c_0^8 d_0^2 + 887066103253056573440 G^6 H_0^9 c_0^8 d_0^3 \\
&+1995898732319377290240 G^6 H_0^8 c_0^8 d_0^4 + 3193437971711003664384 G^6 H_0^7 c_0^8 d_0^5 \\
&+3725677633662837608448 G^6 H_0^6 c_0^8 d_0^6 + 3193437971711003664384 G^6 H_0^5 c_0^8 d_0^7 \\
&+1995898732319377290240 G^6 H_0^4 c_0^8 d_0^8 + 887066103253056573440 G^6 H_0^3 c_0^8 d_0^9 \\
&+26119830975916972032 G^6 H_0^2 c_0^8 d_0^{10} + 48385423813803085824 G^6 H_0 c_0^8 d_0^{11} \\
&+4032118651150257152 G^6 c_0^8 d_0^{12} - 6532826241815068672 G^5 H_0^{10} c_0^{10} \\
&-6532826241815068672 0 G^5 H_0^9 c_0^{10} d_0 - 29397718088167809024 0 G^5 H_0^8 c_0^{10} d_0^2 \\
&-78393914901780824064 0 G^5 H_0^7 c_0^{10} d_0^3 - 137189351078116442112 0 G^5 H_0^6 c_0^{10} d_0^4 \\
&-164627221293739730534 4 G^5 H_0^5 c_0^{10} d_0^5 - 137189351078116442112 0 G^5 H_0^4 c_0^{10} d_0^6 \\
&-78393914901780824064 0 G^5 H_0^3 c_0^{10} d_0^7 - 29397718088167809024 0 G^5 H_0^2 c_0^{10} d_0^8 \\
&-6532826241815068672 0 G^5 H_0 c_0^{10} d_0^9 - 6532826241815068672 G^5 c_0^{10} d_0^{10} \\
&+5266104159343345664 G^4 H_0^8 c_0^{12} + 42128833274746765312 G^4 H_0^7 c_0^{12} d_0 \\
&+147450916461613678592 G^4 H_0^6 c_0^{12} d_0^2 + 294901832923227357184 G^4 H_0^5 c_0^{12} d_0^3 \\
&+368627291154034196480 G^4 H_0^4 c_0^{12} d_0^4 + 294901832923227357184 G^4 H_0^3 c_0^{12} d_0^5 \\
&+147450916461613678592 G^4 H_0^2 c_0^{12} d_0^6 + 42128833274746765312 G^4 H_0 c_0^{12} d_0^7 \\
&+5266104159343345664 G^4 c_0^{12} d_0^8 - 1744290367546064896 G^3 H_0^6 c_0^{14} \\
&-10465742205276389376 G^3 H_0^5 c_0^{14} d_0 - 26164355513190973440 G^3 H_0^4 c_0^{14} d_0^2 \\
&-34885807350921297920 G^3 H_0^3 c_0^{14} d_0^3 - 26164355513190973440 G^3 H_0^2 c_0^{14} d_0^4 \\
&-10465742205276389376 G^3 H_0 c_0^{14} d_0^5 - 1744290367546064896 G^3 c_0^{14} d_0^6 \\
&+152409464354373632 G^2 H_0^4 c_0^{16} + 609637857417494528 G^2 H_0^3 c_0^{16} d_0 \\
&+914456786126241792 G^2 H_0^2 c_0^{16} d_0^2 + 609637857417494528 G^2 H_0 c_0^{16} d_0^3 \\
&+152409464354373632 G^2 c_0^{16} d_0^4 - 884560174776320 G H_0^2 c_0^{18} \\
&-1769120349552640 G H_0 c_0^{18} d_0 - 884560174776320 G c_0^{18} d_0^2 + 134217728 c_0^{20})
\end{aligned}$$

References

1. Dadgar-Rad, F.; Imani, A. Theory of gradient-elastic membranes and its application in the wrinkling analysis of stretched thin sheets. *J. Mech. Phys. Solids* **2019**, *132*, 103679. [CrossRef]
2. Suresh, K.; Katara, N. Design and development of circular ceramic membrane for wastewater treatment. *Mater. Today Proc.* **2021**, *43*, 2176–2181. [CrossRef]
3. Zhao, M.H.; Zheng, W.L.; Fan, C.Y. Mechanics of shaft-loaded blister test for thin film suspended on compliant substrate. *Int. J. Solids Struct.* **2010**, *47*, 2525–2532. [CrossRef]
4. Bernardo, P.; Iulianelli, A.; Macedonio, F.; Drioli, E. Membrane technologies for space engineering. *J. Membr. Sci.* **2021**, *626*, 119177. [CrossRef]
5. Sun, H.X.; Bao, S.S.; Zhao, H.R.; Chen, Y.H.; Wang, Y.X.; Jiang, C.; Li, P.; Niu, Q.J. Polyarylate membrane with special circular microporous structure by interfacial polymerization for gas separation. *Sep. Purif. Technol.* **2020**, *251*, 117370. [CrossRef]
6. Tsiatas, C.G.; Katsikadelis, J.T. Large deflection analysis of elastic space membranes. *Int. J. Numer. Meth. Eng.* **2006**, *65*, 264–294. [CrossRef]
7. Li, X.; Sun, J.Y.; Zhao, Z.H.; He, X.T. Large deflection analysis of axially symmetric deformation of prestressed circular membranes under uniform lateral loads. *Symmetry* **2020**, *12*, 1343. [CrossRef]
8. Nguyen, T.N.; Hien, T.D.; Nguyen-Thoi, T.; Lee, J. A unified adaptive approach for membrane structures: Form finding and large deflection isogeometric analysis. *Comput. Method Appl. Mech. Eng.* **2020**, *369*, 113239. [CrossRef]
9. Mei, D.; Sun, J.Y.; Zhao, Z.H.; He, X.T. A closed-form solution for the boundary value problem of gas pressurized circular membranes in contact with frictionless rigid plates. *Mathematics* **2020**, *8*, 1017. [CrossRef]
10. Arthurs, A.M.; Clegg, J. On the solution of a boundary value problem for the nonlinear Föppl-Hencky equation. *Z. Angew. Math. Mech.* **1994**, *74*, 281–284. [CrossRef]
11. Li, X.; Sun, J.Y.; Shi, B.B.; Zhao, Z.H.; He, X.T. A theoretical study on an elastic polymer thin film-based capacitive wind-pressure sensor. *Polymers* **2020**, *12*, 2133. [CrossRef]
12. Dickey, R.W. A boundary value problem for a class of nonlinear ordinary differential equations. *J. Differ. Equ.* **1968**, *4*, 399–407. [CrossRef]
13. Versaci, M.; Angiulli, G.; Fattorusso, L.; Jannelli, A. On the uniqueness of the solution for a semi-linear elliptic boundary value problem of the membrane MEMS device for reconstructing the membrane profile in absence of ghost solutions. *Int. J. Non-Linear Mech.* **2019**, *109*, 24–31. [CrossRef]
14. Hencky, H. Über den Spannungszustand in kreisrunden Platten mit verschwindender Biegungssteifigkeit. *Z. Angew. Math. Phys.* **1915**, *63*, 311–317.
15. Chien, W.Z. Asymptotic behavior of a thin clamped circular plate under uniform normal pressure at very large deflection. *Sci. Rep. Natl. Tsinghua Univ.* **1948**, *5*, 193–208.
16. Alekseev, S.A. Elastic circular membranes under the uniformly distributed loads. *Eng. Corpus* **1953**, *14*, 196–198. (In Russian)
17. Lian, Y.S.; Sun, J.Y.; Zhao, Z.H.; Li, S.Z.; Zheng, Z.L. A refined theory for characterizing adhesion of elastic coatings on rigid substrates based on pressurized blister test methods: Closed-form solution and energy release rate. *Polymers* **2020**, *12*, 1788. [CrossRef]
18. Li, X.; Sun, J.Y.; Zhao, Z.H.; Li, S.Z.; He, X.T. A new solution to well-known Hencky problem: Improvement of in-plane equilibrium equation. *Mathematics* **2020**, *8*, 653. [CrossRef]
19. Ma, Y.; Wang, G.R.; Chen, Y.L.; Long, D.; Guan, Y.C.; Liu, L.Q.; Zhang, Z. Extended Hencky solution for the blister test of nanomembrane. *Extrem. Mech. Lett.* **2018**, *22*, 69–78. [CrossRef]
20. Lian, Y.S.; Sun, J.Y.; Zhao, Z.H.; He, X.T.; Zheng, Z.L. A revisit of the boundary value problem for Föppl–Hencky membranes: Improvement of geometric equations. *Mathematics* **2020**, *8*, 631. [CrossRef]
21. Sun, J.Y.; Rong, Y.; He, X.T.; Gao, X.W.; Zheng, Z.L. Power series solution of circular membrane under uniformly distributed loads: Investigation into Hencky transformation. *Stuct. Eng. Mech.* **2013**, *45*, 631–641. [CrossRef]
22. Sun, J.Y.; Lian, Y.S.; Li, Y.M.; He, X.T.; Zheng, Z.L. Closed-form solution of elastic circular membrane with initial stress under uniformly-distributed loads: Extended Hencky solution. *Z. Angew. Math. Mech.* **2015**, *95*, 1335–1341. [CrossRef]
23. Chien, W.Z.; Wang, Z.Z.; Xu, Y.G.; Chen, S.L. The symmetrical deformation of circular membrane under the action of uniformly distributed loads in its portion. *Appl. Math. Mech.* **1981**, *2*, 653–668.
24. Alekseev, S.A. Elastic annular membranes with a stiff centre under the concentrated force. *Eng. Corpus* **1951**, *10*, 71–80. (In Russian)
25. Chen, S.L.; Zheng, Z.L. Large deformation of circular membrane under the concentrated force. *Appl. Math. Mech.* **2003**, *24*, 28–31.
26. Jin, C.R. Large deflection of circular membrane under concentrated force. *Appl. Math. Mech.* **2008**, *29*, 889–896. [CrossRef]
27. Huang, P.F.; Song, Y.P.; Li, Q.; Liu, X.Q.; Feng, Y.Q. A theoretical study of circular orthotropic membrane under concentrated load: The relation of load and deflection. *IEEE Access* **2020**, *8*, 126127–126137. [CrossRef]
28. Khapin, A.V.; Abdeev, B.M.; Makhiyev, B.E. Optimal size of an axisymmetric perfectly flexible membrane with a rigid centre loaded with a concentrated static force. *IOP Conf. Ser. Mater. Sci. Eng.* **2020**, *775*, 012138. [CrossRef]
29. Strangeways, I. A history of rain gauges. *Weather* **2010**, *65*, 133–138. [CrossRef]

30. Haselow, L.; Meissner, R.; Rupp, H.; Miegel, K. Evaluation of precipitation measurements methods under field conditions during a summer season: A comparison of the standard rain gauge with a weighable lysimeter and a piezoelectric precipitation sensor. *J. Hydrol.* **2019**, *575*, 537–543. [CrossRef]
31. Saad Al-Wagdany, A. Intensity-duration-frequency curve derivation from different rain gauge records. *J. King Saud Univ. Sci.* **2020**, *32*, 3421–3431. [CrossRef]

Article

Hybrid Model for Time Series of Complex Structure with ARIMA Components

Oksana Mandrikova, Nadezhda Fetisova and Yuriy Polozov *

Institute of Cosmophysical Research and Radio Wave Propagation, Far Eastern Branch of the Russian Academy of Sciences, Mirnaya st, 7, Paratunka, 684034 Kamchatskiy Kray, Russia; oksanam1@ikir.ru or oksanam1@mail.ru (O.M.); nv.glushkova@yandex.ru (N.F.)
* Correspondence: polozov@ikir.ru or up_agent@mail.ru

Abstract: A hybrid model for the time series of complex structure (HMTS) was proposed. It is based on the combination of function expansions in a wavelet series with ARIMA models. HMTS has regular and anomalous components. The time series components, obtained after expansion, have a simpler structure that makes it possible to identify the ARIMA model if the components are stationary. This allows us to obtain a more accurate ARIMA model for a time series of complicated structure and to extend the area for application. To identify the HMTS anomalous component, threshold functions are applied. This paper describes a technique to identify HMTS and proposes operations to detect anomalies. With the example of an ionospheric parameter time series, we show the HMTS efficiency, describe the results and their application in detecting ionospheric anomalies. The HMTS was compared with the nonlinear autoregression neural network NARX, which confirmed HMTS efficiency.

Keywords: time series model; wavelet transform; ARIMA model; neural network NARX; ionospheric parameters

1. Introduction

Time series modeling and analysis are important bases for the methods of studying the processes and phenomena of different natures. They are used in various spheres of human activity (physics, biology, medicine, economics, etc.). Methods of data modeling and analysis aimed at detecting and identifying anomalies are of special actuality. The examples are the problems of the recognition of anomalies in geophysical monitoring data, such as the detection of magnetic and ionospheric storms [1–4], earthquakes [5,6], tsunamis [7,8], geological anomalies [9] and other catastrophic natural phenomena. The need to detect anomalies often arises in the medical field, for example, to detect and to identify clinical conditions of patients [10]. An important property of such methods is their ability to adapt, providing the possibility to detect and identify rapid changes in the system or object state, indicating anomaly occurrences.

As a rule, time series of empirical data have a complex non-stationary structure and contain local features of various forms. The methods for the time series analysis include deterministic [11], stochastic [12–14] approaches and their various combinations [15–19]. Traditional methods for data time series modeling and analysis (AR models, ARMA [20,21], exponential smoothing [22], stochastic approximation [13], etc.) do not allow us to describe the time series of complex structure adequately [23]. At present, hybrid approaches [16,17,19,23–28] are widely applied. They make it possible to improve the efficiency of the procedure of data analysis in case of its complicated structure. For example, in [19], on the basis of wavelet decomposition, a technique was developed to estimate the coefficients of turbulent diffusion and power exponents from single Lagrangian trajectories of particles. Wavelet transform is a flexible tool and was applied in the paper [29] to study the relationship between vegetation and climate in India. The 2D empirical wavelet filters developed by the authors of [30] are effective in image processing applications. Currently, neural network methods are also widely used [4,15,23,31]. They allow us to approximate complex nonlinear relationships in data and are easily automated. However, the reliability and accuracy of neural networks depend on data representativity and it is very laborious to adapt them. For example, the authors of the paper [31] proposed a neural network structure, based on the LSTM paradigm, which allowed them to obtain an accurate forecast of time series for web traffic on a limited data set. The authors of the paper [23] considered combinations of wavelet transform with neural networks to analyze hydrological data.

Due to these aspects and despite the intensive development of machine learning methods and their active application in various fields of artificial intelligence, classical models of time series, in particular, ARIMA models [4,15,32,33], are popular. The obvious advantages of ARIMA models are their mathematical validity, a formalized methodology for model identification and verification for its adequacy. However, the ARIMA model construction is based on the assumption that the process has a normal distribution and is stationary (or stationary in differences). If these assumptions are not satisfied, the model accuracy is significantly reduced. In order to improve the ARIMA efficiency, a number of papers [16,17,26,27,34,35] suggested a hybrid approach to the time series analysis. For example, the paper [17] proposed to apply ARIMA together with discrete wavelet transform and neural network LSTM. The authors of the paper [17] showed that the combination of ARIMA and LSTM with a discrete wavelet transform allowed them to improve the accuracy of ARIMA and LSTM models in order to make forecasts of a monthly precipitation time series. A combination of the discrete wavelet transform with ARIMA and neural network was also proposed in [35] to forecast a hydrological time series.

In this paper, we propose a hybrid model for a time series of complex structure (HMTS). The model includes regular and anomalous components. The HMTS identification is based on the combination of function expansion in a wavelet series [36] with ARIMA models [20]. The time series components obtained after expansion have a simpler structure allowing us to identify ARIMA models in the case of components stationarity. This makes it possible to obtain a more accurate ARIMA model for the time series of a complex structure and expands the field of its application. The HMTS anomalous component describes irregular (sporadic) changes in time series. It is identified on the basis of threshold functions. A large dictionary of wavelet bases allows us to identify models for the time series of complex structure [9,36,37], including local features of various forms. The paper describes a method of HMTS identification and suggests algorithms for anomaly detection. The HMTS efficiency is illustrated on the example of an ionospheric parameter time series. The results and their application in detecting ionospheric anomalies of different intensities are presented. The paper also compares the HMTS with the nonlinear autoregressive neural network NARX, which also confirmed the HMTS efficiency.

2. Materials and Methods

2.1. Description of the Method

The time series of a complex structure may be represented as

$$f(t) = A^{REG}(t) + U(t) + e(t) = \sum_{\mu=\overline{1,T}} \alpha_\mu(t) + U(t) + e(t), \quad (1)$$

where $A^{REG}(t) = \sum_{\mu=\overline{1,T}} \alpha_\mu(t)$ is a regular component, which is a linear combination of the components $\alpha_\mu(t)$, μ is the component number; $U(t)$ is the anomalous component including local features of various forms occurring at random times, $e(t)$ is the noise component, t is time.

2.2. Wavelet Series Expansion and Determination of the Model Regular Components

It is assumed that $f \in L^2(R)$ ($L^2(R)$ is Lebesgue space) there is a unique representation [36]

$$f(t) = \ldots + g_{-1}(t) + g_0(t) + g_1(t) + \ldots,$$

where $g_j \in W_j, j \in \mathbb{Z}$ ($\mathbb{Z}$ is the set of integers), $g_j(t) = \sum_k d_{j,k}\Psi_{j,k}(t)$, $\Psi_{j,k} = \{\Psi_{j,k}\}_{k \in Z}$ is the basis of the space W_j, the coefficients $d_{j,k} = \langle f, \Psi_{j,k}\rangle$, $\Psi_{j,k} = 2^{j/2}\Psi(2^j t - k)$ are considered as a result of mapping f into the space W_j with resolution j. If $\Psi \in L^2(\mathbb{R})$ is $\mathcal{R}$-function and the sequence $\{\Psi_{j,k}\}$ is a Riesz basis [37] in $L^2(\mathbb{R})$, space $L^2(\mathbb{R})$ expansion structure generated by the wavelet $\Psi \in L^2(\mathbb{R})$ is

$$L^2(R) = \sum_{j \in Z}^{\bullet} W_j := \ldots \dotplus W_{-1} \dotplus W_0 \dotplus W_1 \dotplus \ldots, \quad (2)$$

where $W_j := clos_{L^2(\mathbb{R})}\left(\Psi_{j,k}; k \in \mathbb{Z}\right)$, the dots above the summation sign and above the plus signs denote the direct sum.

Using expansion (2), we obtain a sequence of nested and closed subspaces $V_j \in L^2(\mathbb{R}), j \in \mathbb{Z}$ defined as

$$V_j = \ldots \dotplus W_{j-2} \dotplus W_{j-1} \quad (3)$$

where the space $V_j = clos_{L^2(\mathbb{R})}(\phi(2^j t - k))$, ϕ is the scaling function. Based on (2) and (3), we obtain space $L^2(\mathbb{R})$ expansion:

$$L^2(\mathbb{R}) = V_j \dotplus W_j \dotplus W_{j+1} \dotplus \ldots,$$

in case of an orthogonal wavelet Ψ, we have

$$L^2(\mathbb{R}) = V_j \oplus W_j \oplus W_{j+1} \oplus \ldots, \quad (4)$$

where $\oplus$ is the orthogonal sum.

Considering the space $V_j = clos_{L^2(\mathbb{R})}(\phi(2^j t - k))$ with $j = 0$ as the base space f, and using (4) m times, we obtain the following expansion [36]:

$$V_0 = W_{-1} \oplus W_{-2} \oplus \ldots \oplus W_{-m} \oplus V_{-m}.$$

In this case, for f_0 we have the following representation:

$$f_0(t) = g_{-1}(t) + g_{-2}(t) + \ldots + g_{-m}(t) + f_{-m}(t) = \sum_{j=-1}^{-m} g_j(t) + f_{-m}(t) \quad (5)$$

where $f_{-m} \in V_{-m}$, $g_j \in W_j$, $f_{-m}(t) = \sum_k c_{-m,k}\phi_{-m,k}(t)$ is the smoothed component, $c_{-m,k} = \langle f_0, \phi_{-m,k}\rangle$, $\phi_{-m,k}(t) = 2^{-m/2}\phi(2^{-m}t - k)$ is the scaling function, $g_j(t) = \sum_k d_{j,k}\Psi_{j,k}(t)$ are the detailing components, $d_{j,k} = \langle f_0, \Psi_{j,k}\rangle$, $\Psi_{j,k}(t) = 2^{\frac{j}{2}}\Psi(2^j t - k)$ is the wavelet.

Note that, when the scaling function ϕ has L zero moments, i.e., $\int_{-\infty}^{+\infty} t^{\vartheta}\phi(t)dt = 0$, $\vartheta = \overline{1,L}$ and $f \in C^L$ (C^L is the space of functions continuously differentiable by L times), then for t near $2^m k$ [38]:

$$c_{-m,k} = \langle f, \phi_{-m,k}\rangle \cong 2^{-m/2} f(2^m k) \quad (6)$$

It follows from (6) that the component $f_{-m} \in V_{-m}$ gives approximation f with resolution 2^m (it approximates the trend). The detailing component g_j has the resolution

2^{-j}, and approximates the local features of the scale j. Figure 1 shows the amplitude–frequency characteristics (AFC) of the scaling function (solid line) and the wavelet (dashed line) for different m, obtained for the 3rd-order Daubechies wavelet.

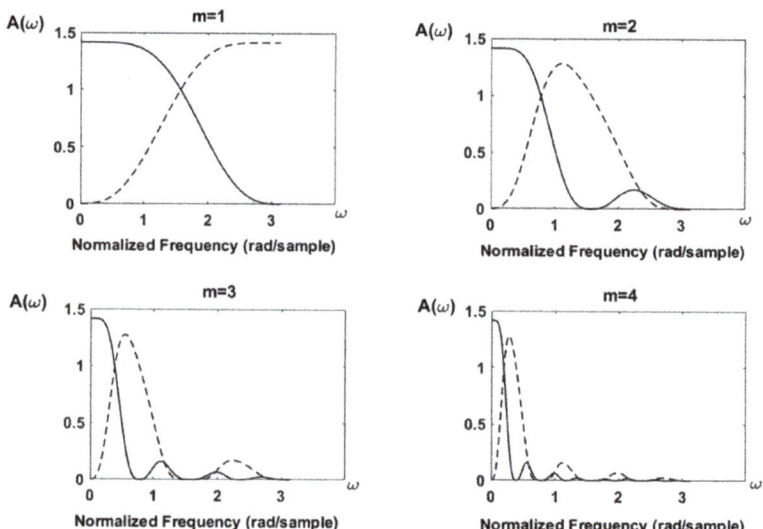

Figure 1. AFC of the scaling function and the wavelet for $m = 1, 2, 3, 4$ obtained for the 3rd-order Daubechies wavelet.

Thus, we can obtain different representations of f_0 in the form (5) for different m. Obviously, it is necessary to determine the level of expansion m^r, for which the component f_{-m^r} is regular. It is natural to assume that the component f_{-m} is regular if it is strictly stationary. In this case, the problem of determining regular components is reduced to the problem of obtaining representation (5) for which the component f_{-m} is strictly stationary. The condition of stationarity of the component f_{-m} will allow us to identify the ARIMA model for it. Following the theory by Box and Jenkins [20], a time series is strictly stationary if its autocorrelation function (ACF) damps rapidly during average and large delays. To determine the model type (AR, MA, ARMA) and the order, ACF and partial ACF (PACF) are studied [20]. Taking into account the fact that the f resolution decreases with the m increase, we define m^r sequentially:

The components f_{-m^r} and g_{j^r} obtained on the basis of Algorithm 1 describe the regular changes of the time series. Then from (1) and (5), we have the representation:

$$f_0(t) = \sum_{\mu=\overline{1,T}} \alpha_\mu(t) + U(t) + e(t) = f_{-m^r}(t) + \sum_{j^r} g_{j^r}(t) + \sum_{j \in P_j} g_j(t), \qquad (7)$$

where $A^{REG}(t) = \sum_{\mu=\overline{1,T}} \alpha_\mu(t) = f_{-m^r}(t) + \sum_{j^r} g_{j^r}(t)$, and we assume that $f_{-m^r}(t) = \alpha_1(t)$, $g_{j^r}(t) = \alpha_\mu(t), \mu = \overline{2,T}$, T is the number of regular components; $P_j = \left\{ j = \overline{-1, -(m^r-1)} \Big| j \neq j^r \right\}$.

Algorithm 1:

1. We map (5) for the expansion level $m = 1$ for f_0: $f_{-m}(t) = \sum_k c_{-m,k} \phi_{-m,k}(t)$, $m = 1$;
2. We check the condition of strict stationarity for the component f_{-m} by estimating the numerical characteristics (analysis of ACF and PACF [20]);
3. In the case of strict stationarity of the component f_{-m}, we assume that it describes regular data changes ($m = m^r$) and go to step 5, otherwise go to step 4;
4. If $m < M$, where M is the maximum level of expansion: $M \leq \log_2 N$ (N is the time series length), we increase the expansion level by 1: $m = m + 1$ and return to step 2; if, $m \geq M$ we terminate the algorithm execution;
5. We check the condition of strict stationarity for the detailing components $g_j(t) = \sum_k d_{j,k} \Psi_{j,k}(t)$, $j = \overline{-1, -m^r}$ by estimating the numerical characteristics (analysis of ACF and PACF [20]). If the condition of strict stationarity is satisfied for the component g_j, we take $j = j^r$ and assume that the component g_{j^r} is regular.

2.3. Estimation of the Parameters for the Model Regular Component

The components f_{-m^r} and g_{j^r} are strictly stationary, thus, we can estimate ARIMA models of order (p, ν, h) for them [20]. Then for the component $f_{-m^r}(t) = \sum_k c_{-m^r,k} \phi_{-m^r,k}(t)$ for brevity, we omit index r and obtain

$$\omega_{-m,k} = \gamma_1^m \omega_{-m,k-1} + \ldots + \gamma_p^m \omega_{-m,k-p} - \theta_1^m a_{-m,k-1} - \ldots - \theta_h^m a_{-m,k-h} \quad (8)$$

where $\omega_{-m,k} = \nabla^\nu c_{-m,k}$, ∇^ν is the difference operator of order ν; p, $\gamma_1^m, \ldots, \gamma_p^m$ are the order and the parameters of autoregression, respectively; h, $\theta_1^m, \ldots, \theta_h^m$ are the order and parameters of the moving average, respectively; $a_{-m,k}$ are residual errors.

In a similar way, for the component $g_{j^r}(t) = \sum_k d_{j^r,k} \Psi_{j^r,k}(t)$ we omit index r and obtain

$$\omega_{j,k}(t) = \gamma_1^j \omega_{j,k-1} + \ldots + \gamma_z^j \omega_{j,k-z} - \theta_1^j a_{j,k-1} - \ldots - \theta_u^j a_{j,k-u} \quad (9)$$

where $\omega_{j,k} = \nabla^{\nu_j} d_{j,k}$, ∇^{ν_j} is the difference operator of order ν_j, z_j, $\gamma_1^j, \ldots, \gamma_z^j$ are the order and the parameters of autoregression, respectively; u, $\theta_1^j, \ldots, \theta_u^j$ are the order and the parameters of the moving average, respectively; $a_{j,k}$ are residual errors.

From (7) to (9) we obtain the representation:

$$A^{REG}(t) = \sum_{\mu=\overline{1,T}} \sum_{k=\overline{1,N^\mu}} s_{j,k}^\mu b_{j,k}^\mu(t), \quad (10)$$

where $s_{j,k}^\mu = \sum_{l=1}^{p^\mu} \gamma_l^\mu \omega_{j,k-l}^\mu - \sum_{n=1}^{h^\mu} \theta_n^\mu a_{j,k-n}^\mu$ is the estimated value of the parameters of a regular μ-th component, p^μ, γ_l^μ are the order and the parameters of autoregression of the μ-th component, h^μ, θ_n^μ are the order and the parameters of the moving average of the μ-th component, $\omega_{j,k}^\mu = \nabla^{\nu_\mu} \delta_{j,k}^\mu$, ν_μ is the order of the μ-th component difference, $\delta_{j,k}^1 = c_{-m,k}$, $\delta_{j,k}^\mu = d_{j,k}$, $\mu = \overline{2,T}$, T is the number of modeled components, $a_{j,k}^\mu$ are the residual errors for the μ-th component model, N^μ is the μ-th component length, $b_{j,k}^1 = \phi_{-m,k}$, ϕ is the scaling function, $b_{j,k}^\mu = \Psi_{j,k}$, $\mu = \overline{2,T}$, Ψ is the wavelet.

The identification of the ARIMA model for the μ-th component requires the determination of the different order ν_μ and the identification of the resulting ARMA process (model order and parameter estimation). The ARIMA model identification is described in detail in [20] and is not presented in the paper.

The diagnostic verification of each of the components f_{-m^r} and g_{j^r} models can be based on the analysis of the model residual errors. Commonly used tests based on the analysis of model residual errors are the cumulative fitting criterion [20] and the cumulative periodogram test [20].

2.4. Anomalous Component of the Model

The anomalous component $U(t)$ of model (1) includes local features of various shapes occurring at random times. Therefore, the application of the parametric approach to identify it is ineffective.

2.4.1. Application of Threshold Functions

In the case of a nonparametric approach, following the results of [37], the function U can be approximated by threshold functions:

$$U(t) = \sum_{j,k} P_j(d_{j,k}) \Psi_{j,k}(t), \qquad (11)$$

$$P_j(d_{j,k}) = \begin{cases} 0, & if\ |d_{j,k}| \leq T_j \\ d_{j,k}, & if\ |d_{j,k}| > T_j \end{cases}$$

In this case, from (7) and (10), we obtain the *hybrid model of time series (HMTS)*

$$f_0(t) = A^{REG}(t) + U(t) + e(t) = \sum_{\mu=\overline{1,T}} \sum_{k=\overline{1,N^\mu}} s_{j,k}^\mu b_{j,k}^\mu(t) + \sum_{j,k} P_j(d_{j,k}) \Psi_{j,k}(t) + e(t), \qquad (12)$$

It was shown in [37] that the mappings (11) allow us to obtain approximations close to optimal ones (by minimizing the minimax risk) for a complex structure function. Moreover, the equivalence of discrete and continuous wavelet expansions [36,38] provides the opportunity to analyze a function on any resolution. In its turn, the increase in the amplitudes of the wavelet coefficients $|d_{j,k}|$ in the vicinity of local features of a function (Jaffard's theorem [39]) will provide, based on (11), their mapping into the component U of model (12).

Obviously, by applying different orthogonal wavelets Ψ we can obtain different representations (12).

We should note that due to the random nature of U, application of any thresholds T_j (see (11)) is inevitably associated with erroneous decisions. In this case, the thresholds can be chosen by minimizing the posteriori risk [40].

The threshold divides the F value space of the function under analysis into two nonintersecting domains F_1 and F_2 determining anomalous and non-anomalous states, respectively. For the specific state h_b, the loss average can be estimated as [40]

$$R_b(f) = \sum_{z=1}^{2} \prod_{bz} P\{f \in F_z | h_b\}, \qquad (13)$$

where $\prod_{bz}$ is the loss function, $P\{f \in F_z | h_b\}$ is the conditional probability of falling within the domain F_z if the state h_b actually exists, $b \neq z$, b, z are the state indices ("|" denotes conditional probability).

Averaging the conditional function of the risk over all the states h_b we obtain the average risk

$$R = \sum_{b=1}^{2} p_b R_b, \qquad (14)$$

where p_b is a priori probability of the state h_b.

If we do not know priori probabilities of the states p_b, then having statistical (priori) data, we can determine posteriori probabilities $P\{h_b|f\}, b = 1,2$. Then, applying a simple loss function

$$\prod_{bz} = \begin{cases} 1, b \neq z, \\ 0, b = z, \end{cases}$$

from (13) and (14), a posteriori risk equals

$$R = \sum_{b \neq z} P\{h_b | f \in F_z\}. \tag{15}$$

2.4.2. Analysis of the Model's Regular Component Errors and Detection of Anomalies

Obviously, during anomalous periods, the residual errors of the model regular component A^{REG} (see (10)) increase. Then *anomaly detection* can be based on the conditional test

$$\varepsilon_j^\mu = \sum_{q=1}^{Q_\mu} \left| a_{j,k+q}^\mu \right| > H_\mu,$$

where $q \geq 1$ is the data lead step, $a_{j,k}^\mu$ are the residual errors of the μ-th component model, Q_μ is the data lead length.

We can estimate the confidence interval of the predicted data [20], which is why it is logical to define the thresholds H_μ as

$$H_\mu(Q_\mu) = \left\{ 1 + \sum_{q=1}^{Q_\mu - 1} \left(\psi_q^\mu \right)^2 \right\}^{1/2} \sigma_{a^\mu}$$

where $\sigma_{a^\mu}^2$ is the variance of residual errors of the μ-th component model; ψ_q^μ are the weighting coefficients of the μ-th component model, they are determined from the equation [20]

$$\left(1 - \varphi_1^\mu B - \varphi_2^\mu B^2 - \ldots - \varphi_{p^\mu + \nu_\mu}^\mu B^{p^\mu + \nu_\mu}\right)\left(1 + \psi_1^\mu B + \psi_2^\mu B^2 + \ldots\right) =$$
$$= \left(1 - \theta_1^\mu B - \theta_2^\mu B^2 - \ldots - \theta_{h^\mu}^\mu B^{h^\mu}\right),$$

where $\varphi_j^\mu = \gamma^\mu(B)(1-B)^{\nu_\mu}$ is the generalized autoregressive operator, B is the back shift operator: $B^l \omega_{j,k}^\mu = \omega_{j,k-l}^\mu$.

It is also possible to use the following probability limits:

$$H_\mu(Q_\mu) = u_{\xi/2} \left\{ 1 + \sum_{q=1}^{Q_\mu - 1} \left(\psi_q^\mu \right)^2 \right\}^{1/2} \sigma_{a^\mu},$$

where $u_{\varepsilon/2}$ is the quantile of the level $(1 - \varepsilon/2)$ of standard normal distribution.

3. Results of the Model Application

3.1. Modeling of Ionospheric Parameter Time Series

The ionosphere is the upper region of the earth's atmosphere. It is located at heights from 70 to 1000 km and higher, and affects radio wave propagation [41]. Ionospheric anomalies occur during extreme solar events (solar flares and particle ejections) and magnetic storms. They cause serious malfunctions in the operation of modern ground and space technical equipment [42]. An important parameter characterizing the state of the ionosphere is the critical frequency of the ionospheric F2-layer (foF2). The foF2 time series have a complex structure and contain seasonal and diurnal components, as well as local features of various shapes and durations occurring during ionospheric anomalies. Intense ionospheric anomalies can cause failures in the operation of technical systems. Therefore, their timely detection is an important applied problem.

In the experiments, we used hourly (1969–2019) and 15-min (2015–2019) foF2 data obtained by the method of vertical radiosonding of the ionosphere at Paratunka station (53.0° N and 158.7° E, Kamchatka, Russia, IKIR FEB RAS). The proposed HMTS was identified separately for foF2 hourly and 15-min data.

To identify HMTS regular components, we used the foF2 data recorded during the periods of absence of ionospheric anomalies. The application of Algorithm 1 showed that the components f_{-3} and g_{-3} are stationary (having damping ACF), thus ARIMA models can be identified for them. Figures 2 and 3 show ACF and PACF of foF2 initial time series, as well as the components f_{-3} and g_{-3}. The results confirm stationarity of the components f_{-3} and g_{-3}. An analysis of PACF shows the possibility to identify the AR models of orders 2 and 3 for the first differences of these components. The results in Figures 2 and 3 also illustrate that foF2 initial time series are non-stationary and, therefore, it is impossible to approximate them by ARIMA model without wavelet decomposition operation.

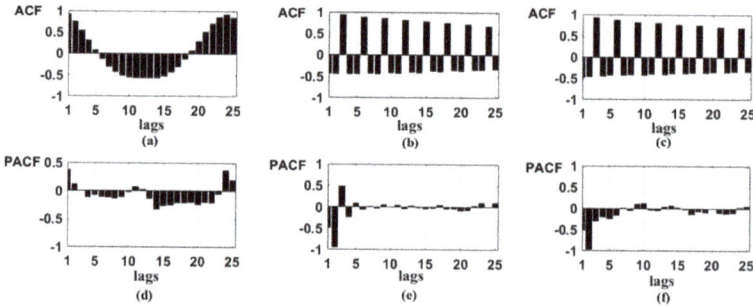

Figure 2. The analyzed period is from 4 January 2014 to 29 January 2014 (high solar activity): (**a**) ACF of the original signal; (**b**) ACF of the component f_{-3}; (**c**) ACF of the component g_{-3}; (**d**) PACF of the 1st difference of the original signal; (**e**) PACF of the 1st difference of the component f_{-3}; (**f**) PACF of the 1st difference of the component g_{-3}.

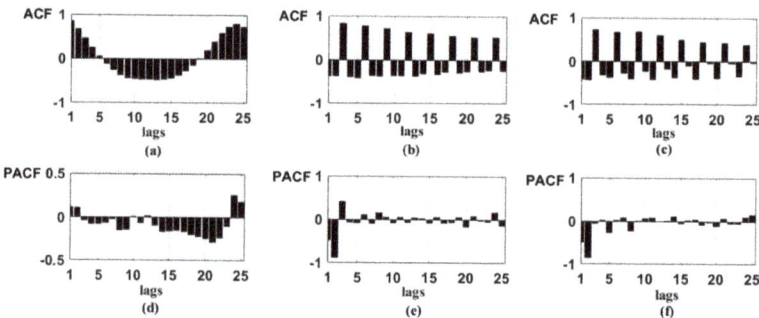

Figure 3. The analyzed period is from 9 February 2008 to 27 February 2008 (low solar activity): (**a**) ACF of the original signal; (**b**) ACF of the component f_{-3}; (**c**) ACF of the component g_{-3}; (**d**) PACF of the 1st difference of the original signal; (**e**) PACF of the 1st difference of the component f_{-3}; (**f**) PACF of the 1st difference of the component g_{-3}.

According to ratio (10) and based on the PACF of the first differences of the components f_{-3} and g_{-3} (Figure 3e,f), we obtain the HMTS regular component

$$A^{REG}(t) = f_{-3}(t) + g_{-3}(t) = \sum_k c_{-3,k}\phi_{-3,k}(t) + \sum_k d_{-3,k}\Psi_{-3,k}(t) = \sum_{\mu=1,2}\sum_{k=\overline{1,N^\mu}} s^\mu_{-3,k} b^\mu_{-3,k}(t),$$

where $s^\mu_{-3,k} = \sum_{l=1}^{p^\mu} \gamma^\mu_l \omega^\mu_{-3,k-l}$, $\mu = 1, 2$, $\omega^1_{-3,k} = \nabla c_{-3,k}$, $\omega^2_{-3,k} = \nabla d_{-3,k}$, $b^1_{-3,k} = \phi_{-3,k}$, $b^2_{-3,k} = \Psi_{-3,k}$. Estimated parameters for $s^1_{-3,k}$ and $s^2_{-3,k}$ are presented in Table 1. The parameters were estimated separately for different seasons and different levels of solar activity.

Table 1. HMTS regular component parameters.

Period	Solar Activity	Parameters of $s^1_{-3,k}$			Parameters of $s^2_{-3,k}$	
		γ^1_1	γ^1_2	γ^1_3	γ^2_1	γ^2_2
winter	low and high	−0.6	−0.6	0.4	−0.9	−0.9
summer	low	−0.8	−0.7	−	−0.9	−0.9
	high	−0.5	−0.6	−	−0.9	−0.8

Based on the data from Table 1 we obtain
(1) for wintertime:
$s^1_{-3,k} = -0.6\omega^1_{-3,k-1} - 0.6\omega^1_{-3,k-2} + 0.4\omega^1_{-3,k-3} + a^1_{-3,k}$,
$s^2_{-3,k} = -0.9\omega^2_{-3,k-1} - 0.9\omega^2_{-3,k-2} + a^2_{-3,k}$,
(2) for summertime and high solar activity:
$s^1_{-3,k} = -0.5\omega^1_{-3,k-1} - 0.6\omega^1_{-3,k-2} + a^1_{-3,k}$,
$s^2_{-3,k} = -0.9\omega^2_{-3,k-1} - 0.8\omega^2_{-3,k-2} + a^2_{-3,k}$,
(3) for summertime and low solar activity:
$s^1_{-3,k} = -0.8\omega^1_{-3,k-1} - 0.7\omega^1_{-3,k-2} + a^1_{-3,k}$,
$s^2_{-3,k} = -0.9\omega^2_{-3,k-1} - 0.9\omega^2_{-3,k-2} + a^2_{-3,k}$.

Figure 4 shows the modeling results for HMTS regular components (f_{-3} and g_{-3}) during the absence of ionospheric anomalies. The model errors do not exceed the confidence interval that indicates their adequacy.

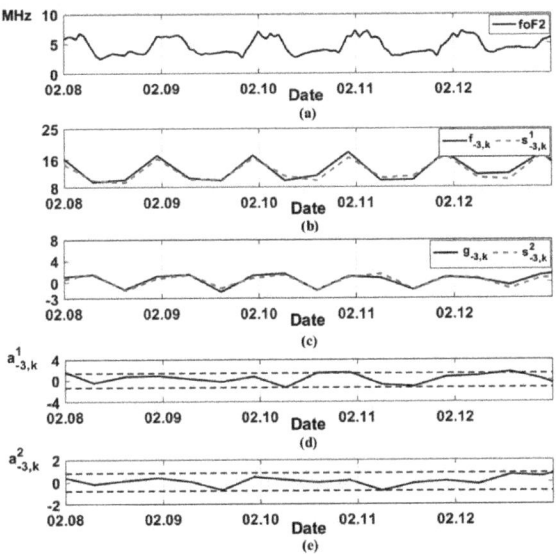

Figure 4. Modeling of the components f_{-3} and g_{-3}: (**a**) foF2 data (8 February 2011–12 February 2011); (**b**) component f_{-3} (black) and its model values $s^1_{-3,k}$ (blue dashed line); (**c**) component g_{-3} (black) and its model values $s^2_{-3,k}$ (blue dashed line); (**d**) errors of $s^1_{-3,k}$; (**e**) errors of $s^2_{-3,k}$. On the graphs (**d**,**e**) the dashed lines show 70% confidence intervals.

Tables 2 and 3, and Figure 5 show the results of validation tests for the obtained models. The tests were carried out for the foF2 data that were not used at the stage of model identification. In order to verify the models, we used the cumulative fitting criterion (Tables 2 and 3), analysis of model residual error ACF (Figure 5a,b) and normalized cumulative periodogram (Figure 5c,d).

Table 2. Cumulative fitting criterion for the winter season.

Periods	Y for s^1_{-3}	Table Value $\chi_{0.1}{}^2/\chi_{0.05}{}^2$	Y for s^2_{-3}	Table Value $\chi_{0.1}{}^2/\chi_{0.05}{}^2$
12.15.1970–12.29.1970	18.36		28.44	
02.07.2002–02.25.2002	22.08		26.40	
01.30.2012–02.11.2012	16.20	24.8/27.6	13.50	26.0/28.9
02.04.2013–02.18.2013	25.90		23.76	
02.19.2016–03.05.2016	19.50		21.06	

Table 3. Cumulative fitting criterion for the summer season.

Periods	Y for s^1_{-3}	Table Value $\chi_{0.1}{}^2/\chi_{0.05}{}^2$	Y for s^2_{-3}	Table Value $\chi_{0.1}{}^2/\chi_{0.05}{}^2$
06.03.1971–06.22.1971	27.26		17.39	
07.11.1990–07.27.1990	16.92	26.0/28.9	18.33	26.0/28.9
08.03.2002–08.17.2002	24.84		23.76	
06.15.2016–06.27.2016	20.70		21.90	

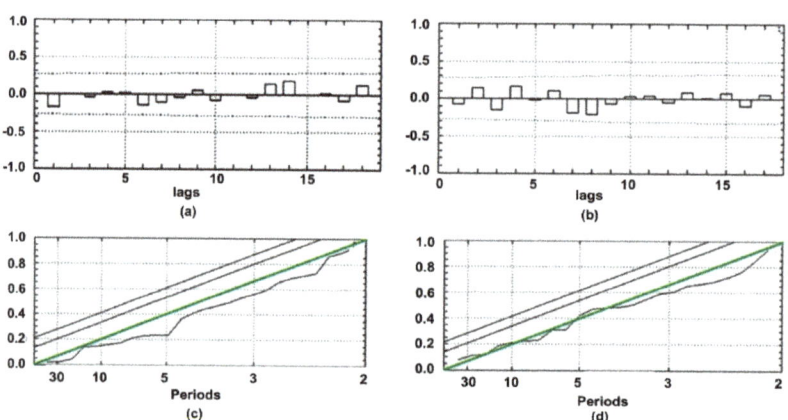

Figure 5. Results of model verification: ACF of residual errors: (**a**) $a^1_{-3,k}$; (**b**) $a^2_{-3,k}$; cumulative periodogram of residual errors: (**c**) $a^1_{-3,k}$; (**d**) $a^2_{-3,k}$.

Based on the cumulative fitting criterion [20], the fitted model is satisfactory if

$$Y = n \sum_{z=1}^{Z} y_z^2(a)$$

is distributed approximately as $\chi^2(Z - p - h)$, where Z are the considered first autocorrelations of model errors, p is the AR model order, h is the MA model order, $y_z(a)$ are the autocorrelations of model error series, $n = N - \nu$, N is the series length, ν is the model difference order.

According to the criterion, if the model is inadequate, the average Y grows. Consequently, the model adequacy can be verified by comparing Y with the table of χ^2 distribution. The results in Tables 2 and 3 show that the Y values of the estimated models, at a significance level $\alpha = 0.05$, do not exceed the table χ^2 values. The model adequacy is also confirmed by the analysis of residual error ACF (Figure 5a,b) and the normalized cumulative periodogram (Figure 5c,d).

Figure 6a,b shows the results of modeling of the hourly foF2 data during the magnetic storm on 18 and 19 December 2019. Figure 6c shows the geomagnetic activity index K (K-index), which characterizes geomagnetic disturbance intensity. The K-index represents the values from 0 to 9, estimated for the three-hour interval. It is known that during increased geomagnetic activity (K > 3), anomalous changes are observed in ionospheric parameters [43]. The analysis of the results in Figure 6 shows an increase in the model errors during the increase in K-index and magnetic storm occurrence (Figure 6b). This indicates ionospheric anomaly occurrences. The results show that the HMTS allows us to detect ionospheric anomalies successfully.

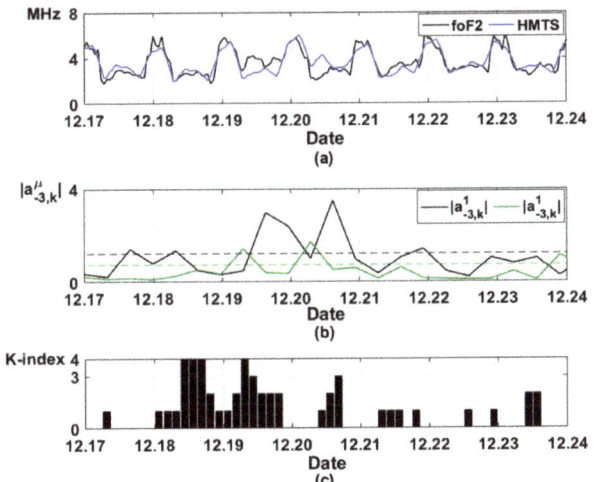

Figure 6. Modeling of foF2 data for the period from 17 December 2019 to 24 December 2019. (**a**) foF2 data (black), HMTS (blue); (**b**) errors of $s^1_{-3,k}$ (black) and $s^2_{-3,k}$ (green), dashed lines show 70% confidence intervals; (**c**) K-index values.

Figure 7 shows the results of the application of operation (11) to 15-min foF2 data during the same magnetic storm. Based on operation (11), ionospheric anomaly occurrences are determined by the threshold function $P_j(d_{j,k})$ with the thresholds T_j.

In this paper, we used the thresholds

$$T_j = V * \sqrt{\frac{1}{\Phi-1} \sum_{k=1}^{\Phi} \left(d_{j,k} - \overline{d_{j,k}}\right)^2}$$

where the coefficient $V = 2.3$ was estimated by minimizing a posteriori risk (ratio (15)), $\overline{d_{j,k}}$ is the average value calculated in a moving time window with the length $\Phi = 480$ (it corresponds to the interval of 5 days).

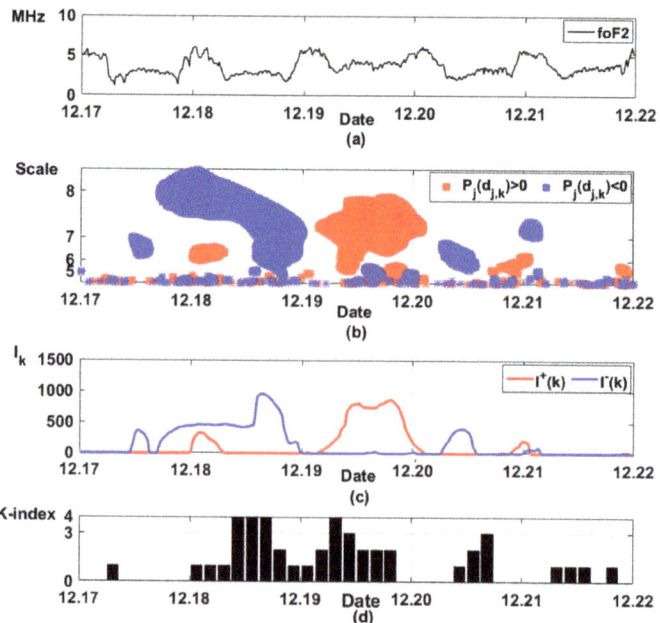

Figure 7. Modeling of foF2 data for the period from 17 December 2019 to 22 December 2019. (**a**) 15-minute data of foF2; (**b**) positive (red) and negative (blue) ionospheric anomalies; (**c**) ionospheric anomaly intensity; (**d**) K-index values.

Positive $(P_j(d_{j,k}) > 0)$ and negative $(P_j(d_{j,k}) < 0)$ anomalies were considered separately. Positive anomalies (shown in red in Figure 7b) characterize the anomalous increase in foF2 values. Negative anomalies (shown in blue in Figure 7b) characterize anomalous decrease in foF2 values. To evaluate the intensity of ionospheric anomalies we used the value

$$I_k = \sum_j P_j(d_{j,k})$$

Assessment of the intensity of positive I_k^+ $(P_j(d_{j,k}) > 0)$ and negative I_k^- $(P_j(d_{j,k}) < 0)$ ionospheric anomalies is shown in Figure 7c, positive anomalies are shown in red, negative ones are shown in blue. Figure 7d shows the K-index values. The results show the occurrence of a negative ionospheric anomaly during the initial and the main phases of the magnetic storm (18 December 2019), and a positive ionospheric anomaly during the recovery phase of the storm (19 December 2019). The observed dynamics of the ionospheric parameters are characteristic of the periods of magnetic storms [43]. The results show the efficiency of HMTS application for detecting ionospheric anomalies of different intensities.

3.2. Comparison of HMTS with NARX Neural Network

To evaluate the HMTS efficiency, we compared it with the NARX neural network [44]. The NARX network is a non-linear autoregressive neural network, and it is often used to forecast time series [44–47]. The architectural structure of recurrent neural networks can take different forms. There are NARX with a Series-Parallel Architecture (NARX SPA) and NARX with a Parallel Architecture (NARX PA) [44,45].

The dynamics of the NARX SPA model is described by the equation

$$y(k+1) = F[x(k), x(k-1), \ldots, x(k-l_x), y(k), y(k-1), \ldots, y(k-l_y)],$$

where $F(\cdot)$ is the neural network display function, $y(k+1)$ is the neural network output, $x(k), x(k-1), \ldots, x(k-l_x)$ are neural network inputs, $y(k), y(k-1), \ldots, y(k-l_y)$ are past values of the time series.

In NARX PA, the network input takes the network outputs $\hat{y}_i = \hat{y}(i)$ instead of the past values of the time series $y_i = y(i)$, $i = \overline{k, k - l_y}$.

The neural networks were trained separately for different seasons and different levels of solar activity. During the training, we used the data for the periods without ionospheric anomalies. We obtained the networks with delays $l_x = l_y = 2$ and $l_x = l_y = 5$ for each season. The results of the networks are shown in Figure 8. Table 4 shows the standard deviations of errors (SD) of networks, which were determined as

$$S = \sqrt{\frac{1}{n}\sum_{i=1}^{n}(y_i - \hat{y}_i)^2}.$$

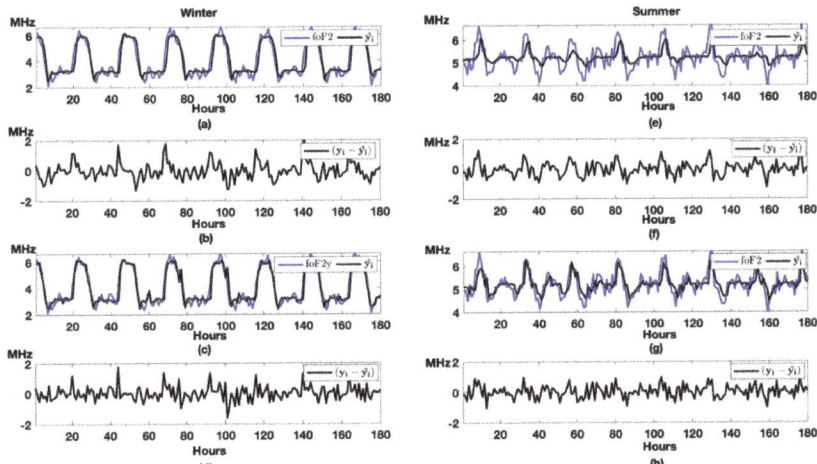

Figure 8. Network errors: (**a,e**) foF2 data (blue), NARX PA output (black); (**b,f**) NARX PA errors; (**c,g**) foF2 data (blue), NARX SPA output (black); (**d,h**) NARX SPA errors.

Table 4. Standard deviations of neural network errors.

	SD of NARX SPA		SD of NARX PA	
Season	Delays: $l_x=l_y=2$	Delays: $l_x=l_y=5$	Delays: $l_x=l_y=2$	Delays: $l_x=l_y=5$
Winter (low solar activity)	0.48	0.43	0.57	0.49
Summer (low solar activity)	0.41	0.36	0.46	0.39
Winter (high solar activity)	0.49	0.48	0.78	0.74
Summer (high solar activity)	0.42	0.36	0.45	0.36

The analysis of the results (Figure 8, Table 4) shows that the NARX SPA predicts the data with fewer errors than the NARX PA. Sending the past time series values to the NARX SPA network input (rather than network outputs) made it possible to obtain a more accurate data prediction. The comparison results of the NARX SPA with the HMTS are presented below.

Figure 9 shows the results of ionospheric data modeling based on HMTS and NARX SPA during the periods of absence of ionospheric anomalies. The results show that the model errors have similar values for the winter and summer seasons, and vary within the interval of $[-1,1]$, both for HMTS and NARX SPA.

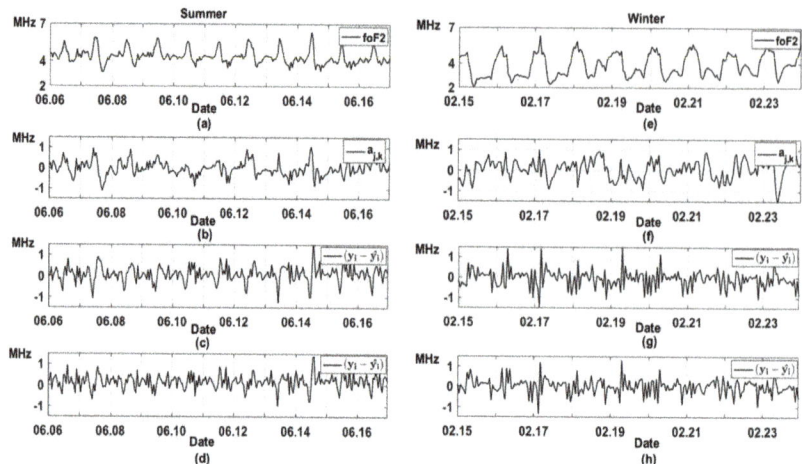

Figure 9. Errors of HMTS and NARX SPA for summer (from 6 June 2019 to 16 June 2019) and winter (from 15 February 2019 to 23 February 2019) seasons: (**a**,**e**) foF2 data; (**b**,**f**) HMTS errors; (**c**,**g**) NARX SPA errors (network delays $l_x = l_y = 2$); (**d**,**h**) NARX SPA errors (network delays $l_x = l_y = 5$).

Figure 10 shows the results of the application of HMTS and NARX SPA for hourly foF2 data during magnetic storms that occurred on 21–22 November 2017 and 5–6 August 2019. NARX SPA errors were calculated in a 3-h moving time window: $\varepsilon_i = \sum_{i=i-1}^{i+1} |y_i - \hat{y}_i|$.

Figure 10e,j shows the geomagnetic activity Dst-index, which characterizes geomagnetic disturbance intensity during magnetic storms. Dst-index takes negative values during magnetic storms. The increases in HMTS and NARX SPA errors during the analyzed magnetic storms (Figure 10b–d,g–i) indicate ionospheric anomaly occurrences. The results show that HMTS and NARX SPA allow us to detect ionospheric anomalies successfully. However, an increase in NARX SPA errors is also observed in wintertime on the eve and after the magnetic storm (Figure 10c,d). This shows the presence of false alarms.

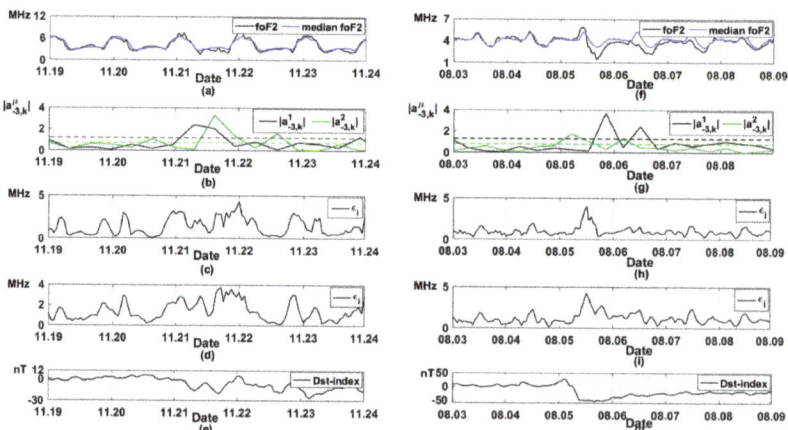

Figure 10. Modeling of hourly foF2 data: (**a**,**f**) recorded foF2 data (black), foF2 median (blue); (**b**,**g**) errors of $s^1_{-3,k}$ (black) and $s^2_{-3,k}$ (green), dashed lines show 70% confidence intervals; (**c**,**h**) NARX SPA errors (network delays $l_x = l_y = 2$); (**d**,**i**) NARX SPA errors (network delays $l_x = l_y = 5$); (**e**,**j**) Dst-index of geomagnetic activity.

The results of detecting ionospheric anomalies based on HMTS and NARX SPA are shown in Tables 5–8. The estimates were based on statistical modeling. The HMTS results are shown for the 90% confidence interval. The analysis of the results shows that NARX SPA efficiency exceeds that for HMTS during high solar activity. However, the frequency of false alarms for HMTS is significantly less than that for NARX SPA.

Table 5. Results for wintertime and high solar activity.

Signal/Noise	HMTS		NARX SPA	
	Detected/False		Detected/False	
	Component f_{-3}	Component g_{-3}	Delays: $l_x=l_y=2$	Delays: $l_x=l_y=5$
1.3	95%/0%	78%/2%	86%/5%	96%/4%
1	92%/0%	74%/7%	76%/8%	94%/9%
0.8	85%/3%	74%/11%	75%/12%	84%/12%

Table 6. Results for wintertime and low solar activity.

Signal/Noise	HMTS		NARX SPA	
	Detected/False		Detected/False	
	Component f_{-3}	Component g_{-3}	Delays: $l_x=l_y=2$	Delays: $l_x=l_y=5$
1.3	97%/0%	90%/5%	81%/1%	89%/2%
1	96%/2%	89%/12%	73%/12%	84%/12%
0.8	85%/6%	89%/17%	70%/19%	82%/18%

Table 7. Results for summertime and high solar activity.

Signal/Noise	HMTS		NARX SPA	
	Detected/False		Detected/False	
	Component f_{-3}	Component g_{-3}	Delays: $l_x=l_y=2$	Delays: $l_x=l_y=5$
1.3	79%/0%	80%/2%	79%/5%	81%/7%
1	70%/0%	65%/4%	71%/15%	72%/14%
0.8	55%/1%	63%/10%	64%/18%	64%/17%

Table 8. Results for summertime and low solar activity.

Signal/Noise	HMTS		NARX SPA	
	Detected/False		Detected/False	
	Component f_{-3}	Component g_{-3}	Delays: $l_x=l_y=2$	Delays: $l_x=l_y=5$
1.3	94%/0%	83%/3%	92%/2%	93%/3%
1	90%/0%	80%/9%	90%/6%	91%/9%
0.8	86%/2%	80%/13%	85%/11%	84%/15%

4. Conclusions

The paper proposes a hybrid model of time series of complex structure. The model is based on the combination of function expansions in a wavelet series with ARIMA models. Ionospheric critical frequency data were used to estimate the HMTS efficiency. The estimates showed:

1. The HMTS regular component adequately describes ionospheric parameter time series during the periods without ionospheric anomalies. Application of wavelet decomposition allows us to detect regular components of ionospheric parameter time series and to use the ARIMA model;
2. Analysis of HMTS regular component errors allows us to detect ionospheric anomalies during a magnetic storm;
3. The HMTS anomalous component allows us to detect ionospheric anomalies of different intensities by threshold functions.

Comparison of HMTS with NARX with Series-Parallel Architecture confirmed the HMTS efficiency to detect anomalies in the ionospheric critical frequency data. The results of the experiments showed that the efficiency of the NARX neural network slightly exceeds that of HMTS (about 2–3%) during high solar activity. However, the frequency of false alarms in NARX is significantly higher (about 15%). During the periods of low solar activity, the efficiency of HMTS exceeds that of NARX.

The HMTS can be used for modeling and analysis of time series of complex structure, including seasonal components and local features of various forms.

Author Contributions: Conceptualization, O.M.; methodology, O.M. and N.F.; software, N.F. and Y.P.; formal analysis, O.M., Y.P. and N.F.; project administration, O.M. All authors have read and agreed to the published version of the manuscript.

Funding: The work was carried out according to the Subject AAAA-A21-121011290003-0 "Physical processes in the system of near space and geospheres under solar and lithospheric influences" IKIR FEB RAS.

Institutional Review Board Statement: Not applicable.

Informed Consent Statement: Not applicable.

Data Availability Statement: Not applicable.

Acknowledgments: The work was carried out by the means of the Common Use Center "North-Eastern Heliogeophysical Center" CKP_558279", "USU 351757. The authors are grateful to the Institutes that support the ionospheric stations data that were used in the work. We would like to thank anonymous Reviewers for their greatly appreciated efforts helping to improve the paper.

Conflicts of Interest: The authors declare no conflict of interest.

References

1. Song, R.; Zhang, X.; Zhou, C.; Liu, J.; He, J. Predicting TEC in China based on the neural networks optimized by genetic algorithm. *Adv. Space Res.* **2018**, *62*, 745–759. [CrossRef]
2. Bailey, R.L.; Leonhardt, R. Automated detection of geomagnetic storms with heightened risk of GIC. *Earth Planets Space* **2016**, *68*, 99. [CrossRef]
3. Mandrikova, O.V.; Solovev, I.S.; Zalyaev, T.L. Methods of analysis of geomagnetic field variations and cosmic ray data. *Earth Planets Space* **2014**, *66*, 148. [CrossRef]
4. Tang, R.; Zeng, F.; Chen, Z.; Wang, J.-S.; Huang, C.-M.; Wu, Z. The Comparison of Predicting Storm-Time Ionospheric TEC by Three Methods: ARIMA, LSTM, and Seq2Seq. *Atmosphere* **2020**, *11*, 316. [CrossRef]
5. Perol, T.; Gharbi, M.; Denolle, M. Convolutional neural network for earthquake detection and location. *Sci. Adv.* **2018**, *4*, e1700578. [CrossRef]
6. Tronin, A.A. Satellite Remote Sensing in Seismology. A Review. *Remote Sens.* **2009**, *2*, 124–150. [CrossRef]
7. Chierici, F.; Embriaco, D.; Pignagnoli, L. A new real-time tsunami detection algorithm. *J. Geophys. Res. Ocean.* **2017**, *122*, 636–652. [CrossRef]
8. Kim, S.-K.; Lee, E.; Park, J.; Shin, S. Feasibility Analysis of GNSS-Reflectometry for Monitoring Coastal Hazards. *Remote Sens.* **2021**, *13*, 976. [CrossRef]
9. Alperovich, L.; Eppelbaum, L.; Zheludev, V.; Dumoulin, J.; Soldovieri, F.; Proto, M.; Bavusi, M.; Loperte, A. A new combined wavelet methodology: Implementation to GPR and ERT data obtained in the Montagnole experiment. *J. Geophys. Eng.* **2013**, *10*, 25017. [CrossRef]
10. Amigó, J.M.; Small, M. Mathematical methods in medicine: Neuroscience, cardiology and pathology. *Philos. Trans. R. Soc. A Math. Phys. Eng. Sci.* **2017**, *375*. [CrossRef]

11. Chen, J.; Heincke, B.; Jegen, M.; Moorkamp, M. Using empirical mode decomposition to process marine magnetotelluric data. *Geophys. J. Int.* **2012**, *190*, 293–309. [CrossRef]
12. Chen, L.; Han, W.; Huang, Y.; Cao, X. Online Fault Diagnosis for Photovoltaic Modules Based on Probabilistic Neural Network. *Eur. J. Electr. Eng.* **2019**, *21*, 317–325. [CrossRef]
13. Robbins, H.; Monro, S. A Stochastic Approximation Method. *Ann. Math. Stat.* **1951**, *22*, 400–407. [CrossRef]
14. Vasconcelos, J.C.S.; Cordeiro, G.M.; Ortega, E.M.M.; De Rezende, É.M. A new regression model for bimodal data and applications in agriculture. *J. Appl. Stat.* **2021**, *48*, 349–372. [CrossRef]
15. Valipour, M.; Banihabib, M.E.; Behbahani, S.M.R. Comparison of the ARMA, ARIMA, and the autoregressive artificial neural network models in forecasting the monthly inflow of Dez dam reservoir. *J. Hydrol.* **2013**, *476*, 433–441. [CrossRef]
16. Li, S.; Wang, Q. India's dependence on foreign oil will exceed 90% around 2025—The forecasting results based on two hybridized NMGM-ARIMA and NMGM-BP models. *J. Clean. Prod.* **2019**, *232*, 137–153. [CrossRef]
17. Wu, X.; Zhou, J.; Yu, H.; Liu, D.; Xie, K.; Chen, Y.; Hu, J.; Sun, H.; Xing, F. The Development of a Hybrid Wavelet-ARIMA-LSTM Model for Precipitation Amounts and Drought Analysis. *Atmosphere* **2021**, *12*, 74. [CrossRef]
18. Miljkovic, D.; Shaik, S.; Miranda, S.; Barabanov, N.; Liogier, A. Globalisation and Obesity. *World Econ.* **2015**, *38*, 1278–1294. [CrossRef]
19. Ivanov, L.; Collins, C.; Margolina, T. Reconstruction of Diffusion Coefficients and Power Exponents from Single Lagrangian Trajectories. *Fluids* **2021**, *6*, 111. [CrossRef]
20. Box, G.E.P.; Jenkins, G.M. *Time Series Analysis: Forecasting and Control, Rev. ed.*; Holden-Day Series in Time Series Analysis and Digital Processing; Holden-Day: San Francisco, CA, USA, 1976; ISBN 978-0-8162-1104-3.
21. Liu, J.; Kumar, S.; Palomar, D.P. Parameter Estimation of Heavy-Tailed AR Model with Missing Data via Stochastic EM. *IEEE Trans. Signal Process.* **2019**, *67*, 2159–2172. [CrossRef]
22. Chatfield, C.; Koehler, A.B.; Ord, J.K.; Snyder, R.D. A New Look at Models for Exponential Smoothing. *J. R. Stat. Soc. Ser. D* **2001**, *50*, 147–159. [CrossRef]
23. Estévez, J.; Bellido-Jiménez, J.A.; Liu, X.; García-Marín, A.P. Monthly Precipitation Forecasts Using Wavelet Neural Networks Models in a Semiarid Environment. *Water* **2020**, *12*, 1909. [CrossRef]
24. Mbatha, N.; Bencherif, H. Time Series Analysis and Forecasting Using a Novel Hybrid LSTM Data-Driven Model Based on Empirical Wavelet Transform Applied to Total Column of Ozone at Buenos Aires, Argentina (1966–2017). *Atmosphere* **2020**, *11*, 457. [CrossRef]
25. Mehdizadeh, S.; Fathian, F.; Adamowski, J.F. Hybrid artificial intelligence-time series models for monthly streamflow modeling. *Appl. Soft Comput.* **2019**, *80*, 873–887. [CrossRef]
26. Shishegaran, A.; Saeedi, M.; Kumar, A.; Ghiasinejad, H. Prediction of air quality in Tehran by developing the nonlinear ensemble model. *J. Clean. Prod.* **2020**, *259*, 120825. [CrossRef]
27. Büyükşahin, Ü.Ç.; Ertekin, Ş. Improving forecasting accuracy of time series data using a new ARIMA-ANN hybrid method and empirical mode decomposition. *Neurocomputing* **2019**, *361*, 151–163. [CrossRef]
28. Vivas, E.; Allende-Cid, H.; Salas, R.; Bravo, L. Polynomial and Wavelet-Type Transfer Function Models to Improve Fisheries' Landing Forecasting with Exogenous Variables. *Entropy* **2019**, *21*, 1082. [CrossRef]
29. Sebastian, D.E.; Ganguly, S.; Krishnaswamy, J.; Duffy, K.; Nemani, R.; Ghosh, S. Multi-Scale Association between Vegetation Growth and Climate in India: A Wavelet Analysis Approach. *Remote Sens.* **2019**, *11*, 2703. [CrossRef]
30. Hurat, B.; Alvarado, Z.; Gilles, J. The Empirical Watershed Wavelet. *J. Imaging* **2020**, *6*, 140. [CrossRef]
31. Casado-Vara, R.; del Rey, A.M.; Pérez-Palau, D.; De-La-Fuente-Valentín, L.; Corchado, J. Web Traffic Time Series Forecasting Using LSTM Neural Networks with Distributed Asynchronous Training. *Mathematics* **2021**, *9*, 421. [CrossRef]
32. Moon, J.; Hossain, B.; Chon, K.H. AR and ARMA model order selection for time-series modeling with ImageNet classification. *Signal Process.* **2021**, *183*, 108026. [CrossRef]
33. Yang, S.; Chen, H.-C.; Wu, C.-H.; Wu, M.-N.; Yang, C.-H. Forecasting of the Prevalence of Dementia Using the LSTM Neural Network in Taiwan. *Mathematics* **2021**, *9*, 488. [CrossRef]
34. Phan, T.-T.-H.; Nguyen, X.H. Combining statistical machine learning models with ARIMA for water level forecasting: The case of the Red River. *Adv. Water Resour.* **2020**, *142*, 103656. [CrossRef]
35. Khan, M.H.; Muhammad, N.S.; El-Shafie, A. Wavelet based hybrid ANN-ARIMA models for meteorological drought forecasting. *J. Hydrol.* **2020**, *590*, 125380. [CrossRef]
36. Chui, C.K. *An Introduction to Wavelets*; Wavelet Analysis and Its Applications; Academic Press: Boston, MA, USA, 1992; ISBN 978-0-12-174584-4.
37. Mallat, S.G. *A Wavelet Tour of Signal Processing*; Academic Press: San Diego, CA, USA, 1999; ISBN 978-0-12-466606-1.
38. Daubechies, I. *Ten Lectures on Wavelets*; CBMS-NSF Regional Conference Series in Applied Mathematics; Society for Industrial and Applied Mathematics: Philadelphia, PA, USA, 1992; ISBN 978-0-89871-274-2.
39. Jaffard, S. Pointwise smoothness, two-microlocalization and wavelet coefficients. *Publ. Matemàtiques* **1991**, *35*, 155–168. [CrossRef]
40. Berger, J.O. *Statistical Decision Theory and Bayesian Analysis*, 2nd ed.; Springer Series in Statistics; Springer: New York, NY, USA, 1993; ISBN 978-0-387-96098-2.
41. Hernández-Pajares, M.; Juan, J.M.; Sanz, J.; Aragón-Àngel, À.; García-Rigo, A.; Salazar, D.; Escudero, M. The ionosphere: Effects, GPS modeling and the benefits for space geodetic techniques. *J. Geod.* **2011**, *85*, 887–907. [CrossRef]

42. Ferreira, A.A.; Borries, C.; Xiong, C.; Borges, R.A.; Mielich, J.; Kouba, D. Identification of potential precursors for the occurrence of Large-Scale Traveling Ionospheric Disturbances in a case study during September 2017. *J. Space Weather Space Clim.* **2020**, *10*, 32. [CrossRef]
43. Danilov, A. Ionospheric F-region response to geomagnetic disturbances. *Adv. Space Res.* **2013**, *52*, 343–366. [CrossRef]
44. Diaconescu, E. The use of NARX neural networks to predict chaotic time series. *WSEAS Trans. Comp. Res.* **2008**, *3*, 182–191.
45. Narendra, K.; Parthasarathy, K. Identification and control of dynamical systems using neural networks. *IEEE Trans. Neural Netw.* **1990**, *1*, 4–27. [CrossRef]
46. Ma, Q.; Liu, S.; Fan, X.; Chai, C.; Wang, Y.; Yang, K. A Time Series Prediction Model of Foundation Pit Deformation Based on Empirical Wavelet Transform and NARX Network. *Mathematics* **2020**, *8*, 1535. [CrossRef]
47. Boussaada, Z.; Curea, O.; Remaci, A.; Camblong, H.; Bellaaj, N.M. A Nonlinear Autoregressive Exogenous (NARX) Neural Network Model for the Prediction of the Daily Direct Solar Radiation. *Energies* **2018**, *11*, 620. [CrossRef]

Article

Applied Machine Learning Algorithms for Courtyards Thermal Patterns Accurate Prediction

Eduardo Diz-Mellado [1,†], Samuele Rubino [2,†], Soledad Fernández-García [2], Macarena Gómez-Mármol [2,*], Carlos Rivera-Gómez [1,*] and Carmen Galán-Marín [1]

1 Departamento de Construcciones Arquitectónicas 1, Escuela Técnica Superior de Arquitectura, Universidad de Sevilla, Avda. Reina Mercedes, 2, 41012 Seville, Spain; ediz@us.es (E.D.-M.); cgalan@us.es (C.G.-M.)
2 Departamento de Ecuaciones Diferenciales y Análisis Numérico, Facultad de Matemáticas and Instituto de Matemáticas (IMUS), Universidad de Sevilla, C/Tarfia, s/n, 41012 Seville, Spain; samuele@us.es (S.R.); soledad@us.es (S.F.-G.)
* Correspondence: macarena@us.es (M.G.-M.); crivera@us.es (C.R.-G.)
† E.D.-M. and S.R. contributed equally.

Abstract: Currently, there is a lack of accurate simulation tools for the thermal performance modeling of courtyards due to their intricate thermodynamics. Machine Learning (ML) models have previously been used to predict and evaluate the structural performance of buildings as a means of solving complex mathematical problems. Nevertheless, the microclimatic conditions of the building surroundings have not been as thoroughly addressed by these methodologies. To this end, in this paper, the adaptation of ML techniques as a more comprehensive methodology to fill this research gap, covering not only the prediction of the courtyard microclimate but also the interpretation of experimental data and pattern recognition, is proposed. Accordingly, based on the climate zoning and aspect ratios of 32 monitored case studies located in the South of Spain, the Support Vector Regression (SVR) method was applied to predict the measured temperature inside the courtyard. The results provided by this strategy showed good accuracy when compared to monitored data. In particular, for two representative case studies, if the daytime slot with the highest urban overheating is considered, the relative error is almost below 0.05%. Additionally, values for statistical parameters are in good agreement with other studies in the literature, which use more computationally expensive CFD models and show more accuracy than existing commercial tools.

Keywords: courtyard; climate change; microclimate; Support Vector Regression (SVR); machine learning

Citation: Diz-Mellado, E.; Rubino, S.; Fernández-García, S.; Gómez-Mármol, M.; Rivera-Gómez, C.; Galán-Marín, C. Applied Machine Learning Algorithms for Courtyards Thermal Patterns Accurate Prediction. *Mathematics* **2021**, 9, 1142. https://doi.org/10.3390/math9101142

Academic Editor: Lucas Jódar

Received: 30 March 2021
Accepted: 14 May 2021
Published: 18 May 2021

Publisher's Note: MDPI stays neutral with regard to jurisdictional claims in published maps and institutional affiliations.

Copyright: © 2021 by the authors. Licensee MDPI, Basel, Switzerland. This article is an open access article distributed under the terms and conditions of the Creative Commons Attribution (CC BY) license (https://creativecommons.org/licenses/by/4.0/).

1. Introduction

According to the latest forecasts, two trends will become progressively reinforced over the present century. The first one is the gradual increase in average surface temperatures mainly due to global greenhouse gas emissions [1]. The second is the concentration of the population in cities [2]. This combination of factors, rising temperatures, and high population concentration will accentuate other environmental problems related to human thermal comfort, such as the so-called Urban Heat Island (UHI) effect. Urban Heat Islands (UHIs) are defined as urban areas with higher air temperatures than their surrounding rural areas [3]. The causes of the UHI effect are classified differently by Givoni [4] as due either to meteorological factors or to urban parameters [5].

Several urban dynamics converge to generate this overheating. Apart from domestic and industrial anthropogenic impacts, some of these factors are related to built-up topography and urban features: firstly, the constructed zones offer more surface area for heat absorption, radiating it slowly during the night; secondly, the canyon effect [6], which causes the thermal energy to remain in the ground by the influence of multiple horizontal reflections and absorption of incoming radiation provoked by tall buildings. UHI is also linked to a capsule of city gases that absorbs heat from the sun. In the city, buildings obstruct the wind and the capsule remains in place [7]. Finally, the urban albedo, which could be defined as the aptitude of construction materials to reflect solar radiation [8].

Considering the need to achieve the medium-term goal of nearly zero-energy buildings and cities [9], different passive strategies have been evaluated to counteract this urban overheating without resorting to energy-dependent cooling systems [10]. Like other animal colonies, cities are usually adapted to the climate as kinds of human termite mounds, perforating the urban fabric to regulate direct solar radiation. On a different scale than other public spaces, such as urban canyons and squares, courtyards have traditionally acted as passive cooling resources in cities around the world and not exclusively in hot and warm climates. One study on low-rise housing in the Netherlands shows how courtyards improve the energy efficiency of the building [11]. Previous research performed on courtyards in Spain has quantified the courtyard tempering effect, which enables improving thermal comfort and helping to reduce cooling energy consumption in buildings [12]. Due to the growing interest in strategies capable of achieving more climate-resilient cities, many studies have examined the microclimatic performance of the courtyard. Furthermore, several literature reviews compiling research on this topic have been published [13–15]. The courtyard microclimate can be explained in terms of the thermodynamic effects that occur within it, i.e., convection, radiation, stratification, and flow patterns. Among the different parameters affecting these microclimatic conditions, most of the studies emphasize the importance of courtyard geometry [13–29], in many cases considered the Aspect Ratio (AR), which is the ratio between the height and the width of the courtyard.

$$AR = \frac{Height}{Width} \qquad (1)$$

Courtyard location, implying climatic conditions and specifically outdoor temperature ranges, is another key factor that is becoming commonplace in a large number of publications [14–21,25–30].

The perforation of the urban block with courtyards responds to light, ventilation, and thermal needs. Different field monitoring campaigns in the existing literature have proved the thermal tempering potential of courtyards to lower the outdoor temperature, in some cases by up to 15 °C [23]. Many simulation methods and tools are currently available for the thermal performance modeling of indoor spaces [22]. Notwithstanding, the alternatives for simulating outdoor ones are more limited. This is mainly due to the complexity of these outdoor spaces' thermodynamics, which involve multiple variables and entails enormous challenges to be modeled with enough accuracy. However, new software means have emerged in recent years that are capable, to some extent, of simulating their microclimatic conditions [31]. One of the most widely used tools is ENVI-met, based on CFD simulation [32]. Other outdoor modeling software alternatives are Urban Weather Generator (UWG), based on energy conservation principles [33]; SOLWEIG, which can simulate spatial variations of 3D radiation fluxes and mean radiant temperatures [34]; Open FOAM, which has been used in previous research to simulate urban wind flows [35]; FreeFem++, employed to perform courtyard microclimate modelling [17,36]; and ANSYS Fluent, which has been applied for the simulation of wind flows in outdoor spaces [17,36]. Most of these tools present adequate accuracy for predicting urban outdoor microclimates, but they tend to show a larger error range when they are used to model the microclimate of smaller-scale spaces, such as courtyards, with greater dependence on the built environment [12,36].

Consequently, in this work, a new tool is proposed to predict this specific microclimate inside courtyards based on Machine Learning (ML) techniques [37,38]. In the computers and information era, a large amount of data are being generated in many different fields, such as science, finance, engineering, and industry. Thus, statistical problems have grown in size and complexity and the statistical analysis tries to understand these data. This is what is called learning from data or ML. Some examples of ML problems are the following:

predict the price of a stock for 6 months from now, based on company performance measures and economic data; identify the numbers in a handwritten ZIP code from a digitized image, or estimate the amount of glucose in the blood of a diabetic person from the infrared absorption spectrum of that person's blood [38]. ML models have been shown previously to be useful for predicting and assessing structural performance [39].

ML problems are categorized as supervised or unsupervised. In supervised learning, the aim is to predict the value of an outcome measure based on a certain number of input measures (also known as features, attributes, or covariates). It is called supervised because of the presence of the outcome variable to guide the learning process. In unsupervised learning, there is no outcome measure, and the goal is to describe the associations and patterns among a set of input measures. Mathematical optimization has played a crucial role in supervised learning [37–40]. Support Vector Machine (SVM) and Support Vector Regression (SVR) are some of the main applications of mathematical optimization for supervised learning [41–46]. These are geometrical optimization problems that can be written as convex quadratic optimization problems with linear constraints, solvable by some nonlinear optimization procedure.

The present paper's main goal is to implement the ML methodology as a suitable and accurate system for predicting courtyard thermal patterns. To achieve this, the most relevant features regarding courtyards' thermoregulatory performance according to the literature, i.e., geometry and outdoor temperature, have been considered. The advantages of using ML techniques over conventional modeling tools are twofold: on the one hand, they allow the identification of the fundamental variables, simplifying the calculation processes; on the other hand, they are perfectible methodologies that make it possible to increase the accuracy of predictions by providing feedback from monitored data by increasing the size of the training dataset. In fact, despite presenting work based on an extensive set of field-monitoring campaigns, the case studies monitored could be considered an initial limitation of the study. Nevertheless, the proposed methodology achieves an accuracy level comparable to, and in some cases superior to, other outdoor thermal modeling methods. The overall structure of the paper can be framed in a three-phase procedure. Firstly, the case studies used to validate the thermal predictions are selected and monitored. Secondly, simulations based on SVR and correlated employing MATLAB interpolations are performed. Finally, different error ranges are verified and compared with other tools simulation errors in the thermal patterns' prediction of the courtyard microclimate. Note that the interpolation technique is applied when characteristic parameters are within an appropriate range, defined by training data. Outside of this range, other prediction techniques are needed.

2. Materials and Methods

Regarding specifically the application of the ML methodology, it was sequenced into four steps. First, the reference study cases are defined and characterized. Second, the field monitoring campaigns are characterized. Third, the problem setup is detailed. Fourth, the SVR method is described. In particular, the following variables are considered: time (hours), outside courtyard measured temperature (CMT), wind speed and direction, with the aim of searching for a function that provides the temperature inside the courtyard all along the week. This problem was solved using the statistical software R. Finally, using the library of predicted data obtained from the ML method, the measured temperature inside a given courtyard is predicted, based on its climate zone, year´s season, and ARs. This will be done in two phases by an interpolation technique implemented in the scientific software MATLAB.

2.1. Location, Climate and Cases Study

In this research, the thermal performance of 22 selected courtyards in a total of 12 different locations in three different Thermal Ranges (TR) are analyzed as case studies.

The study was carried out in Mérida (Badajoz, Spain), Córdoba (Córdoba, Spain) and Seville (Seville, Spain), located in south-western Spain. All three cities are characterized by a hot climate in summer and a mild climate in winter. The specific Spanish regulations CTE-DB-HE [47], characterize them as C4, B4, and B4, respectively. The letter (A–D) represents the winter climate severity ranging from A for mild temperatures to D for very cold climates, and the number (1–4) represents the summer climate severity, being 1 for mild climates and 4 for very hot climates.

The selected case studies are intended to be analyzed in the warm season, so they all belong to the same climatic zone in summer. According to the Köppen classification, the selected cities are defined as Csa, with dry summers with low rainfall and very hot summers. Many case studies were analyzed over an extended period, always exceeding the minimum two-week monitoring period established by previous research [48].

Previous studies have shown the influence of outdoor temperature and geometry on the thermal tempering potential of the courtyard [49] and their thermal sensation [50], so for this research, a selection of case studies with different outdoor temperatures and different AR (Equation (1), defined in (1)) are analyzed. Therefore, two values, ARI and ARII, are defined. In Table 1, the main characteristics of the case studies selected for this research are shown, including the longitude, latitude and meters above sea level (MASL) of each case study.

Table 1. Location of monitored courtyards in this work.

Courtyard	TR	Location	Longitude	Latitude	MASL	Climate Zone	Floor Area (m²)	Dimensions (m)		Height (m)	AR I	AR II
CS1	TR1	M	38.90°	6.35°	229	C4	34.1	5.2	6.5	5.9	1.14	0.91
CS2	TR1	M	38.92°	6.93°	229	C4	25.2	4.1	6.1	6.3	1.54	1.04
CS3	TR1	M	38.91°	6.34°	229	C4	34.7	5.2	6.7	6.3	1.21	0.95
CS4	TR1	M	38.92°	6.35°	229	C4	36.5	3.2	11.5	10.8	3.41	0.94
CS5	TR2	S	37.39°	5.96°	16	B4	35.9	6.9	5.2	5	0.72	0.96
CS6	TR2	C	37.88°	4.78°	106	B4	14.6	4.3	3.4	6.3	1.47	1.85
CS7	TR2	S	37.40°	6.00°	16	B4	101.2	11	9.2	12	1.1	1.3
CS8	TR2	S	37.39°	5.96°	16	B4	35.9	6.9	5.2	5	0.72	0.96
CS9	TR2	S	37.36°	5.99°	16	B4	48.2	7.3	6.6	14	1.92	2.12
CS10	TR2	S	37.40°	6.00°	16	B4	22.4	5.6	4	8.5	1.5	2.12
CS11	TR2	S	37.39°	5.96°	16	B4	35.9	6.9	5.2	5	0.72	0.96
CS12	TR2	S	37.36°	5.99°	16	B4	48.2	7.3	6.6	14	1.92	2.12
CS13	TR2	S	37.28°	5.92°	16	B4	75.9	11	6.9	8.9	0.81	1.29
CS14	TR2	S	37.39°	5.96°	16	B4	35.9	6.9	5.2	5	0.72	0.96
CS15	TR2-3	S	37.28°	5.93°	16	B4	99	13.2	7.5	10.7	0.81	1.43
CS16	TR3	C	37.88°	4.78°	106	B4	65.5	8.4	7.8	6.8	0.81	0.87
CS17	TR3	C	37.88°	4.77°	106	B4	14.6	4.3	3.4	6.3	1.47	1.85
CS18	TR3	S	37.36°	5.99°	16	B4	48.2	7.3	6.6	14	1.92	2.12
CS19	TR3	C	37.88°	4.78°	106	B4	65.5	8.4	7.8	6.8	0.81	0.87
CS20	TR3	S	37.35°	5.99°	16	B4	48.2	7.3	6.6	14	1.92	2.12
CS21	TR3	C	37.88°	4.77°	106	B4	14.6	4.3	3.4	6.3	1.47	1.85
CS22	TR3	C	37.89°	4.78°	106	B4	65.5	8.4	7.8	6.8	0.81	0.87

2.2. Field Monitoring Campaign

As previously mentioned, in this research, numerous monitoring campaigns have been carried out in courtyards with diverse geometries (AR) and with different outdoor temperatures (TR). For both boundary conditions, AR and TR, the selected ranges are based on previous studies [23].

Some campaigns were carried out over several months to select similar outdoor temperature ranges in all case studies. One week was selected as a representative sample for each courtyard. During the monitoring campaigns, outdoor climatological parameters were analyzed, and simultaneously, the temperature inside the courtyards was recorded. According to the U.S. National Weather Service [51], dry-bulb temperature (DBT), can be measured using a normal thermometer freely exposed to the air but shielded from solar radiation and moisture. The thermometer will be affected by thermal radiation from the courtyard walls, so we will refer to the DBT as the Courtyard Measured Temperature (CMT) rather than the air temperature. In the case of outdoor environment analysis, portable weather stations model PCE-FWS 20 were used, the technical data of which are shown in Table 2. The weather station was located on the roof of the building, fully exposed, with no nearby high-rise buildings that could affect data collection. Data, such as courtyard measured temperature and wind speed and direction, were recorded with a measurement interval of 15 min.

Table 2. Technical data of the measurement equipment.

Sensor	Situation	Variable	Resolution	Range	Accuracy
PCE-FWS 20	Outdoor	Wind	-	0–180 km/h	±1 m/s
		Dry bulb Temp	0.1 °C	−40 to +65 °C	±1 °C
		RH	1%	12–99%	±5%
TESTO 174H/T	Courtyard	Dry bulb Temp	0.1 °C	−20 to +70 °C	±0.5 °C
		RH	2%	0–100%	±0.1%

Simultaneously, the temperatures in the courtyards of the selected case studies were recorded with sensors' model TESTO 174 H and TESTO 174 T, whose technical data are shown in Table 2. The sensors were placed vertically suspended from the roof of the building on the north-facing façade of the courtyard so that solar radiation would not influence the results. In addition, they were protected with a reflective shield to prevent overheating and to allow ventilation of the measuring equipment (Figure 1). As the sensors' measured temperature would vary throughout the courtyard due to several factors, including stratification and infrared radiation, all the sensors were placed at +1.00 m and +2.00 m, referring to the height of the courtyard inhabited by users.

(a) (b) (c)

Figure 1. Location of the measurement instruments: (a) Weather station PCE-FWS 20; (b,c) sensors TESTO 174.

2.3. Problem Setup

In this article, the variables selected to predict the value of the temperature inside a courtyard are the two most relevant according to the literature review, namely, TR—considering climate location zone and year´s season, and AR—as a numerical parameter synthesizing courtyard geometry. To perform the modeling, two stages were considered.

In the first stage, work was accomplished with the data from 22 monitored courtyards in every hour of one week, from different periods of the year, in various courtyards located in the Spanish cities of Badajoz, Córdoba and Seville.

The SVM method was used to create the library with some of these training data along one week in different courtyards. After that, we consider courtyards with different characteristic parameters, such as ARI and ARII, which are not included in the training data and use interpolation techniques to obtain the prediction for a week.

2.4. Support Vector Regression Method

Support Vector Machines (SVMs) were introduced in the 90s by Vapnik and his collaborators [45] in the framework of statistical learning theory. Although originally, SVMs were thought to solve binary classification problems, they are currently used to solve various types of problems, for example, regression problems [44], on which this research has focused.

In this first stage, the predicted value of the measured temperature inside a courtyard using some information related to it has been obtained. In particular, the time (hour of the day), the outside CMT, the wind speed and direction have been considered. More specifically, the following has been considered: $x = (x^1, x^2, x^3, x^4)$, where,

- x^1 = time (hour of the day);
- x^2 = outside temperature;
- x^3 = wind speed;
- x^4 = wind direction.

Searching for a function $f : R^4 \to R$, such that $y = f(x)$ provides the temperature inside the courtyard was the goal in this step.

To find this function f for each courtyard, the m-collection of experimental data associated with it was used. The idea of the SVR method [6] is to obtain a function such that for every sample (x_i, y_i), $i = 1, \ldots, m$, it is satisfied that $|f(x_i) - y_i| \leq \varepsilon$, for some $\varepsilon > 0$ small. Concretely, given ε, γ and $C > 0$, the following optimization problem is considered:

$$\max\left\{-\frac{1}{2}\sum_{i,j=1}^{m}(\alpha_i - \alpha_i^*)(\alpha_j - \alpha_j^*)\exp\left(-\gamma \| x_i - x_j \|^2\right) - \varepsilon \sum_{i,j=1}^{m}(\alpha_i + \alpha_i^*) + \sum_{i,j=1}^{m} y_i(\alpha_i - \alpha_i^*)\right\},$$

subject to $\sum_{i,j=1}^{m}(\alpha_i - \alpha_i^*) = 0$, for $\alpha_i, \alpha_i^* \in [0, C]$.

This problem was solved using the statistical software R. In particular, we used the E1071 library [52], a software package designed to solve classification and regression problems, using Support Vector Machines, which can be easily installed in R. The solution provides a possible candidate function as follows:

$$f(x) = \sum_{i=1}^{m}(\alpha_i - \alpha_i^*)\exp\left(-\gamma \| x_i - x \|^2\right) + b,$$

where the constant $b \in R$ can be computed by forcing the Karush–Kuhn–Tucker (KKT) condition [53]. The function $K(x, x') = \exp(-\gamma \| x - x' \|^2)$ is called the radial basis kernel. It holds that $|f(x_i) - y_i| \leq \varepsilon$, for all $i = 1, \ldots, m$. The quality of function f depends on the choice of the parameters ε, γ and C. In order to select the best parameters, Cross-Validation (CV) technique was used to obtain the parameter values: $\gamma = 0.1$, $C = 10$ and $\varepsilon = 0.1$, with a CV error around 1% for all test cases.

2.5. Predicted Temperature of a Courtyard

In the first stage, through the monitoring data and the SVR method, a library of predicted temperatures inside various courtyards located in different cities of the south of Spain was obtained. In this second stage, by using this library, the predicted temperature inside a given courtyard was obtained.

In this section, given that the definition of AR is two-dimensional, two ARs were measured: the first one, ARI was defined as the relation between the width and the height, and the second one, ARII was defined as the relation between the length and the height, as follows:

$$\text{ARI} = h_{max}/W \text{ and } \text{ARII} = h_{max}/L,$$

where h_{max} is the maximum height, W represents the width and L the length of the courtyard.

Once both ARs were fixed, the predicted temperature inside a given courtyard in two different ways was performed. First, the courtyards library was classified considering three different TRs, depending on the range of temperatures of the courtyards and an interpolation technique to predict the temperature inside a courtyard of the same class by using ARs data was used, as it is explained in Section 2.5.1.

Second, the courtyards library was classified into different groups, depending on the courtyards AR range and an interpolation technique to predict the temperature inside a courtyard of the same class by using the maximum and minimum temperature data was used. Two cases (AR.1 and AR.2) were considered: first, the classification by considering ARI, and second, ARII was performed.

2.5.1. Fixed Temperature Range, Interpolation Using the ARs

In this case, the courtyards library was classified into three different TR, depending on the range of temperatures inside the courtyard. These TRs correspond to statistical climatic records in the locations where case studies are placed. The first group corresponds to the hottest days of spring or autumn, the second, to a typical summer season and, the third, to a summer heatwave.

TR1: $(15°, 35°)$.
TR2: $(20°, 40°)$.
TR3: $(25°, 45°)$.

In the following Table 3, the courtyards are classified within these different TR. Note that some courtyards are in more than one TR because the temperature range in the courtyard changed from one week to another. This is because the courtyard, as a thermal tempering device, performs differently depending on the outdoor temperature.

Table 3. Classification of courtyards within temperature range classes.

Thermal Range	Courtyards
TR1	CS1, CS2, CS3, CS4
TR2	CS5, CS6, CS7, CS8, CS9, CS10, CS11, CS12, CS13, CS14
TR3	CS16, CS17, CS18, CS19, CS20, CS21, CS22

For a given courtyard, its range of temperature is first estimated, being classified as TR1, TR2 or TR3, and its AR, being classified as ARI or ARII.

Once courtyards are classified, the temperature prediction is verified through the SVR method; by an interpolation technique, it can be obtained a prediction of the temperature inside a courtyard of the same class. To achieve these data, MATLAB function *scatteredInterpolant* was used, which performs interpolation on a 2D dataset of scattered data. In particular, it returns the interpolant F for the given dataset such that we can evaluate F at a set of query points in 2D to produce interpolated values $T_q = F(\text{ARI}_q, \text{ARII}_q)$, obtaining the temperature inside the courtyard T_q.

2.5.2. Fixed the ARs, Interpolation Using Minimum and Maximum Temperatures

In this section, two different cases, depending on whether we fix ARI or ARII, are considered.

First, the courtyards library was classified into two different classes, depending on *ARI*:

ARI.1: $(0, 1)$.

ARI.2: $(1, 2)$.

In the following Table 4, the courtyards are classified within these different classes. Note that CS4 has not been taken into account, as its ARI is out of the considered ranges (3.41).

Table 4. Classification of courtyards within ARI range class.

ARI Range Class	Courtyards
ARI.1	CS5, CS8, CS11, CS13, CS14, CS15, CS16, CS19, CS22
ARI.2	CS1, CS2, CS3, CS6, CS7, CS9, CS10, CS12, CS17, CS18, CS20, CS21

Thus, for a given courtyard, we measure the ARI and classify it into ARI.1 or ARI.2.

Then, given the minimum and maximum temperature, T_{min} and T_{max}, respectively, of some courtyards in the same class and their corresponding predicted temperatures through the SVR method, by an interpolation technique implemented in the scientific software MATLAB, it can be obtained a prediction of the temperature inside a courtyard of the same class. To do the interpolation, we have used again the MATLAB function *scatteredInterpolant*, which performs interpolation on a 2D dataset of scattered data. In this case, we obtained $T_q = F(T_{min,q}, T_{max,q})$, obtaining the temperature inside the courtyard T_q.

Second, we classified the courtyards library into two different classes, depending on *ARII*:

ARII.1: $(0, 1)$.

ARII.2: $(1, 2.5)$.

In the following Table 5, we classify the courtyards within these different classes:

Table 5. Classification of courtyards within ARII range class.

ARII Range Class	Courtyards
ARII.1	CS1, CS3, CS4, CS5, CS8, CS11, CS14, CS16, CS19, CS22
ARII.2	CS2, CS6, CS7, CS9, CS10, CS12, CS13, CS15, CS17, CS18, CS20, CS21

Thus, for a given courtyard, first it was measured the ARII, being classified as ARII.1 or as ARII.2. To do the interpolation, the same procedure as in the case of ARI was followed, using now ARII instead.

3. Results

3.1. Fixed Temperature Range, Interpolation Using AR

In this section, it is shown the predicted temperature obtained by the method proposed in Section 2.5.1 in one courtyard of each temperature range class. The predicted temperature in comparison to the monitored temperature inside the courtyard, as well as the outdoor temperature, are both represented. In addition, a quantitative analysis was carried out. On the one hand, it was evaluated the relative error of the predicted temperature with respect to the monitored temperature in different discrete norms:

$$L^1\,(\%) = \frac{\sum_{i=1}^{N} \left| T_{monit.} - T_{pred.} \right|(t_i)}{\sum_{i=1}^{N} T_{monit.}(t_i)} \cdot 100, L^2\,(\%) = \left[\frac{\sum_{i=1}^{N} (T_{monit.} - T_{pred.})^2 (t_i)}{\sum_{i=1}^{N} T^2_{monit.}(t_i)} \right]^{1/2} \cdot 100,$$

where it is denoted by $T_{monit.}(t_i)$ (resp, $T_{pred.}(t_i)$), the monitored temperature (resp., the predicted temperature) at time $t_i, i = 1, \ldots, N$ (hours, (h)). Moreover, the percentage in time for which the obtained absolute error within the predicted and the monitored temperature is less than or equal to a fixed tolerance $tol = 2\,°C$ was evaluated. On the other hand, the following statistical parameters were computed: the correlation coefficient

R, the Root Mean Square Error ($RMSE$) and the Mean Absolute Percentage Error ($MAPE$). The formulas for these parameters are as follows:

$$R = \frac{\sum_{i=1}^{N}(T_{monit.}(t_i) - \overline{T}_{monit.})(T_{pred.}(t_i) - \overline{T}_{pred.})}{\left[\sum_{i=1}^{N}(T_{monit.}(t_i) - \overline{T}_{monit.})^2 \sum_{i=1}^{N}(T_{pred.}(t_i) - \overline{T}_{pred.})^2\right]^{1/2}},$$

$$RMSE\,(^{o}C) = \left[\frac{\sum_{i=1}^{N}(T_{monit.} - T_{pred.})^2(t_i)}{N}\right]^{1/2},$$

$$MAPE\,(\%) = \frac{1}{N}\sum_{i=1}^{N}\frac{\left|T_{monit.} - T_{pred.}\right|(t_i)}{T_{monit.}(t_i)} \cdot 100,$$

where, in the formula for the correlation coefficient R, the mean monitored temperature (resp., the mean predicted temperature) is denoted by $\overline{T}_{monit.}$ (resp, $\overline{T}_{pred.}$). The values of the relative and absolute errors and the statistical parameters are shown in Tables 6 and 7 for the CMT in each selected courtyard of each temperature range class.

Table 6. Example 3.0.1. Relative and absolute errors for the courtyard measured temperature in each selected courtyard of each TR.

Thermal Range	L^1 (%)	L^2 (%)	Absolute Error ≤tol (%)
TR1	5.09	6.10	91.67
TR2	4.93	6.62	84.28
TR3	3.50	4.31	89.88

Table 7. Example 3.0.1. Statistical parameters for the courtyard measured temperature in each selected courtyard of each TR.

Thermal Range	R	RMSE (°C)	MAPE (%)
TR1	0.96	1.24	5.17
TR2	0.88	1.62	4.82
TR3	0.89	1.21	3.52

For the class TR1, the courtyard CS1, located in Badajoz was considered. The prediction is performed for the dates 20 to 26 May. In the graph (Figure 2), simulation versus monitoring results of this courtyard with mild and very irregular temperatures is shown. There is hardly any thermal gap, and the prediction shows good accuracy. The obtained results are specified in Tables 6 and 7 (first row).

For the class TR2, the courtyard CS5, located in Seville, was considered. The prediction is performed for the date 7 to 13 September. The obtained results are represented in Figure 3 and Tables 6 and 7 (second row). Note that the prediction for the second half of the last day is not represented in this plot. This is due to the fact that some of the training data used for this prediction had fewer points than the 168 needed for the whole week. However, to be consistent with the other cases, we decided to keep the whole week in this plot.

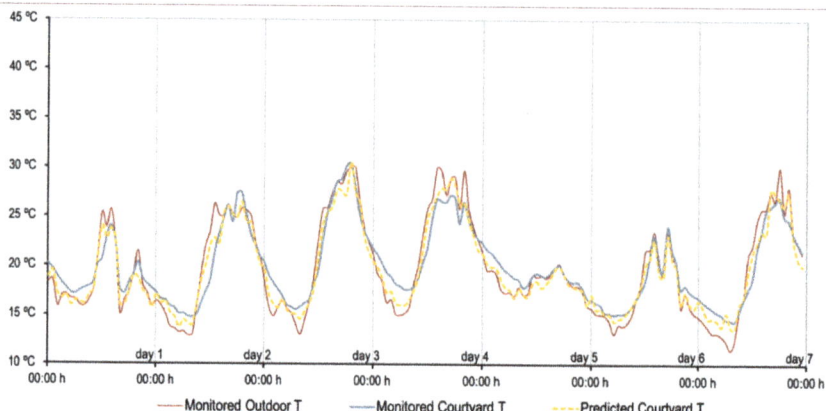

Figure 2. Example 3.0.1. Predicted temperature versus monitored and outdoor temperatures inside a TR1 courtyard.

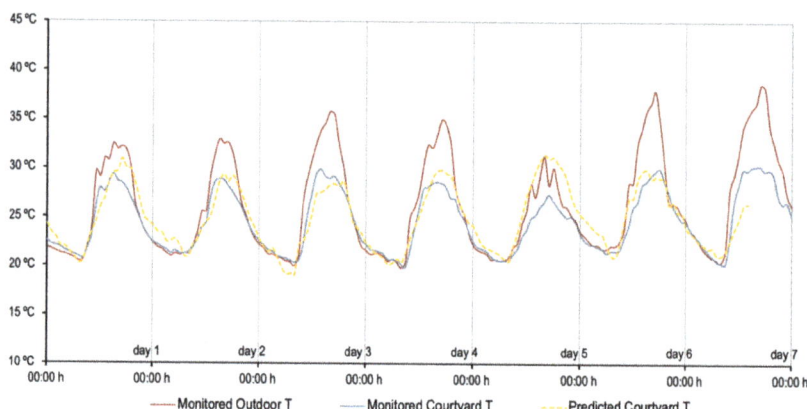

Figure 3. Example 3.0.1. Predicted temperature versus monitored and outdoor temperatures inside a TR2 courtyard.

For the class TR3, the courtyard CS17, located in Córdoba, was considered. The prediction is performed for the date 26 July to 1 August. Unlike the previous case shown in Figure 2, in this one (Figure 4), the outside temperature is higher and there is a large thermal gap. The predicted results show similarly good accuracy, particularly on days of maximum outdoor temperature. The obtained results are detailed in Tables 6 and 7 (third row).

3.2. Fixed AR, Interpolation Using Minimum and Maximum Temperature

In this section, the predicted temperature obtained by the method proposed in Section 2.5.2 in one courtyard of each AR range class is shown. First, the ARI range class is considered, representing the predicted temperature in comparison to the monitored temperature inside the courtyard, as well as the outdoor temperature. Additionally, a quantitative analysis was carried out. On the one hand, the relative error of the predicted temperature with respect to the monitored temperature in different discrete norms (L^1 and L^2) was evaluated, as done in Section 3.1. Moreover, the percentage in time for which the obtained absolute error within the predicted and the monitored temperature is less than or equal to a fixed tolerance $tol = 2\,°C$ was also evaluated. On the other hand, the following statistical parameters were computed: R, $RMSE$ and $MAPE$. The values of the relative and absolute errors and the statistical parameters are shown in Tables 8 and 9 for the courtyard measured temperature in each selected courtyard of each ARI range class.

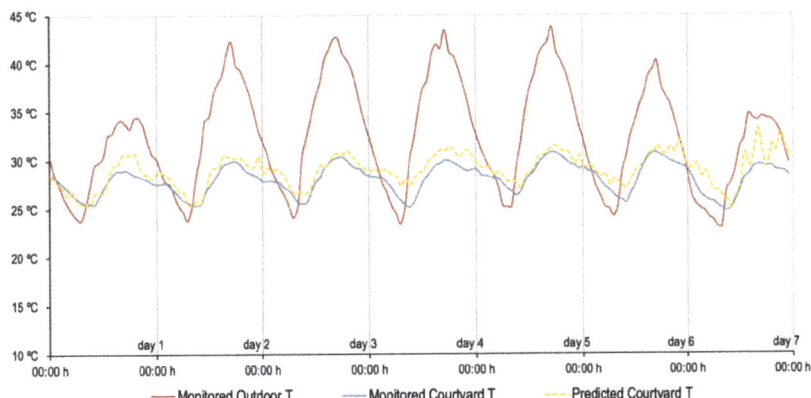

Figure 4. Example 3.0.1. Predicted temperature versus monitored and outdoor temperatures inside a TR3 courtyard.

Table 8. Example 3.0.1. Statistical parameters for the courtyard measured temperature in each selected courtyard of each temperature range class.

ARI Range Class	L^1 (%)	L^2 (%)	Absolute Error $\leq tol$ (%)
ARI.1	5.79	7.05	61.45
ARI.2	5.09	6.10	91.67

Table 9. Example 3.0.2. Statistical parameters for the courtyard measured temperature in each selected courtyard of each ARI range class.

ARI Range Class	R	RMSE (°C)	MAPE (%)
ARI.1	0.92	2.30	5.99
ARI.2	0.96	1.24	5.17

For the ARI.1, the courtyard CS16, located in Córdoba, was considered. The prediction is performed for the date 26 July to 1 August. The obtained results are represented in Figure 5 and Tables 7 and 8 (first row).

For the class ARI.2, the courtyard CS1, located in Badajoz, was considered. The prediction is performed for the date 20 to 26 May. The obtained results are represented in Figure 6 and Tables 7 and 8 (second row).

Finally, the ARII range class was considered and the predicted temperature was represented in comparison to the monitored temperature inside the courtyard as well as the outdoor temperature. As before, a quantitative analysis was carried out. On the one hand, the relative error of the predicted temperature with respect to the monitored temperature in different discrete norms (L^1 and L^2) was evaluated, as done in Section 3.1. Moreover, the percentage in time for which the obtained absolute error within the predicted and the monitored temperature is less than or equal to a fixed tolerance $tol = 2\,°C$ was also evaluated. On the other hand, the following statistical parameters were computed: R, $RMSE$ and $MAPE$. The values of the relative and absolute errors, and the statistical parameters are shown in in Tables 10 and 11 for the courtyard measured temperature in each selected courtyard of each ARII range class.

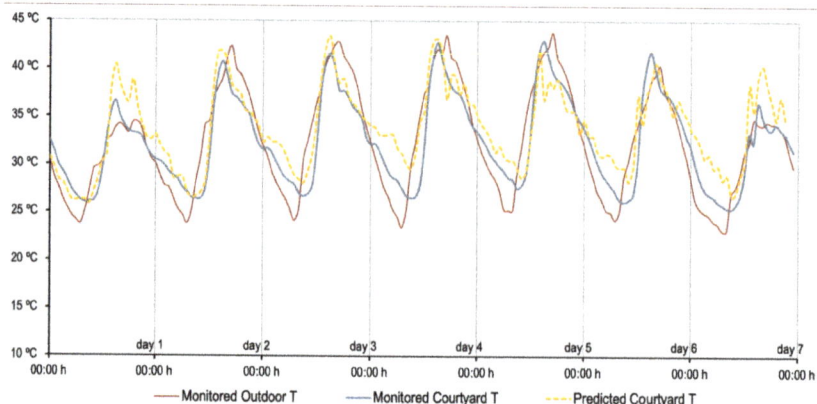

Figure 5. Example 3.0.2. Predicted temperature versus monitored and outdoor temperatures inside an ARI.1 courtyard.

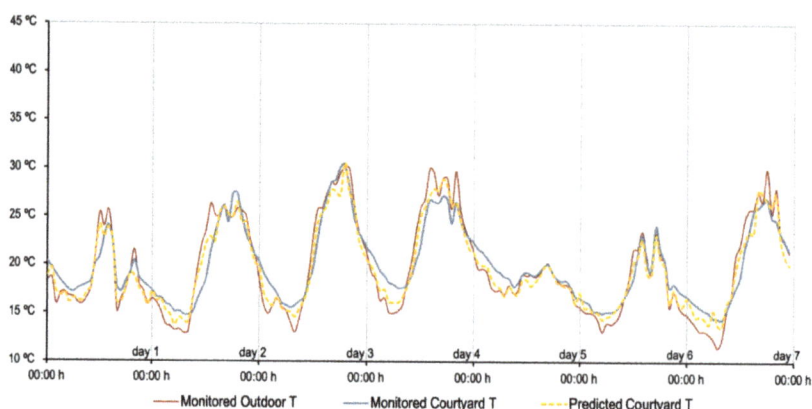

Figure 6. Example 3.0.2 Predicted temperature versus monitored and outdoor temperatures inside an ARI.2 courtyard.

Table 10. Example 3.0.2. Relative and absolute errors for the courtyard measured temperature in each selected courtyard of each ARII range class.

ARII Range Class	L^1 (%)	L^2 (%)	Absolute Error $\leq tol$ (%)
ARII.1	10.10	12.69	58.33
ARII.2	4.80	5.87	80.72

Table 11. Example 3.0.2. Statistical parameters for the courtyard measured temperature in each selected courtyard of each ARII range class.

ARII Range Class	R	RMSE (°C)	MAPE (%)
ARII.1	0.64	2.65	9.78
ARII.2	0.79	1.46	4.86

For the class ARII.1, the courtyard CS4, located in Badajoz, was considered. The prediction is performed for the date 20 to 26 May. In this case, as can be seen in Figure 7, the courtyard has a different thermal performance than in the previously described case studies. This is mainly due to the overheating that occurs in the early morning hours due to the low AR. The predicted results do not show such a tight accuracy under these conditions. The obtained results are detailed in Tables 9 and 10 (first row).

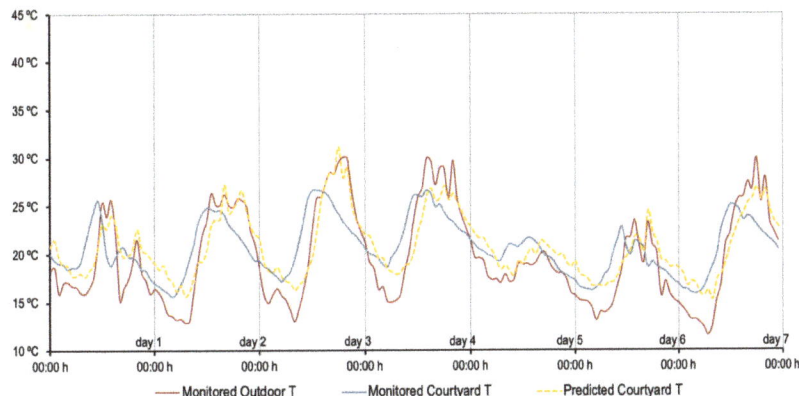

Figure 7. Example 3.0.2. Predicted temperature versus monitored and outdoor temperatures inside a ARII.1 courtyard.

For the class ARII.2, the courtyard CS9, located in Seville, was considered. The prediction is performed for the date 4 to 10 September. The obtained results are represented in Figure 8 and Tables 9 and 10 (second row).

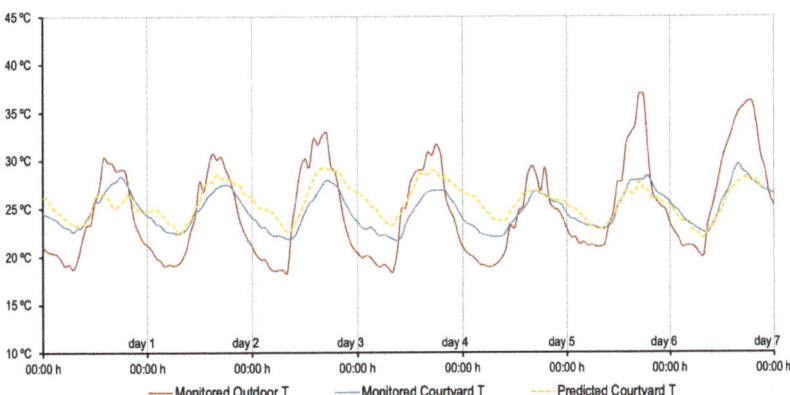

Figure 8. Example 3.0.2. Predicted temperature versus monitored and outdoor temperatures inside an ARII.2 courtyard.

3.3. Relative Errors Calculation

The main goal of this work is the accurate thermal modeling of the courtyard for its optimization as a resilient strategy against climate change and urban overheating. Therefore, the specific performance of courtyard thermodynamics was considered for the evaluation of the model errors. The courtyard´s thermal tempering performance increases as a function of the Thermal Gap (from now on, TG), that is, the difference between the exterior monitored temperature and the monitored temperature inside the courtyard. TG usually increases as the outside temperature rises. Accordingly, in this section, relative errors from two different and representative case studies with different TRs are selected, comparing statistical parameters more in detail.

The first selected case study corresponds to the predicted temperature inside the TR1 courtyard CS1 (Figure 2), and the second case corresponds to the predicted temperature inside the TR3 courtyard CS17 (Figure 4). These cases were selected since, in the first case, monitored and predicted temperatures inside the courtyard are rather close to the exterior monitored one, while in the second case, monitored and predicted temperatures inside the courtyard are quite far from the exterior monitored one.

Conversely, the relative and absolute errors, as well as the statistical parameters considered in the previous section, were computed. Then, the daily computations all along the week were performed. The obtained results are given in Tables 12 and 13.

Table 12. Relative error CS1.

Errors + Stat. Param.\Day	1	2	3	4	5	6	7
L^1 (%)	5.70	5.55	4.77	5.67	4.57	3.62	5.55
L^2 (%)	6.79	6.95	5.62	6.40	5.30	4.42	6.52
Absolute error $\leq tol$ (%)	95.83	83.33	87.50	91.67	95.83	100	87.50
R	0.92	0.95	0.98	0.97	0.91	0.97	0.96
$RMSE$ (°C)	1.30	1.41	1.25	1.44	1.03	0.80	1.32
$MAPE$ (%)	5.82	5.84	4.91	5.90	4.54	3.57	5.60

Table 13. Relative Error CS17.

Errors + Stat. Param.\Day	1	2	3	4	5	6	7
L^1 (%)	2.16	2.68	2.71	5.10	2.40	3.23	5.95
L^2 (%)	3.26	3.14	3.03	5.48	2.65	3.92	7.02
Absolute error $\leq tol$ (%)	95.83	95.83	100	83.33	100	95.83	58.33
R	0.91	0.96	0.96	0.92	0.97	0.91	0.89
$RMSE$ (°C)	0.90	0.87	0.85	1.54	0.76	1.12	1.93
$MAPE$ (%)	2.42	2.65	2.74	5.18	2.42	3.30	5.91

On the other hand, bearing in mind the obtained results, the best predicted day in each week was selected. In the first case, the day that gives better performances is the 6th one, while in the second case, it is the 5th one. Then, the relative error of the predicted temperature was computed hourly and represented in two ways. In the first way, with respect to the TG and in the second way, with respect to the monitored temperature. The graphics corresponding to CS1 and CS17 are included in Figure 9a,b, respectively. The graphic represented in Figure 9a corresponds to the relative error of the predicted temperature with respect to TG, and the graphic in Figure 9b corresponds to the relative error of the predicted temperature with respect to the monitored temperature inside the courtyard. In addition, the segment of the day in which critical urban overheating is concentrated is indicated in each graph. These hours, according to climate records [54], are between 13:00 and 19:00. On the left, considering CS1 plotted in Figure 9a, it can be observed that the relative error with respect to TG is always below 3%, except for two peaks, corresponding to time slots where the exterior temperature and the monitored temperature inside the courtyard almost coincide. Considering CS17 in Figure 9a, it was obtained a very low relative error with respect to TG in the central time slot of the day, that is, between 13:00 and 19:00, where TG is large. On the right, regarding the relative error with respect to the courtyard measured temperature (CMT) in Figure 9b, it can be observed that the plotted relative error for CS1 is below 0.1%, while for CS17, this relative error is always below 0.05%. For both case studies, if the daytime slot with the highest urban overheating is considered, the relative error is always below 0.05%.

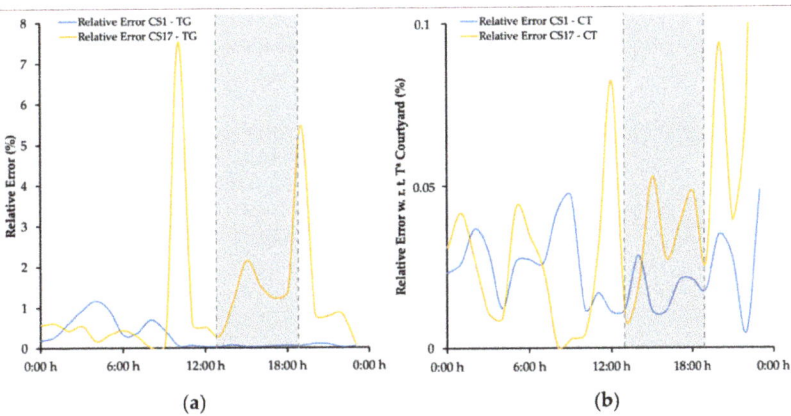

Figure 9. Relative errors CS1 and CS17 according to: (**a**) TG (**b**) CT.

4. Discussion

In this section, the results that were obtained in Section 3 are discussed. Regarding the results obtained in Sections 3.1 and 3.2, on the one hand, it can be appreciated in Tables 6, 8 and 10 that the values for the relative errors in different discrete norms are around 5% and in almost all cases are below 10%, and the percentage in time for which the obtained absolute error w.r.t. the CMT is less than or equal to $tol = 2\,°C$ is superior to 80%, except for the cases of Example 3.0.2, the ARI.1 range class and the ARII.1 range class. For the first critical case, reasonable values for the relative errors in different discrete norms (within 5% and 8%) were obtained, and the percentage in time for which the obtained absolute error w.r.t. the CMT is less than or equal to $tol = 2\,°C$ is 61.45%. However, that case is rather special since it can be observed at relatively high temperatures w.r.t. the other experiments. In any case, if the tolerance parameter is increased to $tol = 3\,°C$ for that case, a higher percentage of up to 80.72%, can be obtained. For the second critical case, the relative errors in different discrete norms are within 10% and 13%, and the percentage in time for which the obtained absolute error w.r.t. the CMT is less than or equal to $tol = 2\,°C$ is 58.33%. In any case, if the tolerance parameter is increased to $tol = 3\,°C$ for that case, we obtain the higher percentage 74.40%. On the other hand, the values for the statistical parameters that indicate that the simulation is accurate are $R \to 1$, $RMSE \to 0$, $MAPE \to 0$ [36,42–44,50–52]. The values of these parameters for the courtyard measured temperature in the present courtyards for each simulation confirm that the used strategy is rather accurate. In particular, in Tables 7, 9 and 11 it can be observed that the correlation coefficient R is quite close to 1 for all range classes (superior to 0.85, except for the cases Example 3.0.2 ARII.1 and ARII.2 range classes for which it is within 0.6 and 0.8). The $RMSE$ values are around $1.5\,°C$ and the $MAPE$ values are around 5%, except for the critical cases identified above for which the $RMSE$ values are around $2.5\,°C$ and the $MAPE$ values are within 5% and 10%.

Finally, in Section 3.3, relative and absolute errors as well as the statistical parameters in two selected cases were computed daily. The case where the predicted CMT inside the courtyard is rather close to the exterior one, and the case where the predicted CMT inside the courtyard is quite far from the exterior one were chosen. The obtained results are given in Tables 12 and 13, respectively. It can be observed that the values for the relative errors in different discrete norms are around 6% in the first case, and around 3% in the second case, and in almost all cases are below 7%. Moreover, the percentage in time for which the obtained absolute error w.r.t. the CMT is less than or equal to $tol = 2\,°C$ is superior to 80% all the days except for the 7th day of the second case, arriving to 100% in the 6th day of the first case and on the 3rd and 5th day of the second case. With respect to statistical parameters, the correlation coefficient R is quite close to 1, being larger than 0.89 in all cases. The $RMSE$ values are around $1.25\,°C$ in the first case and $1\,°C$ in the second one, and the $MAPE$ values are around 5% in the first case and 3% in the second case. Thus, mostly, the results obtained in Section 3.3 daily in these selected cases improve the global results computed for the whole week in Sections 3.1 and 3.2.

In brief, apart from the critical cases identified above, the values of the statistical parameters considered are in a similar range than those obtained in [36] for a similar problem. In that work, the authors performed a very accurate courtyard thermal simulation based upon a Computational Fluid Dynamics (CFD) FreeFEM 3D model, which is much more computationally expensive than the ML technique SVR used in this work. In particular, the computation of one-week temperature through the SVR method takes around one minute, while the CFD method takes around four minutes per one day of simulation.

5. Conclusions

In the present work, the applicability of a supervised ML model as a suitable tool for predicting microclimatic performance inside courtyards has been evaluated. For this purpose, among the ML models developed as supervised learning, Support Vector Machines (SVM) were selected. The model was fed and validated with empirical data from 22 case studies in southern Spain.

The results provided by this strategy showed good accuracy when compared to monitored data. In particular, we selected two representative and highly meaningful case studies with different TGs. The final results for both cases showed that, when the daytime slot with the highest urban overheating is considered, the relative error is almost below 0.05%. Additionally, values for statistical parameters are in good agreement with other studies in the literature that use more computationally expensive CFD models and show more accuracy than existing commercial tools. Indeed, the present strategy shows a Root Mean Square Error (RMSE) around 1 °C for the two representative case studies selected, which is in a similar range to the values obtained in [36] for a similar problem by a more computationally expensive CFD model, while corresponding values for existing commercial software are typically around 3 °C.

Based on the results obtained, it can be stated that the new application proposed for the ML method is useful for the development of design and measurement tools capable of modeling the complex microclimate of courtyards. Furthermore, the accuracy of the predictions for the analyzed case studies increases as a function of the courtyard thermal tempering potential linked to the intensification of the outdoor temperature.

The enhancement of the proposed methodology with the inclusion of other complementary microclimatic strategies, such as shading devices or vegetation as new ML features as well as establishing a balance between an over fitted and under fitted ML model considering the optimal number of training data, can be considered as future ways to develop this research.

Author Contributions: S.F.-G., M.G.-M. and S.R. contributed to develop the mathematical software and error analysis performed in this work. E.D.-M., C.R.-G., C.G.-M. contributed to define research conceptualization, field monitoring campaign design and performance, writing, review, and editing. M.G.-M., C.G.-M. and C.R.-G. contributed to the research funding acquisition. E.D.-M. and S.R., as the main authors of the paper, contributed equally. All authors have read and agreed to the published version of the manuscript.

Funding: Proyecto (RTI2018-093521-B-C31), financiado por: FEDER/Ministerio de Ciencia e Innovación—Agencia Estatal de Investigación (Researchers: Macarena Gómez-Mármol, Samuele Rubino and Soledad Fernández-García). Proyecto (RTI2018-093521-B-C33), financiado por: FEDER/Ministerio de Ciencia e Innovación—Agencia Estatal de Investigación (Researchers: Carmen Galán-Marín y Carlos Rivera-Gómez). Pre-doctoral contract granted to Eduardo Diz-Mellado (FPU 18/04783).

Institutional Review Board Statement: Not applicable.

Informed Consent Statement: Not applicable.

Acknowledgments: The authors gratefully acknowledge AEMET (State Meteorological Agency) for the data supplied and also want to thank Carlos Constantino Oitaven (University of Seville) for the basic version of the R code used as a starting point in the present simulations.

Conflicts of Interest: The authors declare no conflict of interest.

Nomenclature

UHI	Urban Heat Island
AR	Aspect Ratio
ML	Machine Learning
SVM	Support Vector Machine
SVR	Support Vector Regression
CS	Case Study
TR	Thermal Range
CFD	Computational Fluids Dynamics
TG	Thermal Gap
CMT	Courtyard Measured Temperature

References

1. IPCC. Proposed Outline of the Special Report in 2018 on the Impacts of Global Warming of 1.5 °C above Pre-Industrial Levels and Related Global Greenhouse Gas Emission Pathways, in the Context of Strengthening the Global Response to the Threat of Climate cha. Ipcc—Sr15. Available online: www.environmentalgraphiti.org (accessed on 10 March 2021).
2. United Nations, Department of Economic and Social Affairs PD. *The World's Cities in 2018*; United Nations: New York, NY, USA, 2018; p. 34.
3. Bombardelli, F.A. *Handbook of Environmental Fluid Dynamics*; Apple Academic Press: Cambridge, MA, USA, 2012; Volume 2.
4. Taleb, D.; Abu-Hijleh, B. Urban heat islands: Potential effect of organic and structured urban configurations on temperature variations in Dubai, UAE. *Renew. Energy* **2013**, *50*, 747–762. [CrossRef]
5. Climate Considerations in Building and Urban Design | Wiley. Available online: https://www.wiley.com/en-us/Climate+Considerations+in+Building+and+Urban+Design-p-9780471291770 (accessed on 10 March 2021).
6. Vardoulakis, S.; Fisher, B.E.; Pericleous, K.; Gonzalez-Flesca, N. Modelling air quality in street canyons: A review. *Atmos. Environ.* **2003**, *37*, 155–182. [CrossRef]
7. Ulpiani, G. On the linkage between urban heat island and urban pollution island: Three-decade literature review towards a conceptual framework. *Sci. Total Environ.* **2020**, *751*, 141727. [CrossRef] [PubMed]
8. Carpio, M.; González, Á.; González, M.; Verichev, K. Influence of pavements on the urban heat island phenomenon: A scientific evolution analysis. *Energy Build.* **2020**, *226*, 110379. [CrossRef]
9. Villa-Arrieta, M.; Sumper, A. Economic evaluation of Nearly Zero Energy Cities. *Appl. Energy* **2019**, *237*, 404–416. [CrossRef]
10. ZEBRA 2020—NEARLY ZERO-ENERGY BUILDING STRATEGY 2020 Strategies for a nearly Zero-Energy Building Market Transition in the Euro-Pean Union. 2020. Available online: https://zebra2020.eu/ (accessed on 10 May 2021).
11. Taleghani, M.; Tenpierik, M.; Dobbelsteen, A.V.D. Energy performance and thermal comfort of courtyard/atrium dwellings in the Netherlands in the light of climate change. *Renew. Energy* **2014**, *63*, 486–497. [CrossRef]
12. López-Cabeza, V.; Galán-Marín, C.; Rivera-Gómez, C.; Roa-Fernández, J. Courtyard microclimate ENVI-met outputs deviation from the experimental data. *Build. Environ.* **2018**, *144*, 129–141. [CrossRef]
13. Al-Masri, N.; Abu-Hijleh, B. Courtyard housing in midrise buildings: An environmental assessment in hot-arid climate. *Renew. Sustain. Energy Rev.* **2012**, *16*, 1892–1898. [CrossRef]
14. Taleghani, M. Outdoor thermal comfort by different heat mitigation strategies—A review. *Renew. Sustain. Energy Rev.* **2018**, *81*, 2011–2018. [CrossRef]
15. Zamani, Z.; Heidari, S.; Hanachi, P. Reviewing the thermal and microclimatic function of courtyards. *Renew. Sustain. Energy Rev.* **2018**, *93*, 580–595. [CrossRef]
16. Xu, X.; Luo, F.; Wang, W.; Hong, T.; Fu, X. Performance-Based Evaluation of Courtyard Design in China's Cold-Winter Hot-Summer Climate Regions. *Sustainability* **2018**, *10*, 3950. [CrossRef]
17. Rojas-Fernández, J.; Galán-Marín, C.; Rivera-Gómez, C.; Fernández-Nieto, E.D. Exploring the Interplay between CAD and FreeFem++ as an Energy Decision-Making Tool for Architectural Design. *Energies* **2018**, *11*, 2665. [CrossRef]
18. Taleghani, M.; Tenpierik, M.; Dobbelsteen, A.V.D.; Sailor, D.J. Heat mitigation strategies in winter and summer: Field measurements in temperate climates. *Build. Environ.* **2014**, *81*, 309–319. [CrossRef]
19. Shahidan, M.F.; Jones, P.J.; Gwilliam, J.; Salleh, E. An evaluation of outdoor and building environment cooling achieved through combination modification of trees with ground materials. *Build. Environ.* **2012**, *58*, 245–257. [CrossRef]
20. Muhaisen, A.S. Shading simulation of the courtyard form in different climatic regions. *Build. Environ.* **2006**, *41*, 1731–1741. [CrossRef]

21. Taleghani, M.; Kleerekoper, L.; Tenpierik, M.; Dobbelsteen, A.V.D. Outdoor thermal comfort within five different urban forms in the Netherlands. *Build. Environ.* **2015**, *83*, 65–78. [CrossRef]
22. Choi, J.-H. Investigation of the correlation of building energy use intensity estimated by six building performance simulation tools. *Energy Build.* **2017**, *147*, 14–26. [CrossRef]
23. Rivera-Gómez, C.; Diz-Mellado, E.; Galán-Marín, C.; López-Cabeza, V. Tempering potential-based evaluation of the courtyard microclimate as a combined function of aspect ratio and outdoor temperature. *Sustain. Cities Soc.* **2019**, *51*. [CrossRef]
24. Huang, L.; Hamza, N.; Lan, B.; Zahi, D. Climate-responsive design of traditional dwellings in the cold-arid regions of Tibet and a field investigation of indoor environments in winter. *Energy Build.* **2016**, *128*, 697–712. [CrossRef]
25. Ghaffarianhoseini, A.; Berardi, U.; Ghaffarianhoseini, A. Thermal performance characteristics of unshaded courtyards in hot and humid climates. *Build. Environ.* **2015**, *87*, 154–168. [CrossRef]
26. Soflaei, F.; Shokouhian, M.; Abraveshdar, H.; Alipour, A. The impact of courtyard design variants on shading performance in hot-arid climates of Iran. *Energy Build.* **2017**, *143*, 71–83. [CrossRef]
27. Rodríguez-Algeciras, J.; Tablada, A.; Chaos-Yeras, M.; De la Paz, G.; Matzarakis, A. Influence of aspect ratio and orientation on large courtyard thermal conditions in the historical centre of Camagüey-Cuba. *Renew. Energy* **2018**, *125*, 840–856. [CrossRef]
28. Nasrollahi, N.; Hatami, M.; Khastar, S.R.; Taleghani, M. Numerical evaluation of thermal comfort in traditional courtyards to develop new microclimate design in a hot and dry climate. *Sustain. Cities Soc.* **2017**, *35*, 449–467. [CrossRef]
29. Guedouh, M.S.; Zemmouri, N. Courtyard Building's Morphology Impact on Thermal and Luminous Environments in Hot and Arid Region. *Energy Procedia* **2017**, *119*, 153–162. [CrossRef]
30. Rojas-Fernández, J.; Galán-Marín, C.; Roa-Fernández, J.; Rivera-Gómez, C. Correlations between GIS-Based Urban Building Densification Analysis and Climate Guidelines for Mediterranean Courtyards. *Sustainability* **2017**, *9*, 2255. [CrossRef]
31. Mauree, D.; Naboni, E.; Coccolo, S.; Perera, A.; Nik, V.M.; Scartezzini, J.-L. A review of assessment methods for the urban environment and its energy sustainability to guarantee climate adaptation of future cities. *Renew. Sustain. Energy Rev.* **2019**, *112*, 733–746. [CrossRef]
32. ENVI-Met—Decode Urban Nature with ENVI-Met Software. Available online: http://www.envi-met.com (accessed on 10 March 2011).
33. Bueno, B.; Norford, L.; Hidalgo, J.; Pigeon, G. The urban weather generator. *J. Build. Perform. Simul.* **2013**, *6*, 269–281. [CrossRef]
34. Lindberg, F.; Holmer, B.; Thorsson, S. SOLWEIG 1.0—Modelling spatial variations of 3D radiant fluxes and mean radiant temperature in complex urban settings. *Int. J. Biometeorol.* **2008**, *52*, 697–713. [CrossRef]
35. Kastner, P.; Dogan, T. A cylindrical meshing methodology for annual urban computational fluid dynamics simulations. *J. Build. Perform. Simul.* **2019**, *13*, 59–68. [CrossRef]
36. López-Cabeza, V.; Carmona-Molero, F.; Rubino, S.; Rivera-Gómez, C.; Fernández-Nieto, E.; Galán-Marín, C.; Chacón-Rebollo, T. Modelling of surface and inner wall temperatures in the analysis of courtyard thermal performances in Mediterranean climates. *J. Build. Perform. Simul.* **2021**, *14*, 181–202. [CrossRef]
37. Kondarasaiah, M.H.; Ananda, S. Kinetic and Mechanistic Study of Ru(III)-Nicotinic Acid Complex Formation by Oxidation of Bromamine-T in Acid Solution. *Oxid. Commun.* **2004**, *27*, 140–147.
38. Grant, P. Assessment and Selection. *The Business of Giving*. 2014. Available online: https://www.amazon.co.uk/Business-Giving-Philanthropy-Grantmaking-Investment-ebook/dp/B009AQUSHU (accessed on 10 March 2021).
39. Sun, H.; Burton, H.V.; Huang, H. Machine learning applications for building structural design and performance assessment: State-of-the-art review. *J. Build. Eng.* **2021**, *33*, 101816. [CrossRef]
40. Shawe-Taylor, J.; Cristianini, N. *Kernel Methods for Pattern Analysis*; Cambridge University Press: Cambridge, UK, 2004.
41. Fletcher, R. *Practical Methods of Optimization*; John Wiley & Sons: Hoboken, NJ, USA, 2013. [CrossRef]
42. Moguerza, J.M.; Muñoz, A. Support Vector Machines with Applications. *Stat. Sci.* **2006**, *21*, 322–336. [CrossRef]
43. Chen, H.-F. In SilicoLogPPrediction for a Large Data Set with Support Vector Machines, Radial Basis Neural Networks and Multiple Linear Regression. *Chem. Biol. Drug Des.* **2009**, *74*, 142–147. [CrossRef]
44. Vapnik, V.; Golowich, S.E.; Smola, A. Support vector method for function approximation, regression estimation, and signal processing. *Adv. Neural Inf. Process. Syst.* **1997**, 281–287.
45. Baumann, S.; Groß, S.; Voigt, L.; Ullrich, A.; Weymar, F.; Schwaneberg, T.; Dörr, M.; Meyer, C.; John, U.; Ulbricht, S. Pitfalls in accelerometer-based measurement of physical activity: The presence of reactivity in an adult population. *Scand. J. Med. Sci. Sports* **2017**, *28*, 1056–1063. [CrossRef]
46. Platt, J.C. Fast training of support vector machines using sequential minimal optimization. *Adv. Kernel Methods* **1999**, 185–208.
47. Documento, B.E. Introducción I Objeto. 2017. Available online: https://www.codigotecnico.org/images/stories/pdf/ahorroEnergia/DBHE.pdf (accessed on 10 March 2021).
48. Diz-Mellado, E.; López-Cabeza, V.P.; Rivera-Gómez, C.; Roa-Fernández, J.; Galán-Marín, C. Improving School Transition Spaces Microclimate to Make Them Liveable in Warm Climates. *Appl. Sci.* **2020**, *10*, 7648. [CrossRef]

49. Diz-Mellado, E.M.; Galán-Marín, C.; Rivera-Gómez, C.; López-Cabeza, V.P. Facing climate change overheating in cities through multiple ther-moregulatory courtyard potential case studies appraisal. In *REHABEND*; University of Cantabria, Santander (Spain)—Building Technology R&D Group: Santander, Spain, 2020; pp. 1645–1652.
50. Callejas, I.A.; Durante, L.C.; Diz-Mellado, E.; Galán-Marín, C. Thermal Sensation in Courtyards: Potentialities as a Passive Strategy in Tropical Climates. *Sustainability* **2020**, *12*, 6135. [CrossRef]
51. National Weather Service. Available online: https://www.weather.gov/source/zhu/ZHU_Training_Page/definitions/dry_wet_bulb_definition/dry_wet_bulb.htm (accessed on 6 May 2021).
52. Hornik, K.; Weingessel, A.; Leisch, F. Davidmeyerr-Projectorg, M.D.M. Package 'E1071'. 2020. Available online: https://cran.r-project.org/web/packages/e1071/e1071.pdf (accessed on 6 May 2021).
53. Luenberger, G.; Mateos, M.L. *Programación Lineal y no Lineal. Number 90C05 LUEp*; Addison-Wesley Iberoamericana: Madrid, Spain, 1989.
54. Revista espasa.planetasaber. Condiciones Atmosféricas De Un Lugar Clima. Available online: http://espasa.planetasaber.com/theworld/gats/article/default.asp?pk=793&art=59 (accessed on 6 May 2021).

Article

Advances in the Approximation of the Matrix Hyperbolic Tangent

Javier Ibáñez [1], José M. Alonso [1], Jorge Sastre [2], Emilio Defez [3] and Pedro Alonso-Jordá [4,*]

[1] Instituto de Instrumentación para Imagen Molecular, Universitat Politècnica de València, Av. dels Tarongers, 14, 46011 Valencia, Spain; jjibanez@dsic.upv.es (J.I.); jmalonso@dsic.upv.es (J.M.A.)
[2] Instituto de Telecomunicación y Aplicaciones Multimedia, Universitat Politècnica de València, Ed. 8G, Camino de Vera s/n, 46022 Valencia, Spain; jsastrem@upv.es
[3] Instituto de Matemática Multidisciplinar, Universitat Politècnica de València, Ed. 8G, Camino de Vera s/n, 46022 Valencia, Spain; edefez@imm.upv.es
[4] Department of Computer Systems and Computation, Universitat Politècnica de València, Ed. 1F, Camino de Vera s/n, 46022 Valencia, Spain
* Correspondence: palonso@upv.es

Abstract: In this paper, we introduce two approaches to compute the matrix hyperbolic tangent. While one of them is based on its own definition and uses the matrix exponential, the other one is focused on the expansion of its Taylor series. For this second approximation, we analyse two different alternatives to evaluate the corresponding matrix polynomials. This resulted in three stable and accurate codes, which we implemented in MATLAB and numerically and computationally compared by means of a battery of tests composed of distinct state-of-the-art matrices. Our results show that the Taylor series-based methods were more accurate, although somewhat more computationally expensive, compared with the approach based on the exponential matrix. To avoid this drawback, we propose the use of a set of formulas that allows us to evaluate polynomials in a more efficient way compared with that of the traditional Paterson–Stockmeyer method, thus, substantially reducing the number of matrix products (practically equal in number to the approach based on the matrix exponential), without penalising the accuracy of the result.

Keywords: matrix functions; matrix hyperbolic tangent; matrix exponential; Taylor series; matrix polynomial evaluation

1. Introduction and Notation

Matrix functions have been an increasing focus of attention due to their applications to new and interesting problems related, e.g., to statistics [1], Lie theory [2], differential equations (the matrix exponential function e^{At} can be considered as a classical example for its application in the solution of first order differential systems $Y'(t) = AY(t)$ with $A \in \mathbb{C}^{n \times n}$ and for its use in the development of exponential integrators for nonlinear ODEs and PDEs, see [3] for example), approximation theory, and many other areas of science and engineering [4].

There are different ways to define the notion of the function $f(A)$ of a square matrix A. The most common are via the Jordan canonical form, via the Hermite interpolation, and via the Cauchy integral formula. The equivalence among the different definitions of a matrix function can be found in [5]. Several general methods have been proposed for evaluating matrix functions, among which, we can highlight the Taylor or Padé approximations and methods based on the Schur form of a matrix [4].

Among the most well-known matrix functions, we have the matrix hyperbolic cosine and the matrix hyperbolic sine functions, respectively defined in terms of the matrix exponential function e^A by means of the following expressions:

$$\cosh(A) = \frac{1}{2}\left(e^A + e^{-A}\right), \quad \sinh(A) = \frac{1}{2}\left(e^A - e^{-A}\right). \quad (1)$$

These matrix functions are applied, e.g., in the study of the communicability analysis in complex networks [6], or to construct the exact series solution of coupled hyperbolic systems [7]. Precisely due to their applicability, the numerical computation of these functions has received remarkable and growing attention in recent years. A set of state-of-the-art algorithms to calculate these functions developed by the authors can be found in [8–11].

On the other hand, we have the matrix hyperbolic tangent function, defined as

$$\tanh(A) = \sinh(A)(\cosh(A))^{-1} = (\cosh(A))^{-1} \sinh(A), \quad (2)$$

and used, for instance, to give an analytical solution of the radiative transfer equation [12], in the heat transference field [13,14], in the study of symplectic systems [15,16], in graph theory [17], and in the development of special types of exponential integrators [18,19].

In this work, we propose and study two different implementations that compute the matrix hyperbolic tangent function: the first uses the matrix exponential function whereas the second is based on its Taylor series expansion. In addition, for the second approach, we use and compare two different alternatives to evaluate the matrix polynomials involved in the series expansion.

1.1. The Matrix Exponential Function-Based Approach

This first option is derived from the matrix hyperbolic tangent function definition as expressed in Equations (1) and (2), from which the following matrix rational expression is immediately deduced:

$$\tanh(A) = \left(e^{2A} - I\right)\left(e^{2A} + I\right)^{-1} = \left(e^{2A} + I\right)^{-1}\left(e^{2A} - I\right), \quad (3)$$

where I denotes the identity matrix with the same order as A. Equation (3) reduces the approximation of the matrix hyperbolic tangent function to the computation of the matrix exponential e^{2A}.

There exists profuse literature (see e.g., [4,20]) about the approximation of the matrix exponential function and the inconveniences that its calculation leads to [21]. The most competitive methods used in practice are either those based on polynomial approximations or those based on Padé rational approaches, with the former, in general, being more accurate and with lower computational costs [22].

In recent years, different polynomial approaches to the matrix exponential function have been proposed, depending on the type of matrix polynomial used. For example, some approximations use the Hermite matrix polynomials [23], while others derived from on Taylor polynomials [22,24]. More recently, a new method based on Bernoulli matrix polynomials was also proposed in [25].

All these methods use the scaling and squaring method based on the identity

$$e^A = \left(e^{2^{-s}A}\right)^{2^s},$$

which satisfies the matrix exponential. In the scaling phase, an integer scaling factor s is taken, and the approximation of $e^{2^{-s}A}$ is computed using any of the proposed methods so that the required precision is obtained with the lowest possible computational cost. In the squaring phase, we obtain e^A by s repeated squaring operations.

1.2. The Taylor Series-Based Approach

The other possibility for computing the matrix hyperbolic tangent function is to use its Taylor series expansion

$$\tanh(z) = \sum_{k \geq 1} \frac{2^{2k}(2^{2k}-1)\mathcal{B}_{2k}}{(2k)!} z^{2k-1}, |z| < \frac{\pi}{2},$$

where $\mathcal{B}_{2k}$ are the Bernoulli's numbers.

As in the case of the matrix exponential, it is highly recommended to use the scaling and squaring technique to reduce the norm of the matrix to be computed and, thus, to obtain a good approximation of the matrix hyperbolic tangent with an acceptable computational cost. Due to the double angle formula for the matrix hyperbolic tangent function

$$\tanh(2A) = 2\left(I + \tanh^2(A)\right)^{-1} \tanh(A), \tag{4}$$

which is derived from the scalar one

$$\tanh(2z) = \frac{2\tanh(z)}{1 + \tanh^2(z)},$$

it is possible to compute $T_s = \tanh(A)$ by using the following recurrence:

$$\begin{aligned} T_0 &= \tanh(2^{-s}A), \\ T_i &= 2\left(I + T_{i-1}^2(A)\right)^{-1} T_{i-1}(A), i = 1, \ldots, s. \end{aligned} \tag{5}$$

Throughout this paper we will denote by $\sigma(A)$ the set of eigenvalues of matrix $A \in \mathbb{C}^{n \times n}$ and by I_n (or I) the matrix identity of order n. In addition, $\rho(A)$ refers to the spectral radius of A, defined as

$$\rho(A) = \max\{|\lambda|; \lambda \in \sigma(A)\}.$$

With $\lceil x \rceil$, we denote the value obtained by rounding x to the nearest integer greater than or equal to x, and $\lfloor x \rfloor$ is the value obtained rounding x to the nearest integer less than or equal to x. The matrix norm $||\cdot||$ will stand for any subordinate matrix norm and, in particular, $||\cdot||_1$ denotes the 1-norm.

This work is organised as follows. First, Section 2 incorporates the algorithms corresponding to the different approaches previously described for approximating the matrix hyperbolic tangent and for computing the scaling parameter and the order of the Taylor polynomials. Next, Section 3 details the experiments carried out to compare the numerical properties of the codes to be evaluated. Finally, in Section 4, we present our conclusions.

2. Algorithms for Computing the Matrix Hyperbolic Tangent Function

2.1. The Matrix Exponential Function-Based Algorithm

The first algorithm designed, called Algorithm 1, computes the matrix hyperbolic tangent by means of the matrix exponential according to Formula (3). In Steps 1 and 2, Algorithm 2 from [26] is responsible for computing $e^{2^{-s}B}$ by means of the Taylor approximation of order m_k, being $B = 2A$. In Step 3, $T \simeq \tanh(2^{-s}B)$ is worked out using Formula (3). In this phase, T is computed by solving a system of linear equations, with $\left(e^{2^{-s}B} + I\right)$ being the coefficient matrix and $\left(e^{2^{-s}B} - I\right)$ being the right hand side term. Finally, through Steps 4–8, $\tanh(A)$ is recovered from T by using the squaring technique and the double angle Formula (5).

Algorithm 1: Given a matrix $A \in \mathbb{C}^{n \times n}$, this algorithm computes $T = \tanh(A)$ by means of the matrix exponential function.

1. $B = 2A$
2. Calculate the scaling factor $s \in \mathbb{N} \cup \{0\}$, the order of Taylor polynomial $m_k \in \{2, 4, 6, 9, 12, 16, 20, 25, 30\}$ and compute $e^{2^{-s}B}$ by using the Taylor approximation /* Phase I (see Algorithm 2 from [26]) */
3. $T = \left(e^{2^{-s}B} + I\right)^{-1}\left(e^{2^{-s}B} - I\right)$ /* Phase II: Work out $\tanh(2^{-s}B)$ by (3) */
4. **for** $i = 1$ **to** s **do** /* Phase III: Recover $\tanh(A)$ by (5) */
5. $B = I + T^2$
6.   Solve for X the system of linear equations $BX = 2T$
7. $T = X$
8. **end**

Algorithm 2: Given a matrix $A \in \mathbb{C}^{n \times n}$, this algorithm computes $T = \tanh(A)$ by means of the Taylor approximation Equation (8) and the Paterson–Stockmeyer method.

1. Calculate the scaling factor $s \in \mathbb{N} \cup \{0\}$, the order of Taylor approximation $m_k \in \{2, 4, 6, 9, 12, 16, 20, 25, 30\}$, $2^{-s}A$ and the required matrix powers of $4^{-s}B$ /* Phase I (Algorithm 4) */
2. $T = 2^{-s} A P_{m_k}(4^{-s}B)$ /* Phase II: Compute Equation (8) */
3. **for** $i = 1$ **to** s **do** /* Phase III: Recover $\tanh(A)$ by Equation (5) */
4. $B = I + T^2$
5.   Solve for X the system of linear equations $BX = 2T$
6. $T = X$
7. **end**

2.2. Taylor Approximation-Based Algorithms

Let

$$f(z) = \sum_{n \geq 1} \frac{2^{2n}(2^{2n} - 1)\mathcal{B}_{2n}}{(2n)!} z^{2n-1} \tag{6}$$

be the Taylor series expansion of the hyperbolic tangent function, with the radius of convergence $r = \pi/2$, where $\mathcal{B}_{2n}$ are the Bernoulli's numbers, defined by the recursive expression

$$\mathcal{B}_0 = 1, \quad \mathcal{B}_k = -\sum_{i=0}^{k-1} \binom{k}{i} \frac{\mathcal{B}_i}{k+1-i}, k \geq 1.$$

The following proposition is easily obtained:

Proposition 1. *Let $A \in \mathbb{C}^{n \times n}$ be a matrix satisfying $\rho(A) < \pi/2$. Then, the matrix hyperbolic tangent $\tanh(A)$ can be defined for A as the Taylor series*

$$\tanh(A) = f(A) = \sum_{n \geq 1} \frac{2^{2n}(2^{2n} - 1)\mathcal{B}_{2n}}{(2n)!} A^{2n-1}. \tag{7}$$

From [4] Theorem 4.7, this series in Equation (7) converges if the distinct eigenvalues $\lambda_1, \lambda_2, \cdots, \lambda_t$ of A satisfy one of these conditions:

1. $|\lambda_i| < \pi/2$.
2. $|\lambda_i| = \pi/2$ and the series $f^{(n_i - 1)}(\lambda)$, where $f(z)$ is given by Equation (6) and n_i is the index of λ_i, is convergent at the point $\lambda = \lambda_i, i = 1, \ldots, t$.

To simplify the notation, we denote with

$$\tanh(A) = \sum_{k\geq 0} q_{2k+1} A^{2k+1},$$

the Taylor series (7), and with

$$T_{2m+1}(A) = \sum_{k=0}^{m} q_{2k+1} A^{2k+1} = A \sum_{k=0}^{m} p_k B^k = A P_m(B), \qquad (8)$$

the Taylor approximation of order $2m+1$ of $\tanh(A)$, where $B = A^2$.

There exist several alternatives that can be applied to obtain $P_m(B)$, such as the Paterson–Stockmeyer method [27] or the Sastre formulas [28], with the latter being more efficient, in terms of matrix products, compared with the former.

Algorithm 2 works out $\tanh(A)$ by means of the Taylor approximation of the scaled matrix $4^{-s}B$ Equation (8). In addition, it uses the Paterson–Stockmeyer method for the matrix polynomial evaluation, and finally it applies the recurrence Equation (5) for recovering $\tanh(A)$.

Phase I of Algorithm 2 is in charge of estimating the integers m and s so that the Taylor approximation of the scaled matrix B is computed accurately and efficiently. Then, in Phase II, once the integer m_k has been chosen from the set

$$\mathbb{M} = \{2, 4, 6, 9, 12, 16, 20, 25, 30, \dots\},$$

and powers B^i, $2 \leq i \leq q$ are calculated, with $q = \lceil \sqrt{m_k} \rceil$ or $q = \lfloor \sqrt{m_k} \rfloor$ as an integer divisor of m_k, the Paterson–Stockmeyer method computes $P_{m_k}(B)$ with the necessary accuracy and with a minimal computational cost as

$$\begin{aligned}
P_{m_k}(B) = &(((p_{m_k} B^q + p_{m_k-1} B^{q-1} + p_{m_k-2} B^{q-2} + \cdots + p_{m_k-q+1} B + p_{m_k-q} I) B^q \\
&+ p_{m_k-q-1} B^{q-1} + p_{m_k-q-2} B^{q-2} + \cdots + p_{m_k-2q+1} B + p_{m_k-2q} I) B^q \\
&+ p_{m_k-2q-1} B^{q-1} + p_{m_k-2q-2} B^{q-2} + \cdots + p_{m_k-3q+1} B + p_{m_k-3q} I) B^q \\
&\cdots \\
&+ p_{q-1} B^{q-1} + p_{q-2} B^{q-2} + \cdots + p_1 B + p_0 I.
\end{aligned} \qquad (9)$$

Taking into account Equation (9), the computational cost of Algorithm 2 is $O\left(\left(2k + 4 + \frac{8s}{3}\right) n^3\right)$ flops.

Finally, in Phase III, the matrix hyperbolic tangent of matrix A is recovered by squaring and repeatedly solving a system of linear equations equivalent to Equation (4).

With the purpose of evaluating $P_m(B)$ in Equation (8) in a more efficient way compared with that offered by the Paterson–Stockmeyer method, as stated in the Phase II of Algorithm 2, the formulas provided in [28] were taken into consideration into the design of Algorithm 3. Concretely, we use the evaluation formulas for Taylor-based matrix polynomial approximations of orders $m = 8$, $14+$, and $21+$ in a similar way to the evaluation described in [22] (Sections 3.1–3.3) for the matrix exponential function. Nevertheless, the Paterson–Stockmeyer method is still being used for orders equal to $m = 2$ and $m = 4$.

Following the notation given in [22] (Section 4) for an order m, the suffix "+" in $m = 14+$ and $m = 21+$ means that these Taylor approximations are more accurate than those approximations of order $m = 14$ and $m = 21$, respectively, since the former will be composed of a few more polynomial terms. The coefficients of these additional terms will be similar but not identical to the corresponding traditional Taylor approximation ones. It is convenient to clarify that we have used the order $m = 14+$, instead of the order $m = 15+$, because we have not found a real solution for the coefficients of the corresponding evaluation formula with order $m = 15+$.

The evaluation formulas for the order $m = 8$ that comprise the system of non-linear equations to be solved for determining the unknown coefficients $c_i, i = 1, \ldots, 6$, are:

$$\begin{aligned}
B &= -A^2, \\
B_2 &= B^2, \\
y_0(B) &= B_2(c_1 B_2 + c_2 B), \\
y_1(B) &= A((y_0(B) + c_3 B_2 + c_4 B)(y_0(B) + c_5 B_2) + c_6 y_0(B) \\
&\quad + 2B_2/15 + B/3 + I),
\end{aligned} \qquad (10)$$

where

$$y_1(B) = T_{17}(A) = AP_8(B),$$

and $T_{17}(A)$, or $AP_8(B)$, refers to the Taylor polynomial of order 17 of function $\tanh(A)$ given by Equation (8).

Algorithm 3: Given a matrix $A \in \mathbb{C}^{n \times n}$, this algorithm computes $T = \tanh(A)$ by means of the Taylor approximation Equation (8) and the Sastre formulas.

1 Calculate the scaling factor $s \in \mathbb{N} \cup \{0\}$, the order of Taylor approximation $m_k \in \{2, 4, 8, 14, 21\}$, $2^{-s}A$ and the required matrix powers of $4^{-s}B$ /* Phase I (Algorithm 5) */
2 $T = 2^{-s} A P_{m_k}(4^{-s}B)$ /* Phase II: Compute Equation (8) */
3 **for** $i = 1$ **to** s **do** /* Phase III: Recover $\tanh(A)$ by Equation (5) */
4 $\quad B = I + T^2$
5 $\quad$ Solve for X the system of linear equations $BX = 2T$
6 $\quad T = X$
7 **end**

Regarding the non-linear equations for order $m = 14+$ and its unknown coefficients $c_i, i = 1, \ldots, 13$, we have

$$\begin{aligned}
y_0(B) &= B_2(c_1 B_2 + c_2 B), \\
y_1(B) &= (y_0(B) + c_3 B_2 + c_4 B)(y_0(B) + c_5 B_2) + c_6 y_0(B), \\
y_2(B) &= A((y_1(B) + c_7 y_0(B) + c_8 B_2 + c_9 B)(y_1(B) + c_{10} B_2 + c_{11} B) \\
&\quad + c_{12} y_1 + c_{13} B_2 + B/3 + I),
\end{aligned} \qquad (11)$$

where

$$y_2(B) = A(P_{14} + b_{15} B^{15} + b_{16} B^{16}),$$

and AP_{14} represents the Taylor polynomial of order 29 of function $\tanh(A)$ given by Equation (8). If we denote as p_{15} and p_{16} the Taylor polynomial coefficients corresponding to the powers B^{15} and B^{16}, respectively, the relative error of coefficients b_{15} and b_{16} with respect to them, with two decimal digits, are:

$$\begin{aligned}
|(b_{15} - p_{15})/p_{15}| &= 0.38, \\
|(b_{16} - p_{16})/p_{16}| &= 0.85.
\end{aligned}$$

Taking

$$B_3 = B_2 B,$$

the evaluation formulas related to the system of non-linear equations for order $m = 21+$ with the coefficients $c_i, i = 1, \ldots, 21$ to be determined are

$$\begin{aligned}
y_0(B) &= B_3(c_1 B_3 + c_2 B_2 + c_3 B), \\
y_1(B) &= (y_0(B) + c_4 B_3 + c_5 B_2 + c_6 B)(y_0(B) + c_7 B_3 + c_8 B_2) \\
&\quad + c_9 y_0(B) + c_{10} B_3, \\
y_2(B) &= A((y_1(B) + c_{11} B_3 + c_{12} B_2 + c_{13} B)(y_1(B) + c_{14} y_0(B) \\
&\quad + c_{15} B_3 + c_{16} B_2 + c_{17} B) + c_{18} y_1 + c_{19} y_0(B) + c_{20} B_3 \\
&\quad + c_{21} B_2 + B/3 + I),
\end{aligned} \quad (12)$$

where

$$y_2(B) = A(P_{21} + b_{22} B^{22} + b_{23} B^{23} + b_{24} B^{24}),$$

and AP_{21} stands for the Taylor polynomial of order 43 of the function $\tanh(A)$ given by Equation (8). With two decimal digits of accuracy, the relative error made by the coefficients b_{22}, b_{23}, and b_{24} with respect to their corresponding Taylor polynomial coefficients p_{22}, p_{23}, and p_{24} that accompany their respective powers B^{22}, B^{23}, and B^{24}, are the following:

$$\begin{aligned}
|(b_{22} - p_{22})/p_{22}| &= 0.69, \\
|(b_{23} - p_{23})/p_{23}| &= 0.69, \\
|(b_{24} - p_{24})/p_{24}| &= 0.70.
\end{aligned}$$

Similarly to [22] (Sections 3.1–3.3), we obtained different sets of solutions for the coefficients in Equations (10)–(12) using the vpasolve function (https://es.mathworks.com/help/symbolic/vpasolve.html, accessed on 7 March 2020) from the MATLAB Symbolic Computation Toolbox with variable precision arithmetic. For the case of $m = 21+$, the random option of vpasolve has been used, which allowed us to obtain different solutions for the coefficients, after running it 1000 times. From all the sets of real solutions provided, we selected the most stable ones according to the stability check proposed in [28] (Ex. 3.2).

2.3. Polynomial Order m and Scaling Value s Calculation

The computation of m and s from Phase I in Algorithms 2 and 3 is based on the relative forward error of approximating $\tanh(A)$ by means of the Taylor approximation Equation (8). This error, defined as $E_f = \left\| \tanh(A)^{-1}(I - T_{2m+1}) \right\|$, can be expressed as

$$E_f = \sum_{k \geq 2m+2} c_k A^k,$$

and it can be bounded as (see Theorem 1.1 from [29]):

$$E_f = \left\| \sum_{k \geq 2m+2} c_k A^k \right\| = \left\| \sum_{k \geq m+1} \bar{c}_k B^k \right\| \leq \sum_{k \geq m+1} |\bar{c}_k| \beta_m^k \equiv h_m(\beta_m),$$

where $\beta_m = \max\left\{ \|B^k\|^{1/k} : k \geq m+1, \bar{c}_{m+1} \neq 0 \right\}$.

Let Θ_m be

$$\Theta_m = \max\left\{ \theta \geq 0 : \sum_{k \geq m+1} |\bar{c}_k| \theta^k \leq u \right\},$$

where $u = 2^{-53}$ is the unit roundoff in IEEE double precision arithmetic. The values of Θ_m can be computed with any given precision by using symbolic computations as is shown in Tables 1 and 2, depending on the polynomial evaluation alternative selected.

Algorithm 4 provides the Taylor approximation order $m_k \in \mathbb{M}$, $lower \leq k \leq upper$, where m_{lower} and m_{upper} are, respectively, the minimum and maximum order used, the scaling factor s, together with $2^{-s}A$, and the necessary powers of $4^{-s}B$ for computing T in

Phase II of Algorithm 2. To simplify reading this algorithm, we use the following aliases: $\beta_k \equiv \beta_{m_k}$ and $\Theta_k \equiv \Theta_{m_k}$.

Algorithm 4: Given a matrix $A \in \mathbb{C}^{n \times n}$, the values Θ from Table 1, a minimum order $m_{lower} \in \mathbb{M}$, a maximum order $m_{upper} \in \mathbb{M}$, with $\mathbb{M} = \{2, 4, 6, 9, 12, 16, 20, 25, 30\}$, and a tolerance tol, this algorithm computes the order of Taylor approximation $m \in \mathbb{M}$, $m_{lower} \leq m_k \leq m_{upper}$, and the scaling factor s, together with $2^{-s}A$ and the necessary powers of $4^{-s}B$ for computing $P_{m_k}(4^{-s}B)$ from (9).

1 $B_1 = A^2; k = lower; q = \lceil \sqrt{m_k} \rceil; f = 0$
2 **for** $j = 2$ **to** q **do**
3 $\quad B_j = B_{j-1} B_1$
4 **end**
5 Compute $\beta_k \approx \|B^{m_k+1}\|^{1/(m_k+1)}$ from B_1 and B_q ; /* see [30] */
6 **while** $f = 0$ **and** $k < upper$ **do**
7 $\quad k = k + 1$
8 $\quad$ **if** mod $(k, 2) = 1$ **then**
9 $\quad\quad q = \lceil \sqrt{m_k} \rceil$
10 $\quad\quad B_q = B_{q-1} B_1$
11 $\quad$ **end**
12 $\quad$ Compute $\beta_k \approx \|B^{m_k+1}\|^{1/(m_k+1)}$ from B_1 and B_q ; /* see [30] */
13 $\quad$ **if** $|\beta_k - \beta_{k-1}| < tol$ **and** $\beta_k < \Theta_k$ **then**
14 $\quad\quad f = 1$
15 $\quad$ **end**
16 **end**
17 $s = \max\left(0, \left\lceil \frac{1}{2} \log_2(\beta_k / \Theta_k) \right\rceil\right)$
18 **if** $s > 0$ **then**
19 $\quad s_0 = \max\left(0, \left\lceil \frac{1}{2} \log_2(\beta_{k-1} / \Theta_{k-1}) \right\rceil\right)$
20 $\quad$ **if** $s_0 = s$ **then**
21 $\quad\quad k = k - 1$
22 $\quad\quad q = \lceil \sqrt{m_k} \rceil$
23 $\quad$ **end**
24 $\quad A = 2^{-s}A$
25 $\quad$ **for** $j = 1$ **to** q **do**
26 $\quad\quad B_j = 4^{-sj} B_j$
27 $\quad$ **end**
28 **end**
29 $m = m_k$

In Steps 1–4 of Algorithm 4, the required powers of B for working out $P_{m_k}(B)$ are computed. Then, in Step 5, β_k is obtained by using Algorithm 1 from [30].

As $\lim_{t \to \infty} \|B^t\|^{1/t} = \rho(B)$, where ρ is the spectral radius of matrix B, then $\lim_{m \to \infty} |\beta_m - \beta_{m-1}| = 0$. Hence, given a small tolerance value tol, Steps 6–16 test if there is a value β_k such that $|\beta_k - \beta_{k-1}| < tol$ and $\beta_k < \Theta_k$. In addition, the necessary powers of B for computing $P_{m_k}(B)$ are calculated. Next, the scaling factor $s \geq 0$ is provided in Step 17:

$$s = \max\left(0, \left\lceil \frac{1}{2} \log_2 \frac{\beta_k}{\Theta_k} \right\rceil\right).$$

With those values of m_k and s, we guarantee that:

$$E_f(2^{-s}A) \leq h_{m_k}(4^{-s}\beta_k) < h_{m_k}(\Theta_k) < u, \qquad (13)$$

i.e., the relative forward error of $T_{2m_k+1}(2^{-s}A)$ is lower than the unit roundoff u.

Step 18 checks whether the matrices A and B should be scaled or not. If so, the algorithm analyses the possibility of reducing the order of the Taylor polynomial, but at the same time ensuring that Equation (13) is verified (Steps 19–23). For this purpose, the scaling value corresponding to the order of the Taylor polynomial immediately below the one previously obtained is calculated as well. If both values are identical, the polynomial order reduction is performed. Once the optimal scaling parameter s has been determined, the matrices A and B are scaled (Steps 24–27).

Algorithm 5 is an adaptation of Algorithm 4, where the orders in the set $\mathbb{M} = \{2, 4, 8, 14, 21\}$ are used. Steps 1–15 of Algorithm 5 are equivalent to Steps 1–16 of Algorithm 4. Both values β_1 and β_2 are computed in the same way in both algorithms while values β_3, β_4, and β_5 are worked out in Algorithm 5 for the polynomial orders $m_3 = 8$, $m_4 = 14$, and $m_5 = 21$, respectively. Steps 16–31 of Algorithm 5 correspond to Steps 17–29 of Algorithm 4.

Algorithm 5: Given a matrix $A \in \mathbb{C}^{n \times n}$, the values Θ from Table 2, $\mathbb{M} = \{2, 4, 8, 14, 21\}$ and a tolerance tol, this algorithm computes the order of Taylor approximation $m_k \in \mathbb{M}$ and the scaling factor s, together with $2^{-s}A$ and the necessary powers of $4^{-s}B$ for computing $2^{-s}AP_{m_k}(4^{-s}B)$ from (10), (11) or (12).

1 $B_1 = -A^2;\ B_2 = B_1^2$
2 Compute $\beta_1 \approx \|B^3\|^{1/3}$ from B_1 and B_2
3 $f = 0;\ k = 1$
4 **while** $f = 0$ **and** $k < 5$ **do**
5     $k = k + 1$
6   **if** $k < 5$ **then**
7   Compute $\beta_k \approx \|B^{m_{k+1}}\|^{1/(m_{k+1})}$ from B_1 and B_2
8   **else**
9        $B_3 = B_1 B_2$
10   Compute $\beta_5 \approx \|B^{22}\|^{1/22}$ from B_1 and B_3
11   **end**
12   **if** $|\beta_k - \beta_{k-1}| < tol$ **and** $\beta_k < \Theta_k$ **then**
13       $f = 1;\ s = 0$
14   **end**
15 **end**
16 **if** $f = 0$ **then**
17    $s = \max\left(0, \left\lceil \frac{1}{2} \log_2(\beta_k / \Theta_k) \right\rceil\right)$
18 **end**
19 **if** $s > 0$ **then**
20    $s_0 = \max\left(0, \left\lceil \frac{1}{2} \log_2(\beta_{k-1} / \Theta_{k-1}) \right\rceil\right)$
21   **if** $s_0 = s$ **then**
22       $k = k - 1$
23   **end**
24    $A = 2^{-s}A$
25    $B_1 = 4^{-s}B_1$
26    $B_2 = 16^{-s}B_2$
27   **if** $k = 5$ **then**
28       $B_3 = 64^{-s}B_3$
29   **end**
30 **end**
31 $m = m_k$

Table 1. Values of Θ_{m_k}, $1 \leq k \leq 9$, for polynomial evaluation by means of the Paterson–Stockmeyer method.

$m_1 = 2$	$1.1551925093100 \times 10^{-3}$
$m_2 = 4$	$2.8530558816082 \times 10^{-2}$
$m_3 = 6$	$9.7931623314428 \times 10^{-2}$
$m_4 = 9$	$2.3519926145338 \times 10^{-1}$
$m_5 = 12$	$3.7089935615781 \times 10^{-1}$
$m_6 = 16$	$5.2612365603423 \times 10^{-1}$
$m_7 = 20$	$6.5111831924355 \times 10^{-1}$
$m_8 = 25$	$7.73638541973549 \times 10^{-1}$
$m_9 = 30$	$8.68708923627294 \times 10^{-1}$

Table 2. Values of Θ_{m_k}, $1 \leq k \leq 5$, for polynomial evaluation using the Sastre formulas.

$m_1 = 2$	$1.1551925093100 \times 10^{-3}$
$m_2 = 4$	$2.8530558816082 \times 10^{-2}$
$m_3 = 8$	$1.88126704493647 \times 10^{-1}$
$m_4 = 14+$	$4.65700446893510 \times 10^{-1}$
$m_5 = 21+$	$6.84669656651721 \times 10^{-1}$

3. Numerical Experiments

The following codes have been implemented in MATLAB to test the accuracy and the efficiency of the different algorithms proposed:

- `tanh_expm`: this code corresponds to the implementation of Algorithm 1. For obtaining $m \in \{2, 4, 6, 9, 12, 16, 20, 25, 30\}$ and s and computing e^{2A}, it uses function `exptaynsv3` (see [26]).
- `tanh_tayps`: this development, based on Algorithm 2, incorporates Algorithm 4 for computing m and s, where m takes values in the same set than the `tanh_expm` code. The Paterson–Stockmeyer method is considered to evaluate the Taylor matrix polynomials.
- `tanh_pol`: this function, corresponding to Algorithm 3, employs Algorithm 5 in the m and s calculation, where $m \in \{2, 4, 8, 14, 21\}$. The Taylor matrix polynomials are evaluated by means of Sastre formulas.

Three types of matrices with distinct features were used to build a battery of tests that enabled us to compare the numerical performance of these codes. The MATLAB Symbolic Math Toolbox with 256 digits of precision was used to compute the *"exact"* matrix hyperbolic tangent function using the vpa (variable-precision floating-point arithmetic) function. The test battery featured the following three sets:

(a) Diagonalizable complex matrices: one hundred diagonalizable 128×128 complex matrices obtained as the result of $A = V \cdot D \cdot V^{-1}$, where D is a diagonal matrix (with real and complex eigenvalues) and matrix V is an orthogonal matrix, $V = H/\sqrt{n}$, being H a Hadamard matrix and n is the matrix order. As 1-norm, we have that $2.56 \leq \|A\|_1 \leq 256$. The matrix hyperbolic tangent was calculated *"exactly"* as $\tanh(A) = V \cdot \tanh(D) \cdot V^T$ using the vpa function.

(b) Non-diagonalizable complex matrices: one hundred non-diagonalizable 128×128 complex matrices computed as $A = V \cdot J \cdot V^{-1}$, where J is a Jordan matrix with complex eigenvalues whose modules are less than 5 and the algebraic multiplicity is randomly generated between 1 and 4. V is an orthogonal random matrix with elements in the interval $[-0.5, 0.5]$. As 1-norm, we obtained that $45.13 \leq \|A\|_1 \leq 51.18$. The *"exact"* matrix hyperbolic tangent was computed as $\tanh(A) = V \cdot \tanh(J) \cdot V^{-1}$ by means of the vpa function.

(c) Matrices from the Matrix Computation Toolbox (MCT) [31] and from the Eigtool MATLAB Package (EMP) [32]: fifty-three matrices with a dimension lower than or equal to 128 were chosen because of their highly different and significant characteristics from each other. We decided to scale these matrices so that they had 1-norm not exceeding 512. As a result, we obtained that $1 \leq \|A\|_1 \leq 489.3$. The *"exact"* matrix hyperbolic tangent was calculated by using the two following methods together and the vpa function:

- Find a matrix V and a diagonal matrix D so that $A = VDV^{-1}$ by using the MATLAB function eig. In this case, $T_1 = V\tanh(D)V^{-1}$.
- Compute the Taylor approximation of the hyperbolic tangent function (T_2), with different polynomial orders (m) and scaling parameters (s). This procedure is finished when the obtained result is the same for the distinct values of m and s in IEEE double precision.

The *"exact"* matrix hyperbolic tangent is considered only if

$$\frac{\|T_1 - T_2\|}{\|T_1\|} < u.$$

Although MCT and EMP are really comprised of 72 matrices, only 53 matrices of them, 42 from MCT and 11 from EMP, were considered for our purposes. On the one hand, matrix 6 from MCT and matrix 10 from EMP were rejected because the relative error made by some of the codes to be evaluated was greater or equal to unity. This was due to the ill-conditioning of these matrices for the hyperbolic tangent function. On the other hand, matrices 4, 12, 17, 18, 23, 35, 40, 46, and 51 from MCT and matrices 7, 9, 16, and 17 from EMP were not generated because they did not satisfy the described criterion to obtain the *"exact"* matrix hyperbolic tangent. Finally, matrices 8, 13, 15, and 18 from EMP were refused as they are also part of MCT.

For each of the three previously mentioned sets of matrices, one test was respectively and independently carried out, which indeed corresponds to an experiment to analyse the numerical properties and to account for the computational cost of the different implemented codes. All these experiments were run on an HP Pavilion dv8 Notebook PC with an Intel Core i7 CPU Q720 @1.60 Ghz processor and 6 GB of RAM, using MATLAB R2020b. First, Table 3 shows the percentage of cases in which the normwise relative errors of tanh_expm are lower than, greater than, or equal to those of tanh_tayps and tanh_pol. These normwise relative errors were obtained as

$$\text{Er} = \frac{\|\tanh(A) - \widetilde{\tanh}(A)\|_1}{\|\tanh(A)\|_1},$$

where $\tanh(A)$ represents the exact solution and $\widetilde{\tanh}(A)$ stands for the approximate one.

Table 3. The relative error comparison, for the three tests, between tanh_expm vs. tanh_tayps and tanh_expm vs tanh_pol.

	Test 1	Test 2	Test 3
Er(tanh_expm)<Er(tanh_tayps)	32%	0%	22.64%
Er(tanh_expm)>Er(tanh_tayps)	68%	100%	77.36%
Er(tanh_expm)=Er(tanh_tayps)	0%	0%	0%
Er(tanh_expm)<Er(tanh_pol)	44%	0%	30.19%
Er(tanh_expm)>Er(tanh_pol)	56%	100%	69.81%
Er(tanh_expm)=Er(tanh_pol)	0%	0%	0%

As we can appreciate, from the point of view of the accuracy of the results, tanh_tayps outperformed tanh_expm in 68% of the matrices for Test 1 and 100% and 77.36% of them for Tests 2 and 3. On the other hand, tanh_pol obtained slightly more modest results with improvement percentages equal to 56%, 100%, and 69.81% for Tests 1, 2, and 3, respectively.

Table 4 reports the computational complexity of the algorithms. This complexity was expressed as the number of matrix products required to calculate the hyperbolic tangent of the different matrices that make up each of the test cases. This number of products includes the number of matrix multiplications and the cost of the systems of linear equations that were solved in the recovering phase, by all the codes, together with one more in Step 2 of Algorithm 1 by tanh_expm.

The cost of each system of linear equations with n right-hand side vectors, where n denotes the size of the square coefficient matrices, was taken as 4/3 matrix products. The cost of other arithmetic operations, such as the sum of matrices or the product of a matrix by a vector, was not taken into consideration. As can be seen, the lowest computational cost corresponded to tanh_expm, followed by tanh_pol and tanh_tayps. For example, tanh_expm demanded 1810 matrix products to compute the matrices belonging to Test 1, compared to 1847 by tanh_pol and 2180 by tanh_tayps.

Table 4. Matrix products (P) corresponding to the tanh_expm, tanh_tayps, and tanh_pol functions for Tests 1–3.

	Test 1	Test 2	Test 3
P(tanh_expm)	1810	1500	848
P(tanh_tayps)	2180	1800	1030
P(tanh_pol)	1847	1500	855

Respectively, for the three experiments, Figures 1–3 illustrate the normwise relative errors (a), the performance profiles (b), the ratio of the relative errors (c), the lowest and highest relative error rate (d), the ratio of the matrix products (e), and the polynomial orders (f) for our three codes to be evaluated.

Figures 1a, 2a and 3a correspond to the normwise relative error. As they reveal, the three methods under evaluation were numerically stable for all the matrices that were computed, and all of them provided very accurate results, where the relative errors incurred were always less than 10^{-11}. The solid line that appears in these figures traces the function $k_{\tanh} u$, where $k_{\tanh}$ (or $cond$) stands for the condition number of the matrix hyperbolic tangent function [4] (Chapter 3), and u represents the unit roundoff.

It is clear that the errors incurred by all the codes were usually quite close to this line for the three experiments and even below it, as was largely the case for Tests 1 and 3. For the sake of readability in the graphs, normwise relative errors lower than 10^{-20} were plotted with that value in the figures. Notwithstanding, their original quantities were maintained for the rest of the results.

Figures 1b, 2b and 3b depict the performance profile of the codes. In them, the α coordinate on the x-axis ranges from 1 to 5 in steps equal to 0.1. For a concrete α value, the p coordinate on the y-axis indicates the probability that the considered algorithm has a relative error lower than or equal to α-times the smallest relative error over all the methods on the given test.

The implementation of tanh_tayps always achieved the results with the highest accuracy, followed closely by tanh_pol. For Tests 1 and 2, Figures 1b and 2b indicate that the results provided by them are very similar, although the difference in favour of tanh_tayps is more remarkable in Test 3. As expected from the percentages given in Table 3, tanh_expm computed the hyperbolic tangent function with the worst accuracy, most notably for the matrices from Tests 2 and 3.

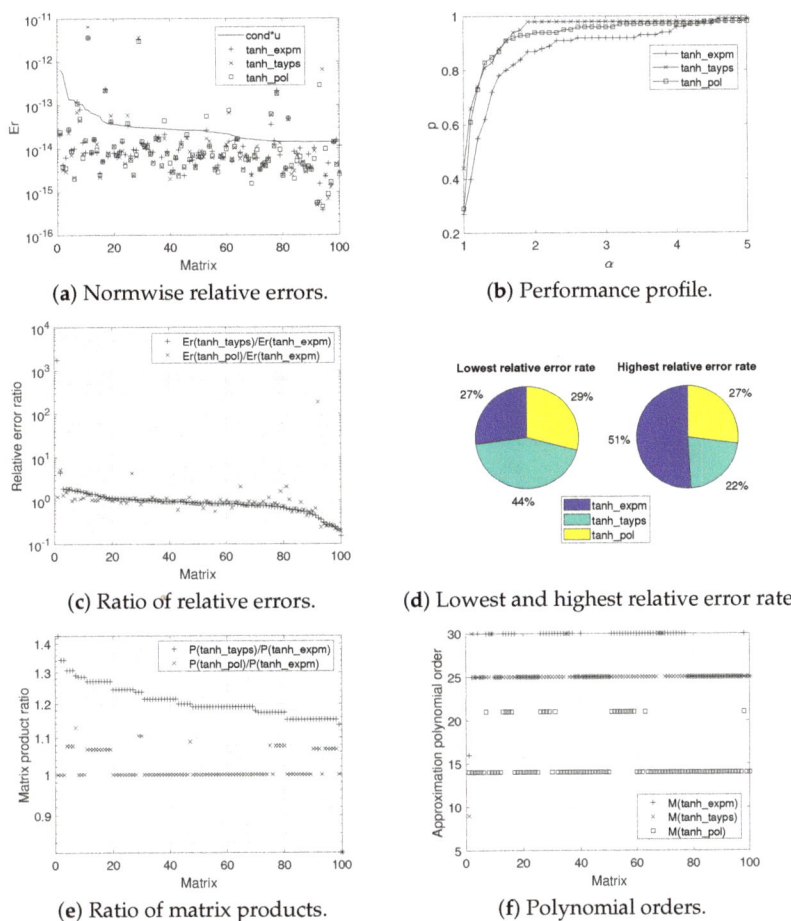

Figure 1. Experimental results for Test 1.

Precisely, the relationship among the normwise relative errors incurred by the codes to be examined is displayed in Figures 1c, 2c and 3c. All these ratios are presented in decreasing order with respect to Er(tanh_tayps)/Er(tanh_expm). This factor is less than 1 for the great majority of the matrices, which indicates that tanh_tayps and tanh_pol are more accurate codes than tanh_expm for estimating the hyperbolic tangent function.

These data are further corroborated by the results shown in Figures 1d, 2d and 3d. These graphs report the percentages of the matrices, for each of the tests, in which each code resulted in the lowest or highest normwise relative error among the errors provided by all of them. Thus, tanh_tayps exhibited the smallest relative error in 44% of the matrices for Test 1 and in 47% of them for Test 3, followed by tanh_pol in 29% and 36% of the cases, respectively. For all the other cases, tanh_expm was the most reliable method. For Test 2, tanh_pol was the most accurate code in 53% of the matrices, and tanh_tayps was the most accurate in 47% of them. In line with these results, tanh_expm was found to be the approach that led to the largest relative errors in 51% of the cases in Test 1, in 100% of them in Test 2, and in 64% for Test 3.

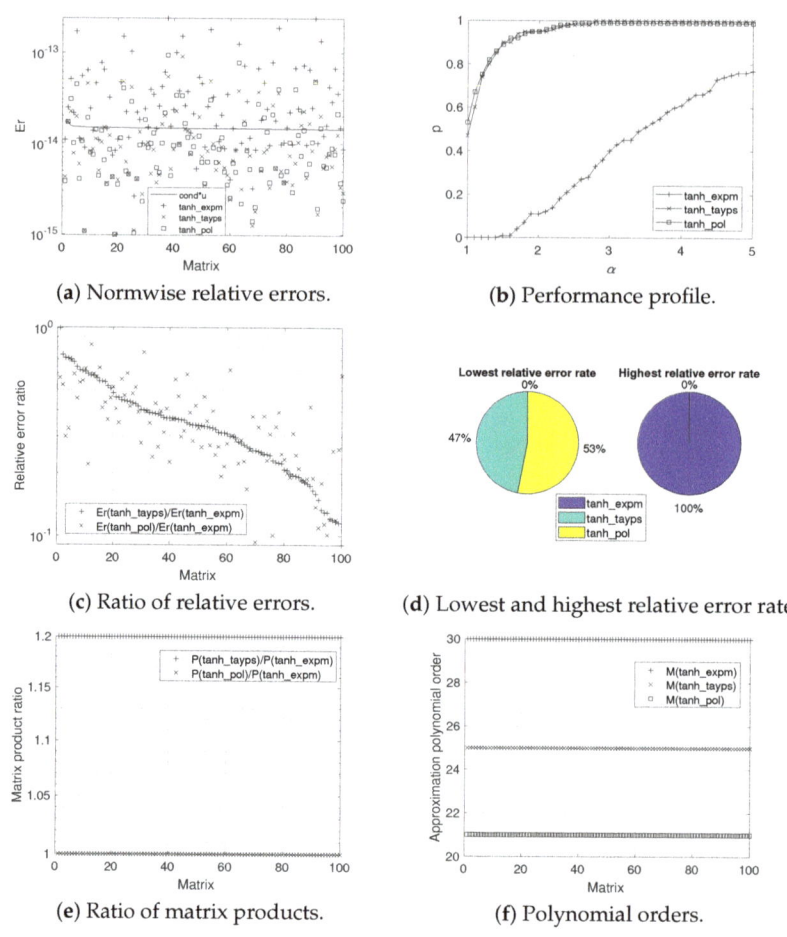

Figure 2. Experimental results for Test 2.

In contrast, although tanh_expm proved to be the most inaccurate code, it is also evident that its computational cost was usually the lowest one, as Table 4 and Figures 1e, 2e and 3e reported. The ratio between the number of tanh_tayps and tanh_expm matrix products ranged from 0.82 to 1.43 for Test 1, was equal to 1.20 for Test 2, and ranged from 0.82 to 1.8 for Test 3. Regarding tanh_pol and tanh_expm, this quotient varied from 0.82 to 1.13 for Test 1, from 0.68 to 1.2 for Test 3, and was equal to 1 for Test 2.

Table 5 lists, in order for Tests 1, 2, and 3, the minimum, maximum, and average values of the degree of the Taylor polynomials m and the scaling parameter s employed by the three codes. Additionally, and in a more detailed way, Figures 1f, 2f and 3f illustrate the order of the polynomial used in the calculation of each of the matrices that compose the testbed. The value of m allowed to be chosen was between 2 and 30 for tanh_expm and tanh_tayps and between 2 and 21 for tanh_pol. As we can see, the average value of m that was typically used varied from 25 and 30 for tanh_expm. It was around 25 for tanh_tayps, and it ranged from 14 to 21 for tanh_pol.

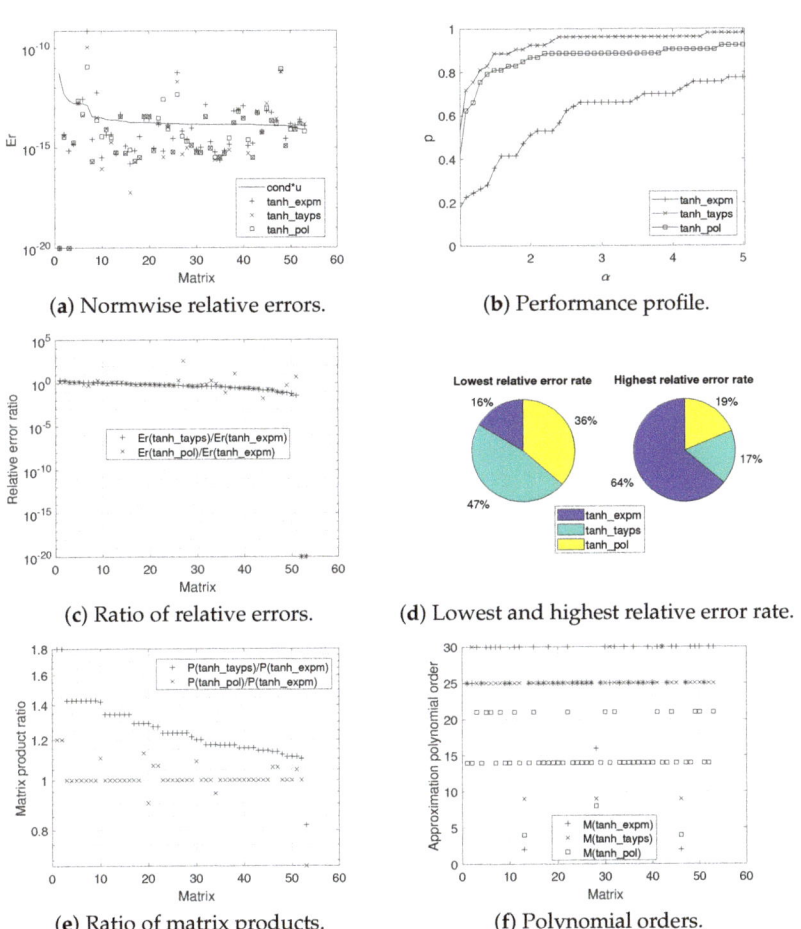

(a) Normwise relative errors.
(b) Performance profile.
(c) Ratio of relative errors.
(d) Lowest and highest relative error rate.
(e) Ratio of matrix products.
(f) Polynomial orders.

Figure 3. Experimental results for Test 3.

Table 5. The minimum, maximum, and average parameters m and s employed for Tests 1–3, respectively.

	m			s		
	Min.	Max.	Average	Min.	Max.	Average
tanh_expm	16	30	27.41	0	5	3.55
tanh_tayps	9	30	25.09	0	6	4.65
tanh_pol	14	21	15.47	0	6	4.83
tanh_expm	30	30	30.00	2	2	2.00
tanh_tayps	25	25	25.00	3	3	3.00
tanh_pol	21	21	21.00	3	3	3.00
tanh_expm	2	30	26.23	0	8	2.77
tanh_tayps	9	30	24.38	0	9	3.74
tanh_pol	4	21	15.36	0	9	3.87

Having concluded the first part of the experimental results, we continue by comparing tanh_tayps and tanh_pol, the Taylor series-based codes that returned the best accuracy in the results. Table 6 presents the percentage of cases in which tanh_tayps gave place to relative errors that were lower than, greater than, or equal to those of tanh_pol. According to the exposed values, both methods provided very similar results, and the percentage of cases in which each method was more accurate than the other was approximately equal to 50% for the different tests.

Table 6. The relative error comparison for the three tests between tanh_tayps vs. tanh_pol.

	Test 1	Test 2	Test 3
Er(tanh_tayps)<Er(tanh_pol)	56%	47%	50.94%
Er(tanh_tayps)>Er(tanh_pol)	44%	53%	45.28%
Er(tanh_tayps)=Er(tanh_pol)	0%	0%	3.77%

Figures 4–6 incorporate the normwise relative errors (a), the ratio of relative errors (b), and the ratio of matrix products (c) between tanh_tayps and tanh_pol. The graphs corresponding to the performance profiles and the polynomial orders are not included now since the results match with those already shown in the previous figures. All this information is also complemented by Table 7, which collects, respectively for each test, the minimum, maximum, and average relative errors incurred by both methods to be analysed, together with the standard deviation.

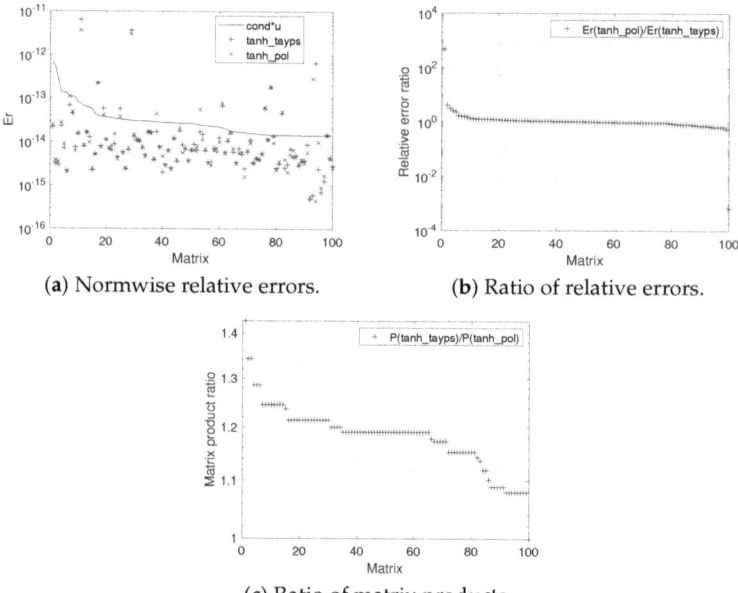

(a) Normwise relative errors.

(b) Ratio of relative errors.

(c) Ratio of matrix products.

Figure 4. Experimental results for Test 1.

As Table 6 details, for Tests 1 and 3, tanh_tayps slightly improved on tanh_pol in the percentage of matrices in which the relative error committed was lower, although it is true that the difference between the results provided by the two methods was small in most cases. However, when such a difference occurred, it was more often in favour of tanh_tayps than the other way around, in quantitative terms.

With all this, we can also appreciate that the mean relative error and the standard deviation incurred by tanh_pol were lower than those of tanh_tayps. For matrices that were part of Test 2, the numerical results achieved by both methods were almost identical, and the differences between them were not significant.

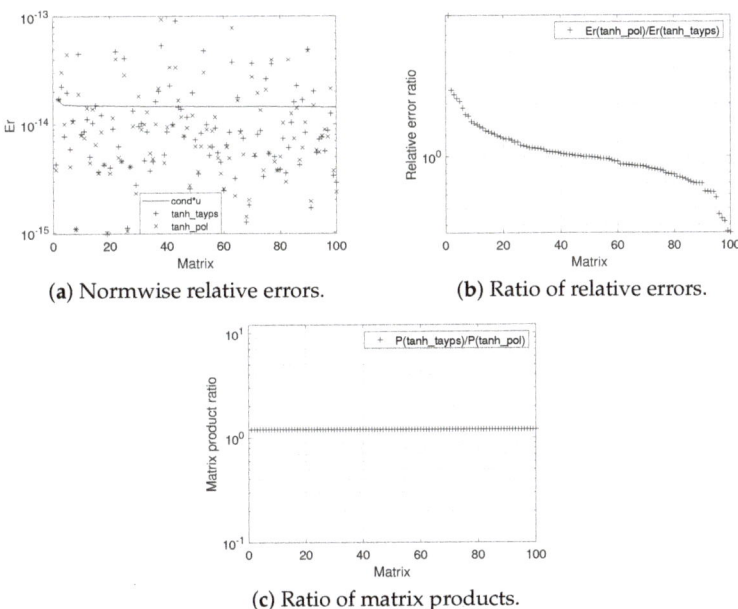

Figure 5. Experimental results for Test 2.

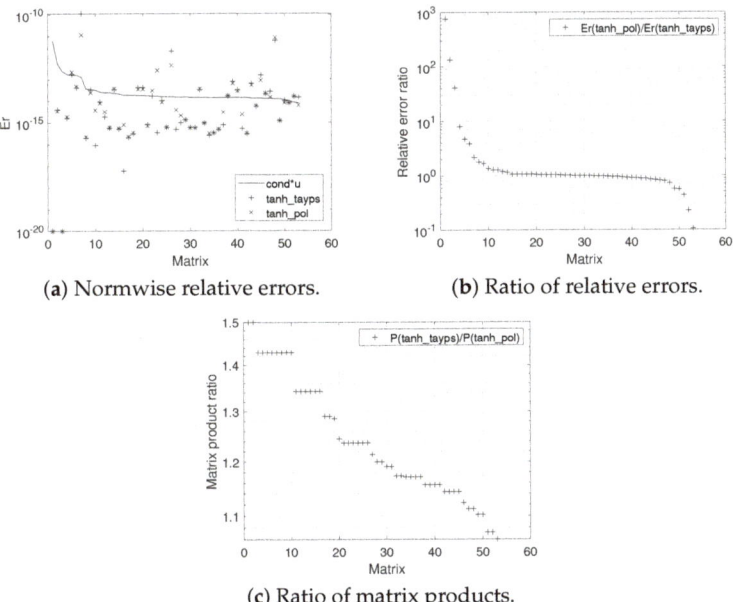

Figure 6. Experimental results for Test 3.

To conclude the analysis and with regard to the computational cost of both codes, such as depicted in Figures 4c, 5c and 6c, it simply remains to note that `tanh_tayps` performed between 1 and 1.43 more matrix products compared with `tanh_pol` for Test 1, 1.2 more for Test 2, and between 1.06 and 1.5 more for Test 3. Therefore, `tanh_pol` computed the matrix tangent function with a very similar accuracy in the results compared to `tanh_tayps` but with a considerably lower computational cost.

Table 7. The minimum, maximum, and average values and the standard deviation of the relative errors committed by `tanh_tayps` and `tanh_pol` for Tests 1–3.

	Minimum	Maximum	Average	Standard Deviation
`tanh_tayps`	4.84×10^{-16}	6.45×10^{-12}	1.22×10^{-13}	7.40×10^{-13}
`tanh_pol`	4.55×10^{-16}	3.64×10^{-12}	8.48×10^{-14}	4.69×10^{-13}
`tanh_tayps`	9.71×10^{-16}	9.06×10^{-14}	1.25×10^{-14}	1.41×10^{-14}
`tanh_pol`	9.92×10^{-16}	9.35×10^{-14}	1.26×10^{-14}	1.47×10^{-14}
`tanh_tayps`	1.09×10^{-254}	1.10×10^{-10}	2.26×10^{-12}	1.52×10^{-11}
`tanh_pol`	1.09×10^{-254}	1.16×10^{-11}	4.10×10^{-13}	1.96×10^{-12}

4. Conclusions

Two alternative methods to approximate the matrix hyperbolic tangent were addressed in this work. The first was derived from its own definition and reduced to the computation of a matrix exponential. The second method deals with its Taylor series expansion. In this latter approach, two alternatives were developed and differed on how the evaluation of the matrix polynomials was performed. In addition, we provided algorithms to determine the scaling factor and the order of the polynomial. As a result, we dealt with a total of three MATLAB codes (`tanh_expm`, `tanh_tayps`, and `tanh_pol`), which were evaluated on a complete testbed populated with matrices of three different types.

The Taylor series-based codes reached the most accurate results in the tests, a fact that is in line with the recommendation suggested in [33] of using a Taylor development against other alternatives whenever possible. However, codes based on Taylor series can be computationally expensive if the Paterson–Stockmeyer method is employed to evaluate the polynomial, as we confirmed with the `tanh_tayps` implementation. One idea to reduce this problem is to use Sastre formulas, as we did in our `tanh_pol` code, resulting in an efficient alternative that significantly reduced the number of matrix operations without affecting the accuracy.

The results found in this paper demonstrated that the three codes were stable and provided acceptable accuracy. However, and without underestimating the other two codes, the `tanh_pol` implementation proposed here offered the best ratio of accuracy/computational cost and proved to be an excellent method for the computation of the matrix hyperbolic tangent.

Author Contributions: Conceptualization, E.D.; methodology, E.D., J.I., J.M.A. and J.S.; software, J.I., J.M.A. and J.S.; validation, J.M.A.; formal analysis, J.I. and J.S.; investigation, E.D., J.I., J.M.A. and J.S.; writing—original draft preparation, E.D., J.I., J.M.A. and J.S.; writing—review and editing, P.A.-J. All authors have read and agreed to the published version of the manuscript.

Funding: This research was funded by the Spanish Ministerio de Ciencia e Innovación under grant number TIN2017-89314-P.

Institutional Review Board Statement: Not applicable.

Informed Consent Statement: Not applicable.

Data Availability Statement: Not applicable.

Conflicts of Interest: The authors declare no conflict of interest.

References

1. Constantine, A.; Muirhead, R. Partial differential equations for hypergeometric functions of two argument matrices. *J. Multivar. Anal.* **1972**, *2*, 332–338. [CrossRef]
2. James, A.T. Special functions of matrix and single argument in statistics. In *Theory and Application of Special Functions*; Academic Press: Cambridge, MA, USA, 1975; pp. 497–520.
3. Hochbruck, M.; Ostermann, A. Exponential integrators. *Acta Numer.* **2010**, *19*, 209–286. [CrossRef]
4. Higham, N.J. *Functions of Matrices: Theory and Computation*; Society for Industrial and Applied Mathematics: Philadelphia, PA, USA, 2008; pp. 425.
5. Rinehart, R. The Equivalence of Definitions of a Matrix Function. *Am. Math. Mon.* **1955**, *62*, 395–414. [CrossRef]
6. Estrada, E.; Higham, D.J.; Hatano, N. Communicability and multipartite structures in complex networks at negative absolute temperatures. *Phys. Rev. E* **2008**, *78*, 026102. [CrossRef] [PubMed]
7. Jódar, L.; Navarro, E.; Posso, A.; Casabán, M. Constructive solution of strongly coupled continuous hyperbolic mixed problems. *Appl. Numer. Math.* **2003**, *47*, 477–492. [CrossRef]
8. Defez, E.; Sastre, J.; Ibáñez, J.; Peinado, J.; Tung, M.M. A method to approximate the hyperbolic sine of a matrix. *Int. J. Complex Syst. Sci.* **2014**, *4*, 41–45.
9. Defez, E.; Sastre, J.; Ibáñez, J.; Peinado, J. Solving engineering models using hyperbolic matrix functions. *Appl. Math. Model.* **2016**, *40*, 2837–2844. [CrossRef]
10. Defez, E.; Sastre, J.; Ibáñez, J.; Ruiz, P. Computing hyperbolic matrix functions using orthogonal matrix polynomials. In *Progress in Industrial Mathematics at ECMI 2012*; Springer: Berlin/Heidelberg, Germany, 2014; pp. 403–407.
11. Defez, E.; Sastre, J.; Ibánez, J.; Peinado, J.; Tung, M.M. On the computation of the hyperbolic sine and cosine matrix functions. *Model. Eng. Hum. Behav.* **2013**, *1*, 46-59.
12. Efimov, G.V.; Von Waldenfels, W.; Wehrse, R. Analytical solution of the non-discretized radiative transfer equation for a slab of finite optical depth. *J. Quant. Spectrosc. Radiat. Transf.* **1995**, *53*, 59–74. [CrossRef]
13. Lehtinen, A. Analytical Treatment of Heat Sinks Cooled by Forced Convection. Ph.D. Thesis, Tampere University of Technology, Tampere, Finland, 2005.
14. Lampio, K. Optimization of Fin Arrays Cooled by Forced or Natural Convection. Ph.D. Thesis, Tampere University of Technology, Tampere, Finland, 2018.
15. Hilscher, R.; Zemánek, P. Trigonometric and hyperbolic systems on time scales. *Dyn. Syst. Appl.* **2009**, *18*, 483.
16. Zemánek, P. New Results in Theory of Symplectic Systems on Time Scales. Ph.D. Thesis, Masarykova Univerzita, Brno, Czech Republic, 2011.
17. Estrada, E.; Silver, G. Accounting for the role of long walks on networks via a new matrix function. *J. Math. Anal. Appl.* **2017**, *449*, 1581–1600. [CrossRef]
18. Cieśliński, J.L. Locally exact modifications of numerical schemes. *Comput. Math. Appl.* **2013**, *65*, 1920–1938. [CrossRef]
19. Cieśliński, J.L.; Kobus, A. Locally Exact Integrators for the Duffing Equation. *Mathematics* **2020**, *8*, 231. [CrossRef]
20. Golub, G.H.; Loan, C.V. *Matrix Computations*, 3rd ed.; Johns Hopkins Studies in Mathematical Sciences; The Johns Hopkins University Press: Baltimore, MD, USA, 1996.
21. Moler, C.; Van Loan, C. Nineteen Dubious Ways to Compute the Exponential of a Matrix, Twenty-Five Years Later. *SIAM Rev.* **2003**, *45*, 3–49. [CrossRef]
22. Sastre, J.; Ibáñez, J.; Defez, E. Boosting the computation of the matrix exponential. *Appl. Math. Comput.* **2019**, *340*, 206–220. [CrossRef]
23. Sastre, J.; Ibáñez, J.; Defez, E.; Ruiz, P. Efficient orthogonal matrix polynomial based method for computing matrix exponential. *Appl. Math. Comput.* **2011**, *217*, 6451–6463. [CrossRef]
24. Sastre, J.; Ibáñez, J.; Defez, E.; Ruiz, P. New scaling-squaring Taylor algorithms for computing the matrix exponential. *SIAM J. Sci. Comput.* **2015**, *37*, A439–A455. [CrossRef]
25. Defez, E.; Ibáñez, J.; Alonso-Jordá, P.; Alonso, J.; Peinado, J. On Bernoulli matrix polynomials and matrix exponential approximation. *J. Comput. Appl. Math.* **2020**, 113207. [CrossRef]
26. Ruiz, P.; Sastre, J.; Ibáñez, J.; Defez, E. High perfomance computing of the matrix exponential. *J. Comput. Appl. Math.* **2016**, *291*, 370–379. [CrossRef]
27. Paterson, M.S.; Stockmeyer, L.J. On the Number of Nonscalar Multiplications Necessary to Evaluate Polynomials. *SIAM J. Comput.* **1973**, *2*, 60–66. [CrossRef]
28. Sastre, J. Efficient evaluation of matrix polynomials. *Linear Algebra Appl.* **2018**, *539*, 229–250. [CrossRef]
29. Al-Mohy, A.H.; Higham, N.J. A New Scaling and Squaring Algorithm for the Matrix Exponential. *SIAM J. Matrix Anal. Appl.* **2009**, *31*, 970–989. [CrossRef]
30. Higham, N.J. FORTRAN Codes for Estimating the One-norm of a Real or Complex Matrix, with Applications to Condition Estimation. *ACM Trans. Math. Softw.* **1988**, *14*, 381–396. [CrossRef]
31. Higham, N.J. The Matrix Computation Toolbox. 2002. Available online: http://www.ma.man.ac.uk/~higham/mctoolbox (accessed on 7 March 2020).

32. Wright, T.G. Eigtool, Version 2.1. 2009. Available online: http://www.comlab.ox.ac.uk/pseudospectra/eigtool (accessed on 7 March 2020).
33. Corwell, J.; Blair, W.D. Industry Tip: Quick and Easy Matrix Exponentials. *IEEE Aerosp. Electron. Syst. Mag.* **2020**, *35*, 49–52. [CrossRef]

MDPI
St. Alban-Anlage 66
4052 Basel
Switzerland
Tel. +41 61 683 77 34
Fax +41 61 302 89 18
www.mdpi.com

Mathematics Editorial Office
E-mail: mathematics@mdpi.com
www.mdpi.com/journal/mathematics

www.ingramcontent.com/pod-product-compliance
Lightning Source LLC
LaVergne TN
LVHW070251100526
838202LV00015B/2205